NEUROINFLAMMATION

Contemporary Neuroscience

NEUROINFLAMMATION
Mechanisms and Management

SECOND EDITION

Edited by

PAUL L. WOOD

Oxon Medica
South San Francisco, CA

HUMANA PRESS
TOTOWA, NEW JERSEY

BS

Cover Illustration: Figure 4 from Chapter 16, "The Neuroinflammatory Components of the Trimethyltin (TMT) Model of Hippocampal Neurodegeneration," by G. Jean Harry and Christian Lefebvre d'Hellencourt.

Production Editor: Jessica Jannicelli.
Cover design by Patricia F. Cleary.

For additional copies, pricing for bulk purchases, and/or information about other Humana titles, contact Humana at the above address or at any of the following numbers: Tel.: 973-256-1699; Fax: 973-256-8341; E-mail: humana@humanapr.com or visit our website: http://humanapress.com

This publication is printed on acid-free paper. ⊗
ANSI Z39.48-1984 (American National Standards Institute) Permanence of Paper for Printed Library Materials.

Printed in the United States of America. 10 9 8 7 6 5 4 3 2 1

Library of Congress Cataloging-in-Publication Data

Neuroinflammation : mechanisms and management / edited by Paul L. Wood.-- 2nd ed.
    p. ; cm. -- (Contemporary neuroscience)
  Includes bibliographical references and index.
  ISBN 1-58829-002-6 (alk. paper) 1-59259-297-X (e-book)
    1. Nervous system--Degeneration--Immunological aspects. 2. Inflammation. 3. Neuritis.
4. Inflammation--Mediators. 5. Nervous system--Pathophysiology. I. Wood, Paul L. II.
Series.
  [DNLM: 1. Neurodegenerative Diseases--immunology. 2. Anti-Inflammatory
Agents--pharmacology. 3. Inflammation--immunology. 4. Nerve
Degeneration--immunology. 5. Neurodegenerative Diseases--drug therapy. WL 359
N49483 2003]
  RC363 .N48 2003
  616.8'0479--dc21
                                                                    2002038760

6/23/04

PREFACE

The first edition of *Neuroinflammation: Mechanisms and Management* was the first book to organize the early concepts of neuroinflammatory mechanisms and the role of these processes in complex neurodegenerative diseases. The field is unique in the neuroscience area in that it has required the skills and experimental analyses of an extremely diverse array of scientific and clinical research groups. This field includes publications from neurologists, psychiatrists, pathologists, clinical imaging groups, neurophysiologists, neurochemists, immunologists, molecular biologists, anatomists, biochemists, and pharmacologists. This field has also generated excitement in both academic and pharmaceutical research arenas, and since the last edition of this book, has resulted in the introduction of two novel inhibitors of neuroinflammation into clinical trials. These include CEP-1347 for Parkinson's disease and CPI-1189 for Alzheimer's disease. Both compounds are currently in Phase II clinical trials, and pivotal efficacy data should be available within the next 3 years.

In the second edition, we have included extensive updates of new knowledge of the mediators produced by activated microglia and their role in neuroinflammatory-induced neuronal lysis. In addition, we have increased the coverage of animal models used in the study of neuroinflammatory mechanisms and in the new imaging methods that allow the noninvasive evaluation of microglial activation in human neurodegenerative disorders. These imaging techniques have demonstrated that microglial activation and the associated neuroinflammation precedes neuronal degeneration in a number of clinical conditions.

Another important aspect of neuroinflammation that has evolved since the first edition of this book is the role of neuroinflammation in amyloid-dependent neuronal lysis. Both in vitro and in vivo data indicate that amyloid is unlikely to be directly neurotoxic, but that amyloid deposition activates neuroinflammatory processes that lead to neuronal degeneration.

In summary, the field of neuroinflammation is evolving rapidly and advancing new potential therapeutics into clinical trials. When scientific concepts result in drugs with clinical utility, a research field has achieved significant maturity and productivity. I hope that this maturity, and its benefit to the treatment of devastating neurological disorders, is solidly in place for the next edition of *Neuroinflammation: Mechanisms and Management*.

Paul L. Wood

Contents

III. ALZHEIMER'S DISEASE

IV. MULTIPLE SCLEROSIS

V. PARKINSON'S AND HUNTINGTON'S DISEASES

CONTRIBUTORS

STEVEN ACKERLEY • *Departments of Neuroscience and Neurology, Institute of Psychiatry, Kings College London, Denmark Hill, London, UK*

CRAIG S. ATWOOD • *Institute of Pathology, Case Western Reserve University, Cleveland, OH*

HIROKO BABA • *Department of Physiology, Nagoya City University School of Medicine, Nagoya, Japan*

RICHARD B. BANATI • *Clinical Sciences Centre, PET-Neurology, Hammersmith Hospital; and Division of Neuroscience and Psychological Medicine, Department of Neuropathology (Molecular Neuropsychiatry), Faculty of Medicine, Imperial College, London, UK*

KATHLEEN M. K. BOJE • *Department of Pharmaceutical Sciences, SUNY Buffalo School of Pharmacy, Buffalo, NY*

TARA BRENNAN • *Neuroinflammation Research Laboratories, Department of Psychiatry, Mount Sinai School of Medicine, New York, NY*

JANET BROWNLEES • *Departments of Neuroscience and Neurology, Institute of Psychiatry, Kings College London, Denmark Hill, London, UK*

W. BRÜCK • *Department of Neuropathology, Charité, Humboldt University, Berlin, Germany*

ASHLEY I. BUSH • *Laboratory for Oxidation Biology, Genetics and Aging Unit, Department of Psychiatry, Harvard Medical School, Boston, MA*

ANNACHIARA CAGNIN • *Clinical Sciences Centre, PET-Neurology, Hammersmith Hospital, London, UK; and Department of Neurological Sciences, Padua University, Padua, Italy*

PAK H. CHAN • *Neurological Laboratories, Stanford University, Stanford, CA*

JEAN-CHRISTOPHE COPIN • *Neurological Laboratories, Stanford University, Stanford, CA; and Divisions of Surgical and Medical Critical Care, Geneva University Hospital, Geneva, Switzerland*

CHRISTIAN LEFEBVRE D'HELLENCOURT • *Neurotoxicology Group, National Institute of Environmental Health Sciences, Research Triangle Park, NC*

ROBERT A. FLOYD • *Free Radical Biology and Aging Research Program, Oklahoma Medical Research Foundation; and Department of Biochemistry and Molecular Biology, University of Oklahoma Health Sciences Center, Oklahoma City, OK*

YVAN GASCHE • *Neurological Laboratories, Stanford University, Stanford, CA; and Divisions of Surgical and Medical Critical Care, Geneva University Hospital, Geneva, Switzerland*

ALEXANDER GERHARD • *Clinical Sciences Centre, PET-Neurology, Hammersmith Hospital, London, UK*

ANDRZEJ R. GLABINSKI • *Department of Neurology, Medical University of Lodz, Lodz, Poland*

MARCIA N. GORDON • *Department of Pharmacology, Alzheimer's Research Laboratory, University of South Florida, Tampa, FL*

ANDREW J. GRIERSON • *Academic Neurology Unit, The Medical School, University of Sheffield, Sheffield, UK*

G. JEAN HARRY • *Neurotoxicology Group, National Institute of Environmental Health Sciences, Research Triangle Park, NC*

HANS-PETER HARTUNG • *Department of Neurology, Heinrich-Heine-University, Düsseldorf, Germany*

BEATRICE HAUSS-WEGRZYNIAK • *Division of Neural Systems, Memory and Aging, University of Arizona, Tucson, AZ*

MICHAEL T. HENEKA • *Department of Neurology, University of Bonn, Bonn, Germany*

KENNETH HENSLEY • *Free Radical Biology and Aging Research Program, Oklahoma Medical Research Foundation, Oklahoma City, OK*

HIDEKI HIDA • *Department of Physiology, Nagoya City University School of Medicine, Nagoya, Japan*

LAP HO • *Neuroinflammation Research Laboratories, Department of Psychiatry, Mount Sinai School of Medicine, New York, NY*

KURT A. JELLINGER • *Institute of Clinical Neurobiology, University of Vienna, Vienna, Austria*

BERND C. KIESEIER • *Department of Neurology, Heinrich-Heine-University, Düsseldorf, Germany*

JARI KOISTINAHO • *A. I. Virtanen Institute of Molecular Science, University of Kuopio, Kuopio, Finland*

CARL J. KOVELOWSKI • *Roberts Center, Sun Health Research Institute, Sun City, AZ*

GARY E. LANDRETH • *Department of Neurosciences and Neurology, Alzheimer Research Laboratory, Case Western Reserve University, Cleveland, OH*

HANS LASSMANN • *Department of Neuroimmunology, Institute of Brain Research, University of Vienna, Vienna, Austria*

CLAUDIA F. LUCCHINETTI • *Department of Neurology, Mayo Clinic, Rochester, MN*

LIH-FEN LUE • *Sun Health Research Institute, Sun City, AZ*

RALPH N. MARTINS • *Sir James McCusker Alzheimer's Disease Research Unit, Department of Surgery, University of Western Australia, Perth, Western Australia*

CHRISTOPHER C. J. MILLER • *Departments of Neuroscience and Neurology, Institute of Psychiatry, Kings College London, Denmark Hill, London, UK*

DAVE MORGAN • *Department of Pharmacology, Alzheimer's Research Laboratory, University of South Florida, Tampa, FL*

HITOO NISHINO • *Department of Physiology, Nagoya City University School of Medicine, Nagoya, Japan*

KENJI OKAJIMA • *Department of Laboratory Medicine, Kumamoto University, Kumamoto, Japan*

GIULIO MARIA PASINETTI • *Neuroinflammation Research Laboratories, Department of Psychiatry, Mount Sinai School of Medicine, New York, NY*

GEORGE PERRY • *Institute of Pathology, Case Western Reserve University, Cleveland, OH*

PATRICK POMPL • *Neuroinflammation Research Laboratories, Department of Psychiatry, Mount Sinai School of Medicine, New York, NY*

RICHARD M. RANSOHOFF • *Department of Neurosciences, Cleveland Clinic Foundation, Cleveland, OH*

JOSEPH ROGERS • *Roberts Center, Sun Health Research Institute, Sun City, AZ*

ALEX E. ROHER • *Haldeman Laboratory for Alzheimer Disease Research, Sun Health Research Institute, Sun City, AZ*

MARK A. SMITH • *Institute of Pathology, Case Western Reserve University, Cleveland, OH*

SOPHIA SUNDARARAJAN • *Department of Neurology, Case Western Reserve University, Cleveland, OH*

RUDOLPH E. TANZI • *Genetics and Aging Unit, Department of Neurology, Harvard Medical School, Boston, MA*

YUJI TAOKA • *Department of Laboratory Medicine, Kumamoto University, Kumamoto; and Department of Orthopedic Surgery, School of Medicine, University of Tokushima, Tokushima, Japan*

PAUL THORNHILL • *Department of Biochemistry and Molecular Biology, University of Leeds, Leeds, UK*

DOUGLAS G. WALKER • *Roberts Center, Sun Health Research Institute, Sun City, AZ*

GARY L. WENK • *Division of Neural Systems, Memory and Aging, University of Arizona, Tucson, AZ*

PAUL L. WOOD • *Oxon Medica, South San Francisco, CA*

JUHA YRJÄNHEIKKI • *Cerebricon Ltd., Kuopio, Finland*

I
Neuroinflammatory Mechanisms

Microglia

Roles of Microglia in Chronic Neurodegenerative Diseases

Paul L. Wood

1. INTRODUCTION

Over the last decade, the neuroinflammatory hypothesis of neurodegeneration has become well established. Our increased scientific understanding of the cascade(s) of events involved in chronic, slowly progressing neurodegenerative diseases has established the foundations for the first mechanistic drug discovery programs for the pharmacological treatment of these devastating diseases. An exciting feature of the neuroinflammatory hypothesis is that a target cell, microglia, has been identified for pharmacological approaches to halting or preventing slowly progressing neurodegenerative diseases.

Microglia are the resident macrophage cell population within the entire neuroaxis and represent the primary immunocompetent cells to deal with invasions by infectious agents and tumors and to remove cellular debris. These cells are present in large numbers representing 10–20% of the glial cell population in the brain and, in the case of perivascular microglia, may play a role in antigen recognition and processing at the level of the blood-brain barrier.

In the normal "resting" state, microglia demonstrate a ramified shape with extended pseudopodia. In the resting state, these cells clearly demonstrate suppressed genomic activity. Upon cellular activation, by a diverse array of stimuli, microglia downregulate surface-bound keratan sulfate proteoglycans and assume an amoeboid shape characteristic of the "activated" and the "phagocytic" stages of microglial cellular activation. In this state, genomic upregulation occurs, leading to the production of a large number of potentially neurotoxic mediators. These mediators are crucial to the normal "housekeeping" activities of microglia and are downregulated once these housekeeping functions have been completed. However, it now appears that in a number of clinical conditions and in a number of preclinical models, microglia remain in an activated state for extended periods and may contribute to neuronal lysis by the direct cytotoxic actions of some microglial mediators and via what has been termed "bystander lysis." This term merely reflects that neurons in the local vicinity of a *sustained* inflammatory response initiated by microglia are prone to lytic attack. The stimuli that act to elicit microglial activation are numerous and are related to the normal "housekeeping" functions of microglia. As immunocompetent cells, a number of chemotactic factors are potential modulators of microglial migration and activation. Receptors on microglia have been demonstrated for a number

From: *Neuroinflammation, 2nd Edition: Mechanisms and Management*
Edited by: P. L. Wood © Humana Press Inc., Totowa, NJ

of known mediators of chemotaxis in inflammatory responses. These include receptors for platelet activating factor, interleukin-8, C5a anaphylotoxin, and the bacterial *N*-formyl peptides as defined by the f-Met-Leu-Phe (FMLP) receptor.

It is the purpose of this chapter to review the basic biochemical characteristics of microglia, with specific attention to the potential neurotoxic mediators involved in various neurodegenerative diseases and in the associated preclinical models of these diseases. The diverse array of potentially neurotoxic mediators secreted by activated microglia are not neurotoxic under the conditions of *transient* microglial activation associated with normal "housekeeping" activities. The neurotoxic actions of such mediators are presumably "buffered" by an equally diverse array of inactivation and cytoprotective mechanisms. However, local microenvironmental compromises including genetic factors and frank tissue insult will dramatically affect the degree of "buffering" capacity available to remove different cytotoxic mediators.

2. MICROGLIA IN NEURODEGENERATIVE DISEASES

There are currently a number of different approaches under clinical and preclinical investigation for inhibiting neuroinflammation. The potential clinical impact of drugs that can provide such a pharmacologic action with minimal side effects will be enormous. The enormity of this impact is best evaluated by tabulating the vast array of neurodegenerative conditions in which activated microglia have been demonstrated. With regard to regional neuronal losses in neuroinflammatory diseases, experiments with basal forebrain mixed neuronal/glial/microglial cultures have shown that lipopolysaccharide (LPS) activation of microglia results in selective losses of cholinergic neurons, demonstrating their selective susceptibility to the toxic actions of activated microglia. These data demonstrate that not all neurons are equally susceptible to the mediators produced by activated microglia and suggest a basis for the losses of only selected neuronal populations in inflammatory-mediated neurodegeneration, like that seen in Alzheimer's disease.

2.1. Alzheimer's Disease

In Alzheimer's disease (AD), the widespread activation of microglia (*see* Chapter 15) in cortical regions affected by AD neuropathology (i.e., plaques and tangles) has consistently been demonstrated with autopsy studies of late-stage AD *(1–32)*. These data led to the original hypothesis that microglial activation might lead to a local neuroinflammatory response that is sustained over long periods (i.e., decades) and that as the local buffering capacity (i.e., inactivation mechanisms) is saturated and/or eroded, local neurons are killed via "bystander lysis." This hypothesis has been validated by more recent studies with in vivo imaging of activated microglia, utilizing [^{11}C]PK-11195. These studies *(32)* have demonstrated that microglial activation in the cortex happens early in the disease process. [^{11}C]PK-11195 is a high-affinity ligand for peripheral benzodiazepine receptors that are concentrated in microglia in the central nervous system (CNS) *(33)* and are upregulated in activated microglia in vivo *(34)*.

Activation of microglia has also been reported in a number of experimental models of AD. Increased microglial activation in cholinergic cell body regions has been demonstrated in the nucleus basalis with local injections of excitotoxins *(35)*, immunotoxins *(36)*, and LPS *(37)* and in the septum with local excitotoxic lesions *(38,39)*. Similarly, hippocampal excitotoxic lesions *(40–44)* and cortical infusions of tumor necrosis factor-α

(TNF-α) *(45)* result in local microglial activation. Cortical microglial activation also occurs in a number of the transgenic mouse models of amyloid deposition, including the Tg2576 *(46–48)*, APP23 *(49)*, PPV717F *(50)*, and PS/APP *(51)* strains.

2.2. Parkinson's Disease

In Parkinson's disease (PD), there are two parallel neurodegenerative processes occurring. These include the events that lead to the degeneration of the nigrostriatal dopaminergic pathway, resulting in motor disturbances and degeneration of the nucleus basalis–cortical cholinergic projection and cortical degeneration, resulting in AD-like neuropathology and cognitive decline. Microglial activation has been demonstrated in the substantia nigra (*see* Chapter 21) around the degenerating dopaminergic neurons *(52–54)* and in the cerebral cortex *(55–57)*. Autopsy studies of younger patients have shown that microglial activation precedes AD-like pathology in the cortex of late-stage PD patients with dementia *(57)*.

Activation of microglia is also seen in the substantia nigra in preclinical models of PD. The substantia nigra is characterized by large numbers of resident microglia *(58)*. Preclinical models of nigral neuroinflammation include local nigral injections of 6-hydroxydopamine *(59,60)* or LPS *(61,62)* and systemic injections of MPTP *(63)*. The LPS model is extremely interesting in that the local inflammation results in the degeneration of dopaminergic neurons but not local GABAergic interneurons. This dopaminergic cell loss appears to be permanent in that it was still evident 1 yr after the LPS lesions were induced.

2.3. Multiple Sclerosis

Activation of microglia is most intense in areas of focal pathology *(64–67)* and is a useful surrogate marker of disease severity utilizing [^{11}C]PK-11195 as an imaging agent *(68)*. MS lesions are also characterized by production of microglial cytokines *(69)* and proteases *(70)*. The association of microglia and microglial toxins with areas of tissue destruction in MS support the evaluation of inhibitors of microglial activation in this devastating disease.

In the experimental allergic encephalitis (EAE) model of MS, microglial activation has also been reported *(66,71,72)*. These cells are also associated with the production of cytokines *(73,74)*, complement *(75)*, and amyloid precursor protein (APP) *(76)*.

2.4. Huntington's Disease

In Huntington's disease (HD), there is microglial activation in the extrapyramidal system early in the disease, with a progressive augmentation of microglial activation during disease progression *(77)*.

Although there is no generally accepted preclinical model of HD, quinolinate lesions of the extrapyramidal system have been utilized to understand the cellular mechanisms that can lead to neuronal degeneration in these clinically relevant brain regions *(78, 79)*. As in the human disease, microglial activation occurs in these excitotoxic models of neurodegeneration.

2.5. Supranuclear Palsy

In progressive supranuclear palsy (SNP), there is robust microglial activation associated with the corticobasal degeneration *(80)*.

2.6. Prion and Viral Dementias

In a vast array of prion diseases that lead to neurodegeneration, there is a consistent activation of large numbers of microglia *(81–85)*. These diseases include Kuru, Creutzfeldt–Jakob disease, Gerstmann–Straussler disease, scrapie, and spongiform encephalopathy.

Viral-mediated neuroinflammation also occurs with acquired immunodeficiency syndrome (AIDS) dementia *(86,87)* and herpes encephalitis *(88)*. Microglial activation is also seen in preclinical models of viral encephalopathy *(89–91)*.

2.7. Age-Related Macular Degeneration

In the dry type of age-related macular degeneration (ARMD), which encompasses 85–90% of ARMD patients, microglial activation is a hallmark feature of the diseased retina in these patients *(92)*. Microglial activation in the eye is also seen in experimental models of diabetic retinopathy *(93,94)*, excitotoxic degeneration *(95)*, and glaucoma *(96,97)*.

2.8. Traumatic Brain Injury

Traumatic brain injury (TBI) to the skull *(98–104)* or spinal cord *(105,106)* results in massive levels of local microglial activation in humans. This activation of microglia (*see* Chapter 10) is also seen in preclinical models of spinal cord contusion *(107–109)* and in TBI to the skull *(110–112)*.

2.9. Amylotrophic Sclerosis

Microglial activation has been demonstrated in the spinal cord of ALS patients *(113)* and in experimental models of ALS *(114)*.

2.10. Neuropathic Pain

Neuropathic pain is a clinical condition in which body areas sense normal stimuli as painful. This supersensitivity to pain is seen in patients with diabetic neuropathy, postherpatic neuralgia, and reflex sympathetic dystrophy. In preclinical models of neuropathic pain, microglial activation in the dorsal horn of the spinal cord, the first synaptic relay point of afferent pain fibers, is seen. These models include formalin-induced hyperalgesia *(115,116)*, neuropathic pain induced by gp120 *(117)*, and chronic constriction injury to the sciatic nerve or spinal dorsal roots *(118–121)*.

3. MICROGLIAL MEDIATORS

3.1. Cytokines

Microglia are the major cellular source, within the CNS, for the proinflammatory cytokines interleukin (IL)-1α, IL-1β, and TNF-α *(122)*. Prior to release, the 26 to 37-kDa precursors of these cytokines undergo posttranslational processing to biologically active 17-kDa products. Direct injections of the cytokines IL-1α, IL-1β, and TNF-α into the CNS result in local inflammatory responses and neuronal degeneration *(123)* and exacerbate EAE-like inflammation *(124)*. These actions are consistent with the potential role of cytokines in neurodegeneration when microglia remain in a sustained activated state and contrast with the potential neurotrophic actions of these cytokines under transient exposure conditions *(125)*. Microglia also produce the inflammatory and chemokinetic cytokines macrophage inflammatory protein-1α (MIP-1α) *(126)*, IL-5 *(127)*, and IL-8 (neutrophil chemotactic peptide) *(128)*. The microglial cytokines IL-1α and TNF-α potently upregulate expression of ICAM-1, VCAM-1, and LFA-1 by microglia *(129)*. Consistent with these observations is the upregulation of microglial CAMs in MS *(130)* and AD *(17)*.

3.2. Proteases

Upon cellular activation, microglia upregulate the synthesis and secretion of a number of proteolytic enzymes that are potentially involved in an equally vast array of functions. Of particular interest to neuroimmune function and neurotoxic potential are the actions of these enzymes in antigen processing for antigen presentation; degradation of the extracellular matrix (ECM) and direct lytic attack of neurons. Proteases that possess the potential to degrade both the ECM and neuronal cells in the vicinity of microglial release include the following:

- Cathepsins B *(131–133)*, L *(132,134–135)*, and S *(134,136)*. Extracellular cathepsin B has been demonstrated to be increased in both AD *(137)* and MS *(138)*.
- The matrix metalloproteinases MMP-1, MMP-2, MMP-3, and MMP-9 *(139–140)*. MMP-1 and MMP-3 are significantly elevated in AD *(141–143)*.
- The metalloprotease-disintegrin ADAM8 (CD156) *(144)*.
- Microglia secrete plasminogen, the prohormone of plasmin, and plasminogen activator, the processing enzyme for this prohormone. Plasmin, once formed outside of microglia, can act to degrade the extracellular matrix and thereby participate in local inflammatory responses *(145–147)*.
- Elastase is another protease secreted by microglia that could have profound detrimental effects on the extracellular matrix *(148)*.

3.3. Prostaglandins

Historically, prostanoid biosynthesis was thought to take place mainly in astrocytes; however, more recent studies indicate that microglia may be a major source of these lipid mediators *(149,150)*. These data, along with the initial observations that chronic treatment with anti-inflammatory agents that gain access to the brain, including dapsone *(151)* and nonsteroidal anti-inflammatory drugs (NSAIDS) *(152)*, decrease the incidence of AD, have led to the hypothesis that inhibitors of CNS inflammatory responses may have clinical utility in treating chronic neurodegenerative diseases *(153–156)*. This has resulted in the evaluation of selective COX-2 inhibitors in AD, based on the decreased potential of these agents to induce ulceration of the gut, relative to classic NSAIDs. Unfortunately, these clinical studies ignored the basic research data that demonstrated that in contrast to rodents, human microglia only express COX-1 and not COX-2 *(157,158)*.

3.4. Acute-Phase Proteins

Experimental studies have demonstrated a temporal relationship for cytokine production after brain injury. Almost immediately, there is enhanced production and release of IL-1β and TNF-α and a delayed but more sustained release of IL-6 into the extracellular space *(159–161)*. Il-6, in turn, is the major trigger *(162)* for the production of both class 1 and class 2 acute-phase proteins *(163)*. The early production of IL-1 and TNF-α, followed by delayed but more sustained increases in IL-6, has been demonstrated in experimental models of closed head injury *(164)*, with excitotoxic brain lesions *(165)*, and with CNS infections *(166)*.

Table 1
Listing of Endogenous Negative Modulators
of Complement Activation Upregulated in AD and MS

Inhibitor	Complement stage inhibited	Cellular source	Disease	Refs.
C1 inhibitor (serpin)	C1	Microglia	AD	*174*
Membrane cofactor protein (CD46)	C3	Astroglia	AD	*175*
Protectin (CD59)	MAC	Astroglia	AD	*176,177*
Clusterin (SGP-2)	MAC	Astroglia	AD	*178,179*
Vitronectin	MAC	Astroglia	AD	*180*
Vitronectin	MAC	Astroglia	MS	*181*

A component of the acute-phase protein response that has come under intense scrutiny is activation of the complement cascade, which ultimately leads to generation of the membrane attack complex (MAC; C5b-9) *(167,168)*. Indeed, both in experimental models of microglial activation and in AD and MS, complement activation has been reported *(169–173)*. Of interest, full complement activation, resulting in the generation of the MAC, has been demonstrated in both AD and MS. Such enhancement of the complement cascade could well contribute to neuronal lysis in these clinical conditions and in any other disorder involving microglial activation. The triggers leading to complement activation presumably are complex and varied.

Complement activation is also under the regulatory control of a number of protein negative modulators *(167)*. It is of interest to note that in AD a number of these factors are also upregulated (Table 1). However, despite this upregulation of intrinsic "buffering" mechanisms to limit complement activation, neuronal destruction still occurs. These data are consistent with the hypothesis that, in AD, neuroinflammatory responses proceed unchecked (i.e., "sustained" microglial activation) and lead to neuronal losses ("bystander lysis").

Microglia produce a number of acute-phase proteins that include chaperone proteins, trophic factors, and protease inhibitors. When overexpressed, these acute-phase proteins can lead to abnormal trophic factor function and dramatic imbalances in protease–protease inhibitor balance in the brain. Under conditions of sustained production, these chaperone proteins may serve a detrimental role in the deposition of amyloid fibrils and have been termed "pathological chaperones" *(182)*.

3.5. Amyloid Precursor Protein

Diffuse plaques in AD contain activated microglia, suggesting that microglia contribute to the ongoing pathological process rather than reacting to neuronal losses because microglia are present around diffuse non-neuritic β-amyloid deposits in areas devoid of neuronal losses *(183)*. Sustained microglial activation has therefore been hypothesized as an essential element of the initiation and support of the progressive pathological cascade leading to neuritic plaque formation *(183)*.

Another consequence of sustained acute-phase protein responses is the deposition of amyloid plaques, the hallmark pathology of AD *(184)*. These plaques are the result of abnormal proteolytic cleavage of membrane-bound APP; however, the cellular source of these soluble degradation products is still a matter of debate. In the case of microglia, all four isoforms of APP have been demonstrated: APP_{695}, APP_{714}, APP_{751}, and APP_{770} *(185)*. Microglia have also been shown to synthesize APP in response to excitotoxic injury *(186)*. Additionally, microglia have been proposed as a possible major source of secreted β-amyloid *(187–189)*. The fact that aggregates of activated microglia are the *sole and consistent* accompaniment of amyloid deposition suggests that they are pivotal in promoting the formation of dense plaque formation in AD *(189)*. β-Amyloid also may act in a feed-forward mechanism to maintain microglial activation because β-amyloid activates microglia directly and stimulates growth factor production by astroglia *(190)*, which, in turn, activate microglia.

3.6. Transmembrane Proteins Involved in Cytotoxicity or Astrogliosis

A number of transmembrane receptors have been characterized for microglia that play roles in microglial toxicity and/or microglial signaling to astrocytes. These include the following:

- Platelet activating factor (PAF) receptors that play a role in the chemotactic response to neuronally released PAF *(191)*.
- CD81 or "target of the antiproliferative antibody" (TAPA). This member of the tetraspanin family is involved reactive gliosis *(192)*.
- Fas ligand, which induces apoptosis in Fas-positive target cells *(193)*.

3.7. Nitric Oxide and Free Radicals

Studies of rat *(194–196)* and murine *(196,197)* microglia have all demonstrated low levels of constitutive nitric oxide synthase (NOS) and a dramatic upregulation of inducible NOS after microglial activation (*see* Chapter 5). This inducible NOS appears to be both cytosolic and membrane bound *(197)*. The concentrations of NO produced by rodent microglial cultures are sufficient to be both bacteriostatic *(198)* and neurotoxic *(194)*. In contrast to these observations with rodent microglia, fetal human microglia appear to possess low levels of inducible NOS *(199–201)*, suggesting that astroglial inducible NOS may play a more pivotal role in human neuroinflammation *(202)*. However, the degree of cell maturation in vitro appears to be important, as a more recent report has shown that human microglia subcultured for 5–6 mo do possess inducible NOS *(203)*. In this regard, NOS mRNA has been demonstrated in the activated microglia associated with MS lesions *(204)*. Augmented production of superoxide has been demonstrated in experimental models of ischemia and TBI *(205)*, in trisomy 16 mice *(206)*, and in autopsy samples of Alzheimer cortex *(207)*.

4. PROGRESS IN THE CLINIC

The evolution of the neuroinflammatory hypothesis combined with the observations that sustained use of NSAIDs for 2 yr or more significantly reduces the risk for onset and progression of AD has led to the hypothesis that an inhibitor of neuroinflammation may have disease-modifying properties that protect against AD neuropathology *(152–156)*. These historical correlations were based on data in individuals taking NSAIDs, which inhibit both COX-1 and COX-2. This lead to a leap of faith by several pharmaceutical companies to enter into clinical trials in AD patients with the new selective COX-2

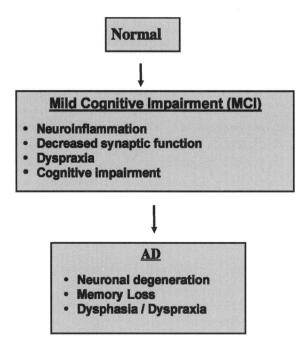

Fig. 1. Proposed sequence of events in neuroinflammatory-dependent loss of cognitive function in Alzheimer's disease.

inhibitors. Unfortunately, these studies were ill conceived based on two key observations. First, human microglia possess COX-1 and not COX-2 *(157,158,208,209)*, so although COX-2 inhibitors would inhibit prostaglandin production by neurons containing COX-2, they would be ineffective in blocking prostaglandin production by microglia, the key cell type responsible for neuroinflammation. Second, these studies of COX-2 inhibitors were performed in early-stage AD patients (ADAS-cog of around 22), whereas the historical data for NSAIDs relates to drug utilization one to two decades prior to this. At that earlier stage of the disease, namely mild cognitive impairment (Fig. 1), an anti-inflammatory action on a single proinflammatory pathway (i.e., COX-1 and COX-2) would provide significant additional buffering capacity against ongoing neuroinflammation. In contrast, in early-stage AD patients, a significant degree of these endogenous buffering systems has deteriorated to the point that neuronal degeneration is occurring at a dramatic rate. Hence, blocking only neuronal COX-2 and leaving a vast array of other inflammatory paths (i.e., cytokines, proteases, acute-phase proteins, nitric oxide, free radicals, and microglial COX-1) intact is probably insufficient at this stage of the disease. Hence, clinical trials in AD patients with the selective COX-2 inhibitor, celecoxib, have failed to date *(210)*.

4.1. Inhibitors of Amyloid Deposition

A hallmark feature of AD is the deposition of amyloid plaques in the neocortex. A number of pharmaceutical companies have developed strategies to inhibit this process and thereby limit the associated neuroinflammation. These include inhibitors of the enzymes responsible for the generation of extracellular amyloid, namely β-secretase and γ-secretase inhibitors *(211)*. Although γ-secretase inhibitors are currently in phase II

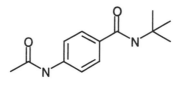

Fig. 2. CPI-1189.

clinical trials in AD, their clinical utility may well be limited by their immunosuppressive properties *(212)*. An alternate strategy, termed the vaccine approach, involves raising antibodies against a fragment of amyloid. This approach has demonstrated efficacy in the transgenic mouse models and is currently in phase II clinical trials in AD *(213)*, but clinical trials have recently been halted as a result of induction of neuroinflammation in some patients.

4.2. Inhibitors of Microglial Activation

Tissue culture studies of microglia have demonstrated that microglial activation by diverse stimuli is dependent on activation of mitogen-activated protein kinase (MAPK) signaling pathways *(214–219)*. MAPK-dependent signal transduction in microglia includes (1) thrombin stimulation of nitric oxide production *(220)*; (2) TGF-β stimulation of caspase 8 inhibitory protein *(221)*; (3) LPS stimulation of TNF-α production *(222)*, and (4) P2 purinergic stimulation of TNF-α production *(223)*.

Additionally, MAPK activation has been demonstrated in microglia in the neocortex of autopsy tissues from AD patients *(224–226)*. Similarly, in the trimethyltin (TMT) model of hippocampal neurotoxicity mediated by microglial activation *(227–229)*, dramatic upregulation of microglial p38^{MAPK} and pJNK^{MAPK} occurs.

Early approaches targeting downregulation of activated microglia in preclinical models focused on immunosuppressants, of which cyclosporin and FK-506 were found to be the most effective *(230)*. More recently, a less toxic compound, the tetracycline derivative minocycline, has demonstrated downregulation of microglial MAPK and neuroprotection against excitotoxic lesions *(231)*. Another approach in this area is CPI-1189 (Fig. 2), a novel signal transduction inhibitor of MAPK activation by cytokine and Toll receptors *(232)*. CPI-1189 protects neurons against TNF-α-induced neurotoxicity both in vitro *(233)* and in vivo *(234–235)*. In the TMT model of neuroinflammatory-induced neuronal cell death in the hippocampus (*see* Chapter 16), CPI-1189 dose-dependently provides significant neuroprotection (Fig. 3) by downregulating MAPK (p38 and pJNK) in activated microglia (Fig. 4). This drug candidate is currently in phase IIb clinical trials in AD. Curcumin, an antioxidant that also potently inhibits MAPK activation has been shown to be neuroprotective in a transgenic mouse model of amyloid deposition *(236)*.

5. SUMMARY

As the resident macrophages of the CNS, microglia are critical in host defense against micro-organisms, against tumors, and in cleanup of cellular debris. However, these are transient "housekeeping" functions that involve profound cellular activation of a normally resting cell population, which returns to the resting state upon completion of these tasks. In contrast, preclinical and clinical observations have demonstrated that when cellular

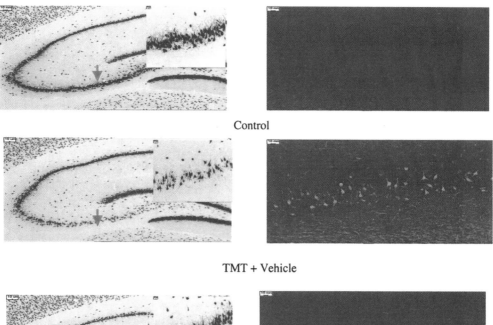

Control

TMT + Vehicle

TMT + CPI-1189 (50 mg/kg)

Fig. 3. Sections of rat hippocampus in animals treated with the toxin trimethyltin ± CPI-1189. The sections on the left are for the neuronal-specific marker NeuN (inserts are 10× magnifications of the CA3 region) and the sections on the right are degenerating neurons stained with fluoro-jade.

activation of microglia is maintained, neuronal injury can occur via multiple mechanisms. The rate of progress of such compromise to neurons will be determined by the capacity of local "buffering" systems involved in the inactivation of toxic microglial mediators. A key question currently being addressed is the potential primary role of microglia in initiating neuronal damage. This may be a facet of a number of clinical conditions; however, even in situations where microglial activation may be a secondary event, pharmacological suppression of this activity should provide clinical utility *(237)*. The potential early and/or primary roles of microglia in neurodegenerative processes are suggested by several preclinical and clinical observations. With the knowledge that MAPK pathways (Fig. 5) regulate the induction of proinflammatory pathways in microglia *(238)*, we now also have specific biochemical targets to pharmacologically modulate in this target cell population.

In summary, sustained and early microglial activation, leading to a chronic inflammatory state, may be a hallmark feature of neurodegenerative disorders and agents that modulate the activity of these cells will represent a new generation of therapeutics, which are much needed in neurology today.

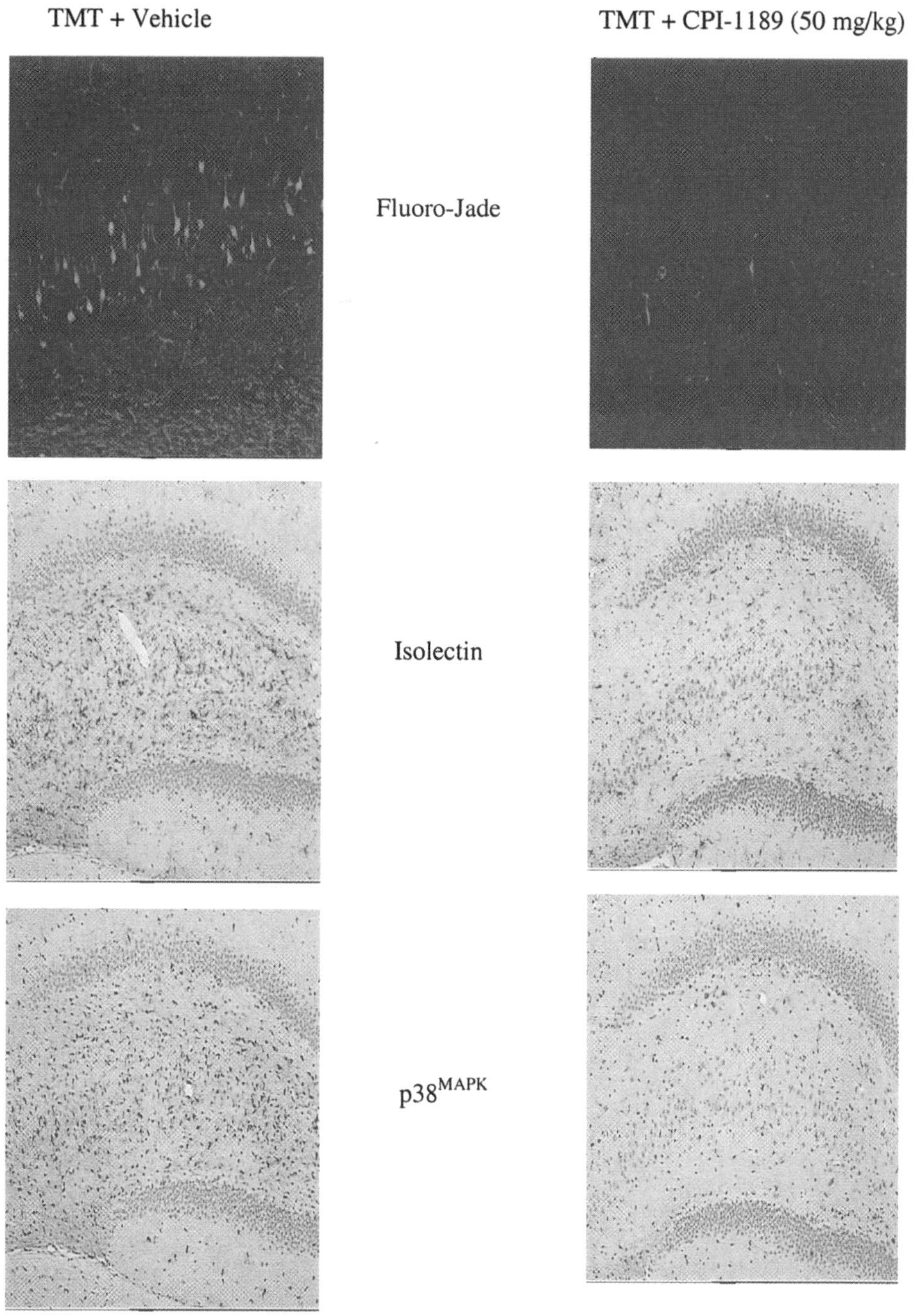

TMT + Vehicle   TMT + CPI-1189 (50 mg/kg)

Fluoro-Jade

Isolectin

p38^{MAPK}

Fig. 4. Sections of rat hippocampus in animals treated with the toxin trimethyltin ± CPI-1189. The sections were stained with fluoro-jade to reveal degenerating neurons, with isolectin for microglial cell counts, and with antibodies to p38 to monitor MAPK upregulation in activated microglia.

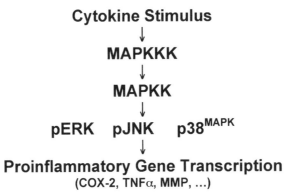

Fig. 5. MAPK signal transduction pathways leading to induction of pro-inflammatory gene expression.

REFERENCES

1. McGeer, P. L., Itagaki, S., Boyes, B. E., and McGeer, E. G. (1988) Reactive microglia are positive for HLA-DR in the substantia nigra of Parkinson's and Alzheimer's disease brains. *Neurology* **38,** 1285–1291.

2. McGeer, P. L., Akiyama, H., Itagaki, S., and McGeer, E. G. (1989) Immune system response in Alzheimer's disease. *Can. J. Neurol. Sci.* **16,** 516–527.

3. McGeer, P. L., Walker, D. G., Akiyama, H., Kawamata, T., Guan, A. L., Parker, C. J., et al. (1991) Detection of the membrane inhibitor of reactive lysis (CD59) in diseased neurons of Alzheimer brain. *Brain Res.* **544,** 315–319.

4. Itagaki, S., McGeer, P. L., and Akiyama, H. (1988) Presence of T-cytotoxic suppressor and leukocyte common antigen positive cells in Alzheimer's disease brain tissue. *Neurosci. Lett.* **91,** 259–264.

5. Itagaki, S., McGeer, P. L., Akiyama, H., Zhu, S., and Selkoe, D. (1989) Relationship of microglia and astrocytes to amyloid deposits of Alzheimer disease. *J. Neuroimmunol.* **24,** 173–182.

6. Griffin, W. S. T., Stanley, L. C., Ling, C., White, L., McLeod, V., Perrot, L. J., et al. (1989) Brain interleukin-1 and S-100 immunoreactivity are elevated in Down syndrome and Alzheimer disease. *Proc. Natl. Acad. Sci. USA* **86,** 7611–7615.

7. Griffin, W. S. T., Sheng, J. G., Roberts, G. W., and Mrak, R. E. (1995) Interleukin-1 expression in different plaque types in Alzheimer's disease: significance in plaque evolution. *J. Neuropathol. Exp. Neurol.* **54,** 276–281.

8. Perlmutter, L. S., Barron, E., and Chui, H. C. (1990) Morphologic association between microglia and senile plaque amyloid in Alzheimer's disease. *Neurosci. Lett.* **119,** 32–36.

9. Perlmutter, L. S., Scott, S. A., Barrón, E., and Chui, H. C. (1992) MHC class II-positive microglia in human brain: association with Alzheimer lesions. *J. Neurosci. Res.* **33,** 549–558.

10. Rozemuller, J. M., Eikelenboom, P., Pals, S. T., and Stam, F. C. (1989) Microglial cells around amyloid plaques in Alzheimer's disease express leukocyte adhesion molecules of the LFA-1 family. *Neurosci. Lett.* **101,** 288–292.

11. Rozemuller, J. M., Van der Valk, P., and Eikelenboom, P. (1992) Activated microglia and cerebral amyloid deposits in Alzheimer's disease. *Res. Immunol.* **143,** 646–649.

12. Tooyama, I., Kimura, H., Akiyama, H., and McGeer, P. L. (1990) Reactive microglia express class I and class II major histocompatibility complex antigens in Alzheimer's disease. *Brain Res.* **523,** 273–280.

13. Masliah, E., Mallory, M., Hansen, L., Alford, M., Albright, T., Terry, R., Shapiro, P., et al. (1991) Immunoreactivity of CD45, a protein phosphotyrosine phosphatase, in Alzheimer's disease. *Acta Neuropathol. (Berl.)* **83,** 12–20.

14. Cras, P., Kawai, M., Siedlak, S., Mulvihill, P., Gambetti, P., Lowery, D., et al. (1990) Neuronal and microglial involvement in β-amyloid protein deposition in Alzheimer's disease. *Am. J. Pathol.* **137,** 241–246.

15. Cras, P., Kawai, M., Siedlak, S., and Perry, G. (1991) Microglia are associated with the extracellular neurofibrillary tangles of Alzheimer disease. *Brain Res.* **558,** 312–314.

16. Johnson, S. A., Lampert-Etchells, M., Pasinetti, G. M., Rozovsky, I., Finch, C. E., et al. (1992) Complement mRNA in the mammalian brain: responses to Alzheimer's disease and experimental brain lesioning. *Neurobiol. Aging* **13,** 641–648.

17. Akiyama, H. and McGeer, P. L. (1990) Brain microglia constituitively express β-2 integrins. *J. Neuroimmunol.* **30,** 81–93.

18. Akiyama, H., Kawamata, T., Dedhar, S., and McGeer, P. L. (1991) Immunohistochemical localization of vitronectin, its receptor and beta-3 integrin in Alzheimer brain tissue. *J. Neuroimmunol.* **32,** 19–28.

19. Akiyama, H., Yamada, T., Kawamata, T., and McGeer, P. L. (1991) Association of amyloid P component with complement proteins in neurologically diseased brain tissue. *Brain Res.* **548,** 349–352.

20. Akiyama, H., Ikeda, K., Kondo, H., Kato, M., and McGeer, P. L. (1993) Microglia express the type 2 plasminogen activator inhibitor in the brain of control subjects and patients with Alzheimer's disease. *Neurosci. Lett.* **164,** 233–235.

21. Akiyama, H., Nishimura, T., Kondo, H., Ikeda, K., Hayashi, Y., and McGeer, P. L. (1994) Expression of the receptor for macrophage colony stimulating factor by brain microglia and its upregulation in brains with Alzheimer's disease and amyotrophic lateral sclerosis. *Brain Res.* **639,** 171–174.

22. Mrak, R. E., Sheng, J. G., and Griffin, W. S. T. (1995) Glial cytokines in Alzheimer's disease: review and pathogenic implications. *Hum. Pathol.* **26,** 816–823.

23. Yamada, T., Kawamata, T., Walker, D. G., and McGeer, P. L. (1992) Vimentin immunoreactivity in normal and pathological human brain tissue. *Acta Neuropathol. (Berl.)* **84,** 157–162.

24. Yamada, T., Miyazaki, K., Koshikawa, N., Takahashi, M., Akatsu, H., and Yamamoto, T. (1995) Selective localization of gelatinase A, an enzyme degrading β-amyloid protein, in white matter microglia and in Schwann cells. *Acta Neuropathol. (Berl.)* **89,** 199–203.

25. Yamada, T., Yoshiyama, Y., Sato, H., Seiki, M., Shinagawa, A., and Takahashi, M. (1995) White matter microglia produce membrane-type matrix metalloprotease, an activator of gelatinase A, in human brain tissues. *Acta Neuropathol. (Berl.)* **90,** 421–424.

26. Akiyama, H., Arai, T., Kondo, H., Tanno, E., Haga, C., and Ikeda, K. (2000) Cell mediators of inflammation in the Alzheimer disease brain. *Alzheimer Dis.* **14,** S47–S53.

27. Arends, Y. M., Duyckaerts, C., Rozemuller, J. M., Eikelenboom, P., and Hauw, J. J. (2000) Microglia, amyloid and dementia in Alzheimer disease—a correlative study. *Neurobiol. Aging* **21,** 39–47.

28. McGeer, P. L., McGeer, E. G., and Yasojima, K. (2000) Alzheimer disease and neuroinflammation. *J. Neural Transm.* **59,** 53–57.

29. Sheffield, L. G., Marquis, J. G., and Berman, N. J. (2000) Regional distribution of cortical microglia parallels that of neurofibrillary tangles in Alzheimer's disease. *Neurosci. Lett.* **285,** 165–168.

30. Togo, T., Akiyama, H., Kondo, H., Ikeda, K., Kato, M., Iseki, E., and Kosaka, K. (2000) Expression of CD40 in the brain of Alzheimer's disease and other neurological diseases. *Brain Res.* **885,** 117–121.

31. Wegiel, J., Wang, K. C., Tarnawski, M., and Lach, B. (2000) Microglial cells are the driving force in fibrillar plaque formation, whereas astrocytes are a leading factor in plaque degradation. *Acta Neuropathol.* **100,** 356–364.

32. Cagnin, A., Brooks, D. J., Kennedy, A. M., Gunn, R. N., Myers, R., Turkheimer, F. E., et al. (2001) In-vivo measurement of activated microglia in dementia. *Lancet* **358,** 461–467.

33. Park, C. H., Carboni, E., Wood, P. L., and Gee, K. W. (1996) Characterization of peripheral benzodiazepine type sites in a cultured murine BV-2 microglial cell line. *Glia* **16,** 65–70.

34. Stephenson, D. T., Schober, D. A., Smalstig, E. B., Mincy, R. E., Gehlert, D. R., and Clemens, J. A. (1995) Peripheral benzodiazepine receptors are colocalized with activated microglia following transient global forebrain ischemia in the rat. *J. Neurosci.* **15,** 5263–5274.

35. Prosperi, C., Scali, C., Pepeu, G., and Casamenti, F. (2001) NO-flurbiprofen attenuates excitotoxin-induced brain inflammation, and releases nitric oxide in the brain. *Jpn. J. Pharmacol.* **86(2),** 230–235.

36. Rossner, S., Härtig, W., Schliebs, R., Brückner, G., Brauer, K., Perez-Polo, J. R., et al. (1995) IgG-saporin immunotoxin-induced loss of cholinergic cells differentially activates microglia in rat basal forebrain nuclei. *J. Neurosci. Res.* **41,** 335–346.

37. Kim, W. G., Mohney, R. P., Wilson, B., Jeohn, G. H., Liu, B., and Hong, J. S. (2000) Regional difference in susceptibility to lipopolysaccharide-induced neurotoxicity in the rat brain: role of microglia. *J. Neurosci.* **20,** 6309–6316.

38. Coffey, P. J., Perry, V. H., Allen, Y., Sinden, J., and Rawlins, J. N. P. (1988) Ibotenic acid induced demyelination in the central nervous system: a consequence of a local inflammatory response. *Neurosci. Lett.* **84,** 178–184.

39. Coffey, P. J., Perry, V. H., and Rawlins, J. N. P. (1990) An investigation into the early stages of the inflammatory response following ibotenic acid-induced neuronal degeneration. *Neuroscience* **35,** 121–132.

40. Bernal, F., Saura, J., Ojuel, J., and Mahy, N. (2000) Differential vulnerability of hippocampus, basal ganglia, and prefrontal cortex to long-term NMDA excitotoxicity. *Exp. Neurol.* **161,** 686–695.

41. Lu, X. R. and Ong, W.Y. (2001) Heme oxgenase-1 is expressed in viable astrocytes and mircoglia but in degenerating pyramidal neurons in the kainate-lesioned rat hippocampus. *Exp. Brain Res.* **137,** 424–431.

42. Finsen, B. R., Jorgensen, M. B., Diemer, N. H., and Zimmer, J. (1993) Microglial MHC antigen expression after ischemic and kainic acid lesions of the adult rat hippocampus. *Glia* **7,** 41–49.

43. Andersson, P.-B., Perry, V. H., and Gordon, S. (1991) The CNS acute inflammatory response to excitotoxic neuronal cell death. *Immunol. Lett.* **30,** 177–182.

44. Jorgensen, M. B., Finsen, B. R., Jensen, M. B., Castellano, B., Diemer, N. H., and Zimmer, J. (1993) Microglial and astroglial reactions to ischemic & kainic acid-induced lesions of the adult rat hippocampus. *Exp. Neurol.* **120,** 70–88.

45. Wright, J. L. and Merchant, R. E. (1992) Histopathological effects of intracerebral injections of human recombinant tumor necrosis factor-α in the rat. *Acta Neuropathol. (Berl.)* **85,** 93–100.

46. Benzing, W. C., Wujek, J. R., Ward, E. K., Shaffer, D., Ashe, K. H., Younkin, S. G., et al. (1999) Evidence for glial-mediated inflammation in aged APP(SW) transgenic mice. *Neurobiol. Aging* **20,** 581–589.

47. Mehlhorn, G., Hollborn, M., and Schliebs, R. (2000) Induction of cytokines in glial cells surrounding cortical beta-amyloid plaques in transgenic Tg2576 mice with Alzheimer pathology. *Int. J. Dev. Neurosci.* **18,** 423–431.

48. Apelt, J. and Schliebs, R. (2001) β-Amyloid-induced glial expression of both pro- and anti-inflammatory cytokines in cerebral cortex of aged transgenic Tg2576 mice with Alzheimer plaque pathology. *Brain Res.* **894,** 21–30.

49. Bornemann, K. D., Wiederhold, K. H., Pauli, C., Ermini, F., Stalder, M., Schnell, L., et al. (2001) Aβ-Induce inflammatory processes in microglia cells of APP23 transgenic mice. *Am. J. Pathol.* **158,** 63–73.
50. Murphy, G. M., Zhao, F. F., Yang, L., and Cordell, B. (2000) Expression of macrophage colony-stimulating factor receptor is increased in the A beta PPV717F transgenic mouse model of Alzheimer's disease. *Am. J. Pathol.* **157,** 895–904.
51. Matsuoka, Y., Picciano, M., Malester, B., LaFrancois, J., Zehr, C., Daeschner, J. M., et al. (2001) Inflammatory responses to amyloidosis in a transgenic mouse model of Alzheimer's disease. *Am. J. Pathol.* **158,** 1345–1354.
52. McGeer, P. L., Kawamata, T., Walker, D. G., Akiyama, H., Tooyama, I., and McGeer, E. G. (1993) Microglia in degenerative neurological disease. *Glia* **7,** 84–92.
53. Knott, C., Stern, G., and Wilkin, G. P. (2000) Inflammatory regulators in Parkinson's disease: iNOS, lipocortin-1, and cyclooxygenases-1 and-2. [review]. *Mol. Cell. Neurosci.* **16,** 724–739.
54. Mirza, B., Hadberg, H., Thomsen, P., and Moos, T. (2000) The absence of reactive astrocytosis is indicative of a unique inflammatory process in Parkinson's disease. *Neuroscience* **95(2),** 425–432.
55. Mackenzie, I. A. (2000) Activated microglia in dementia with Lewy bodies. *Neurology* **55,** 132–134.
56. Jellinger, K. A. (2000) Cell death mechanisms in Parkinson's disease. [review]. *J. Neural Transm.* **107,** 1–29.
57. Mackenzie, I. A. (2000) Activated microglia in dementia with Lewy bodies. *Neurology* **55,** 132–134.
58. Lawson, L. J., Perry, V. H., Dri, P., and Gordon, S. (1990) Heterogeneity in distribution and morphology of microglia in the normal adult mouse brain. *Neuroscience* **39,** 151–170.
59. Akiyama, H. and McGeer, P. L, (1989) Microglial response to 6-hydroxydopamine-induced substantia nigra lesions. *Brain Res.* **489,** 247–253.
60. Haruhiko, A. and McGeer, P. L. (1989) Microglial response to 6-hydroxydopamine-induced substantia nigra lesions. *Brain Res.* **489,** 247–253.
61. Kim, W. G., Mohney, R. P., Wilson, B., Jeohn, G. H., Liu, B., and Hong, J. S. (2000) Regional difference in susceptibility to lipopolysaccharide-induced neurotoxicity in the rat brain: role of microglia. *J. Neurosci.* **20,** 6309–6316.
62. Herrera, A. J., Castano, A., Venero, J. L., Cano, J., and Machado, A. (2000) The single intranigral injection of LPS as a new model for studying the selective effects of inflammatory reactions on dopaminergic system. *Neurobiology* **7,** 429–447.
63. Grunblatt, E., Mandel, S., and Youdim, M. H. (2000) MPTP and 6-hydroxydopamine-induced neurodegeneration as models for Parkinson's disease: neuroprotective strategies. *J. Neurol.* **247,** 95–102.
64. Li, H., Newcombe, J., Groome, P., and Cuzner, M. L. (1993) Characterization and distribution of phagocytic macrophages in multiple sclerosis plaques. *Neuropathol. Appl. Neurobiol.* **19,** 214–223.
65. Bö, L., Mork, S., Kong, P. A., Nyland, H., Pardo, C. A., and Trapp, B. D. (1994) Detection of MHC class II-antigens on macrophages and microglia, but not on astrocytes and endothelia in active multiple sclerosis lesions. *J. Neuroimmunol.* **51,** 135–146.
66. Brosnan, C. F., Cannella, B., Battistini, L., and Raine, C. S. (1995) Cytokine localization in multiple sclerosis lesions: correlation with adhesion molecule expression and reactive nitrogen species. *Neurology* **45(Suppl. 6),** S16–S21.
67. Boyle, E. A. and McGeer, P. L. (1990) Cellular immune response in multiple sclerosis plaques. *Am. J. Pathol.* **137,** 575–584.
68. Banati, R. B., Newcombe, J., Gunn, R. N., Cagnin, A., Turkheimer, F., Heppner, F., G. et al. (2000) The peripheral benzodiazepine binding site in the brain in multiple sclerosis

—quantitative in vivo imaging of microglia as a measure of disease activity. *Brain* **123,** 2321–2337.

69. Bitsch, A., Kuhlmann, T., Da Costa, C., Bunkowski, S., Polak, T., and Bruck, W. (2000) Tumour necrosis factor alpha mRNA expression in early multiple sclerosis lesions: correlation with demyelinating activity and oligodendrocyte pathology. *Glia* **29,** 366–375.

70. Bever, C. T. Jr. and Garver, D. W. (1995) Increased cathepsin B activity in multiple sclerosis brain. *J. Neurol. Sci.* **131,** 71–73.

71. Matsumoto, Y., Ohmori, K., and Fujiwara, M. (1992) Microglial and astroglial reactions to inflammatory lesions of experimental autoimmune encephalomyelitis in the rat central nervous system. *J. Neuroimmunol.* **37,** 23–33.

72. Bauer, J., Sminia, T., Wouterlood, F. G., and Dijkstra, C. D. (1994) Phagocytic activity of macrophages and microglial cells during the course of acute and chronic relapsing experimental autoimmune encephalomyelitis. *J. Neurosci. Res.* **38,** 365–375.

73. Renno, T., Krakowski, M., Piccirillo, C., Lin, J., and Owens, T. (1995) TNF-α expression by resident microglia and infiltrating leukocytes in the central nervous system of mice with experimental allergic encephalomyelitis: regulation by Th1 cytokines. *J. Immunol.* **154,** 944–953.

74. Bauer, J., Berkenbosch, F., VanDam, A.-M., and Dijkstra, C. D. (1993) Demonstration of interleukin-1β in Lewis rat brain during experimental allergic encephalomyelitis by immunocytochemistry at the light and ultrastructural level. *J. Neuroimmunol.* **48,** 13–22.

75. Linington, C., Lassmann, H., Morgan, B. P., and Compston, D. A. S. (1989) Immunohistochemical localization of terminal complement component C9 in experimental allergic encephalomelitis. *Acta Neuropathol.* **79,** 78–85.

76. Banati, R. B., Gehrmann, J., Lannes-Vieira, J., Wekerle, H., and Kreutzberg, G. W. (1995) Inflammatory reaction in experimental autoimmune encephalomyelitis (EAE) is accompanied by a microglial expression of the βA4-amyloid precursor protein (APP). *Glia* **14,** 209–215.

77. Sapp, E., Kegel, K. B., Aronin, N., Hashikawa, T., Uchiyama, Y., Tohyama, K., P. G., et al. (2001) Early and progressive accumulation of reactive microglia in the Huntington disease brain. *J. Neuropathol. Exp. Neurol.* **60,** 161–172.

78. Töpper, R., Gehrmann, J., Schwarz, M., Block, F., Noth, J., and Kreutzberg, G. W. (1993) Remote microglial activation in the quinolinic acid model of Huntington's disease. *Exp. Neurol.* **123,** 271–283.

79. Bresjanac, M. and Antauer, G. (2000) Reactive astrocytes of the quinolinc acid-lesioned rat striatum express GFR alpha 1 as well as GDNF in vivo. *Exp. Neurol.* **164,** 53–59.

80. Ishizawa, K. and Dickson, D. W. (2001) Microglial activation parallels system degeneration in progressive supranuclear palsy and corticobasal degeneration. *J. Neuropathol. Exp. Neurol.* **60,** 647–657.

81. Aoki, T., Kobayashi, K., and Isaki, K. (1999) Microglial and astrocytic change in brains of Creutzfeldt–Jakob disease: an immunocytochemical and quantitative study. *Clin. Neuropathol.* **18,** 51–60.

82. Muhleisen, H., Gehrmann, J., and Meyermann, R. (1995) Reactive microglia in Creutzfeldt–Jakob disease. *Neuropathol. Appl. Neurobiol.* **21,** 505–517.

83. Rezaie, P. and Lantos, P. L. (2001) Microglia and the pathogenesis of spongiform encephalopathies. [review]. *Brain Res. Rev.* **35,** 55–72.

84. Bugiani, O., Giaccone, G., Piccardo, P., Morbin, M., Tagliavini, F., and Ghetti, B. (2000) Neuropathology of Gerstmann–Straussler–Scheinker disease. *Microscopy* **50,** 10–15.

85. Brown, D. R., Schmidt, B., and Kretzschmar, H. A. (1996) Role of microglia and host prion protein in neurotoxicity of a prion protein fragment. *Nature* **380,** 345–347.

86. Glass, J. D. and Wesselingh, S. L. (2001) Microglia in HIV-associated neurological diseases. *Microsc. Res. Tech.* **54,** 95–105.

87. Kure, K., Weidenheim, K. M., Lyman, W. D., and Dickson, D. W. (1990) Morphology and distribution of HIV-1 gp41-positive microglia in subacute AIDS encephalitis. Pattern of involvement resembling a multisystem degeneration. *Acta Neuropathol. (Berl.)* **80,** 393–400.

88. Cagnin, A., Myers, R., Gunn, R. N., Lawrence, A. D., Stevens, T., Kreutzberg, G. W., et al. (2001) In vivo visualization of activated glia by [11C] (R)-PK11195-PET following herpes encephalitis reveals projected neuronal damage beyond the primary focal lesion. *Brain* **124,** 2014–2027.

89. Lynch, W. P., Czub, S., McAtee, F. J., Hayes, S. F., and Portis, J. L. (1991) Murine retrovirus-induced spongiform encephalopathy: productive infection of microglia and cerebellar neurons in accelerated CNS disease. *Neuron* **7,** 365–379.

90. Lynch, W. P., Robertson, S. J., and Portis, J. L. (1995) Induction of focal spongiform neurodegeneration in developmentally resistant mice by implantation of murine retrovirus-infected microglia. *J. Virol.* **69,** 1408–1419.

91. Gillespie, J. S., Cavanagh, H. M. A., Behan, W. M. H., Morrison, L. J. A., McGarry, F., and Behan, P. O. (1993) Increased transcription of interleukin-6 in the brains of mice with chronic enterovirus infection. *J. Gen. Virol.* **74,** 741–743.

92. Penfold, P. L., Madigan, M. C., Gillies, M. C., and Provis, J. M. (2001) Immunological and aetiological aspects of macular degeneration. *Prog. Retin. Eye Res.* **20,** 385–414.

93. Rungger-Brandle, E., Dosso, A. A., and Leuenberger, P. M. (2000) Glial reactivity, an early feature of diabetic retinopathy. *Invest. Ophthalmol. Vis. Sci.* **41,** 1971–1980.

94. Zeng, X. X., Ng, Y. K., and Ling, E. A. (2000) Neuronal and microglial response in the retina of streptozotocin-induced diabetic rats. *Vis. Neurosci.* **17,** 463–471.

95. Shin, D. H., Lee, H. Y., Lee, H. W., Lee, K. H., Lim, H. S., Jeon, G. S., et al. (2000) Activation of microglia in kainic acid induced rat retinal apoptosis. *Neurosci. Lett.* **292,** 159–162.

96. Neufeld, A. H. (1999) Microglia in the optic nerve head and the region of parapapillary chorioretinal atrophy in glaucoma. *Arch. Ophthalmol.* **117,** 1050–1056.

97. Wang, X., Tay, S. W., and Ng, Y. K. (2000) An immunohistochemical study of neuronal and glial cell reactions in retinae of rats with experimental glaucoma. *Exp. Brain Res.* **132,** 476–484.

98. Beschorner, R., Engel, S., Mittelbronn, M., Adjodah, D., Dietz, K., Schluesener, H. J., et al. (2000) Differential regulation of the monocytic calcium-binding peptides macrophage-inhibiting factor related protein-8 (MRP8/S100A8) and allograft inflammatory factor-1 (AIF-1) following human traumatic brain injury. *Acta Neuropathol.* **100,** 627–634.

99. Beschorner, R., Schluesener, H. J., Nguyen, T. D., Magdolen, V., Luther, T., Pedal, I., et al. (2000) Lesion-associated accumulation of uPAR/CD87-expressing infiltrating granulocytes, activated microglial cells/macrophages and upregulation by endothelial cells following TBI and FCI in humans. *Acta Neuropathol.* **100,** 627–634.

100. Engel, S., Schluesener, H., Mittelbronn, M., Seid, K., Adjodah, D., Wehner, H. D., et al. (2000) Dynamics of microglial activation after human traumatic brain injury are revealed by delayed expression of macrophage-related proteins MRP8 and MRP14. *Acta Neuropathol.* **100,** 313–322.

101. Beschorner, R., Adjodah, D., Schwab, J. M., Mittelbronn, M., Pedal, I., Mattern, R., et al. (2000) Long-term expression of heme oxygenase-1 (HO-1, HSP-32) following focal cerebral infarctions and traumatic brain injury in humans. *Acta Neuropathol.* **100,** 377–384.

102. Aihara, N., Hall, J. J., Pitts, L. H., Fukuda, K., and Noble, L. J. (1995) Altered immunoexpression of microglia and macrophages after mild head injury. *J. Neurotrauma* **12,** 53–64.

103. Griffin, W. S., Sheng, J. G., Gentleman, S. M., Graham, D. I., Mrak, R. E., and Riberts, G. W. (1994) Microglial interleukin-1α expression in human head injury: correlations with neuronal and neuritic beta-amyloid precursor protein expression. *Neurosci. Lett.* **176,** 133–136.

104. Takeuchi, A., Miyaishi, O., Kiuchi, K., and Isobe, K. (2001) Macrophage colony-stimulating factor is expressed in neuron and microglia after focal brain injury. *J. Neurosci. Res.* **65,** 38–44.

105. Mautes, A. E. M. and Noble, L. J. (2000) Co-induction of HSP70 and heme oxygenase-1 in macrophages and glia after spinal cord contusion in the rat. *Brain Res.* **883,** 233–237.

106. Schwab, J. M., Brechtel, K., Nguyen, T. D., and Schluesener, H. J. (2000) Persistent accumulation of cyclooxygenase-1 (COX-1) expressing microglia/macrophages and upregulation by endothelium following spinal cord injury. *J. Neuroimmunol.* **111,** 122–130.

107. Dijkstra, S., Geisert, E. E., Gispen, W. H., Bar, P. R., and Joosten, E. J. (2000) Up-regulation of CD81 (target of the antiproliferative antibody; TAPA) by reactive microglia and astrocytes after spinal cord injury in the rat. *J. Comp. Neurol.* **428,** 266–277.

108. Lee, Y. L., Shih, K., Bao, P., Ghirnikar, R. S., and Eng, L. F. (2000) Cytokine chemokine expression in contused rat spinal cord. *Neurochem. Int.* **36,** 417–425.

109. Saito, N., Yamamoto, T., Watanabe, T., Abe, Y., and Kumagai, T. (2000) Implications of p53 protein expression in experimental spinal cord injury. *J. Neurotrauma* **17,** 173–182.

110. Otto, V. I., Stahel, P. F., Rancan, M., Kariya, K., Shohami, E., Yatsiv, I., et al. (2001) Regulation of chemokines and chemokine receptors after experimental closed head injury. *Neuroreport* **12,** 2059–2064.

111. Beer, R., Franz, G., Srinivasan, A., Hayes, R. L., Pike, B. R., Newcomb, J. K., et al. (2000) Temporal profile and cell subtype distribution of activated caspase-3 following experimental traumatic brain injury. *J. Neurochem.* **75,** 1264–1273.

112. Beer, R., Franz, G., Schopf, M., Reindl, M., Zelger, B., Schmutzhard, E., et al. (2000) Expression of Fas and Fas ligand after experimental traumatic brain injury in the rat. *J. Cereb. Blood Flow Metab.* **20,** 669–677.

113. Sitte, H. H., Wanschitz, J., Budka, H., and Berger, M. L. (2001) Autoradiography with [^{3}H]PK11195 of spinal tract degeneration in amyotrophic lateral sclerosis. *Acta Neuropathol. (Berl.)* **101,** 75–78.

114. Almer, G., Vukosavic, S., Romero, N., and Przedborski, S. (1999) Inducible nitric oxide synthase up-regulation in a transgenic mouse model of familial amyotrophic lateral sclerosis. *J. Neurochem.* **72,** 2415–2425.

115. Nomura, H., Furuta, A., Suzuki, S. O., and Iwaki, T. (2001) Dorsal horn lesion resulting from spinal root avulsion leads to the accumulation of stress-responsive proteins. *Brain Res.* **893,** 84–94.

116. Fu, K. Y., Light, A. R., and Maixner, W. (2000) Relationship between nociceptor activity, peripheral edema, spinal microglial activation and long-term hyperalgesia induced by formalin. *Neuroscience* **101,** 1127–1135.

117. Herzberg. U. and Sagen, J. (2001) Peripheral nerve exposure to HIV viral envelope protein gp120 induces neuropathic pain and spinal gliosis. *J. Neuroimmunol.* **116,** 29–39.

118. Fu, K. Y., Light, A. R., and Maixner, W. (2000) Relationship between nociceptor activity, peripheral edema, spinal microglial activation and long-term hyperalgesia induced by formalin. *Neuroscience* **101,** 1127–1135.

119. Winkelstein, B. A., Rutkowski, M. D., Sweitzer, S. M., Pahl, J. L., and DeLeo, J. A. (2001) Nerve injury proximal or distal to the DRG induces similar spinal glial activation and selective cytokine expression but differential behavioral responses. *J. Comp. Neurol.* **439,** 127–139.

120. Colburn, R. W., Rickman, A. J., and DeLeo, J. A. (1999) The effect of site and type of nerve injury on spinal glial activation and neuropathic pain behavior. *Exp. Neurol.* **157,** 289–304.

121. Stuesse, S. L., Cruce, W. R., Lovell, J. A., McBurney, D. L., and Crisp, T. (2000) Microglial proliferation in the spinal cord of aged rats with a sciatic nerve injury. *Neurosci. Lett.* **287,** 121–124.

122. McGeer, P. L. and McGeer, E. G. (1995) The inflammatory response system of brain: implications for therapy of Alzheimer and other neurodegenerative diseases. *Brain Res. Rev.* **21,** 195–218.

123. Wright, J. L. and Merchant, R. E. (1992) Histopathological effects of intracerebral injections of human recombinant tumor necrosis factor-α in the rat. *Acta Neuropathol. (Berl.)* **85,** 93–100.

124. Simmons, R. D. and Willenborg, D. O. (1990) Direct injection of cytokines into the spinal cord causes autoimmune encephalomyelitis-like inflammation. *J. Neurol. Sci.* **100,** 37–42.

125. Rothwell, N. J. and Strijbos, P. J. L. M. (1995) Cytokines in neurodegeneration and repair. *Int. J. Dev. Neurosci.* **13,** 179–185.

126. Murphy, G. M. Jr., Jia, X.-C., Song, Y., Ong, E., Shrivastava, R., Bocchini, V., et al. (1995) Macrophage inflammatory protein 1-α mRNA expression in an immortalized microglial cell line and cortical astrocyte cultures. *J. Neurosci. Res.* **40,** 755–763.

127. Ringheim, G. E. (1995) Mitogenic effects of interleukin-5 on microglia. *Neurosci. Lett.* **201,** 131–134.

128. Atanassov, C. L., Muller, C. D., Dumont, S., Rebel, G., Poindron, P., and Seiler, N. (1995) Effect of ammonia on endocytosis and cytokine production by immortalized human microglia and astroglia cells. *Neurochem. Int.* **27,** 417–424.

129. Sebire, G., Hery, C., Peudenier, S., and Tardieu, M. (1993) Adhesion proteins on human microglial cells and modulation of their expression by IL1-alpha and TNF-alpha. *Res. Virol.* **144,** 47–52.

130. Cannella, B. and Raine, C. S. (1995) The adhesion molecule and cytokine profile of multiple sclerosis lesions. *Ann. Neurol.* **37,** 424–435.

131. Ryan, R., Sloane, B. F., Sameni, M., and Wood, P. L. (1995) Microglial cathepsin B: an immunological examination of cellular and secreted species. *J. Neurochem.* **65,** 1035–1045.

132. Banati, R. B., Rothe, G., Valet, G., and Kreutzberg, G. W. (1993) Detection of lysosomal cysteine proteinases in microglia: flow cytometric measurement and histochemcial localization of cathepsin B and L. *Glia* **7,** 183–191.

133. Kingham, P. J. and Pocock, J. M. (2001) Microglial secreted cathepsin B induces neuronal apoptosis. *J. Neurochem.* **76,** 1475–1484.

134. Gresser, O., Weber, E., Hellwig, A., Riese, S., Regnier-Vigouroux, A. (2001) Immunocompetent astrocytes and microglia display major differences in the processing of the invariant chain and in the expression of active cathepsin L and cathepsin S. *Eur. J. Immunol.* **31,** 1813–1824.

135. Yoshiyama, Y., Arai, K., Oki, T., and Hattori, T. (2000) Expression of invariant chain and pro-cathepsin L in Alzheimer's brain. *Neurosci. Lett.* **290,** 125–128.

136. Liuzzo, J. P., Petanceska, S. S., and Devi, L. A. (1999) Neurotrophic factors regulate cathepsin S in macrophages and microglia: a role in the degradation of myelin basic protein and amyloid beta peptide. *Mol. Med.* **5,** 334–343.

137. Cataldo, A. M. and Nixon, R. A. (1990) Increased cathepsin B activity in multiple sclerosis brain. *Proc. Natl. Acad. Sci. USA* **87,** 3861–3865.

138. Bever, C. T. Jr. and Garver, D. W. (1995) Increased cathepsin B activity in multiple sclerosis brain. *J. Neurol. Sci.* **131,** 71–73.

139. Ihara, M., Tomimoto, H., Kinoshita, M., Oh, J., Noda, M., Wakita, H., et al. (2001) Chronic cerebral hypoperfusion induces MMP-2 but not MMP-9 expression in the microglia and vascular endothelium of white matter. *J. Cereb. Blood Flow Metab.* **21,** 828–834.

140. Gottschall, P. E., Yu, X., and Bing, B. (1995) Increased production of gelatinase B (matrix metalloproteinase-9) and interleukin-6 by activated rat microglia in culture. *J. Neurosci. Res.* **42,** 335–342.

141. Yoshiyama, Y., Sato, H., Seiki, M., Shinagawa, A., Takahashi, M., and Yamada, T. (1998) Expression of the membrane-type 3 matrix metalloproteinase (MT3-MMP) in human brain tissues. *Acta Neuropathol. (Berl.)* **96,** 347–350.

142. Yoshiyama, Y., Asahina, M., and Hattori, T. (2000) Selective distribution of matrix metalloproteinase-3 (MMP-3) in Alzheimer's disease brain. *Acta Neuropathol. (Berl.)* **99,** 91–95.

143. Leake, A., Morris, C. M., and Whateley, J. (2000) Brain matrix metalloproteinase 1 levels are elevated in Alzheimer's disease. *Neurosci. Lett.* **291,** 201–203.

144. Schlomann, U., Rathke-Hartlieb, S., Yamamoto, S., Jockusch, H., and Bartsch, J. W. (2000) Tumor necrosis factor a induces a metalloprotease-disintegrin, ADAM8 (CD 156): implications for neuron-glia interactions during neurodegeneration. *J. Neurosci.* **20,** 7964–7971.

145. Nakajima, K., Tsuzaki, N., Takemoto, N., and Kohsaka, S. (1992) Production and secretion of plasminogen in cultured rat brain microglia. *FEBS Lett.* **308,** 179–182.

146. Nakajima, K., Tsuzaki, N., Shimojo, M., Hamanoue, M., and Kohsaka, S. (1992) Microglia isolated from rat brain secrete a urokinase-type plasminogen activator. *Brain Res.* **577,** 285–292.

147. Behrendt, N., Ronne, E., and Dane, K. (1993) Binding of the urokinase-type plasminogen activator to its cell surface receptor is inhibited by low doses of suramin. *J. Biol. Chem.* **268,** 5985–5989.

148. Nakajima, K., Shimojo, M., Hamanoue, M., Ishiura, S., Sugita, H., and Kohsaka, S. (1992) Identification of elastase as a secretory protease from cultured rat microglia. *J. Neurochem.* **58,** 1401–1408.

149. Matsuo, M., Hamasaki, Y., Fujiyama, F., and Miyazaki, S. (1995) Eicosanoids are produced by microglia, not by astrocytes, in rat glial cell cultures. *Brain Res.* **685,** 201–204.

150. Minghetti, L. and Levi, G. (1995) Induction of prostanoid biosynthesis by bacterial lipopolysaccharide and isoproterenol in rat microglial cultures. *J. Neurochem.* **65,** 2690–2698.

151. McGeer, P. L., Harada, N., Kimura, H., McGeer, E. G., and Schulzer, M. (1992) Prevalence of dementia amongst elderly Japanese with leprosy: apparent effect of chronic drug therapy. *Dementia* **3,** 146–149.

152. Breitner, J. C. S., Welsh, K. A., Helms, M. J., Gaskell, P. C., Gau, B. A., Roses, A. D., et al. (1995) Delayed onset of Alzheimer's disease with nonsteroidal anti-inflammatory and histamine H2 blocking drugs. *Neurobiol. Aging* **16,** 523–530.

153. McGeer, P. L. and Rogers, J. (1992) Anti-inflammatory agents as a therapeutic approach to Alzheimer's disease. *Neurology* **42,** 447–449.

154. Hull, M., Lieb, K., and Fiebich, B. L. (2000) Anti-inflammatory drugs: a hope for Alzheimer's disease? [review]. *Expert Opin.* **9,** 671–683.

155. Blain, H., Jouzeau, J.Y., Blain, A., Terlain, B., Trechot, P., Touchon, J., et al. (2000) Alzheimer's disease and non steroidal anti-inflammatory drugs with selectivity for cyclooxygenase-2: rationale and perspectives. *Presse Med.* **29,** 267–273 (in French).

156. Halliday, G. M., Shepherd, C. E., McCann, H., Reid, W. J., Grayson, D. A., Broe, G. A., et al. (2000) Effect of anti-inflammatory medications on neuropathological findings in Alzheimer disease. *Arch. Neurol.* **57,** 831–836.

157. Hoozemans, J. M., Rozemuller, A. M., Janssen, I., De Groot, C. A., Veerhuis, R., Eikelenboom, P. (2001) Cyclooxygenase expression in microglia and neurons in Alzheimer's disease and control brain. *Acta Neuropathol.* **101,** 2–8.

158. Yermakova, A. V., Rollins, J., Callahan, L. M., Rogers, J., and O'Banion, M. K. (1999) Cyclooxygenase-1 in human Alzheimer and control brain: quantitative analysis of expression by microglia and CA3 hippocampal neurons. *J. Neuropathol. Exp. Neurol.* **58,** 1135–1146.

159. Woodroofe, M. N., Sarna, G. S., Wadhwa, M., Hayes, G. M., Loughlin, A. J., Tinker, A., et al. (1991) Detection of interleukin-1 and interleukin-6 in adult rat brain, following mechanical injury, by in vivo microdialysis: evidence of a role for microglia in cytokine production. *J. Neuroimmunol.* **33,** 227–236.

160. Bauer, J., Ganter, U., Strauss, S., Stadtmüller, G., Frommberger, U., Bauer, H., et al. (1992) The participation of interleukin-6 in the pathogenesis of Alzheimer's disease. *Res. Immunol.* **143,** 650–657.

161. Yan, H. Q., Banos, M. A., Herregodts, P., Hooghe, R., and Hooghe-Peters, E. L. (1992) Expression of interleukin (IL)-1β, IL-6 and their respective receptors in the normal rat brain and after injury. *Eur. J. Immunol.* **22,** 2963–2971.

162. Castrell, J. V., Andus, T., Kunz, D., and Heinrich, P. C. (1989) Interleukin-6: the major regulator of acute-phase protein synthesis in man and rat. *Ann. NY Acad. Sci.* **557,** 87–101.

163. Wood, J. A., Wood, P. L., Ryan, R., Graff-Radford, N. R., Pilapil, C., Robitaille, Y., and Quirion, R. (1993) Cytokine indices in Alzheimer's temporal cortex: no changes in mature IL-1β or IL-1RA but increases in the associated acute phase proteins IL-6, α2-macroglobulin and C-reactive protein. *Brain Res.* **625,** 245–252.

164. Shohami, E., Novikov, M., Bass, R., Yamin, A., and Gailly, R. (1994) Closed head injury triggers early production of TNFalpha and IL-6 by brain injury. *J. Cereb. Blood Flow Metab.* **14,** 615–619.

165. Minami, M., Kuraishi, Y., and Satoh, M. (1991) Effects of kainic acid on messenger RNA levels of IL-1beta, IL-6, TNFalpha and LIF in the rat brain. *Biochem. Biophys. Res. Commun.* **176,** 593–598.

166. Gillespie, J. S., Cavanagh, H. M. A., Behan, W. M. H., Morrison, L. J. A., McGarry, F., and Behan, P. O. (1993) Increased transcription of interleukin-6 in the brains of mice with chronic enterovirus infection. *J. Gen. Virol.* **74,** 741–743.

167. McGeer, P. L. and McGeer, E. G. (1992) Complement proteins and complement inhibitors in Alzheimer's disease. *Res. Immunol.* **143,** 621–624.

168. Ishii, T. and Haga, S. (1992) Complements, microglial cells and amyloid fibril formation. *Res. Immunol.* **143,** 614–616.

169. McGeer, P. L., Kawamata, T., Walker, D. G., Akiyama, H., Tooyama, I., and McGeer, E. G. (1993) Microglia in degenerative neurological disease. *Glia* **7,** 84–92.

170. Hofman, F. M., Hinton, D. R., Johnson, K., and Merrill, J. E. (1989) Tumor necrosis factor identified in multiple sclerosis brain. *J. Exp. Med.* **170,** 607–612.

171. Walker, D. G. and McGeer, P. L. (1992) Complement gene expression in human brain: comparison between normal and Alzheimer disease cases. *Mol. Brain Res.* **14,** 109–116.

172. Brachova, L., Lue, L, Schultz, J., el Rashidy, T., and Rogers, J. (1993) Association cortex, cerebellum, and serum concentrations of C1q and factor B in Alzheimer's disease. *Mol. Brain Res.* **18,** 329–334.

173. Walker, D. G., Kim, S. U., and McGeer, P. L. (1995) Complement and cytokine gene expression in cultured microglia derived from postmortem human brains. *J. Neurosci. Res.* **40,** 478–493.

174. Walker, D. G., Yasuhara, O., Patston, P. A., McGeer, E. G., and McGeer, P. L. (1995) Complement C1 inhibitor is produced by brain tissue and is cleaved in Alzheimer disease. *Brain Res.* **675,** 75–82.

175. Gordon, D. L., Sadlon, T. A., Wesselingh, S. L., Russell, S. M., Johnstone, R. W., and Purcell, D. F. J. (1992) Human astrocytes express membrane cofactor protein (CD46), a regulator of complement activation. *J. Neuroimmunol.* **36,** 199–208.

176. McGeer, P. L., Walker, D. G., Akiyama, H., Kawamata, T., Guan, A. L., Parker, C. J., Okada, N., and McGeer, E. G. (1991) Detection of the membrane inhibitor of reactive lysis (CD59) in diseased neurons of Alzheimer brain. *Brain Res.* **544,** 315–319.

177. Gordon, D. L., Sadlon, T., Hefford, C., and Adrian, D. (1993) Expression of CD50, a regulator of the membrane attack complex of complement, on human astrocytes. *Mol. Brain Res.* **18,** 335–338.

178. McGeer, P. L., Kawamata, T., and Walker, D. G. (1992) Distribution of clusterin in Alzheimer brain tissue. *Brain Res.* **579,** 337–341.

179. Bertrand, P., Poirier, J., Oda, T., Finch, C. E., and Pasinetti, G. M. (1995) Association of apolipoprotein E genotype with brain levels of apolipoprotein E and apolipoprotein J (clusterin) in Alzheimer disease. *Mol. Brain Res.* **33,** 174–178.

180. Akiyama, H., Kawamata, T., Dedhar, S., and McGeer, P. L. (1991) Immunohistochemical localization of vitronectin, its receptor and beta-3 integrin in Alzheimer brain tissue. *J. Neuroimmunol.* **32,** 19–28.

181. Sobel, R. A., Chen, M., Maeda, A., and Hinojoza, J. R. (1995) Vitronectin and integrin vitronectin receptor localization in multiple sclerosis lesions. *J. Neuropathol. Exp. Neurol.* **54,** 202–213.

182. Wisniewski, T. and Frangione, B. (1992) Apolipoprotein E: a pathological chaperone protein in patients with cerebral and systemic amyloid. *Neurosci. Lett.* **135,** 235–238.

183. Griffin, W. S. T., Sheng, J. G., Roberts, G. W., and Mrak, R. E. (1995) Interleukin-1 expression in different plaque types in Alzheimer's disease: significance in plaque evolution. *J. Neuropathol. Exp. Neurol.* **54,** 276–281.

184. McGeer, P. L., Akiyama, H., Kawamata, T., Yamada, T., Walker, D. G., and Ishii, T. (1992) Immunohistochemical localization of beta-amyloid precursor protein sequences in Alzheimer and normal brain tissue by light and electron microscopy. *J. Neurosci. Res.* **31,** 428–442.

185. Konig, G., Monning, U., Czech, C., Prior, R., Banatis, R., Schreiter-Gasser, U., et al. (1992) Identification and differential expression of a novel alternative splice isoform of the βA4 amyloid precursor protein (APP) mRNA in leukocytes and brain microglial cells. *J. Biol. Chem.* **267,** 10,804–10,809.

186. Töpper, R., Gehrmann, J., Banati, R., Schwarz, M., Block, F., Noth, J., et al. (1995) Rapid appearance of β-amyloid precursor protein immunoreactivity in glial cells following excitotoxic brain injury. *Acta Neuropathol.* **89,** 23–28.

187. Perlmutter, L. S., Barron, E., and Chui, H. C. (1990) Morphologic association between microglia and senile plaque amyloid in Alzheimer's disease. *Neurosci. Lett.* **119,** 32–36.

188. Frackowiak, J., Wisniewski, H. M., Wegiel, J., Merz, G. S., Iqbal, K., and Wang, K. L. (1992) Ultrastructure of the microglia that phagocytose amyloid and the microglia that produce β-amyloid fibrils. *Acta Neuropathol. (Berl.)* **84,** 225–233.

189. Mackenzie, I. R. A., Hao, C. H., and Munoz, D. G. (1995) Role of microglia in senile plaque formation. *Neurobiol. Aging* **16,** 797–804.

190. Araujo, D. M. and Cotman, C. W. (1992) β-Amyloid stimulates glial cells in vitro to produce growth factors that accumulate in senile plaques in Alzheimer's disease. *Brain Res.* **569,** 141–145.

191. Aihara, N., Ishii, S., Kume, K., and Shimizu, T. (2000) Interaction between neurone and microglia mediated by platelet-activating factor. *Genes Cells* **5,** 397–406.

192. Dijkstra, S., Geisert, E. E., Gispen, W. H., Bar, P. R., and Joosten, E. J. (2000) Up-regulation of CD81 (target of the antiproliferative antibody; TAPA) by reactive microglia and astrocytes after spinal cord injury in the rat. *J. Comp. Neurol.* **428,** 266–277.

193. Frigerio, S., Silei, V., Ciusani, E., Massa, G., Lauro, G. M., and Salmaggi, A. (2000) Modulation of Fas-Ligand (Fas-L) on human microglial cells: an in vitro study. *J. Neuroimmunol.* **105,** 109–114.

194. Boje, K. M. and Arora, P. K. (1992) Microglial-produced nitric oxide and reactive nitrogen oxides mediate neuronal cell death. *Brain Res.* **587,** 250–256.

195. Chao, C. C., Hu, S., Molitor, T. W., Shaskan, E. G., and Peterson, P. K. (1992) Activated microglia mediate neuronal cell injury via a nitric oxide mechanism. *J. Immunol.* **149,** 2736–2741.

196. Colton, C. A. and Gilbert, D. L. (1993) Microglia, an *in vivo* source of reactive oxygen species in the brain. *Adv. Neurol.* **59,** 321–326.

197. Wood, P. L., Choksi, S., and Bocchini, V. (1994) Inducible microglial nitric oxide synthase: a large membrane pool. *Neuroreport* **5,** 977–980.

198. Chao, C. C., Hu, S., Close, K., Choi, C. S., Molitor, T. W., Novick, W. J., et al. (1992) Cytokine release from microglia: differential inhibition by pentoxifylline and dexamethasone. *J. Infect. Dis.* **166,** 847–853.

199. Peterson, P. K., Hu, S., Anderson, W. R., and Chao, C. C. (1994) Nitric oxide production and neurotoxicity mediated by activated microglia from human versus mouse brain. *J. Infect. Dis.* **170,** 457–460.

200. Bö, L., Dawson, T. M., Wesselingh, S., Mörk, S., Choi, S., Kong, P. A., et al. (1994) Induction of nitric oxide synthase in demyelinating regions of multiple sclerosis brains. *Ann. Neurol.* **36,** 778–786.

201. Walker, D. G., Kim, S. U., and McGeer, P. L. (1995) Complement and cytokine gene expression in cultured microglia derived from postmortem human brains. *J. Neurosci. Res.* **40,** 478–493.

202. Sherman, M. P., Griscavage, J. M., and Ignarro, L. J. (1992) Nitric oxide-mediated neuronal injury in multiple sclerosis. *Med. Hypotheses* **39,** 143–146.

203. Colasanti, M., Persichini, T., Di Pucchio, T., Gremo, F., and Lauro, G. M. (1995) Human ramified microglial cells produce nitric oxide upon *Escherichia coli* lipopolysaccharide and tumor necrosis factor a stimulation. *Neurosci. Lett.* **200,** 144–146.

204. Bagasra, O., Michaels, F. H., Zheng, Y. M., Bobroski, L. E., Spitsin, S. V., Fu, Z. F., et al. (1995) Activation of the inducible form of nitric oxide synthase in the brains of patients with multiple sclerosis. *Proc. Natl. Acad. Sci. USA* **92,** 12,041–12,045.

205. Kontes, H. K. and Wei, E. P. (1986) Superoxide production in experimental brain injury. *J. Neurosurg.* **64,** 803–807.

206. Colton, C. A., Yao, J., Gilbert, D., and Oster-Granite, M. L. (1990) Enhanced production of superoxide anion by microglia from trisomy 16 mice. *Brain Res.* **519,** 236–242.

207. Zhou, Y., Richardson, J. S., Mombourquette, M. J., and Weil, J. A. (1995) Free radical formation in autopsy samples of Alzheimer and control cortex. *Neurosci. Lett.* **195,** 89–92.

208. Schwab, J. M., Brechtel, K., Nguyen, T. D., and Schluesener, H. J. (2000) Persistent accumulation of cyclooxygenase-1 (COX-1) expressing microglia/macrophages and upregulation by endothelium following spinal cord injury. *J. Neuroimmunol.* **111,** 122–130.

209. Schwab, J. M., Nguyen, T. D., Postler, E., Meyermann, R., and Schluesener, H. J. (2000) Selective accumulation of cyclooxygenase-1-expressing microglial cells/macrophages in lesions of human focal cerebral ischemia. *Acta Neuropathol.* **99,** 609–614.

210. Advances in Alzheimer Therapy. Sixth International Stockholm/Springfield Symposium, 2000.

211. Wood, P. L. (1998) Disease-modifying drugs for Alzheimer's disease?: inhibitors of apoptosis, APP processing and prostaglandin formation. *IDrugs* **1,** 675–677.

212. Hadland, B. K., Manley, N. R., Su, D. M., Longmore, G. D., Moore, C. L., Wolfe, M. S., et al. (2001) Gamma-secretase inhibitors repress thymocyte development. *Proc. Natl. Acad. Sci. USA* **98,** 7487–7491.

213. Brayden, D. J., Templeton, L., McClean, S., Barbour, R., Huang, J. P., Nguyen, M., et al. (2001) Encapsulation in biodegradable microparticles enhances serum antibody response to parenterally-delivered beta-amyloid in mice. *Vaccine* **19,** 4185–4193.

214. Akama, K. T. and Van Eldik, L. J. (2000) β-Amyloid stimulation of inducible nitric-oxide synthase in astrocytes is interleukin-1 beta- and tumor necrosis factor-alpha (TNF alpha)-dependent, and involves a TNF alpha receptor-associated factor- and NF kappa B-inducing kinase-dependent signaling mechanism. *J. Biol. Chem.* **275,** 7918–7924.

215. Justicia, C., Gabriel, C., and Planas, A. M. (2000) Activation of the JAK/STAT pathway following transient focal cerebral ischemia: signaling through Jak1 and Stat3 in astrocytes. *Glia* **30,** 253-270.

216. Tan, J., Town, T., and Mullan, M. (2000) CD45 inhibits CD40L-induced microglial activation via negative regulation of the Src/p44/42 MAPK pathway. *J. Biol. Chem.* **275,** 37,224–37,231.

217. Tan, J., Town, T., Mori, T., Wu, Y. J., Saxe, M., Crawford, F., et al. (2000) CD45 opposes beta-amyloid peptide-induced microglial activation via inhibition of p44/42 mitogen-activated protein kinase. *J. Neurosci.* **20,** 7587–7594.

218. Hasegawa, H., Nakai, M., Tanimukai, S., Taniguchi, T., Terashima, A., Kawamata, T., et al. (2001) Microglial signaling by amyloid beta protein through mitogen-activated protein kinase mediating phosphorylation of MARCKS. *Neuroreport* **12,** 2567–2571.
219. Fabrizi, C., Silei, V., Menegazzi, M., Salmona, M., Bugiani, O., Tagliavini, F., et al. (2001) The stimulation of inducible nitric-oxide synthase by the prion protein fragment 106-126 in human microglia is tumor necrosis factor-alpha-dependent and involves p38 mitogen-activated protein kinase. *J. Biol. Chem.* **276,** 25,692–25,696.
220. Ryu, J., Pyo, H., Jou, I., and Joe, E. H. (2000) Thrombin induces NO release from cultured rat microglia via protein kinase C, mitogen-activated protein kinase, and NF-kappa B. *J. Biol. Chem.* **275,** 29,955–29,959.
221. Schlapbach, R., Spanaus, K. S., Malipiero, U., Lens, S., Tasinato, A., Tschopp, J., et al. (2000) TGF-beta induces the expression of the FLICE-inhibitory protein and inhibits Fas-mediated apoptosis of microglia. *Eur. J. Immunol.* **30,** 3680–3688.
222. Lee, Y. B., Schrader, J. W., and Kim, S. U. (2000) p38 map kinase regulates TNF-alpha production in human astrocytes and microglia by multiple mechanisms. *Cytokine* **12,** 874–880.
223. Hide, I., Tanaka, M., Inoue, A., Nakajima, K., Kohsaka, S., Inoue, K., et al. (2000) Extracellular ATP triggers tumor necrosis factor-alpha release from rat microglia. *J. Neurochem.* **75,** 965–972.
224. Zhu, X. W., Rottkamp, C. A., Boux, H., Takeda, A., Perry, G., and Smith, M. A. (2000) Activation of p38 kinase links tau phosphorylation, oxidative stress, and cell cycle-related events in Alzheimer disease. *J. Neuropathol. Exp. Neurol.* **59,** 880–888.
225. Zhu, X., Rottkamp, C. A., Hartzler, A., Sun, Z., Takeda, A., Boux, H., et al. (2001) Activation of MKK6, an upstream activator of p38, in Alzheimer's disease. *J. Neurochem.* **79,** 311–318.
226. Hensley, K., Floyd, R. A., Zheng, N. Y., Nael, R., Robinson, K. A., Nguyen, X., et al. (1999) p38 kinase is activated in the Alzheimer's disease brain. *J. Neurochem.* **72,** 2053–2058.
227. Monnet-Tschudi, F., Zurich, M. G., Pithon, E., Van Melle, G., and Honegger, P. (1995) Microglial responsiveness as a sensitive marker for trimethyltin (TMT) neurotoxicity. *Brain Res.* **690,** 8–14.
228. McCann, M. J., O'Callaghan, J. P., Martin, P. M., Bertram, T., and Streit, W. J. (1996) Differential activation of microglia and astrocytes following trimethyl tin-induced neurodegeneration. *Neuroscience* **72,** 273–281.
229. Maier, W. E., Brown, H. W., Tilson, H. A., Luster, M. I., and Harry, G. J. (1995) Trimethyltin increases interleukin (IL)-1α, IL-6 and tumor necrosis factor α mRNA levels in rat hippocampus. *J. Neuroimmunol.* **59,** 65–75.
230. Mogi, M., Togari, A., Tanaka, K., Ogawa, N., Ichinose, H., and Nagatsu, T. (2000) Increase in level of tumor necrosis factor-alpha in 6-hydroxydopamine-lesioned striatum in rats is suppressed by immunosuppressant FK506. *Neurosci. Lett.* **289,** 165–168.
231. Tikka, T., Fiebich, B. L., Goldsteins, G., Keinanen, R., and Koistinaho, J. (2001) Minocycline, a tetracycline derivative, is neuroprotective against excitotoxicity by inhibiting activation and proliferation of microglia. *J. Neurosci.* **21,** 2580–2588.
232. Hensley, K., Robinson, K. A., Pye, Q. N., Floyd, R. A., Cheng, I., Garland, W. A., et al. (2000) CPI-1189 inhibits interleukin 1beta-induced p38-mitogen-activated protein kinase phosphorylation: an explanation for its neuroprotective properties? *Neurosci. Lett.* **281,** 179–182.
233. Pulliam, L., Irwin, I., Kusdra, L., Rempel, H., Flitter, W. D., and Garland, W. A. (2001) CPI-1189 attenuates effects of suspected neurotoxins associated with AIDS dementia: a possible role for ERK activation. *Brain Res.* **893,** 95–103.
234. Bjugstad, K. B., Flitter, W. D., Garland, W. A., Su, G. C., and Arendash, G. W. (1998) Preventive actions of a synthetic antioxidant in a novel animal model of AIDS dementia. *Brain Res.* **795,** 349–357.

235. Bjugstad, K. B., Flitter, W. D., Garland, W. A., Philpot, R. M., Kirstein, C. L., and Arendash, G. W. (2000) CPI-1189 prevents apoptosis and reduces glial fibrillary acidic protein immunostaining in a TNF-alpha infusion model for AIDS dementia complex. *J. Neurovirol.* **6,** 478–491.

236. Lim, G. P., Chu, T., Yang, F. S., Beech, W., Frautschy, S. A., and Cole, G. M. (2001) The curry spice curcumin reduces oxidative damage and amyloid pathology in an Alzheimer transgenic mouse. *J. Neurosci.* **21,** 8370–8377.

237. Wood, P. L. (1994) Microglia: a possible cellular target for pharmacological approaches to neurodegenerative disorders. *Drug News Perspect.* **7,** 138–157.

238. Lee, Y. B., Schrader, J. W., and Kim, S. U. (2000) p38 map kinase regulates TNF-alpha production in human astrocytes and microglia by multiple mechanisms. *Cytokine* **12,** 874–880.

2

Apoptosis vs Nonapoptotic
Mechanisms in Neurodegeneration

Kurt A. Jellinger

1. INTRODUCTION

Neurodegenerative disorders are morphologically featured by progressive cell loss in specific vulnerable neuronal populations of the brain and/or spinal cord, often associated with typical cytoskeletal protein changes forming intracytoplasmic and/or intranuclear inclusions and gliosis. In Alzheimer's disease (AD), the most common type of dementia in advanced age, loss of cortical neurons and synapses is accompanied by extracellular deposition of Aβ4 amyloid peptide (Aβ) in senile plaques and cerebral vessels. Paired helical filaments containing hyperphosphorylated microtubule-associated tau protein forming neurofibrillary tangles (NFT), neuropil threads, and neuritic plaques are other histopathological hallmarks of AD (1). In Parkinson's disease (PD), the most frequent extrapyramidal movement disorder in adults, neuron loss in substantia nigra (SN) and other subcortical nuclei is associated with widespread occurrence of intracytoplasmic Lewy bodies (LB) formed from fibrillary α-synuclein and hyperphosphorylated neurofilament protein (2–5). Frequent cortical LBs occur in dementia with Lewy bodies (DLB), which is the second most frequent type in adult age dementia (6), but small numbers are also seen in both PD and AD (7,8). In Pick disease (PiD), a rare presenile type of frontotemporal dementia (FTD), progressive cortical degeneration is associated with Pick bodies, intracytoplasmic accumulations of hyperphosphorylated tau protein (9–11). In multisystem atrophy (MSA), progressive supranuclear palsy (PSP) and corticobasal degeneration (CBD), which are rigid-akinetic extrapyramidal disorders, multisystemic neuronal degeneration is accompanied by glial cytoplasmic inclusions (GCI) containing tau protein that differs from that in AD and PiD (2,10,11) and α-synuclein (12). In Huntington disease (HD), an autosomal dominantly inherited hyperkinetic extrapyramidal disorder resulting from mutation of the IT-15 gene on chromosome 4p16.3 with expanded polyglutamine CAG repeats, neurodegeneration in man and transgenic mice is related to neuronal intranuclear inclusions containing huntingtin and ubiquitin (13,14). Similar inclusions are seen in other rare autosomal dominant ataxic polyglutamine disorders (e.g., dentatorubropallidoluysian atrophy [DRPLA]), suggesting that these protein aggregates are a common feature of the pathogenesis of glutamine repeat neurodegenerations (13,15). In amyotrophic lateral sclerosis (ALS), which is an adult neurodegenerative

From: *Neuroinflammation, 2nd Edition: Mechanisms and Management*
Edited by: P. L. Wood © Humana Press Inc., Totowa, NJ

disease of both lower and upper motor neurons, skein-like ubiquitin-containing inclusions are seen *(16)*. The nature, time-course, and molecular causes of cell death and their relation to abnormal cytoskeletal protein aggregations, in these and other neurodegenerative disorders, are a matter of considerable controversy. Recent studies have provided new insights into cell death programs and their roles in neurodegeneration that will be discussed in this chapter.

2. BASIC ROUTES OF CELL DEATH

2.1. Major Modes of Cell Death

Currently, three major mechanisms of neuronal demise are discussed in neurodegeneration: apoptosis, (oncotic) necrosis, and autophagy. These cell death types are different, frequently divergent, but sometimes overlapping cascades of cellular breakdown. Because of modulation of these cascades by cellular available energy, cells may use diverging executive pathways of demise *(17)*.

Apoptosis, coined from its Greek equivalent *(18)*, is a specific form of gene-directed programmed cell death (PCD) to execute removal of unnecessary (superflual), aged, or damaged cells and plays a central role in metazoan development and homeostasis. Based on distinct morphologic criteria recognized by Flemming as long ago as 1885 *(19)* and biochemical features, a clear distinction has been made between apoptosis and other types of cell death, such as necrosis or autophagic degeneration *(20–22)*. The major morphologic features of the three major types of cell death are summarized in Table 1.

Apoptosis is carried out by an intrinsic suicidal program of the cell and can be triggered by environmental stimuli, including irradiation, lack of mitochondrial DNA gene expression leading to DNA damage, oxidative stress, toxins, viruses, withdrawal of neurotrophic support, and so forth *(22–37)*. Distinct morphologic features of apoptosis, representing the nuclear type of cell death *(21)*, include cell shrinking, plasma membrane blebbing, nuclear and cytoplasmic condensation, clumping of chromatin at nuclear membrane, fragmentation of chromosomal DNA at internucleosomal intervals, loss of ribosomes from the rough endoplasmic reticulum and from polysomes, preservation of convoluted membranes forming blebs, fragmentation of the cell into apoptotic bodies, and degradation of the apoptotic body by lysosomal enzymes. The earliest changes occur in the nucleus and consist of the formation of spherical chromatin masses and flocculent densities, likely associated with subtle complex changes in the nucleolus. These changes occur while the nuclear envelope remains intact, and cytoplasmic organelles are relatively unaltered, with the exception that a few mitochondria may develop a defect in their outer limiting membrane, and the mitochondrial matrix becomes edematous in the vicinity of the membrane defect. Typically, these early changes are followed in the mid- and end stages by progressive fragmentation of the nuclear envelope, intermixing of nucleoplasmic and cytoplasmic contents, gradual condensation of the entire cell, and gradual deteriorative changes in cytoplasmic organelles. In the late stages, apoptotic bodies consisting of one or more balls of clumped nuclear chromatin and a contingent of cytoplasm, all enclosed within a limiting membrane, are sometimes extruded within a limiting membrane and sometimes extruded from the cell into the neuropil, where they become surrounded by glial processes *(38)*. This process has no adverse consequences for neighboring cells in terms of inflammation or the release of potentially dangerous genetic

Table 1
Basic Features of Different Forms of Cell Death

Apoptosis	Necrosis	Autophagic degeneration
Cell and cytoplasmic shrinkage	Cell and mitochondrial swelling	Endocytosis and blebbing
Membrane budding	Disintegration of membranes (blebbing)	Pyknosis; part of nucleus may bleb or segregate
Nuclear condensation; chromatin aggregation	Random aggregation of DNA	
Margination of condensed chromatin at nuclear membrane	Breaks of membranes	
Nuclear DNA fragmentation/DNA laddering on agarose gel	DNA smear in agarose gel	
Formation of apoptotic bodies (nuclear fragment surrounded by small rim of cytoplasm)	Lysis of the cell	Abundant autophagic vacuoles
Organelles and membranes remain intact	Disintegration of organelles and plasma membranes	Dilatation of endoplasmic reticulum, mitochondria, and Golgi
Loss of ribosomes from rough endoplasmic reticulum		
Partial maintainance of ion homeostasis	Disturbed ion homeostasis	
Enzymatic process/caspase-3 activation, release of cytochrome-c		
Production of "death proteins"	Cessation of protein synthesis	
Energy-dependent maintainance of ATP levels	Energy-independent decrease of ATP	
Mitochondrial membrane permeability transition		
Well-controlled cell death (immediate clearing from tissue)	Insult-induced spontaneous cell death	
Heterophagic elimination	No heterophagic elimination	Occasional and late heterophagic elimination
Phagocytoses by macrophages or glial cells	Phagocytoses by macrophages	
No inflammatory response *in situ*	Inflammatory reaction	

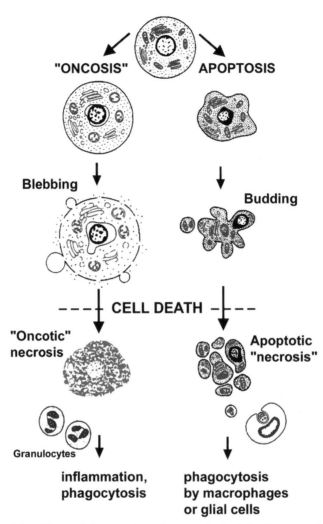

"ONCOSIS"   APOProsis

Blebbing

Budding

- - -| CELL DEATH -|- - -

"Oncotic" necrosis

Apoptotic "necrosis"

Granulocytes

inflammation, phagocytosis

phagocytosis by macrophages or glial cells

Fig. 1. Relationship of "oncotic" necrosis and apoptosis, both leading to cell death. Modified from ref. *29* with permission from Dr. Levin.

material, as apoptosis is generally associated with heterophagic elimination (Fig. 1). The main criterion of apoptosis, the presence of intranucleosomal DNA cleavage, seen as "DNA laddering" on gel electrophoresis, is considered a relatively late event *(22)*. However, cell death is preceded in many instances by fragmentation of DNA by $Ca^{2+}Mg^{2+}$-dependent DNAse into 180 and 200 base pair (bp) fragments with endonuclease activation occurring early in the process of cell death *(27–35)*.

Necrosis means mortification of tissue, but since 1980, it has been given a particular cellular sense with the claim that virtually all cell deaths can be classified dichotomously as either apoptosis or necrosis *(20–25)*. Necrosis or cytoplasmic cell death *(21)* is a passive pathological event—killing of the cell—arising from spontaneous insults (e.g., stroke or trauma) *(21,29)*. It is characterized by cell, organelle, and mitochondrial swelling or dilatation, cytoplasmic vacuolation, breaks of cell membranes, disintegration of organelles, and a final cell bursting. During its development, the nucleus remains intact, at least at the beginning of the process. There is no heterophagic elimination, but phago-cytosis by macrophages with inflammatory reaction in the surrounding tissue (Fig. 1).

Another cell death process is *excitotoxic neurodegeneration* that has been described either as a necrotic or an apoptotic process *(38)*, as sometimes one or the other, depending on intensity of the stimulus, or as a hybrid mixture of both processes on a continuum. Ultrastructurally, the sequence and types of changes that characterize apoptotic cell death have been shown to be strikingly different from those in excitotoxic cell death, whereas DNA fragmentation analysis is not a reliable means of identifying an apoptotic process in that these tests were positive in both PCD and excitotoxic cell death (ECD) *(38,39)*. On the other hand, comparison between ECD and necrosis *(18,22)* suggests that these also are two separate and distinct phenomena, because, in contrast to classical "necro-sis," ECD often entails massive nuclear chromatin clumping that remains conspicuously even after the dead cell has been phagozyted *(38)* and, thus, does not meet the Kerr/Wyllie criteria for either apoptosis or necrosis. Although excitotoxic and apoptotic nuclear changes sometimes look very similar, they occur in different temporal sequence. More-over, fragmentation of the nuclear membrane, a fundamental characteristic of neuronal apoptosis, does not occur in excitotoxic neurodegeneration *(38,39)*.

Modern research on apoptosis has been concentrated mainly in the areas of cancer and immunology *(22–26,29,40)*, but a growing number of diseases, including neurodegen-erative disorders, have been linked to inappropriate apoptosis *(27,30,31,33,37,41–49)*. In the Medline database (October 9, 2002), of almost 61,000 papers mentioning apop-tosis, 6900 in the nervous system, 875, i.e., 9.13%, deal with neurodegeneration, and most assume "apoptotic" neurons (and "necrotic" ones when mentioned) to have the main characteristics of apoptosis (or necrosis) as elucidated in the fields of cancer and immunology *(21,40)*. Although the term *apoptosis* has often been used interchange-ably with PCD, they are not the same, because apoptosis is only *one* specific form of PCD, and several authors have emphasized that the proposed dichotomy between apop-tosis and necrosis is an oversimplification *(21,25,33,36–38,51–53,53a)*.

Other modes of cell death include *autophagic degeneration (47)*, characterized by the formation of numerous autophagic vacuoles, endocytosis, enlargement of the Golgi appa-ratus as a source of lysosomes providing hydrolytic enzymes for the autophagic vacu-oles, vacuolization of the endoplasmic reticulum, and moderate condensation of nuclear chromatin that may ultimately leave the pyknotic nuclei and is destroyed by autolyso-somes of the same cell, with occasional and late heterophagic elimination (Table 1).

It appears doubtful whether these modes of cell death really represent distinct entities. They would rather appear to form part of a continuum between apoptosis and necrosis, depending on the severity of the insult, such as Ca^{2+} levels, intracellular energy (ATP) levels, mitochondrial function, glutamate receptor stimulation, oxidative stress, nitric oxide (NO) release, and so forth *(17,21,25,30,31,54–59)*. Although most dying neurons in development fall into one of the three main categories, some combine features of more than one of them or do not match any, whereas such exceptions are more common under abnormal conditions (e.g., in genetic mutants) *(21)*.

Recently, several alternative nonapoptotic forms of PCD that are distinct from apop-tosis by the criteria of morphology, biochemistry, and response to apoptosis inhibitors have been discussed *(38,60)*. Morphologically, this alternative form of PCD that is

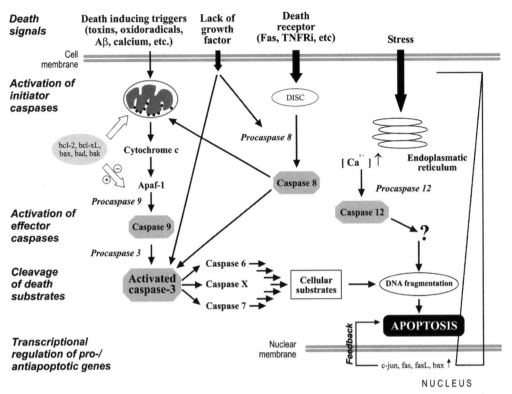

Fig. 2. Markers for apoptosis in tissue and their localization in the apoptotic cascade: at least four distinct signaling pathways, such as apoptotic triggers, death receptors, lack of neurotrophic factors, and stress, induce complex caspase activation. All these different events may have the activation of caspase-3 in common, finally leading to DNA fragmentation and apoptotic cell death. Modified from ref. *53.*

driven by an alternative caspase-9 activity (*see* Subheading 2.2.) appears during development and in some cases of neurodegeneration.

2.2. Molecular Players of Cell Death Cascades

The principal molecular players of the apoptotic program include a number of apoptosis-inducing or death receptors (e.g., Apo-1/Fas [CD95/Fas Receptor], Apaf-1 [apoptotic protease-activating factor-1] and other apoptosis-initiating factors [AIFs], small proteins released from damaged mitochondria into the cytoplasm in early stages of cell death, aspartate-specific cystein proteases of the caspase/calpain family, the Bcl-2 and p53 oncogene families, and mitogen-activated protein kinase pathways regulated by neurotrophins) (Fig. 2). Once activated, many, but not all, of them induce proteolysis of specific cellular substructures and consequently amplify the death signal cascade *(30–35, 53,57,59,61–66)*. Extensive in vitro studies indicate that caspases are required as effectors of proapototic signaling pathways. They consist of a family of cysteine proteases that are expressed primarily as single chain proenzymes *(61,67,68)*. These proteases are structurally similar, with the inactive forms consisting of an N-terminal prodomain, a large and a small subunit. The three domains are released by limited proteolysis, and the active protease is formed as a tetramer made up of two small and two large subunits.

Caspases with long prodomains are classified as upstream initiator proteases (e.g., caspase-9), whereas caspases containing short prodomaines (e.g., caspase-3) function as downstream effectors of the apoptotic process. Based on their substrate specifities and DNA sequence homologs, the 14 currently identified caspases may be divided into three families: apoptotic initiators, apoptotic executioners, and inflammatory mediators *(66)*. They are located in the cytosol as inactive zymogens (procaspases) that can be proteolytically activated either by autoprocessing (oligomerization) or by other caspases (cleavage). Caspases are activated through two principal pathways known as the "extrinsic pathway," which is initiated by cell surface death receptor ligation, and the "intrinsic pathway," which arises from mitochondria (reactive oxygen species production, membrane depolarization, and so forth), and there can be crosstalk between these two *(34)*. The process occurs after aspartic acid residues at substrate recognition consensus sites, leading to removal of an N-terminal prodomain and generation of heterodimeric catalytic domains. These catalytic subunits, in turn, cleave a variety of vital nuclear (DNA-PK, PK-γ, chromatin, poly/ADP-ribose/polymerase/PARP, etc.) and extranuclear (gelsolin, lamins, β-actin, α-fodrin, etc.) constituents, also called cell death substances, as well as proteins involved in DNA repair. This contributes to the disassembly of cell structures, to the reorganization of the cytoskeleton, to deficits in DNA repair and replication, and to the cleavage of DNA into 180 bp fragments *(31,63)*. A well established sequel of DNA damage is the activation PARP and the ensuing addition of poly(ADP-ribose) (PAR) to nuclear proteins. PARP is activated by free DNA ends generated during the apoptotic process, and its consumption of energetic substrates needed for PAR-synthesis, i.e., NAD^+, may lead to energy depletion and contributes to cell death *(30,32)*. Increased poly(ADP-ribosyl)ation has been reported in extracts of apoptotic cells, and detection of PAR has been used as an early marker for apoptotis associated DNA damage (*see* refs. *63* and *66*). Depending on the cell type and the apoptotic stimulus, different caspases are preferentially activated which, in turn, have different substrate specificities *(31,66)*. Central players of PCD are upstream instigator caspases (#8, #9, and #10), activating other caspases, named effector caspases (#3, #6, #7, and #12), important regulators of postmitotic neuronal homeostasis *(31,35,69–71)*. Processing of the key protease, caspase-3, depends on the activation of caspase-9 but not caspase-8. Selective peptide inhibitors of caspase-9 block processing of caspase-3 and -8 and inhibit apoptosis, whereas a selective inhibitor of caspase-8 blocks neither processing of caspase-3 nor cell death. On the other hand, neuronal apoptosis triggered by potassium deprivation is death-receptor-independent but involves the mitochondrial pathway of the predominant upstream mediators of caspase activation are cytokines of the tumor necrosis factor (TNF) family, such as Fas-L, Apo-3L, TNF-α, and tumor necrosis factor receptor 1 (TNF-RI), which may induce trimerization of their respective receptors (Fas, TNF-RI, Apo-E) or binding (Fig. 2). The Fas (CD95, APO-1) receptor has an extracellular domain for ligand binding and an intracellular death domain. Following binding to its specific ligand, Fas-L trimerization of Fas recruits the Fas-associated death domain (FADD) via interaction between the death domains of Fas and FADD. Caspase-8 activation, by binding of its death effector domain to the FADD domain, initiates the caspase caseade. The Fas/Fas-L signaling system plays a role in the control of cell death and survival of lymphocytes, in the regulation of the immune system, and in the progression of autoimmune diseases. Although it appears not important in regulating neuronal cell loss in neurodegenerative disorders, its activation in reactive astrocytes in damaged brains remains to be elucidated *(72)*. Through adaptor proteins, such as FADD and TRADD, caspase-8 is autoproteolytically activated, which, in turn, may either cleave caspase-3 directly or amplify the death signal through translocation of BID, a mitochondrial protein of the Bcl-2 family, into the mitochondria and subsequent release of cytochrome c from the mitochondrial intermembrane space into the cytosol *(65)*. Recent studies suggest that extracellular signals that lead to oxidative damage and mitochondrial dysfunction with subsequent cytochrome-c release have the potential to activate the caspase-3 apoptotic pathway *(73)*. Such downstream caspases and their proteolytic products are recognized markers of apoptosis, their activation representing an irreversible step in the cell death cascade, and cells expressing these enzymes are prone to death PCD finally culminates in an execution phase where key cellular structures undergo proteolytic digestions. Caspase-3 appears to be one of the main executioner enzymes. Caspase-3, caspase-9 and caspase-12 (–/–) mutant mice show decreased apoptosis, resulting in altered nervous system development *(74,75)*.

Other molecular changes associated with caspase-independent pathways of apoptosis include induction of the expression of proapoptotic proteins such as prostate apoptosis response-4 (Par-4) *(62,76)* and certain members of the proto-oncogene Bcl-2 family, which includes a highly homogenous group of mitochondrial proteins that act either to enhance (Bax, Bid, Bad, Bak, *Bcl-x*) or prevent (*Bcl-2, Bcl-xL*, Mcl-1) apoptosis by forming homotypic or heterotypic dimers, which may affect the formation of a permeability transition pore in the mitochondrial membrane and the release of cytochrome-c *(30,31,77)*. Opening of these pores seems to be closely controlled by members of the Bcl-family, which sit on the external mitochondrial membrane where they could act as ionic channels *(78)*. Cytochrome-c release from mitochondria intermembrane space into the cytosol leads to the formation of a cytoplasmic complex, including Apaf-1 and caspase-9, which results in caspase-9 activation. Subsequently, caspase-9 cleaves and activates downstream caspase-3. Alternatively, the release of AIF, a protease, is capable of directly inducing nuclear fragmentation and caspase-3 activation that is independent of APA1/caspase-9 *(79)*. AIF-dependent cell death displays structural features of apoptosis and can be genetically uncoupled from Apaf1 and caspase-9 expression *(80)*. Conversely, antiapoptotic members of the Bcl-family, such as Bcl-2 and *BCl-xL*, appear to inhibit the opening of the mitochondrial permeability transition pores *(81)*. Bcl-2 inhibiting both apoptotic and necrotic cell death, presumably by interfering with reactive oxygen molecules *(82)*, is upregulated in cells with DNA fragmentation *(83)*. Bcl-x, an antiapoptotic member of the *Bcl-2* family, acts by holding the proapoptotic Apaf-1/caspase-9 complex inactively bound to the mitochondial membrane *(31,84)*. Bax, a proapoptotic homo-log of *Bcl-2* also bound to mitochondria, which induces the opening of the mitochondrial permeability transition pores and subsequent release of cytochrome-c by heterodimerizing with Bcl-xL and displaying it from the inactive Apaf-1/caspase-3 complex, activates caspase-3, which cleaves other caspases in the death cascade *(69,72,85)*. It also promotes neuronal death in response to cytotoxic injury with a key role for p53 activation and of the Bcl-2 family *(86)*. Cell death in primary cultures induced by oxidative glutamate toxicity or glutamate-mediated excitotoxicity T is not altered in the Bax–/– homozygous knockout animals. In contrast, there is a 50% inhibition of spontaneous cell death which suggests that a classical Bax-dependent apoptotic pathway contributes to the spontaneous cell death when nerve cells are initially exposed to culture conditions, while a Bax-dependent RCD pathway is not utilized in oxidative glutamate toxicity and NMDA receptor mediated excitotoxicity following brief exposure to low concentrations of glutamate *(87)*. Cells overexpressing both Bcl-2 and Bax show no signs of caspase activation and survive even though they have significant amounts of cytochrome-c in the cytoplasm, indicating that Bcl-2 can prevent Bax-induced apoptosis by other mechanisms *(85)*. The abundance of proapoptotic and antiapoptotic Bcl-2 family members—most likely via the formation of homodimers and/or heterodimers—serves as an important regulatory instance *(77)*. The p53 family of oncogenes are transcription factors playing a major role in determining the cell fate in response to DNA damage. P53 induces Fas/Apo1, DRS, Bax, and other cell death-related proteins and represses survival mediators, such as IGF-15 *(88)* The recently reported involvement of the p53 family in regulating apoptotisis in both developing and injured mature neuronas systems *(88)* has not been observed in human neurodegenerative disorders *(52,53)*.

In neuronal apoptosis, several lines of evidence indicate that caspase-3, previously designated CPP32/YamalApopain, a 32-kDa protein that can be cleaved into active p17 and p12 fragments, is a major effector of intracellular death signals. Its involvement in apoptotic neuronal death evoked by DNA damage *(89)*, and neuronal death in several chronic neurodegenerative disorders has been associated with activation of caspases (e.g., cleavage by caspase-3 of presenilins and amyloid precursor protein [APP] in AD) *(90–92)*, and of huntingtin and atrophin-1 in HD and DRPLA, respectively *(13,93,94)*, and its increased activation in the SN of PD brain *(55,95,96)*, whereas neuronal apoptosis induced by Aβ is mediated by multiple caspases *(97,97a)*. Caspase-8 is an effector in neuronal apoptosis induced by Aβ 1-42 *(97b)* and in apopoptotic death of dopaminergic neurons induced by 1,2,3,6-tetrahydropyridine in mice, whereas specific caspase-8 inhibitors in tissue cultures did not result in neuroprotection but seemed to trigger a switch from apoptotis to necrosis *(96)*. Caspase activation and cytochrome-c release from mitochondria play an important role in hypoxia-induced neuronal apoptosis *(98)*, in delayed neurodegeneration in neonatal rat thalamus after hypoxia-ischemia appears to be apoptosis-mediated by death receptor activation (e.g., Fas death receptor, Bax, cytochrome-c, activation of caspase-8, abnormalities in mitochondria that precede the activation of caspase-3, and the appearance of neuronal apoptosis) *(99)*, and in dopamine-induced cell death.

Because of the genetically driven nature of apoptosis, nuclear transcription factors, specifically immediate early genes (IEG), may influence the initition and execution of apoptosis. The proto-oncogenes c-Jun and c-Fos, members of the IEG family, have a putative role in transcriptional regulation of apoptosis-associated genes (e.g., p53) *(100)*. C-Jun is expressed for extended periods, whereas the expression of c-Fos appears to be a late event in apoptotis, occurring just before chromatin condensation. Grand et al. *(101)* described an IEG-encoded protein, c-Jun/Ap1 and Ab-2, appearing rather late in the process of apoptosis. Necrotic cells are not stained with these antibodies *(102)* that define cellular responses following excitotoxicity *(103)*. Increasing evidence suggests that the regulation of neuronal cell death is complex, utilizing multiple pathways that are depending on the neurotoxic insult and are also influenced by subtle differences among neuronal cell phenotypes *(104)*. Similar death-signaling pathways might be activated in neurodegenerative disorders by proteins belonging to abnormal subcellular structures and cytoplasmic or nuclear inclusions, such as Aβ peptide, tau protein, parkin, huntingtin, α-synuclein, etc. *(30,33,105–107,107a)*.

Endogenous inhibitors of the apoptosis protein (IAP) family, e.g., neuronal inhibitor of apoptosis protein (NIAP), X-linked inhibitor of apoptosis protein (X-IAP), survivin, c-IAP1, c-IAP2 *(108)*, and vital antiapoptotic proteins such as CrmA, Iaps, and p35 *(109)*, modulate caspase activity at various points within these pathways. The reason for this additional safety check in activation of the caspase-3 apoptotic cascade was unclear until the recent discovery of Smac/DIABLO, a mitochondrial protein that is released with cytochrome-c; it displaces XIAP from caspase-9 and -3, allowing the former to be activated through its interaction with Apaf-1 *(110,111)*. Apoptosis of the cell body can be delayed by members of the Bcl-2 family that may elicit protective effects by preventing the release of Apaf-1/caspase-3 from the mitochondria *(31)* Recently, heat shock protein 70 has been added to the list of molecules with strong antiapoptotic properties *(112)*. However, blockage of the caspase execution machinery may only temporarily rescue damaged neurons, which does not necessarily warrant cell survival, and classical apoptotic features can still appear in caspase-inhibited neurons *(113)*. In some cases, proteolysis activates caspase substrates, in others, it inactivates or destroys them, but the crucial substrate proteins that coordinate cell death are as yet uncharacterized (Fig. 3). Caspase inhibition is considered a potential therapeutic strategy in neurodegenerative diseases. Although the various molecular players in apoptotic cell death are becoming increasingly well known, whether or not they are involved or may interact with one another in neuronal cell death cascade, in various neurodegenerative processes, is just beginning to be understood.

2.3. Methods for Detection of Apoptosis

As a result of constantly expanding insights into cell biology, detection of apoptotic cells that has been previously been based on morphology and the detection of DNA fragmentation by biochemical or histochemical techniques is now advancing to more specific methods *(63,66,115,116)*. Detection of *in situ* DNA fragmentation, currently used as one of the most frequently used techniques to highlight apoptotic cells in tissue, is not specific for apoptosis, but detects DNA damage in a variety of cell injury and cell death paradigms. DNA fragmentation into multiples of 180–200 bp is considered a hallmark of PCD *(18,22,23)*, but large fragments in the range of 50–150 kbp are also generated, and other forms of PCD that lack the prototypical internucleosomal pattern of DNA fragmentation have also been reported. In contrast, necrotic cell death is accompanied by late and random DNA fragmentation owing to release of lysosomal DNAse, although this notion has been challenged recently *(see ref. 63)*. Besides the characteristic, oligonucleosomal fragmentation, it is mainly the enormous amount of DNA strand breaks that is the basis for the relative specificity of *in situ* DNA fragmentation techniques of apoptotic cell death.

Enzymatic histochemical techniques for the *in situ* detection of fragmented DNA include the widely used TUNEL reaction (terminal deoxynucleotidyl transferase [TdT])-mediated dUTP-biotin nick-end labeling) *(116)* or *in situ* nick-end labeling by digoxigenine-antidigoxigenin *(117)*. These enzymatic methods to detect free three ends generated by endonuclease cleavage of genuine DNA into 180–200 bp fragments during the apoptotic process have undergone many variations and improvements to serve different tissue

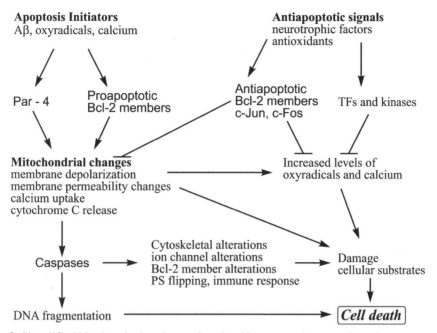

Fig. 3. Simplified biochemical pathways involved in neuronal apoptosis and neuroprotection (modified from ref. *53*). Apoptotic and antiapoptotic signals can be activated locally. Initiating signals, such as Aβ-peptide, oxyradicals, and calcium, lead to induction of Par-4 and Bcl-2 family members. Par-4 and proapoptotic Bcl-2 family members induce mitochondrial alterations, which, in turn, lead to caspase activation and alterations in cellular calcium homeostasis and free radical metabolism. Caspases cleave substrates that modulate synaptic plasticity and mediate proteolysis of substrates leading to nuclear chromatin condensation and fragmentation. Antiapoptotic pathways can be activated by neurotrophic factors and other signals. Such pathways lead to activation of transcription factors (TF) that induce expression of neuroprotective proteins such as antioxidant enzymes and calcium-regulating proteins. In addition, antiapoptotic signals may modulate cell death pathways by activating kinases that phosphorylate substrates such as ion channels and Bcl-2 family members that influence the cell death process.

and cell culture needs *(63)*. However, the techniques to detect endonuclease-mediated double-strand cuts in nuclear DNA can also label nonspecific DNA single-strand cuts, thus signaling both apoptosis and necrosis *(21,24,63,118,119)* as well as for ECD *(38,39)*.

Because these and related methods cannot always distinguish between the different modes of cell death, efforts have been made to characterized them according to their ability to prefernetially detect apoptotic vs necrotic cells. In general, necrotic cells are usually stained very faintly by TdT-based assays and more consistently labeled by DNA polymerase-based techniques given the amplification inherent in nick translation *(117)*. However, none of the DNA fragmentation techniques is currently considered specific for either necrosis or apoptosis. Therefore, these methods have to be supplemented by morphological (ultrastructural) evidence or by detection of apoptosis-related proteins allowing the unequivocal identification of dying cells as apoptotic *(21,27,31,63,66,96, 117)*. This type of studies can be problematic in human postmortem tissue, because tissue factors, such as autolysis, fixation, archival length in buffered formalin, embedding, cutting, blocking of endogenous peroxidase with H_2O_2, various pretreatments, low tissue pH owing to ante mortem hypoxia, and so forth, have been shown to affect DNA fragmentation *(63,120–122)*. However, in general, no relationship between postmortem delay and TUNEL labeling has been found *(63,122–125)*.

RNA transcription and DNA repair may also be detected with the ISEL technique *(126)*. The level of DNA fragmentation detected depends on technical parameters, e.g., concentration, type and activity of enzymes used, duration of exposure, timing of immunohistochemical color reaction, and pretreatment procedures. For this purpose, the use of appropriate positive and negative control tissue is indispensible for validation of the technique, and a multiparametric approach for the identification of apoptotic cells, e.g., using TUNEL/ISEL and morphological criteria or immunohistochemistry for apoptosis-related proteins and activated caspase-3 (e.g., using the CM-1 antiserum/IDUN Pharmaceuticals, La Jolla, CA), specifically recognizing the large (p18) subunit of processed caspase-3 *(127)* and, thus, being specific for the cleaved and activated form of caspase-3) is particularly useful *(63,128)*. Detection of activated caspase-3 is a valuable tool to identify dying cells even before all features of apoptosis (e.g., DNA fragmentation) are present, whereas no activation of the caspase cascade has been found in necrotic cell death. Other studies have simultaneously used BODIPY-Texas-red-dUTP for ISEL technique and the YOYO method (fluorescent DNA binding with the dimeric cyanid dye YOYO-1 to stain condensed chromatin) *(119)*, whereas Annexin V, used for flow-cytometric detection of apoptotic cells in vivo is not useful for its detection in tissue sections *(63)*. In experimental models of neurodegeneration, the combined evaluation of the form of nuclear damage (karyorrhexis, pyknosis), the presence or absence of activated caspase-3 or, alternatively, caspase-cleaved actin (fractin) or MAP-2, and the extent of the damage to cell cytoskeleton, allows for precise assessment of the extent of injury and the mode of cell death (apoptosis, oncosis) for individual neurons *(64)*.

3. EVIDENCE FOR PROGRAMMED CELL DEATH FROM ANIMAL MODELS AND TISSUE CULTURES

Implication for apoptosis as a general mechanism in many neurodegenerative disorders has largely been supported by evidence from animal models and tissue cultures.

3.1. Alzheimer's Disease

Mutation of the presenilin (PS) and the amyloid precursos protein (APP) genes of familial early onset AD (FAD) have been shown to sensitize neuronal cells to apoptosis in both cell cultures and transgenic mice *(129–131)*, with AD-linked mutant PSs being less effective at inducing apoptosis in *Drosophila melanogaster* than wild-type PS *(132)*. Cleavage of PS 1 and PS 2 proteins generates antiapoptotic C-terminal fragments *(133)* and PS-1 protects against apoptosis caused by interacting protein PAG *(133a)*. However, neurons overexpressing mutant PS-1 are more sensitive to apoptosis, and mutant PS-2 gene enhances neuronal death, decreases Bcl-2 expression and triggers p-53-dependent apoptosis *(133b,133c)*. Transfection of neuronal cells with mutants of the APP gene causes DNA fragmentation, and a novel APP mutation increases Aβ peptide levels and induces apoptosis *(134)*, whereas intracellular accumulation of wild-type APP is an intrinsic activator of caspase-3-mediated cell death machinery in postmitotic neurons *(135)*. APP is cleaved by caspases at the C-terminal inducing neurotoxicity *(135a)*. Aβ peptide, the amyloidogenic cleavage product of APP in neuronal cultures, causes selective increase of cellular Aβ-42, which is related to apoptosis but not necrosis *(136)*. Aβ peptide is neurotoxic and can produce apoptosis by (1) activating proapoptotic genes or inhibiting antiapoptotic factors, (2) signaling through G-proteins, (3) promoting superoxide and oxidative free radical production, (4) synergetically acting with other neurotoxic or excitotoxic agents, (5) disrupting intracellular ion homeostasis, or (6) interacting with mutated presenilin to perturb cellular calcium regulation and promote oxidative stress *(see* refs. *33,59,131)*. It has also been shown that exposure of cultured neurons to Aβ and prooxidants induces apoptotic cell death *(97)*. Incubation procedures of neuronal cultures with Aβ-40 and Aβ 25–35 displayed that Aβ toxicity is dependent on aggregation state and that cell death induced by Aβ-40 may be apoptotic *(137)*. Incubation of cultured neurons with cytotoxic concentrations of soluble Aβ-42 that is accumulated inside the endosomal/lysosomal system invokes rapid free radical generation within lysosomes and disruption of lysosomal membrane proton gradient which precedes cell death. This is only specific to the Aβ-42 isoform, whereas incubation of cells with high concentrations of Aβ-40 has no effect on lysosomal hydrolase release into the cytoplasmic compartment *(138)*. Intracellular Aβ 1-42 but not extracellular Aβ peptides induce neuronal death which is accompanied by Aβ accumulation in the endoplasmic reticulum, indicating that intraneuronal deposition of "neurotoxic" Aβ may be an early event in neurodegeneration in AD *(136a)*. Treatment of cells with an antioxidant or lysosomotropic amine (methylamine) partially blocks the release of lysosomal contents suggesting that this is the result of lysosomal membrane oxidation. These data suggest that cell death mediated by the soluble Aβ peptide may be different from cell loss observed following extracellular Aβ deposition *(138)*. Aβ peptide also induces apoptosis-related changes in synapses and dendrites that are potentiated by inhibition of NFκB *(139)*. The preferential activation of the executioner caspase-3, responsible for DNA fragmentation in late stage apoptosis, in APP-related FAD suggests that this enzyme is involved in APP processing and proteolytic Aβ formation in vivo *(92,140,141)*, although Aβ deposition in the APPsw transgenic mouse model does not result in caspase-3 activation which is independent from the neurotoxic effect of Aβ *(142)*. Aβ peptide induced neuronal death in vitro has been found to be Bax-dependent but caspase-independent *(143)* or is mediated by caspase-2 *(144)*, although early neurotoxic events have been shown to be induced by activating caspases other than caspase-3 *(145)*. On the other hand, Aβ and prion peptides even in sublethal concentrations induce a rapid, marked, and prolonged activation of proapoptotic markers, caspase-3, -6, and -8-like activity in neuronal cultures, which may contribute to chronic neuronal dysfunction and increased susceptibility to additional metabolic insults *(146)*. Although ESF1-mediated death of Aβ-treated cortical neurons is independent of p53 and dependent on Bax and caspase-3 *(147)*, Aβ-induced neuronal apoptosis requires c-Jun N-terminal kinase (JNK) activation with downstream activation of caspase-2 and -3, implicating the JNK pathway as a required element in cell death evoked by Aβ peptide *(148)*. Inhibition of the SAFK JNK pathway promotes cell survival end suggests multiple events associated with Aβ-induced cortical neuronal apoptosis *(97,149)*. Although Aβ-induced cerebral endothelial cell death has recently been shown to involve both mitochondrial dysfunction owing to oxidative stress and activation of caspase-3 and -8 *(150)*, a novel Aβ peptide-binding protein (BBP) containing a G protein activation module is suggested to be another target of Aβ-induced neuronal apoptosis *(151)*. Overexpression of *Bcl-xL* in response to subtoxic concentrations of Aβ in neuronal cell cultures as stress response increases neuronal resistance against apoptotic and oxidative injury *(152)*. On the other hand, interaction of AD PS 1 and PS 2 with the antiapoptotic protein *Bcl-X* may play a role in modulating the threshold of cell death *(153)*. Absence of caspase activation in frontotemporal dementia with Parkinsonism related to chromosome 17 (FTDP-17), caused by mutations of the *tau* gene, argue against a role of this enzyme in tau cleavage and tangle formation *(154)*. Although several studies indicated that tau protein is cleaved by proteases other than caspase (e.g., calpain and cathepsin D) *(155)*, recent studies showed that microtubule-associated tau in vitro was selectively cleaved by caspase-3 or calpain, a calcium-activated protease, but not by caspases -1, -8, or -9. Furthermore, tau- and oxidative stress-induced cell death was augmented by expression of APP or Swedish APP mutant, suggesting that activation of caspase-3 and the proapoptotic caspase-3 cleavage product of tau may contribute to progression of neuronal cell death in AD *(156,157)*. Aβ-induced induction of tau phosphorylation has been shown to involve MAP kinases in cultured cells and that tau phosphorylation precedes cell death *(158)*. Induction of neuronal apoptosis by inhibition of phosphatidylinositol 3-kinase results in a rapid increase in the phosphorylation of tau, which is followed by the dephosphorylation and cleavage of the protein. In contrast, necrosis triggered by high salt shock or glutamate treatment leads to a rapid dephosphorylation and an almost complete proteolysis of tau. These data suggest that a transient tau hyperphosphorylation occurs at an early stage of apoptosis, whereas tau is dephosphorylated and cleaved during the late phase of apoptosis as well as in necrotic neurons *(158a)*. Activation of the JNK/p53 pathway in tauopathies, i.e., diseases characterized by tau protein pathology, is related to tau phosphorylation but not to apoptosis *(158a)*. The kinase inhibitor N-(6-aminohexyl)-5-chloro-1-naphtalenfonamide (W 7) prevents tau phosphorylation in Aβ-treated neuroblastoma cell cultures, but it does not prevent cell death *(159)*. The MAP kinases and extracellular signal-regulated kinases (ERKs) may facilitate apoptosis in several paradigms *(160–162)* but are associated with cell survival in others *(163,164)*. In vivo studies have shown that excitotoxic cell damage by systemic administration of kainic acid is associated with activation of JNK-1, c-Jun and reduced expression of ERK and p38 *(165)*, although activation of p38 kinase has been shown to link tau phosphorylation, oxidative stress, and cell cycle-related events in AD *(166)*. However, in human AD brain, no colocalization of phosphorylated ERK immunoreactivity and nuclear DNA fragmentation was seen, suggesting a lack of direct involvement of ERK activation in the cell death cascade in AD *(167)*.

3.2. Parkinson's Disease

In vitro studies have shown that two major neurotoxins used in the induction of Parkinsonian syndromes in animals, 1-methyl-4-phenylpyridinum (MPP+), the active metabolite of 1-methyl-4-phenyl-1,2,3,6-tetrahydropyridine (MPTP), and 6-hydroxydopamine (6-OHDA), are able to kill cells by triggering an apoptotic mechanism. 6-OHDA can lead to apoptosis in neuronally differentiated PC12 cells *(168)*; MPP+ also induces apopototic cell death on neuroblastoma cells and primary dopaminergic cultures *(169–171)*. In these paradigms, direct caspase-3 activation was seen in tissue extracts and cell death can be prevented for at least 72 h by inhibition of caspase-3 *(172)*. The concentration of these substances used were low because higher concentrations induced necrosis, further reinfocing the notion that apoptosis and necrosis may be part of the same continuum.

Natural death of dopaminergic neurons with the morphologic features of apoptosis occurs during normal development *(173)*, particularly in the first postnatal week *(174)*. Its extent can be influenced by excitotoxic injury to the striatum *(175,176)*. PCD/"apoptosis" of dopaminergic cells in vitro and in vivo can be induced by a variety of substrates, e.g., complex I inhibition *(177)*, dopamine and related catecholes/monoamines *(178–183)*, dopamin-melanin *(184)*, dopamine and iron *(185)*, tumor necrosis factor TNFα *(186)*, L-dopa *(187,188)*, MPTP in both high doses *(189)* and low-doses *(187)*, MPP+ *(189,190)*, peroxinitrit and nitric oxide *(191)*, cytochrome-c *(192)*, paraquat *(193)*, and in a number of other models *(see* refs. *194–196)*. There are controversial data about apoptosis induced by MPTP and the proapoptotic properties of MPP+. It has been suggested that MPTP/MPP+ causes apoptosis when its neurotoxic effect is only mild and necrosis when it is stronger *(171)*. In chronic low-dose MPTP-mouse model, apoptotic SN neurons were detected *(187,197)* as was seen following a single bolus injection of MPTP *(198)*. It was accompanied by activation of caspase-9 early and later of caspase-8 in the course of cell demise. These data and the observation that caspase-3 activation was significantly higher after 12 and 24 h after treatment of primary dopaminergic cell cultures with MPP+, whereas con-comitant nuclear fragmentation was observed only at later stages (72 h) of cell demise, suggest that activation of both caspases -3 and -8 precede and are not the consequences of cell death *(199)*. Caspase-3 activation has been shown to be an early and transient phenomenon in apoptototic cell death of DA neurons, and loss of TH immunoreactivity in subchronic MPTP model is the result of cell loss rather than to loss of TH protein expression *(200)*. However, cotreatment of MPP+-intoxicated primary dopaminergic cell cultures with specific caspase-8 inhibitors did not result in neuroprotection but seemed to trigger a switch from apoptosis to necrosis, which may be related to intracellular energy or ATP depletion *(201)*. In adult mice, there was upregulation of Bax and a decrease of Bcl-2 after MPTP administration, whereas mice lacking Bax are significantly more resistent to MPTP neurotoxicity than their wild-type littermates *(202)*. However, despite almost complete survival of the SN cells they show a marked dopamine reduction, which indicates that MPTP induces more than neuronal death. Transgenic mice (over)expressing Bcl-2 are resistant to 6-OHDA and MPTP neurotoxicity *(203)*, and overexpression of Bcl-2 attenuates MPP+ but not 6-OHDA induced neurotoxicity *(204)*. MPP+ treatment of dopaminergic neurons in culture did not cause redistribution of Bax, although cytochrome-c was released from the mitochondria and nuclear condensation/fragmentation was induced, suggesting that Bax may not have a central role in apoptotic death of dopaminergic neurons *(205)*. Recent studies in various models of inducing apoptosis in SN neurons (e.g., 6-OHDA intrastriatal injection, axotomy, and neuN) have shown that expression of cyclin-dependent kinase 5 (cdk5) and of its known activators, particularly p35, is a general feature of apoptotic neuron death in SN in vivo *(206)*. Apoptosis of SN neurons was seen after intrastriatal injection of dopamine *(207)*, and of 6-OHDA in early postnatal stages, whereas in adults *(208)* or in SN in adult murine Weaver mutations both apoptotic and nonapoptotic morphologies of SN neurons were observed *(209)*. In another animal model, intracerebral injection of 6-OHDA caused both apopotitc and necrotic cell death of dopaminergic SN neurons *(210,211)*, whereas 6-OHDA has also been shown to mediate apoptosis via activation of a caspase-3-like protease in neuronal PC 12 cells *(212)*, and protection was observed by broad-spectrum caspase-3 inhibitors *(213–215)*. In some models, activity changes of apoptosis-modulating proteins have been observed (e.g., upregulation of p53 and Bax proteins), but no changes of Bcl-xL, after 6-OHDA treatment of PC 12 cells, suggesting that they could represent relevant markers of apoptosis *(216)*, activation of transcription factor AP-1 in paraquat-induced apoptosis in PC-12 cells *(217)*, of transcription factor E2F1 in 6-OHDA-evoked apoptosis in rat cortical neurons *(218)*, and of c-Jun, Bax, and Bax mRNA in SN neurons in a MPTP mouse model *(219,220)*. These data and inhibition of dopamine- and 6-OHDA induced apoptosis in PC-12 cells by Bcl-2 *(180,182,221)* suggest the existence of some cascade mechanisms of apoptotic and nonapoptotic death of dopaminergic neurons. It can be inhibited by glial cell line derived neurotrophic growth factor (GDNF) *(222–224)*, free radical scavengers *(177)*, antioxidants *(225)*, estradiol *(226)*, inhibition of transcription factor p 53 *(227)*, melatonin *(228)*, selegilin *(191)*, and caspase inhibitors *(221)*. However, another recent study using primary dopaminergic cultures showed that 6-OHDA induced cell death was blocked by caspase inhibitors, whereas MPP$^+$-induced cell death could not be inhibited by them, possibly because a necrotic pathway being engaged *(229)*. At earlier stages, in these models, cell death is regulated by p-35 and Bcl-2 family systems, by NFκB and MAP kinase, whereas in the very end stage, it depends on activated caspase *(229a)*. However, it should be noted that nonprotection by caspase inhibitors does not necessarily mean that these are not activated. Although not all dopaminergic neurotoxins (e.g., MPP$^+$) appear to induce apoptosis *(210)*, partizipation of Par-4 related to Fe 2^+-induced mitochondrial dysfunction in experimental models of PD has been observed *(230)*. Attenuation of 6-OHDA-induced neuronal apoptosis in some models has been shown by overexpression of hemag-glutinin epitope-tagged Bax (HA-Bax) but not Bcl-2 or Bcl-xL *(230)*, and caspase inhibition protects dopaminergic neurons against 6-OHDA-induced apoptosis but not against the loss of their terminals *(231)*. Bcl-2 and GDNF delivered by HSV-mediated gene transfer act additively to protect dopaminergic neurons from 6-OHDA-induced degeneration *(232)*. Bax ablation may prevent dopaminergic neurodegeneration in the MPTP mouse model *(202)*. However, Bcl-xL attenuates MPP$^+$ but not 6-OHDA induced dopaminergic cell death *(205)*, whereas in other models, toxin-induced apoptosis can be prevented by expression of Bcl-2 and Bax, suggesting that Bcl-2 related proteins may show a specific interaction with a distinct partner protein or cell-death pathway, determining its role as a positive or negative modulator of cell death *(51,96,233)*. In NO-mediated neuronal apoptosis produced by astrocytes after hypoxic insult, decreased Bcl-2 and increased Bar expression together with caspase-3 activation have been observed *(234)*. Both MPP$^+$ and other PD-related neurotoxins (e.g., paraquat, dieldrin and salsolinol) induce apoptotic cell death in tissue cultures following initial increase of H$_2$O$_2$-related ROS activity and subsequent activation of NH$_2$-terminal (JNK)1/2MAP kinases. Both MPP$^+$ and the oxidant H$_2$O$_2$ equally induce the ROS-dependent events, although oxidant treatment alone induces similar sequential molecular events: ROS increase, activation of JNK MAP kinases, of PITSLRE kinase, p 110, by both caspase-1 and -8-like activities and apoptotic cell death. A combination of the antioxidant Troloxand and a pan-caspase inhibitor BOC (Asp)-fmk (BAP) exerts significant neuroprotection against ROS-induced dopaminergic cell death *(235)*. In addition, high-throughout cDNA screening using the current model identified downstream response genes, such as heme oxigenase-1, a constitutent of LBs, might be a useful biomarker for the pathological condition of dopaminergic neurons under neurotoxic insult *(235)*. MPP$^+$ causes apoptosis of mesencephalic dopaminergic neurons in culture, combined with overexpression and nuclear accumulation of glyceraldehyde-3-phosphate dehydrogenase (GAPDH), whereas GAPDH knockdown rescues them from MPP$^+$-induced apoptosis *(236).*

Systemic administration of MPTP produces DNA fragmentation with induction of both caspase-3 and -8 activity *(55,95),* whereas inhibition of the downstream cellular substrate of caspase-3, PARP, an enzyme involved in DNA repair, protects against MPTP-mediated neurotoxicity *(55).* A similar activation of caspase-3 was described in the rat SNc following intrastriatal injection of 6-OHDA *(237).* MPTP administration in mice increases nigrostriatal activity of both c-Jun and c-Jun NH2-terminal kinase (JNK), members of the stress-induced protein kinase (SAPK) pathway, which is attenuated by a JNK-specific inhibitor also reducing dopaminergic SN cell loss *(238).* Overexpression of human α-synuclein caused dopamine neuron death in primary human mesencephalic cultures *(238a)* and of transfected astroglia seen in several human α-synucleinopathies *(238b).* Release and aggregation of cytochrome-c and α-synuclein have been shown to be inhibited by certain antiparkinsonian drugs (e.g., talipexole and pramipexole) *(239).*

MPTP is a mitochondrial enzyme that elicits its action first via monoamine oxidase B-catalysed conversion to its metabolite MPP$^+$, which is selectively taken up into dopaminergic SN neurons by the dopamine transporter (DAT). It kills these cells by specific inhibition of mitochondrial complex I which has also been implicated in the pathogenesis of SN cell death in human PD *(see* refs. *48,54,55,233,240,241).* Loss of complex I activity may result in decreased ATP production and increased formation of mitochondrially generated reactive oxygen species (ROS), both of which contribute to neuronal cell death via decreased protein pumping and reduced voltage differential across the inner mitochondrial membrane that would elicit opening of the mitochondrial transition pore and subsequent initiation of apoptosis *(48,54–56,240,242).* Similar inhibition of complex I activity and subsequent mitochondrial impairment is seen with decreased levels of the antioxidant compound glutathion (GSH) that, in PD, is suggested to precede both complexe I and dopamine loss *(48,233,243).* Depletion of intracellular GSH has been shown to induce apoptosis in neuronal cells in vitro *(244,245).* Conversely, apoptosis can be prohibited by overproduction of Bcl-2 owing to increase in GSH levels *(246).*

3.3. Other Neurodegenerations

In Huntington disease transgenic mice expressing exon 1 of the human HD gene containing an expanded polyglutamine, degenerating neurons within those specific areas of the brain known to be affected in HD have not been found to show defining characteristics of apoptosis *(13).* Similarly, no TUNEL-positive neurons were found, suggesting that within dark neurons that have also been observed in postmortem brain from patients with HD, there is no intranucleosomal DNA fragmentation *(247).* In marked contrast to the late onset nonapoptotic form of neuronal death observed in human HD brain and transgenic mouse and *Drosophila* models *(248),* is the rapid apoptotic death observed following expression of high levels of mutant huntingtin in transgenic mice *(249)* or in transfected cells in culture *(250).* This latter form of degeneration does not appear to be dependent on aggregation of the mutant protein, forming intracellular inclusions, and the mechanism of neurodegeneration clearly seems to differ from that seen in striatal neurons induced by 3-nitroproponic acid (3-NP) *(251).* However, recent studies showed DNA fragmentation, strong granular Bax expression and increased Bax/Bcl-2 ratio only in the center of severe lesions in rat striatum following 3-NP administration, while the dark compromised neurons in mild and subtle lesions revealed an equally enhanced expression of both Bax and Bcl-2 but lacked TUNEL-labeling, therefore, are not considered apoptotic *(252)*. This is in line with recent studies in transgenic mice expressing exon 1 of the human HD gene carrying a CAG repeat expansion in which exposure to neurotoxic concentration of dopamine induced neuronal autophagy that colocalized with high levels of oxygen radicals and ubiquitin. These data suggest that the combination of mutant huntingtin and a source of oxyradical stress induces autophagy and may underlie the selective cell death in HD *(253)*. Neuronal cytochrome c immunoreactivity and activated caspase-8 and -3 have been observed in end-stage disease in both human HD and in RG/2 mice, suggesting that apoptosis may play a role in neuronal death in these disorders *(253a,253b)*. Despite considerable progress in our understanding of polyglutamine diseases, there is, however, a big gap of knowledge between the expansion of CAG in huntingtin and the massive but selective death of neurons and glial cells *(254)*. Similar controversial data arguing for or against apoptotic neuronal death have been reported in ALS transgenic mouse models *(255–258)*.

4. IS THERE EVIDENCE FOR APOPTOSIS IN THE HUMAN POSTMORTEM BRAIN?

In human postmortem brain of patients with various neurodegenerative disorders (e.g., AD, PD, CH, ALS, etc.), dying neurons are present that have been reported by some groups to display morphologic features of apoptosis. These include cell shrinkage, chromatin condensation, DNA fragmentation, and increase expression of both proapoptotic (c-Jun, c-Fas, Bas, p53, APO-1/Fax-CD 95, Fas, Fas-L, capsase-8 and -9, activated caspase-3, tumor necrosis factor (R1/p55), and antiapoptotic proteins (Bcl-2, Bal-X) or DNA repair enzymes, such as Ref-1 and the coexpressed GADD 45 *(31,32,41,43–46, 48,55,59,76,95,96,105,120,123–125,194,195,204,257–267)*. However, other groups have observed little or no evidence of apoptotic neuronal death associated with a variety of neurodegenerative disorders *(50–53,268–274)*. These studies have relied on the presence of morphological markers of apoptosis, such as TUNEL/ISEL staining and related methods associated with the expression of ARPs, oncogenes, and other detectable players involved in the cascade of PCD (Table 2). The following summarizes some of the most important findings in different types of human neurodegenerative disorders.

4.1. Alzheimer's Disease

PCD or apoptosis has been implicated as a major mechanism of neuronal cell death in AD *(41–43)*. Histochemical techniques for the demonstration of fragmented DNA revealed large numbers of positive neurons and glial cells in postmortem AD brain *(43, 120,121,123,275–283)*. Compared to controls, DNA fragmentation in AD brain is 50-fold increased in neurons and approx 25-fold in glial cells, microglia and oligodendroglia being mainly affected, whereas only 25% of all degenerating cells represent neurons *(120)*. Apoptosis of astrocytes with enhanced lysosomal activity and of oligodendroglia in white matter lesions in AD have been associated with significant atrophy of the frontal and temporal white matter *(283a)*. Most of the TUNEL-positive neurons are seen in the temporal allocortex, the region initially and most severely involved in AD *(284)*. Only 13–50% (mean 28%) of the degenerating neurons are located within or near Aβ

Table 2
In Situ Tailing and Immunohistochemical Results in PD, PSP, CBD, AD, and Controls (CO)

Method/antibody applied	PD Neurons	PD Microglia	PSP Neurons	PSP Oligos	CBD Neurons	CBD Oligos	AD	CO
In situ tailing (TUNEL)	0	+	+ (~0.1%)	++	0	+	+ (1.1 ± 0.4%)	+ (0.02 ± 0.01%)
Immunohistochemistry (ARPs)								
c-Jun/AP-1 (ASP 1)	+	+	±	+	±	+	±	−
c-Jun	+	+	++	+	+	+	++	++
Bcl-2	±	+	+	++	+	+	++	++
Bcl-X	++	+	++	+	+	+	++	++
p 53	±	±	±	+	NE	NE	+	−
CD 95	−	±	±	+	NE	NE	+	−
Immunohistochemistry (heat shock) and others								
α B-Crystallins	−	−	±	+	++	±	±	±
HSP 27	−	−	−	−	NE	NE	±	±
HSP 65	−	−	++	−	NE	NE	++	++
HSP 70	−	−	+	−	NE	NE	±	±
HSP 90	−	−	++	−	NE	NE	++	++
PHF/ubiquitin	−	−	±	+	++	+	++ (NFT)	± (NFT)
Ubiquitin	−	−	+	−	++	++	++ (NFT)	± (NFT)
PHF-tau (AT-8)	−	−	++	++	++	++	+++	± (NFT)
Activated caspase III (CM1)	−	±	± (<0.1%)	+	±	±	± (0.02-0.05%)	−

Note: (−) No labeling; (±) exceptional cells labeled; (+) few cells labeled; (++) many cells labeled; (NE: not examined; ARPs: apoptosis related factors; NFT: neurofibrillary tangles; Oligos: oligodendrocytes.

deposits (plaques), but these are 5.7 (−0.8 SD)-fold more frequent than neurons without contact to plaques. NFTs involve a mean of 41% (range 18–66%) of all degenerating neurons, which means an approx threefold increased risk of degeneration compared to tangle-free neurons *(120)*. These data suggest that both histological markers of AD, i.e., senile plaques and NFTs, are not necessarily associated with neuronal death, because it frequently occurs at a distance from the Aβ deposits or in cells not affected by NFTs *(120,279)*. On the other hand, recent studies in sporadic AD and PS-1 FAD brains, confirming rare apoptosis of NFT-bearing temporal neurons showed simultaneous intracellular Aβ42 labeling, which suggests that intraneuronal deposition of a neurotoxic form of Aβ42 may be an early event in neurodegeneration in AD *(280)*. In APP and PS-1 double-transgenic mice, intraneuronal Aβ accumulation precedes plaque formation, although intraneuronal Aβ is not detected in brains of aged double-transgeic mice, which correlates with the typical neuropathology of chronic AD patients *(281)*.

Increased expression of pro-apoptotic proteins or gene products, such as c-Jun, c-Fos, Par-4, Bax, Bas, Bad, Bcl-X, p53, APO-1/Fas (DC 95), but also of antiapoptotic protein Bcl-2 or of repair enzymes such as redox factor 1 (Ref-1), a potent activator of p53, have been observed in AD brains *(120,121,131,260,261,265,279–290)*. Recent Western blot studies of human AD brain tissue revealed dysregulation of apoptosis-related proteins, decrease of caspase-3, -8, and -9, of DFF45 (DNA fragmentation factor 45), and PLIP (Fas-associated death domain/FADD/-like interleukin-1β-converting enzyme inhibitory protein) and increase of ARC (apoptotis repressore with caspase recruitment domain), and RICK (receptor interacting protein/RIP)-like interacting CLARP protease, whereas cytochrome-c and Apaf-1 (apoptose protease activating factor 1) were unchanged *(291)*. These data point towards a disturbed balance between proapoptotic and antiapoptotic proteins in AD which is associated with incomplete cell cycle activation in postmitotic AD neurons possibly leading to their elimination via apoptosis *(131, 292)*. This may occur via the death receptor pathway independent of cytochrome-c *(291)*. Cell cycle activation, an obligatory component in the cell death pathway *(293)*, has been demonstrated recently in hippocampal and basal forebrain neurons in AD brain suggesting genetic imbalance as a cause of neuronal loss in AD *(294)*. In addition to activation of c-Jun, the entire JNK/SAPK pathway is activated in AD. It therefore appears that susceptible neurons face the dilemma of proliferation of death, a dilemma in which c-Jun and the JNK/SAPK pathway play an important role *(295)*.

Recent studies suggest that the development of Aβ plaques may cause damage to axons, and the abnormally prolonged stimulation of neurons to this injury ultimately results in profound cytoskeletal alterations that underly neurofibrillary pathology and neurodegeneration *(296)*. In some cells showing DNA damage, coexpression of ARPs like c-Jun *(104)*, Bax and Bcl-2 with decreased levels of Bcl-2 in tangle-bearing neurons *(265,266,297)* and of DNA repair regulating GADD545 protein (T251) have been reported *(290)*. Despite large numbers of cells with DNA fragmentation in the severely involved hippocampus, only exceptional neurons display the morphology of apoptosis or show diffuse cytoplasmic expression of ARPs, although the intensity of Bcl-2, c-Jun, Apo-1/Fas and several stress proteins shows no differences between AD and control brains *(298)*. However, only in one among 2500 to 5650 hippocampal neurons (0.02 to 0.05 %) apoptotic features and expression of activated caspase-3 have been observed that are almost never found in aged controls. These data document the extremely rare occurrence of apoptotic neuronal cell death in AD *(128,298)*. On the other hand, increased expression of Bcl-2 and nonactivated caspase-3 has been observed in AD neurons and glial cells showing DNA fragmentation, with negative correlation to glutamate transporters but related to spliced APP forms *(299)*. This suggests that excitatory injury and imbalance in the ratio of spliced APP may lead to cell death via caspase-3 activation that defines late stages in the cell death cascade *(44)*. Although some studies revealed that neither senile plaques nor NFTs in human brain are associated with activation of caspase-3 despite the ability of Aβ peptide to induce caspase-3 activation and apoptosis in vitro *(273)*, which requires caspase-8, a downstream component of the Fas-FADD pathway *(97,300)*. Other recent studies reported elevated immunoreactivity of activated caspase-3 in neurons, astroglia, and blood vessels with high degree of colocalization with NFTs and SPs, suggesting that activated caspase-3 may play an important role in neuronal cell death and plaque formation in AD *(301)*. Recent studies in human AD brain demonstrating the neo epitope generated by caspase-3 mediated cleavage of APP (αδC^{csp}-APP) immunoreactivity in neurons, glial cells, and dystrophic terminals, colocalized with around 30% of senile plaques suggest that apoptosis may contribute to cell death resulting from amyloidosis and plaque deposition in AD *(301a)*. On the other hand, activated caspase-3 was detected in more than half of the hippocampal neurons showing granulovacuolar degeneration (GVD) in AD brain. It was restricted to the granules which often showed colocalization with tau protein (using antibody AT-8), suggesting a "pretangle" stage *(300)* but no nuclear signs of apoptosis *(273,298)*. Caspase-cleaved APP and activated caspase-3 are colocalized in the granules of CVD in AD and Down's syndrome *(298a)*.

Although neurons with completed NFTs do not express activated caspase-3, recent demonstration of elevated casein kinase-1 (CK-1), a member of protein kinases, in both the matrix of GVD bodies and tangle-bearing cells suggests a link between these two lesions in AD *(302)*. However, the localization of caspase-3 and -6 activity within a select subcellular compartment, such as autophagic granules, does not necessarily indicate a global effective caspase amplification. Moreover, modulation of distal substrates of activated caspase-3 may lead to further modification of this cell death pathway and may explain absent evidence demonstrating acute end stages of apoptotis, including nuclear compensation and blebbing, in AD susceptible neurons *(303)*. The upregulation of individual caspases, though seemingly not leading to apoptosis, could have great significance for the pathogenesis of AD, whereas the lack of some downstream effector caspases in neuronal pathology argues against a specific role in the apoptotic cascade. The Fas/Fas-L signaling system also does not appear to be involved in specific processes in AD *(304)*. Detection of granular precipitates of phosphorylated MAP kinase (ERK 1 and 2) associated with early tau deposition in AD and other tauopathies (PiD, PSP and CBD), suggesting activated Ras as the upstream activator of the phosphorylation-independent MEK 1/ERK pathway of tau phosphorylation, however, was not associated with increased nuclear DNA vulnerability and cell death *(167)*, whereas other recent studies suggest mitochondrial DANN damage as a major mechanism of cell loss in AD *(305)*. The incidence of TUNEL-positive neurons in AD being increased in comparison to age-matched controls is significantly higher than could be expected in a chronic disease with an average duration of 10 yr plus *(303)*. However, the demonstration of very rare hippocampal neurons, displaying both morphologic signs of apoptosis and expression of activated caspase-3, appears fairly realistic given the short duration required for the completion of apoptosis and the protracted course of neurodegeneration in AD *(97a)*. On the other hand, the significantly increased incidence of cells with DNA fragmentation together with a "proapoptotic" environment in AD brain indicate increased susceptibility of AD neurons to metabolic and other noxious factors (e.g., oxidative stress, hypoxia, etc.) *(120, 128,298)*. Incomplete cell cycle activation in postmitotic AD neurons possibly may lead to their elimination by apoptosis *(47,131)*, with autophagy as a possible protective mechanism in early stages of PCD *(51,52,298)*.

4.2. Synucleinopathies

This group of neurodegenerative disorders that are characterized by neuronal or glial inclusions of α-synuclein, includes PD, DLB, and MSA *(12,106,307)*, in which extensive studies of cell death markers have been performed.

4.2.1. Parkinson's Disease

In 1996, the first human post-mortem study described a significantly higher number of TUNEL-positive dopaminergic neurons in SNc of PD brains compared to controls *(308)*. Since then, conflicting results have contributed to the controversy.

The presence of DNA fragmentation, using TUNEL and related techniques, in dopaminergic SNc neurons in most studies is very low, ranging from 0 to approx 12%, with a means of 1–2% of the total dopaminergic SN cell population as compared to 0–1% in controls, whereas other groups did not find them in PD brains (Table 3). Occasionally, a combination of apoptosis and autophagic degeneration of SN neurons has been found in PD *(47)*. A good example of the variability of results obtained in human postmortem samples was provided by a study showing that in control brains, the number of TUNEL-positive neurons was correlated with tissue pH as an index of postmortem hypoxia *(122)*. However, in addition to a percentage of TUNEL-positive SNc and brainstem neurons that was twice as high in PD (2%) as in controls (1%), no correlation between tissue pH and the number of TUNEL-positive neurons was observed in PD brains. This finding was interpreted as reflecting the primary disease process rather than a perimortal effect. Although this study can be classed among those that favor the apoptosis hypothesis in PD, it also provides an interesting explanation for the observation that the number of apoptotic dopaminergic neurons is higher than expected in PD brains. Given the slow rate of neuronal degeneration in PD, particularly in late stages of the disease, it is unlikely that this relatively high percentage of "dying" cells may reflect a primary disease process. It is more likely that SNc neurons in PD brain are already altered in their function and show a higher vulnerability towards deleterious stimuli such as hypoxia and, thus, these increased numbers reflect a perimortem phenomenon to hypoxia or other factors related to the patients' agonal state *(96,124)*.

Although very few SN neurons display a reduction in cell size and clumping of nuclear chromatin resembling apoptosis *(269)*, and show very mild expression of some proapoptotic proteins (c-Jun, AP1, ASP, and Bax) *(50,51)*, increased levels of antiapoptotic proteins Bcl-2 and Bcl-X in basal ganglia and SN but not in cerebral cortex of PD brain were reported *(309)*. However, others have found no differences in the immunoreactivity of Bcl-2 and Bcl-X, and an unchanged expression of Bcl-2 mRNA in SN neurons of PD patients compared with controls *(50,310)*. By contrast, some postmortem data suggest that Bax is expressed in SNc neurons *(202)*, although the intensity of Bax immuno-

Table 3
Incidence of DNA Fragmentation
in Substantia Nigra in Various Neurodegenerative Disorders

Disease	n	Time (h) pm	Method	% Neurons	Refs.
PD	3	?	TU	0	*123*
Juvenile PD	4	3–12.6	TU	0	*46, 308*
Late PD	7	1–5	TU	0–4.2 (m 1.2)	
Co	6	1–3	TU	0	
PD	3	8.3 ± 2.3	EM	3.7	*47*
PD	5	1.7–31	TU	6.9 ± 2.2	*125*
DLB	7	2.5–24.3	TU	11.46 ± 1.3	
AD/PD	4	5.5–20	TU	7.8 ± 2.45	
AD	5	11–16	TU	1.7 ± 0.65	
Co	3	3–8	TU	0.93 ± 0.47	
PD	22	?	TU	"Few" 1/22 brains	*271*
PD	3	7–30	TU	0	*268*
PD	3	?	TU+	1.5	*48*
Co	3	?	YO	0.1	
PD	16	5–3	TU	0–12.8 (m = 2.0)	*124*
MSA	4	8.5–35	TU	0–19.4 (m = 9.0)	
DLB	1	16	TU	9.3	
PSP	1	25.5	TU	0	
Co	14	5.5–48	TU	0–10.5 (m = 1.0)	
PD	3	20–38	TU	0	*270*
Co	4	4–42	TU	0	
PD	3	20–38	TU	2.0 ± 1.2	
Co	4	4–42	(prol)	1.3 ± 1.1	
PD	5	4–12	TU	0/1080	*50*
DLB	2	18–24	TU	0/1010	
PSP	3	12–24	TU	0/1080	
CBD	3	14–24	TU	0/1010	
Co	4	16–24	TU	0	
PD	8	2.75–7	ISEL	9.0 ± 5.0	*55*
Co	4	8–11	YO	0.4 ± 0.1	
DLB	11	15 ± 13	TU	0	*366*
Co	11	18 ± 14	TU	0	

AD = Alzheimer's disease; juv = juvenile; Co = Controls; PD = Parkinson's disease; DLB = dementia with Lewy bodies; MSA = multiple system atrophy; PSP = progressive supranuclear palsy; TU = TUNEL; CBD = corticobasal degeneration; YO = YOYO 1; pm = postmortem.

reactivity was identical in LB and non-LB body containing neurons, leading to the conclusion that Bax does not contribute to LB formation *(311)*.

Bcl-xL that is expressed in a stable manner in adult mammalian CNS *(312)*, was observed to label practically all dopaminergic cells in human SNc and, thus, appears to be a constitutive protein in these neurons. In PD brains, Bcl-xL mRNA was 1.8 times higher than in controls, suggesting that surviving SN neurons in advanced stages of illness might be protected by this anti-apoptotic protein. This was supported by significantly higher expression of Bcl-xL mRNA in LB bearing neurons than in those without LBs *(96)*. Recent studies showed almost double increase of Bcl-X mRNA in dopamine neurons in PD brain compared to controls, which may be indicated by redistribution of Bcl-X from the cytosol to the outer mitochondrial membrane *(310a)*. These findings suggest that the presence of LBs may represent an indicator for dysfunction or damage of the involved neuron, and incidental LB disease is considered a presymptomatic form of PD *(313)*.

On the other hand, immunohistochemical studies in human post-mortem brain observed a positive correlation between the degree of neuronal loss in dopaminergic cell groups affected in the PD mesencephalon and the percentage of caspase-3-positive neurons in these cell groups in control subjects, and a 76% decrease of caspase-3-positive pigmented SNc neurons in PD compared to controls *(95)*. Furthermore, the percentage of caspase-3-positive neurons with LBs was significantly higher than that of caspase-3-positive neurons among all catecholaminergic neurons in PD, as was the percentage of caspase-3-positive neurons among dopaminergic cells. Ultrastructural studies showed activated caspase-3 staining in dopaminergic cells, displaying the morphologic features of increased protein synthesis, whereas in cells with morphologic signs of apoptosis, activated caspase-3 staining was not detectable *(96)*. These data are in accordance with recent studies showing elevated caspase activities and tumor necrosis factor receptor R1 (p53) in SN of PD brains *(240)*, as well as increased caspase-3- and Bax immunoreactivity accompanying nuclear GAPDH (glyceraldehyde-3-phosphate dehydrogenase) translocation and neuronal apoptosis in PD *(55,202,314)*. Other studies showed that activated forms of both caspase-8 and -9, upstream caspases that are known to cleave and activate caspase-3, are present in dopaminergic SN neurons in PD brain *(315)*. Taken together, these results suggest that activation of the caspase cascade may be involved in the degeneration of dopaminergic neurons in PD.

However, LBs, the morphologic markers of PD, are consistently negative for ARPs and activated caspase-3 *(50–53,316)*. LBs in the brainstem of FD and DLB are stained with anti-ERK-2 antibodies, but negative for MAPK-P, SAPKJNK/P and p/38/P, though immunoreactive cytoplasmic granules are found in association with α-synuclein deposits in few neurons. α-Synuclein affects the MAP-kinase pathway by reducing the amount of available MAPK and, thus, may accelerate cell death. These kinases are not expressed in cortical LBs or in cortical neurons with synuclein inclusions in DLB nor in dystrophic Lewy neurites in PD and DLB *(317)*. These data suggest different expression of MAPKs in brainstem and cortical LBs in PD and DLB. Furthermore, no relationship between MAPK-expression and *in situ* labeling of nuclear DANN fragmentation or caspase-3 activation in SN neurons in PD and DLB have been observed that argues against apoptosis of SN neurons in these synucleinopathies *(317)*. Reduced Fas and Fas-L in both LB-bearing and nonbearing SN neurons, but increased Fas and Fas-L immunoreactivity in reactive astrocytesin SN were seen in PD *(72)*. This suggests that the Fas/Fas-L pathway does not play an essential role in regulating dopaminergic cell loss in PD and related neurodegenerative disorders. On the other hand, frequent DNA fragmentation and expression of ARPs and activated caspase-3 have been observed in reactive astrocytes and microglia in SNc of PD brains, indicating their increased turnover in neurodegeneration *(50,268,270)*. This is confirmed by findings in human subjects with Parkinsonism following exposure to MPTP and survival for 3–16 yr, where signs of active ongoing neuron loss, and clustering of microglia around nigral neurons, have been observed *(318)*. In population control of microglia, apoptosis plays an important role, and overactivation-induced apoptosis of microglia may be a fundamental self-regula-

Table 4
Morphology, Iron, and Microglia in Movement Disorders
(Semiquantitative Assessment)

	PD	DLB	PSP	MSA	CBD	Controls
N cases	14	8	5	12	3	6
Age (mean)	64–87 (75)	57–84 (70)	67–71 (70)	51–82 (66)	58–63 (60)	66–97 (80)
SN neuron loss	3+	3+	3+	3+	2–3+	0–0.5
Sc Lewy bodies	3+	3+	0	0	0	0
SNC-Fe(III)	1.5	0–0.5	3+	3+	2–3+	0–0.5
SNC-Microglia[b]	1.5	0–0.5	3+	3+	3+	0–0.5
Putamen Fe(III)[a]	0	0	0.5	3+	1+	0–0.5
GPI-Fe(III)[a]	1+	NE	0.5	3+	1+	0
GPE-Fe(III)[a]	0	NE	0.5	3+	1+	0
LC neuron loss[b]	3+	0	1.2	3+	NE	0
LC Fe(III)[a]	0	0	0	0	NE	0
LC-Microglia	1.1	0	1.5	1.2	NE	0

[a]Perl's stain.
[b]Mainly in involved areas.
Note: NE: not examined; LC: locus ceruleus; 0: absent, 0.5: very mild, 1+: mild, 2+: moderate, 3+: severe.

tory mechanism devised to limit bystander killing of vulnerable neurons *(319)*. On the other hand, activated microglia can trigger neuronal apoptosis that is mediated through the secretion of cathepsin B *(320)*. Large numbers of reactive microglia and macrophages, along with LBs and free melanin, have been found in SNc of PD brain and have been considered a sensitive index of neuropathological activity and of inflammation *(321,322)*. They are accompanied by considerable increase of iron, particularly Fe^{3+}, and ferritin in proliferated microglia, perivascular macrophages, and astrocytes in various subcortical regions not only of PD brain *(323–325)*, but also in other movement disorders, such as PSP, MSA, and CBD, but much less in DLB (unpublished data). Although neither iron nor microglia are detected in the putamen and globus pallidus in PD brain, there are variable amounts of microglia and Fe^{3+} in these nuclei in PSP, and CBD and, in particular, in MSA (Table 4).

In conclusion, the contribution of Bcl family members and caspases in dopaminergic cell death still needs to be better established, but a relationship between mitochondrial dysfunction specific to PD and involvement of apoptosis-related proteins appears possible. Specifically, abnormalities of complex I of the mitochondrial respiratory chain are now well recognized in PD *(see refs. 240,241,326–328)*, and mitochondrial inhibition of cytochrome-c release by reinforcing antiapoptotic or blocking proapoptotic factors (e.g., by caspase inhibition), which may block the apoptotic program but will not reverse the underlying cellular dysfunction, could represent a new challenge in PD therapy *(44,96)*.

4.2.2. Other Synucleinopathies

In *DLB*, similar controversial findings in both SN and cortex have been reported. While increased numbers of TUNEL-positive neurons in SNc, ranging between 9.3 and 11.45–1.3%, were observed by some authors *(124,125)*, our personal studies could neither confirm these data nor show significant differences in the expression of proapoptotic and antiapoptotic proteins in SNp and cortical neurons involved by LBs vs controls or any expression of activated caspase-3 *(50,51)*.

In *MSA*, a Parkinson-like disorder histologically featured by widespread GCIs *(12, 329)*, TUNEL positivity, expression of ARPs, and activated caspase-3 were not seen in neurons, but only in microglia in SNc and in oligodendroglia, often showing α-synuclein positive inclusions *(50,51,329)*, whereas other authors observed occasional DNA fragmenation of SN neurons *(330)*, with an incidence of DNA fragmentation in SN neurons ranging from 0 to 19.4% (mean 9.0%) *(124)*. A distinct cytoplasmic expression of Bcl-2 was seen in oligodendrocytes with coexpression of α-synuclein in about 25% of inclusion-bearing cells *(51,330)*. Since oligodendrocytes are usually Bcl-2 negative, its expression in pathologically altered cells in MSA may represent a final repair mechanism of a sublethally damaged cell to avoid cell death via apoptotis by upregulation of this antiapoptotic protein.

4.3. Other Neurodegenerative Disorders

In *PSP*, a mostly sporadic tauopathies, clinically presenting with atypical parkinsonism and cognitive deficits, morphologically characterized by multisystem neuronal degeneration associated with gliosis and tau-positive inclusions in neurons and glia differing from those in AD *(331–335)*, only single neurons in brainstem tegmentum (about one among 1050 cells) are TUNEL positive, with moderate expression of c-Jun and some heat-shock proteins, much less of ASP-1 and Bcl-2 *(51–53)*, without distinction between tangle-bearing and non-tangle bearing neurons *(336)*. Only one among five such neurons showed coexpression with hyperphosphorylated tau-protein, whereas neurons in SN, basal ganglia and pontine nuclei were all negative, although significant decrease in aconitase activity, cellular ATP level and oxyagen consumption in PSP cybrids suggest a contributary role of impaired mitochondrial energy metabolism *(337)*. In addition, a number of oligodendroglial cells in brainstem tegmentum and pontine basis were TUNEL-positive and expressed both ARPs and activated caspase-3; about 25–30% of these oligos contained tau-positive inclusions (coiled bodies), whereas no signs of apoptosis were detected in astrocytes with tau-positive inclusions *(51,52)*.

In *CBD*, another rare sporadic movement disorder with atypical parkinsonism and cognitive deficits, morphologically featured by asymmetric frontotemporal cortical atrophy with neuronal loss, NFT-like inclusions in ballooned neurons and tau-positive inclusions in astro- and oligoendroglia differing in structure and biochemisty from those in AD and PSP *(333,338–340)*, some of the ballooned neurons in cerebral cortex and in the severely damaged SNc displayed mild expression of Bcl-2 and activated caspase-3, but none of them showed DNA fragmentation nor the morphological phenotype of apoptosis. In SNc and other subcortical nuclei, part of the proliferated microglia were TUNEL-and ARP-positive, as was a moderate number of oligodendrocytes and a few astrocytes, some of them showing Gallyas- and/or tau-positive inclusions *(51,52)*.

In *Pick disease*, a rare form of presenile dementia with severe frontotemporal cortical atrophy, neuronal loss and ballooned neurons (Pick cells) and tau-positive neuronal inclusions (Pick bodies) composed tau proteins forming straight and twisted fibrils that differ from thos in AD and PSP *(10,11)*, only very rare neurons with Pick bodies have been shown to exhibiting DNA fragmentation *(9)*. By contrast, in frontotemporal dementia (FTD) without Pick bodies, DNA damage and activated caspase-3 expression as evidence for apoptosis were observed in neurons and astrocytes *(341)*. Recent studies suggest a relation between degenerating astrocytes and disturbances of cerebral perfusion in FTD *(342)*.

In *Huntington disease*, oligodendroglia but only very rare striatal neurons present DNA fragmentation *(123,343–346)*, possibly resulting from mutant huntingtin that enhances excitotoxic cell death *(13,346)*. Experimental data revealed various ways in which mutant huntingtin, either in soluble or insoluble form, might disrupt neuronal function, but they could not explain the regional selectivity of neurodegeneration. The basic mechanisms of polyglutamine toxicity could apply not only to various types of neurons but also to nonneuronal cells. Thus additional downstream changes, some intrinsic to the susceptible neurons and others triggered by mutant huntingtin, should come into play shaping the pattern of neuronal loss. Particularly strong evidence seems to incriminate glutamate excitotoxicity and related factors, such as disturbed homeostasis of Ca^{2+} and mitochondrial dysfunction *(13)*. Reduced levels of Fas an Fas-L were seen in striatum and striatal neurons with increase in reactive astrocytes of vulnerable regions, suggesting selective attenuation of Fas/Fas-L expression in striatal neurons in HD, although the meaning of their increased expression in astrocytes is still unclear *(72)*. Early and progressive accumulation of reactive microglia was seen in HD brain increasing with the degree of neuronal loss and often in proximate association with degenerating neurons, suggesting microglial response to changes in the neuropil and axons *(347)*. Recent studies have shown that cycle-AMP-responsive element binding protein (CREB), member of transcriptional factors essential for neuronal survival, are disrupted in HD suggesting that neurodegeneration is related to disturbance of CREB-dependent expression of cell-survival factors *(347a)*.

In *motor neuron disorders*, controversial findings for and against apoptotic neuronal death have been reported: Whereas earlier findings in ALS reported increased DNA fragmentation in ALS *(348–352)*, and recent studies demonstrated motor neurons apoptosis in Werdnig-Hoffmann disease *(264)*, and changes in the levels of Bcl-2 family members, activation of caspase-1 and -3, and detection of apoptosis-related molecular La[4] antibodies *(see* ref. *352a)*, others failed to find evidence of apoptosis in ALS spinal cord *(255,257,274,353)*. *In situ* hybridization studies in postmortem samples of ALS showed increased levels of Bax mRNA and significant decrease in Bcl-2 mRNA *(349)* and Bcl-2 protein levels *(352)*, whereas recent research on the subcellular alterations of Bcl-2 (decreased) Bak, Bax, and Bcl-Xl (increased) in ALS spinal cord and cortex revealed a pattern of protein distribution that suggests neuronal death to be apoptotic *(351)*. This was confirmed by increased expression of p53 in motor neurons of spinal cord and motor cortex as well as in astroglia, suggesting that p53 may participate in apoptotic cascade in ALS *(354)*. Delaying caspase activation by Bcl-2 was considered a clue to disease retardation in a transgenic mouse model of ALS *(355)*, while motor neuron degeneration is attenuated in Bax-deficient neurons in vitro *(356)*. However, using different forms of fixation for human postmortem material, recent personal studies revealed only sparse motor neurons in spinal cord and motor cortex being immunoreactive for Bcl-2, Bax, Bak, and activated caspase-3, with no differences in their numbers between ALS patients and controls, thus showing no evidence that apoptosis is a major mechanism of motor neuron cell death in ALS *(255,274)*.

Recent studies of human postmortem brains in major depression, occasional apoptotis in the absence of major pyramidal cell loss was found in hippocampus of both patients treated with corticosteroids and untreated ones, whereas immunohistochemistry for heat-shock protein 70 and NFκB was negative. These data indicate that apoptosis probably only contributes to a minor extent to the volume changes of hippocampus in major depression *(357)*.

5. APOPTOSIS AND TIME-COURSE OF NEURODEGENERATION

Although evidence for a role of apoptosis and related forms of PCD has been repeatedly reported in both experimental animal and cell culture models of various neurodegenerative disorders, and recent studies have shown an increased number of cells with DNA fragmentation in human AD brain and other neurodegenerative diseases, their incidence is significantly higher than could be expected in chronic processes with long preclinical phases and duration of illness *(303)*. Indeed, unlike DNA fragmentation, the hallmark end-stage signs of apoptosis such as nuclear chromatin condensation and apoptotic bodies are extremely rare or almost absent in AD and other neurodegenerative disorders. In addition to the morphologic and biochemical ambiguity of neuronal apoptotic mechanisms, there is a temporal dichotomy between the acuteness of apoptosis and the slowly progressive process in neurodegeneration with cell death occurring over periods of 5–20 yr and more *(303,358)*. In this case, apoptotic events may be rare at any given time and apoptotic bodies may be rapidly phagozytosed by neighboring glial cells without essential inflammatory reaction, and therefore difficult to detect. However, at any time, a small proportion of neurons and other cells appear to undergo conspicuous morphological changes. These include DNA fragmentation and nuclear condensation, hyperchromatic changes in the cytoplasm, and dislocation of the nucleus which are consistent with a rapid transient phase of the neurodegenerative process. The morphology of this phase that contrasts with the classical morphological features of necrosis *(21–24,29–33)* is sufficiently distorted to make normal cell function unlikely. Following these changes, it appears likely that the cellular cytoplasm disappears and abnormal nuclei remain. Such cell withering has been called *apoklesis (269)*. On the other hand, it has been shown that, although initiated, apoptosis does not necessarily progress to its completion. In in vitro models, the activation of caspases signifies apoptosis. However, in chronic neurodegenerative disorders, a different, low-grade process leading to cell dysfunction but not immediate cell death would have to be postulated. One possibility is that such a process might be initiated and mediated by procaspeses that have a much lower catalytic activity *(359–361)*; the proenzyme form of caspase-3 has approx 60-fold less activity than the activated enzyme *(13,62,66)*. However, it is not known whether caspases, especially procaspase-3, may be differentially expressed or regulated in affected and nonaffected brain regions in neurodegenerative disorders *(13)*.

Given that execution of apoptosis requires amplification of caspase-mediated apoptotic signals (Figs. 2 and 3), recent data indicate that in AD and other neurodegenerative disorders, there is a lack of effective apoptotic signal propagation to downstream caspase effectors *(359)*. Although upstream caspases (#8 and #9) are elevated in NFT-containing neurons in AD, downstream caspases, such as #3 and #7, remain at control levels. In addition, caspase-6, also an effector caspase, is restricted to senile plaques and not neurons, suggesting that extracellular pathology, i.e., Aβ deposition, differs from neuronal pathology. While the upstream caspases are restricted specifically to susceptible neurons in AD, the absence of effector caspases indicates a lack of propagation of the initial apoptotic signal. The lack of increased caspase-3 and -7 in AD brain, except within autophagic granules and rare neurons *(298)*, indicate incomplete or absent amplification of downstream events of the caspase cascade *(61)*. Because such downstream caspases and their proteolytic products are recognized as markers of apoptotic irreversibility *(360)*, their avoidance or sporadic appearance in AD indicates an absence of effective distal propagation of the caspase-mediated apoptotic signals. Thus, AD represents the first in vivo situation reported in which the initiation of apoptosis does *not* directly lead to apoptotic cell death. In certain areas in AD brain, while many neurons degenerate, it is unclear whether this is through successful propagation of the apoptotic cascade or by another pathway such as paratosis *(60)*. However, in those surviving neurons, it can be suggested that neuronal viability is, in part, maintained by the lack of distal transmission of the caspase-mediated apoptotic signals. This novel phenomenon of apoptotic avoidance termed *abortive apoptosis* or *abortosis*, may represent an exit from the caspase-induced apoptotic program *(359,361)*, that, given the robust survival of neurons with NFTs in both AD *(362)* and Parkinson-dementia complex of Guam *(363)* may ultimately lead to prolonged neuronal survival. This argues that neurons in AD may have an effective defense to apoptotic death (avoidance) rather than active complications of apoptosis *(363a)*. Alternatively, there may be other cellular mechanisms that limit the activation of the caspase cascade, which would be consistent with evidence that there are many compensatory mechanism in neurons that respond either to one-hit *(364)*, or perhaps accumulating insults, leading to neurodegeneration *(43)*. Thus, in neurodegenerative disorders, there may occur forms of cell death that are neither classical necrosis nor apoptosis *(140)*. Such factors are found in cells with abundant intracellular filaments *(23,24)* or insoluble protein filaments in the cytoplasm (e.g., NFTs, LBs, tau inclusions, Pick bodies, etc.), suggesting that such filamentas may contribute to dysfunction or increased vulnerability of the involved cell *(105,106)* but not necessarily to immediate cell demise.

6. CYTOSKELETAL CHANGES AND CELL DEATH

The relationship between PCD and aggregation of insoluble protein filaments and inclusions in the cytoplasm and, rarely, within nuclei of neurons and glia, in various neurodegenerative disorders remains unclear, although there are some interesting parallels between the neurochemical pathology of some of these inclusions and the biochemical correlates of PCD.

6.1. Alzheimer's Disease

In sporadic AD, DNA fragmentation may accompany tangle formation but is less correlated with the amyloid (plaques) load *(120,365,366)*. On the other hand, recent studies suggest that the development of Aβ plaques in the brain may cause damage to axons, and the abnormally prolonged stimulation of the neurons to this injury ultimately results in cytoskeletal alterations that underly neurofibrillary pathology and neurodegeneration *(291)*. The variable correlation between neuronal loss with both Aβ deposition and NFTs in AD suggests that these two hallmark lesions may not be the only causative factors of neuronal death *(120)*. For comparison, in aged canine brain lacking cytoskeletal (neurofibrillary) abnormalities, DNA damage is related to areas of Aβ deposition and is not considered a consequence of tau pathology or agonal artefacts *(367)*. On the other hand, recent studies showing colocalization of both caspase-cleavage product (CCP) antibody and PHF-1 in hippocampal neurons in AD brain, provide evidence for the activation of death-inducing mechanisms in neurons and suggest a possible association between NFT formation and caspase activation *(368)*. However, induction of tau phosphorylation by Aβ protein has been shown to involve MAP kinase and to precede cell death *(158)*. In AD brain and other tauopathies (PSP, PiD, CBD), expression of phosphorylated MAP kinases is associated with early tau deposition in neurons and glial cells but not with increased nuclear DNA vulnerability and cell death, indicating a lack of direct involvement of the MEK/ERK pathway of tau phosphorylation in the cell death cascade in these neurodegenerative disorders *(167)*. Light- and ultrastructural features of autophagic degeneration (i.e., mild condensation of nuclear chromatin, moderate vacuolation of endoplasmic reticulum, and lysosome-like vacuoles), but normal mitochondria *(21)*, which were observed in hippocampal neurons of AD brain *(128,273)*, and in occasional melanized SNc neurons in PD *(47)*, suggest alternative mechanisms of cell death not necessarily via apoptosis, but rather reflect the combined action of deficient DNA repair and accelerated DNA damage within susceptible cell populations *(43,52,53,298)*. However, it may be related to other underlying abnormalities that render cells more susceptible to oxidative stress and free radical damage that are suggested to represent major contributors to cell loss in AD as well as in other neurodegenerative diseases *(59, 242,369–378)*. Recent studies showed that, in addition to apoptosis and related forms of PCD, mitochondrial DNA damage represents a major pathogenic factor for cell loss and increased susceptibility to neuronal demise in AD *(379)*. The increased mitochondrial DNA damage is associated with reduced mitochondrial mass and enzyme gene expression. Impaired mitochondrial function could enhance neuronal sensitivity to apoptosis and may result in the activation of proapoptotic signaling in AD *(375,376)*.

Oxidative stress and free radical injury have been shown to induce some AD typical molecular abnormalities, including p53-mediated apoptosis, impaired mitochondrial function, and accumulation of hyperphosphorylated tau in neurons *(376,379)*. Furthermore, AD-associated neural thread protein (AD7c-NTP), a novel cDNA isolated from a library prepared with mRNA extracted from AD brain tissue, has recently been demonstrated to mediate some of the important cell death cascades associated with AD neurodegeneration. This suggests a link between overexpression of AD7c-NTP and the accumulation of hyperphosphorylated tau protein in susceptible neurons *(380)*. These changes may promote intracellular oxidative injury, and thus may increase the vulnerability of neurons in AD brain.

6.2. Synucleinopathies

Lewy bodies and degenerating (Lewy) neurites in PD and DLB brain are composed of aggregates of 7–20 nm intermediate filaments associated with a granular electron-dense coating material and vesicular structures, the core of the LB contains densely packed filaments and dense granular material. Cortical LBs are eosinophilic, rounded, angular, or reniform structures without an obvious halo *(3,381)*. Both LBs and "pale bodies"—their precursors *(382)*—are diagnostic hallmarks for both PD and DLB but are not specific for these disorders; they have been described in a variety of conditions as secondary pathology and show a wide distribution within the central and autonomous nervous system *(2–5,307)*.

The precise biochemical composition of LBs is still unknown, but immunohistochemical studies have shown that major components are phosphorylated neurofilament proteins present in both core and periphery, ubiquitin, a heat shock protein targeting proteins for breakdown, enzymes associated with ubiquitin mediated proteolysis and (de)phosphorylation, cytosolic- and microtubule-associated proteins except for tau-protein (only tau that is not bound to microtubules can interact with α-synuclein, whereas α-synuclein may modulate tau function *(107)*, αB-crystalin modulating intermediate filament organization and probably mediating the aggregation of microfilaments *(383)*, synaptophysin, chromogranin A (suggesting that vesicular structures in LBs may represent degenerating nerve endings), α-synuclein, a 143-amino-acid presynaptic, neuron-specific protein *(105,106,384–389)*, and synphylin-1, another protein that associates with α-synuclein and promotes LB formation *(106,399)*. In addition, LBs contain lipids and redox-active iron *(391,392)*. Altered α-synuclein is incorporated into LBs, their precursors ("pale bodies") and dystrophic Lewy neurites before ubiquitination; it is aggregated and fibrillated in vitro, morphologically resembling LB-like fibrils *(388)*. Ultrastructurally, major components of LBs are straight or twisted α-synuclein-positive filaments, 50–700 nm long and 5–10 nm wide, suggested to assemble from 5 nm protofibrils, two of which may interact to form a LB-filament *(385)*. LBs and related cytoplasmic inclusions express cdk5, a proline-directed protein kinase involved in cell cycle regulation that is likely to catalyse the in vivo phosphorylation of neurofilament proteins. The aberrant accumulation or ectopic expression of cdk5 and mitogen-activated protein kinase (MAPK), normally not found in neurons and glia, may lead to the formation of pathologic cytoskeletal inclusions *(393)*. Recent studies on the development of LBs in brainstem suggest an initial intraneuronal appearance of fine dust-like particles related to neuromelanin or lipofuscin, with homogeneous deposition of α-synuclein and ubiquitin with transition into a continuous deposition of α-synuclein in the periphery and of both substances in the central core of the LB *(391,394)*. The later frequent extraneural deposition of LBs may be related to disappearance of the involved neurons *(3)*.

The mechanisms of α-synuclein aggregation in vivo has not yet been fully elucidated, whereas in vitro it is modulated by various factors (Fig. 4). Although α-synuclein is usually unfolded or has an α-helical form, gene mutations, environmental stress, interaction with chaperon molecules Aβ, apolipoprotein E (ApoE) and metal ions can induce a transformation to β-folding *(91,107)*. In this form, it is sensible to self-aggregation in filamentous, amyloid-like fibrils and formation of insoluble intracellular inclusions (Fig. 5). In cultured cells, aggregation of α-synuclein takes place under certain conditions, such as high temperature or low pH *(388)*. Both wild-type and mutant types of α-synuclein may undergo self-aggregation and form insoluble fibrillar aggregates with antiparallel β-sheet structure upon incubation at physiologic temperature, which is accelerated for both hitherto known PD-linked point mutations *(395,396)*. Both α-synuclein mutations linked to autosomal dominant early-onset forms of PD promote the in vitro conversion of the natively unfolded protein into ordered prefibrillar oligomers, suggesting that these protofibrils, rather than the fibril itself, may induce cell death. Protofibrils differ markedly from fibrils with respect to their interactions with synthetic membranes. Protofibrillar α-synuclein, in contrast to the monomeric and the fibrillar forms, binds synthetic vesicles very tightly via a β-sheet rich structure and transiently permeabilizes these vesicles. The destruction of vesicular membranes by protofibrillar α-synuclein was directly observed by atomic force microscopy. The possibility that the toxicity of α-synuclein fibrillization may derive from an oligomeric intermediate, rather than the fibril, may

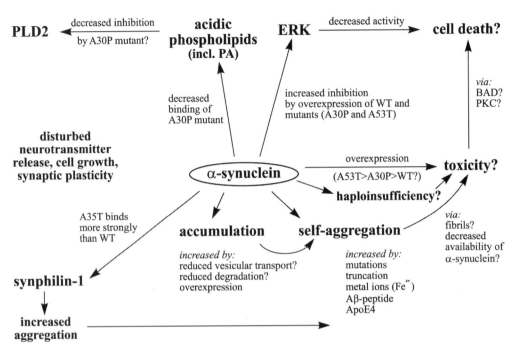

Fig. 4. Putative pathological mechanisms of α-synuclein aggregation (modified from ref. *92*). The binding partners and hypothetical steps of the different pathogenic mechanisms are indicated by arrows. PA, phosphatidic acid; PLD2, phospholipase D2; ERK, extracellular regulated kinase; WT, wild-type; A30P and A53T, described α-synuclein mutations; > possibly more toxic than.

have implications for the therapy of PD *(397)*. Because aggregation has been shown to be a nucleation-dependent process followed by fibril formation, α-synuclein nucleation may be the rate-limiting step for the formation of LB α-synuclein fibrils *(397)*. Because it binds to phospholipids with a certain conformational change of structure *(398)*, the altered composition of membrane lipids in neurodegeneration might be a prerequisite for the aggregation of α-synuclein *(391)*. Sequestration of redox-active iron and aberrant accumulation of ferrric iron causing formation of hydroxyl radicals via the Fenton reaction *(325, 399,400)* suggest that the iron-catalyzed oxidation reaction also plays a significant role in α-synuclein aggregation in vivo *(107)*. This aggregation could be partially blocked by desferoxamine, a high-affinity iron chelator *(401)*. Whether the interaction with ApoE ε3 and 4 that could facilitate the aggregation of Aβ peptide and NAC, the non-Aβ component of APP *(402)*, is important in the pathogenesis of PD still needs to be established. However, a recent association study in patients with sporadic PD suggested that a combination of ApoE and a polymorphism in the α-synuclein promoter may be pathogenic in PD *(403)*. Because α-synuclein interacts with several proteins known to affect cell viability (BAD, PKC, ERK) and overexpression of α-synuclein inhibits ERK, a MAP kinase activated in some cell survival transduction pathways, inhibition of ERK may result in loss of cell viability, and overexpression of the human A53T mutant resulted in dopaminergic cell death in culture *(404)*. In vivo, transgenic mice overexpressing human α-synuclein show marked loss of dopaminergic neurons in different brain regions including SNc and formation of inclusion bodies composed of granular, not fibrillar α-synu-

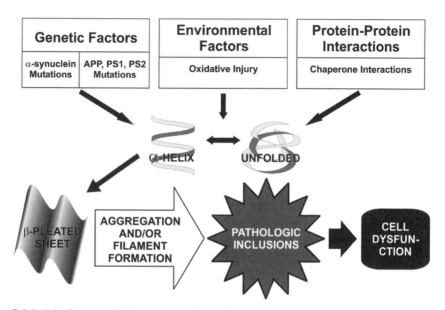

Fig. 5. Model of α-synuclein aggregate formation caused by several factors inducing conversion to a β-pleated sheet conformation. In this conformation, α-synuclein is more susceptible to self-aggregation into filamentous protein fibrils forming intracellular inclusions. Modified from ref. *53.*

clein aggregates which might be possibly correlate with the more toxic nonfilamentous aggregates in vitro *(405)*. In *Drosophila* that overexpress human wild-type or mutant α-synuclein, dopaminergic cell loss, filamentous intraneural inclusions, and age-dependent locomotor dysfunction were observed *(406)*. Overexpression of wild-type and mutant α-synuclein showed different effects on dopaminergic cell susceptibility: overexpression of wild-type-delayed cell death induced by MPP$^+$, while decreases of glutathione (GSH) levels were attenuated by wild-type α-synuclein. Mutant forms increased levels of 8-hydroguanidine, protein carbonyls, lipid peroxidation, and 3-nitrotyrosine, along with markedly accelerating cell death in response to the above insults. The presence of abnormal α-synuclein in SN in PD brain, thus, may increase neuronal vulnerability to a range of toxic agents inducing apoptotic mechanisms *(407)*. On the other hand, decreased α-synuclein mRNA in both rat SN following administration of 6-OHDA *(408)*, and in SN and cortex of human PD brain in the presence of preserved dopamine transporter (DAT) and vesicular VMAT2, suggest that its increase is an early change in the process, leading to neuronal degeneration likely preceding changes in tyrosin hydroxylase and dopamine markers *(409)*. Because immunoreactivity for the heme protein, cytochrome-c, which is located in the intramembrane of mitochondria and released upon apoptotic stimuli into the cytosole, was detected in LBs with defects of cytochrome-c oxidase in SNp neurons in PD *(410)*, aggregation of α-synuclein that is stimulated by cytochrome-c *(411)* may also be closely related to mitochondrial dysfunction *(388)*. α-Synuclein has been shown to produce neuronal death in vitro owing to promotion of mitochondrial deficit, oxidative stress, and effectors of the APK pathway *(412–414)* (*see* Fig. 6).

Crosslinking of α-synuclein by advanced glycation end products (AGEs) thought to play an important role in the pathogenesis of neurodegenerative disorders have been

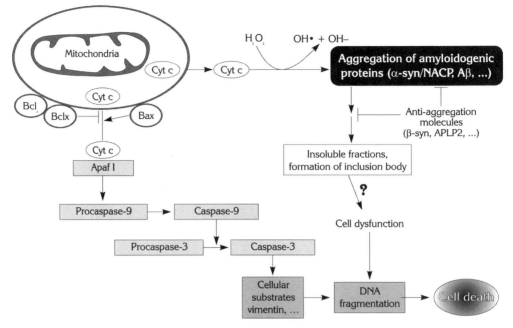

Fig. 6. Possible mechanisms of cell death in synucleinopathies. Modified from ref. *414.*

described not only in advanced stages of PD, but also in very early LBs of cases with incidental Lewy body disease believed to represent preclinical PD *(415).* Because AGEs are both markers of transition metal-induced oxidative stress, and inducers of protein crosslinking and free radical formation, it appears likely that AGE-promoted formation of LBs reflects early disease-specific changes rather than late epiphenoma of PD *(400).*

Although α-synuclein belongs to a family of proteins involved in synaptic function, plasticity, and neurodegeneration *(106,107),* and recent in vivo and in vitro studies suggest toxicity of α-synuclein and its mutant forms, it is still unknown whether neurodegeneration is owing to α-synuclein aggregation, to interference with specific functions of this protein, or whether LBs containing α-synuclein are harmless side products of cell damage. On the other hand, the involvement of the ubiquitin proteolytic system suggests that they may be the structural manifestation of a cytoprotective response designed to eliminate damaged cell elements. The deposits of insoluble proteinaceous fibrils may contribute to dysfunction or death of the involved cells *(106,387).* Furthermore, LBs being the sequelae of frustraneous proteolytic degradation of abnormal cytoskeletal constiuents may represent—similar to other cellular inclusions (e.g., NFTs in AD, Pick bodies, etc.)—end products or reactions to hitherto unknown neuronal degeneration processes that are associated with disturbancs of axonal protein transport, and finally lead to cell death. In PD, SNc cell degeneration is preceded by loss of neurofilament proteins (NFP), neuronal thyrosin hydroxylase (TH) immunoreactivity, TH, DAT, and NFT mRNA, TH- and DAT-proteins, and cytochrome oxidase-c, indicating functional neuronal damage *(4,124,410,416).* It is accompanied by distribution of neuromelanin with uptake into macrophages *(3),* astroglial reaction, and proliferation of MHC class II positive microglia, the latter by releasing cytokins, CD 23, NO, and other

substances mediating inflammatory reactions that may be both inducing factors or sequelae of neuronal death *(319,322,323,417–420)*.

In *MSA*, argyrophilic intracytoplasmic inclusions are found in numerous oligodendrocytes (GCIs) and scattered neurons *(12,106,329,384)*. They contain ubiquitin, αB-crystalin, tau protein, tubulins, and high-molecular-weight aggregates of truncated α-synuclein
(107,108,421–424). Ultrastructurally, they contain 15–30 nm fibrils that are irregular
on their outer surface; some are tubular perhaps because of incorporation of other proteins or entrapment of microtubules in bundles of α-synuclein *(425)*. The GCIs in MSA,
like LBs and NFTs in AD, express cdk5kinase, that is likely to catalyse the in vivo phosphorylation of proteins. The ectopic expression of ckd5 and MAPK in GCIs, both kinases
not normally found in neurons and glial cells, resulting in the abnormal phosphorylation
of microtubular cytoskeletal proteins (e.g., neurofilament and tau-protein), has been suggested to lead to the formation of pathological cytoskeletal inclusions *(426)*. The formation of GCIs in MSA may be the crucial event in this sporadic synucleinopathy through
which the oligodendroglia-myelin-axon pathway causes neurodegeneration *(12,329,423)*.

The selective accumulation of insoluble α-synuclein in LBs, Lewy neurites, GCIs, and
neuronal inclusions in MSA *(425)*, suggests that a reduction in its solubility may induce
filament formation with aggregation into cytoplasmic inclusions. However, recent studies
showed different solubility of α-synuclein between LBs in DLB (insoluble) and GCIs
in MSA oligodendroglia (soluble) possibly resulting from different processing of α-synuclein *(427)*, which might also influence the viability of involved cells in different ways.
In MSA, a distinct cytoplasmic expression of Bcl-2 was seen in oligodendrocytes in an
GCI-like pattern with moderate Bax expression and coexpression of ubiquitin and/or α-
synuclein and Bcl-2 in about a quarter of the GCI-containing cells *(51,330)*.

6.3. Other Neurodegenerative Disorders

In *PSP* and *CBD*, both tauopathies, the histological hallmarks are widespread tau-
positive inclusions in either neurons (globose NFTs) and both oligodendroglia ("coiled
bodies") and "tufted" astrocytes, with additional astrocytic plaques in CBD composed
of 15 nm tubuli containing 64 + 69 kd tau doublet (four R isoforms) that differ from
paired helical filaments in AD *(332–335,338–340)*, while the argyrophilic Pick bodies
in PiD are composed of straight and twisted 12-15 nm fibrils and NFT-like 24 nm fibrils
containing ubiquitin, αB-crystallin, chromogranin A, and 3-repeat tau doublets *(11)*. In
both PSP and CBD, frequent DNA fragmentation and expression of ARP and activated
caspase-3 were observed in oligodendroglia and occasional astroglia, approx 1/4 of which
showed coexpression with tau-positive cytoplasmic inclusions *(51,52)*, suggesting some
causal relationship between these two findings. On the other hand, in PiD, the Pick-
body containing neurons in only less than 1% showed DNA fragmentation or signs of
apoptosis *(9)*.

In *Huntington disease*, the DNA fragmentation observed in oligodendroglia was
suggested to result from either Wallerian degeneration because of neuronal death or a
direct effect of genetic IT-15 huntingtin mutation changes *(344–346)*.

In general, these intracellular inclusions mainly composed of insoluble proteinaceous fibrils may contribute to dysfunction of the involved cells *(105–107,387,388,421)*.
This is in line with the suggestion that neurodegeneration is prone in cells containing
phosphorylated proteins that participate in formation of various cytoskeletal inclusions

and, thus, may be vulnerable to PCD induced by many different causes *(105–107,387, 421,428)*. However, recent demonstration of negative DNA fragmentation in SN neurons containing LBs *(48–52,316)*, in neurons with Pick bodies *(9)*, limited involvement of NFT-bearing AD neurons by apoptosis *(120,128)*, recent calculation that tangle-bearing neurons in Parkinson-dementia complex, and AD may survive for an average of 2.5 yr or even up to 20 yr *(362,363)*, indicates that these inclusions may not predispose a cell to undergo (programmed) death. This is in agreement with recent data after chronic inhibition of protein phosphatases 1 (PP1) and 2 (PPSA), causing dephosphorylation of tau protein and neuronal apoptotis that show different distribution, suggesting that these cytoskeletal changes have no obvious consequences for the viability of the involved neurons *(429)*. Thus, the biological significance of these inclusions, related to mismetabolism of cytoskeletal proteins (often representing morphological hallmarks of a specific neurodegenerative disorder) and also the role they play in neurodegeneration, are still undetermined. Recent studies on the subunit dissociation of oligomeric proteins induced by hydrostatic pressure have shown that interconversion between different conformations occurs within hours or days, suggesting that the proteins represent a persistently heterogeneous population of different long-lived conformers. This may have important implications for our understanding of protein folding and misfolding and biological functions. This deterministic behavior of proteins may be closely related to the genesis of conformational diseases, including AD, transmissible spongiform encephalopathies, and other neurodegenerative disorders *(430)*. Despite considerable progress in the clarification of the molecular mechanisms of PCD *(30–36,44,45,63,96)*, the intracellular cascade leading to neuronal death in chronic progressive disorders remains to be elucidated. Better understanding of these mechanisms and their basic pathobiological causes and routes may lead to future development of protective strategies and novel approaches for the treatment of neurodegenerative disease.

7. CONCLUSIONS

On the basis of the presently available data, it has to be stated that, although many in vitro and in vivo studies favor apoptosis in Alzheimer's and Parkinson's disease and other neurodegenerative disorders, the majority of human brain tissue studies have yielded mixed or controversial results, and there is increasing evidence for alternative mechanisms of neuronal demise in neurodegeneration. Human and experimental data indicate that cells with DNA fragmentation are either sensitive to various noxious factors or are injured cells, although not necessarily undergoing apoptosis or necrosis. Furthermore, the activation of some caspases and other cell-death signals, important for the activation of the apoptotic cascade, apparently does not have a significant role in the wide-spread neuronal death that occurs in most neurodegenerative disoders, although it may contribute to the loss by apoptosis or related forms of programmed cell death of specifically vulnerable neurons. Increased sensitivity of specific neuronal populations may be related to a proapoptotic environment in the brain of patients with AD or other neurodegenerative diseases. The death cascades, however, may be counteracted by other cellular mechanisms that limit the activation of caspases and other apoptotic triggers, suppress oxyradicals, and stabilize calcium homeostasis and mitochondrial function. This would suggest that there may be compensatory mechanisms in neurons that respond to various one-time, chronic, and perhaps accumulating insults in neurodegeneration. Because of the temporal dichotomy between the acuteness of apoptosis and the slowly progressive process of neurodegeneration with cell death over periods of years to decades, demonstration of apoptosis in human brain may be extremely rare and can be influenced by perimortem factors. However, neuronal death in neurodegeneration may often represent a form of demise that is neither classical necrosis nor apoptosis, with occasional autophagic degeneration to be observed. Death of vulnerable cells may be finally triggered during the terminal period of the patient's life. In conclusion, despite considerable progress in our understanding of the causes and routes of programmed cell death, the intracellular cascade leading to neuronal and glial demise in chronic progressive neurodegenerative disorders remains to be elucidated. Understanding of their molecular pathobiology and causative mechanisms may provide better insights into their pathogenesis and lead to development of future protective strategies and novel approaches for the treatment of these disorders.

REFERENCES

1. Jellinger, K. A. and Bancher, C. (1998) Neuropathology of Alzheimer's disease; a critical update. *J. Neural Transm.* **54(Suppl.),** 77–95.
2. Jellinger, K. A. (1998) Neuropathology of movement disorders. *Neurosurg. Clin. North. Am.* **9,** 237–262.
3. Forno, L. S. (1996) Neuropathology of Parkinson's disease. *J. Neuropathol. Exp. Neurol.* **55,** 259–272.
4. Jellinger, K. A. (2001) The pathology of Parkinson's disease. *Adv. Neurol.* **86,** 55–72.
5. Braak, H. and Braak, E. (2000) Pathoanatomy of Parkinson's disease. *J. Neurol.* **247, Suppl. 2,** II3–10.
6. McKeith, I. G., Galasko, D., Kosaka, K., Perry, E. K., Dickson, D. W., Hansen, L. A., et al. (1996) Consensus guidelines for the clinical and pathological diagnosis of dementia with Lewy bodies (DLB), report of the consortium on DLB International Workshop. *Neurology* **47,** 1113–1124.
7. Hughes, A. J., Daniel, S. E., Kilford, L., and Lees, A. J. (1992) Accuracy of clinical diagnosis of idiopathic Parkinson's disease: a clinico-pathological study of 100 cases. *J. Neurol. Neurosurg. Psychiatry* **55,** 181–184.
8. Hamilton, R. L. (2000) Lewy bodies in Alzheimer's disease: a neuropathological review of 145 cases using α-synuclein immunohistochemistry. *Brain Pathol.* **10,** 378–384.
9. Gleckman, A. M., Jiang, Z., Liu, Y., and Smith, T. W. (1999) Neuronal and glial DNA fragmentation in Pick's disease. *Acta Neuropathol.* **98,** 55–61.
10. Buée, L. and Delacourte, A. (1999) Comparative biochemistry of tau in progressive supranuclear palsy, corticobasal degeneration, FTD-17 and Pick's disease. *Brain Pathol.* **9,** 681–693.
11. Dickson, D. W. (1998) Pick's disease: a modern approach. *Brain Pathol.* **8,** 339–354.
12. Dickson, D. W., Lin, W.-I., Liu, W.-K., and Yen, S.-H. (1999) Multiple system atrophy: a sporadic synucleinopathy. *Brain Pathol.* **9,** 721–732.
13. Sieradzan, K. A. and Mann, D. M. (2001) The selective vulnerability of nerve cells in Huntington's disease. *Neuropathol. Appl. Neurobiol.* **27,** 1–21.
14. Ho, L. W., Carmichael, J., Swartz, J., Wyttenbach, A., Rankin, J., and Rubinsztein, D. C. (2001) The molecular biology of Huntington's disease. *Psychol. Med.* **31,** 3–14.
15. Becher, M. W., Kotzuk, J. A., Sharp, A. H., Davies, S. W., Bates, G. P., Price, D. L., and Ross, C. A. (1998) Intranuclear neuronal inclusions in Huntington's disease and dentatorubral and pallidoluysian atrophy: correlation between the density of inclusions and IT15 CAG triplet repeat length. *Neurobiol. Dis.* **4,** 387–397.

16. Julien, J.-P. and Beaulieu, J.-M. (2000) Cytoskeletal abnormalities in amyotrophic lateral sclerosis: beneficial or detrimental effects? *J. Neurol. Sci.* **180,** 7–14.

17. Nicotera, P., Leist, M., Fava, E., Berliocchi, L., and Volbracht, C. (2000) Energy requirement for caspase activation and neuronal cell death. *Brain Pathol.* **10,** 276–282.

18. Kerr, J. F. R., Wyllie, A. H., and Currie, A. R. (1972) Apoptosis, a basic biological phenomenon with wide-ranging implications in tissue kinetics. *Br. J. Cancer* **26,** 239–257.

19. Flemming, W. (1885) Ueber die Bildung von Richtungsfiguren in Säugethiereiern beim Untergang Graaf'scher Follikel. *Arch. Anat. Physiol.* **1885,** 221–224.

20. Clarke, P. G. H. (1990) Developmental cell death: morphological diversity and multiple mechanisms. *Anat. Embryol.* **181,** 195–213.

21. Clarke, P. G. H. (1999) Apoptosis versus necrosis, in *Cell Death and Disease of the Nervous System* (Koliatsosue, M. and Ratan, R. R., eds.), Humana Press, Totowa, NJ, pp. 3–28.

22. Wyllie, A. H., Kerr, J. F. R., and Currie, A. R. (1980) Cell death: the significance of apoptosis. *Int. Rev. Cytol.* **68,** 251–305.

23. Wyllie, A. H. (1997) Apoptosis: an overview. *Br. Med. Bull.* **53,** 451–465.

24. Majno, G. and Joris, I. (1995) Apoptosis, oncosis and necrosis: an overview of cell death. *Am. J. Pathol.* **146,** 3–14.

25. Nicotera, P., Leist, M., and Ferrando-May, E. (1999) Apoptosis and necrosis, different execution of the same death. *Biochem. Soc. Symp.* **66,** 69–73.

26. Thompson, C. B. (1995) Apoptosis in the pathogenesis and treatment of disease. *Science* **267,** 1456–1462.

27. Bredesen, D. E. (1995) Neural apoptosis. *Ann. Neurol.* **38,** 839–851.

28. Savitz, S. I. and Rosenbaum, D. M. (1998) Apoptosis in neurological disease. *Neurosurgery* **42,** 555–574.

29. Levin, S., Bucci, T. J., Cohen, S. M., Fix, A. S., Hardisty, J. F., LeGrand, E. K., et al. (1999) The nomenclature of cell death: recommendations of an ad hoc committee of the Society of Toxicologic Pathologists. *Toxic. Pathol.* **27,** 484–490.

30. Yuan, J. Y. and Yankner, B. A. (2000) Apoptosis in the nervous system. *Nature* **407,** 802–809.

31. Reed, J. C. (2000) Mechanisms of apoptosis. *Am. J. Pathol.* **157,** 1415–1430.

32. Wang, K. K. W. (2000) Calpain and caspase; can you tell the difference? *Trends Neurol. Sci.* **23,** 20–26.

33. Behl, C. (2000) Apoptosis and Alzheimer's disease. *J. Neural Transm.* **107,** 1325–1344.

34. Bredesen, D. E. (2000) Apoptosis: overview and signal transduction pathways. *J. Neurotrauma* **17,** 801–810.

35. Bratton, S. B. and Cohen, G. M. (2001) Apoptotic death sensor: an organelle's alter ego? *Trends Pharmacol. Sci.* **22,** 306–315.

36. Gorman, A. M., Ceccatelli, S., and Orrenius, S. (2000) Role of mitochondria in neuronal apoptosis. *Dev. Neurosci.* **22,** 348–358.

37. Desagher, S. and Martinou, J. C. (2001) Mitochondria as the central control point of apoptosis. *Trends Cell Biol.* **10,** 369–377.

38. Ishimaru, M. J., Ikonomidou, C., Tenkova, T. I., Der, T. C., Dikranian, K., Sesma, M. A., and Olney, J. W. (1999) Distinguishing excitotoxic from apoptotic neurodegeneration in the developing rat brain. *J. Comp. Neurol.* **408,** 461–476.

39. Dikranian, K., Ishimaru, M. J., Tenkova, T., Labruyere, J., Qin, Y. Q., Ikonomidou, C., and Olney, J. W. (2001) Apoptosis in the *in vivo* mammalian forebrain. *Neurobiol. Dis.* **8,** 359–379.

40. Lorenz, H. M., Herrmann, M., Winkler, T., Gaipl, U., and Kalden, J. R. (2000) Role of apoptosis in autoimmunity. *Apoptosis* **5,** 443–449.

41. Cotman, C. W. (1998) Apoptosis decision cascades and neuronal degeneration in Alzheimer's disease. *Neurobiol. Aging* **19(Suppl. 1),** S29–S32.

42. Cotman, C. W. and Su, J. H. (1996) Mechanisms of neuronal cell death in Alzheimer's disease. *Brain Pathol.* **6,** 493–506.

43. Cotman, C. W., Qian, H. Y., and Anderson, A. J. (2000) Cellular signaling pathways in neuronal apoptosis. Role in neurodegeneration and Alzheimer's disease, in *Cerebral Signal Transduction. From First to Fourth Messengers* (Reith, M. E. A., ed.), Humana Press, Totowa, NJ, pp. 175–206.

44. Michel, P. P., Lambeng, N., and Ruberg, M. (1999) Neuropharmacologic aspects of apoptosis: significance for neurodegenerative diseases. *Clin. Neuropharmacol.* **22,** 137–150.

45. Mattson, M. P. (2001) Mechanisms of neuronal apoptosis and excitotoxicity, in *Pathogenesis of Neurodegenerative Disorders* (Mattson, M. P., ed.), Humana Press, Totowa, NJ, pp. 1–20.

46. Mochizuki, H., Mori, H., and Mizuno, Y. (1997) Apoptosis in neurodegenerative disorders. *J. Neural Transm.* **50(Suppl.),** 125–140.

47. Anglade, P., Vyas, S., Javoy-Agid, F., Herrero, M. T., Michel, P. P., Marquez, J., et al. (1997) Apoptosis and autophagy in nigral neurons of patients with Parkinson's disease. *Histol. Histopathol.* **12,** 25–31.

48. Tatton, W. G. and Olanow, C. W. (1999) Apoptosis in neurodegenerative disease: The role of mitochondria. *Biochem. Biophys. Acta* **1410,** 195–214.

49. Gibson, R. M. (2001) Does apoptosis have a role in neurodegeneration? *Br. M. J.* **322,** 1539–1540.

50. Jellinger, K. A. (2000) Cell death mechanisms in Parkinson's disease. *J. Neural Transm.* **107,** 1–29.

51. Jellinger, K. A. and Stadelmann, C. (2000) The enigma of cell death in neurodegenerative disorders. *J. Neural Transm.,* **60(Suppl.),** 365–380.

52. Jellinger, K. A. and Stadelmann, C. (2001) Problems of cell death in neurodegeneration and Alzheimer's disease. *J. Alz. Dis.* **3,** 31–40.

53. Jellinger, K. A. (2001) Cell death mechanisms in neurodegeneration. *J. Cell. Mol. Med.* **5,** 1–17.

53a. Graeber, M. B. and Moran, L. B. (2002) Mechanisms of cell death in neurodegenerative diseases: fashion, fiction, and facts. *Brain Pathol.* **12,** 385–390.

54. Tatton, W. G., Chalmers-Redman, R. M. E., Rideout, H. J., and Tatton, N. A. (1999) Mitochondrial permeability in neuronal death: possible relevance to the pathogenesis of Parkinson's disease. *Parkinsonism Relat. Disord.* **5,** 221–229.

55. Tatton, N. A. (2000) Increased caspase-3 and BAX immunoreactivity accompanying nuclear GAPDH translocation and neuronal apoptosis in Parkinson's disease. *Exp. Neurol.* **166,** 29–43.

56. Vieira, H. L. and Kroemer, G. (1999) Pathophysiology of mitochondrial cell death control. *Cell. Mol. Life Sci.* **56,** 971–976.

57. Tamatani, M., Ogawa, S., Niitsu, Y., and Tohyama, M. (1998) Involvement of bcl-2 family and caspase-3-like protease in NO-mediated neuronal apoptosis. *J. Neurochem.* **71,** 1588–1596.

58. Wang, J., Silva, J. P., Gustafsson, C. M., Rustin, P., and Larsson, N. G. (2001) Increased *in vivo* apoptosis in cells lacking mitochondrial DNA gene expression. *Proc. Natl. Acad. Sci. USA* **98,** 4036–4043.

59. Mattson, M. P. (2001) Inflammation, free radicals, glycation, metabolism and apoptosis, and heavy metals, in *Functional Neurobiology of Aging* (Hof, R. R. and Mobbs, L. C. K., eds.), Academic Press, San Diego, CA, pp. 349–371.

60. Sperandio, S., de Belle, I., and Bredesen, D. E. (2000) An alternative, nonapoptotic form of programmed cell death. *Proc. Natl. Acad. Sci. USA* **97,** 14,376–14,381.

61. Cohen, G. M. (1997) Caspases: the executioners of apoptosis. *Biochem. J.* **326,** 1–16.

62. Chan, S. L. and Mattson, M. P. (1999) Caspase and calpain substrates: roles in synaptic plasticity and cell death. *J. Neurosci. Res.* **58,** 167–190.

63. Stadelmann, C. and Lassmann, H. (2000) Detection of apoptosis in tissue sections. *Cell Tissue Res.* **301,** 19–31.
64. Adamec, E., Yang, F., Cole, G. M., and Nixon, R. A. (2001) Multiple-label immunocytochemistry for the evaluation of nature of cell death in experimental models of neurodegeneration. *Brain Res. Brain Res. Protoc.* **7,** 193–202.
65. Green, D. R. and Reed, J. C. (1998) Mitochondria and apoptosis. *Science* **281,** 1309–1312.
66. Leblanc, A. C., ed. (2002) Apoptosis techniques and protocols. 2nd edition. Humana Press, Totowa, NJ.
67. Wolf, B. B. and Green, D. R. (1999) Suicidal tendencies, apoptotic cell death by caspase family proteinases. *J. Biol. Chem.* **274,** 20,049–20,052.
68. Stennicke, H. R. and Salvesen, G. S. (1997) Biochemical characteristics of caspases-3, -6, -7, and -8. *J. Biol. Chem.* **272,** 25,719–25,723.
69. Roth, K. A., Kuan, C.-Y., Haydar, T. F., D'Sa-Eipper, C., Shindler, K. S., Zheng, T. S., et al. (2000) Epistatic and independent functions of caspase-3 and Bcl-XL in developmental programmed cell death. *Proc. Natl. Acad. Sci. USA* **97,** 466–471.
70. Gerhardt, E., Kugler, S., Leist, M., Beier, C., Berliocchi, L., Volbracht, C., et al. (2001) Cascade of caspase activation in potassium-deprived cerebellar granule neurons: targets for treatment with peptide and protein inhibitors of apoptosis. *Mol. Cell. Neurosci.* **17,** 717–731.
71. Springer, J. E., Nottingham, S. A., McEwen, M. L., Azbill, R. D., and Jin, Y. (2001) Caspase-3 apoptotic signaling following injury to the central nervous system. *Clin. Chem. Lab. Med.* **39,** 299–307.
72. Ferrer, I., Blanco, R., Cutillas, B., and Ambrosio, S. (2000) Fas and Fas-L expression in Huntington's disease and Parkinson's disease. *Neuropathol. Appl. Neurobiol.* **26,** 424–433.
73. Budd, S. L., Tenneti, L., Lishnak, T., and Lipton, S. A. (2000) Mitochondrial and extramitochondrial apoptotic signaling pathways in cerebrocortical neurons. *Proc. Natl. Acad. Sci. USA* **97,** 6161–6166.
74. Kuida, K., Zheng, T. S., Na, S., Kuan, C., Yang, D., Karasuyama, H., et al. (1996) Decreased apoptosis in the brain and premature lethality in CPP32-deficient mice. *Nature* **384,** 368–372.
75. Nakagawa, T., Zhu, H., Morishima, N., Li, E., Xu, J., Yankner, B. A., and Yuan, J. (2000) Caspase-12 mediates endoplasmic reticulum-specific apoptosis and cytotoxicity by amyloid-β. *Nature* **403,** 98–103.
76. Mattson, M. P., Duan, W. Z., Chan, S. L., and Camandola, S. (1999) Par-4: An emerging pivotal player in neuronal apoptosis and neurodegenerative disorders. *J. Molec. Neurosci.* **13,** 17–30.
77. Gross, A., McDonnell, J. M., and Korsmeyer, S. J. (1999) Bcl-2 family members and the mitochondria in apoptosis. *Genet. Dev.* **13,** 1899–1911.
78. Adams, J. M. and Cory, S. (1998) The Bcl-2 protein family: arbiters of cell survival. *Science* **281,** 1322–1326.
79. Susin, S. A., Lorenzo, H. K., Zamzami, N., Marzo, I., Snow, B. E., Brothers, G. M., et al. (1999) Molecular characterization of mitochondrial apoptosis-inducing factor. *Nature* **397,** 441–446.
80. Joza, N., Susin, S. A., Daugas, E., Stanford, W. L., Cho, S. K., Li, C. Y. J., et al. (2001) Essential role of the mitochondrial apoptosis-inducing factor in programmed cell death. *Nature* **410,** 549–554.
81. Finucane, D. M., Bossy-Wetzel, E., Waterhouse, N. J., Cotter, T. G., and Green, D. R. (1999) Bax-induced caspase activation and apoptosis via cytochrome c release from mitochondria is inhibitable by Bcl-xL. *J. Biol. Chem.* **274,** 2225–2233.
82. Sadoul, R. (1998) Bcl-2 family members in the development and degenerative pathologies of the nervous system. *Cell Death Diff.* **5,** 805–815.

83. Hsu, Y. T., Wolter, K. G., and Youle, R. J. (1997) Cytosol to membrane re-distribution of Bax and Bcl-x-l during apoptosis. *Proc. Natl. Acad. Sci. USA* **94,** 3668–3672.

84. Reed, J. C. (1997) Double identity for proteins of the Bcl-2 family. *Nature* **387,** 773–776.

85. Boonman, Z. and Isacson, O. (1999) Apoptosis in neuronal development and transplantation: role of caspases and trophic factors. *Exp. Neurol.* **156,** 1–18.

86. Cregan, S. P., MacLaurin, J. G., Craig, C. G., Robertson, G. S., Nicholson, D. W., Park D. S., and Slack, R. S. (1999) Bax dependent caspase-3 activation is a key determinant in p53-induced apoptosis in neurons. *J. Neurosci.* **19,** 7860–7869.

87. Dargusch, R., Piasecki, D., Tan, S., Liu, Y. B., and Schubert, D. (2001) The role of Bax in glutamate-induced nerve cell death. *J. Neurochem.* **76,** 295–301.

88. Miller, F. D., Pozniak, C. D., and Walsh, G. S. (2000) Neuronal life and death: an essential role for the p53 family. *Cell Death Differ.* **7,** 880–888.

89. Keramaris, E., Stefanis, L., MacLaurin, J., Harada, N., Takaku, K., Ishikawa, T., et al. (2000) Involvement of caspase 3 in apoptotic death of cortical neurons evoked by DNA damage. *Mol. Cell. Neurosci.* **15,** 368–379.

90. Kim, T. W., Pettingell, W. H., Jung, Y. K., Kovacs, D. M., and Tanzi, R. E. (1997) Alternative cleavage of Alzheimer-associated presenilins during apoptosis by a caspase-3 family protease. *Science* **277,** 373–376.

91. Grunberg, J., Walter, J., Loetscher, H., Deuschle, U., Jacobsen, H., and Haass, C. (1998) Alzheimer's disease associated presenilin-1 holoprotein and ist 18-20 kDa C-terminal fragment are death substrates for proteases of the caspase family. *Biochemistry* **37,** 2263–2270.

92. Gervais, F. G., Xu, D., Robertson, G. S., Vaillancourt, J. P., Zhu, Y., Huang, J., et al. (1999) Involvement of caspases in proteolytic cleavage of Alzheimer's amyloid-beta precursor protein and amyloidogenic A beta peptide formation. *Cell* **97,** 395–406.

93. Goldberg, Y. P., Nicholson, D. W., Rasper, D. M., Kalchman, M. A., Koide, H. B., Graham, R. K., et al. (1996) Cleavage of huntingtin by apopain, a proapoptotic cysteine protease, is modulated by the polyglutamine tract. *Nat. Genet.* **13,** 442–449.

94. Miyashita, T., Okamura-Oho, Y., Mito, Y., Nagafuchi, S., and Yamada, M. (1997) Dentatorubral pallidoluysian atrophy (DRPLA) protein is cleaved by caspase-3 during apoptosis. *J. Biol. Chem.* **272,** 29,238–29,242.

95. Hartmann, A., Hunot, S., Michel, P. P., Muriel, M. P., Vyas, S., Faucheux, B. A., et al. (2000) Caspase-3. A vulnerability factor and final effector in apoptotitc death of dopaminergic neurons in Parkinson's disease. *Proc. Natl. Acad. Sci. USA* **97,** 2875–2880.

96. Hartmann, A. and Hirsch, E. C. (2001) Parkinson's disease: the apoptosis hypothesis revisited. *Adv. Neurol.* **86,** 143–153.

97. Allen, J. W., Eldadah, B. A., Huang, X., Knoblach, S. M., and Faden, A. I. (2001) Multiple caspases are involved in β-amyloid-induced neuronal apoptosis. *J. Neurosci. Res.* **65,** 45–53.

97a. Roth, K. A. (2002) Caspases, apoptosis, and Alzheimer disease: causation, correlation, and confusion. *J. Neuropathol. Exp. Neurol.* **60,** 829–838.

97b. Wei, W., Norton, D. D., Wang, X., and Kusiak, J. W. (2002) Aβ 17-42 in Alzheimer's disease activates JNK and caspase-8 leading to neuronal apoptosis. *Brain* **125,** 2036–2043.

98. Araya, R., Uehara, T., and Nomura, Y. (1998) Hypoxia induces apoptosis in human neuroblastoma SK-N-MC cells by caspase activation accompanying cytochrome c release from mitochondria. *FEBS Lett.* **439,** 168–172.

99. Northington, F. J., Ferriero, D. M., Flock, D. L., and Martin, L. J. (2001) Delayed neurodegeneration in neonatal rat thalamus after hypoxia-ischemia is apoptosis. *J. Neurosci.* **21,** 1931–1938.

100. Herdegen, T., Skene, P., and Bähr, M. (1997) The c-Jun transcription factor: bipotential mediator of neuronal death, survival and regeneration. *Trends Neurosci.* **20,** 227–231.

101. Grand, R. J., Milner, A. E., Mustoe, T., Johnson, G. D., Owen, D., Grant, M. L., and Gregory, C. D. (1995) A novel protein expressed in mammalian cells undergoing apoptosis. *Exp. Cell. Res.* **218,** 439–451.

102. Ferrer, I., Planas, A. M., and Pozas, E. (1997) Radiation-induced apoptosis in developing rats and kainic acid-induced excitotoxicity in adult rats are associated with distinctive morphological and biochemical c-Jun/AP-1 (N) expression. *Neuroscience* **80,** 449–458.

103. Pozas, E., Ballabriga, J., Planas, A. M., and Ferrer, I. (1997) Kainic acid-induced excitotoxicity is associated with a complex c-Fos and c-Jun response, which does not preclude either cell death or survival. *J. Neurobiol.* **33,** 232–246.

104. Reimann-Philipp, U., Ovase, R., Weigel, P. H., and Grammas, P. (2001) Mechanisms of cell death in primary cortical neurons and PC12 cells. *J. Neurosci. Res.* **64,** 654–660.

105. Wolozin, B. and Behl, C. (2000) Mechanisms of neurodegenerative disorders. Part I. Protein aggregates. *Arch. Neurol.* **57,** 793–796.

106. Duda, J. E., Lee, V. M. Y., and Trojanowski, J. Q. (2000) Neuropathology of synuclein aggregates. New insights into mechanism of neurodegenerative diseases. *J. Neurosci. Res.* **61,** 121–127.

107. Lucking, C. B. and Brice, A. (2000) α-Synuclein and Parkinson's disease. *Cell. Mol. Life Sci.* **57,** 1894–1908.

107a. Kahns, S., Lykkebo, S., Jakobsen, L. D., Nielsen, M. S., and Jensen, P. H. (2002) Caspase-mediated parkin cleavage in apoptotic cell death. *J. Biol. Chem.* **277,** 15,303–15,308.

108. Miller, L. K. (1999) An exegesis of IAPs: salvation and surprises from BIR motifs. *Trends Cell Biol.* **9,** 323–328.

109. Tschopp, J., Thome, M., Hofmann, K., and Meinl, E. (1998) The fight of viruses against apoptosis. *Curr. Opin. Genet. Dev.* **8,** 82–87.

110. Verhagen, A. M., Ekert, P. G., Pakusch, M., Silke, J., Connolly, L. M., Reid, G. E., et al. (2000) Identification of DIABLO, a mammalian protein that promotes apoptosis by binding to and antagonizing IAP proteins. *Cell* **102,** 43–53.

111. Du, C., Fang, M., Li, Y., Li, L., and Wang, X. (2000) Smac, a mitochondrial protein that promotes cytochrome c-dependent caspase activation by eliminating IAP inhibition. *Cell* **102,** 33–42.

112. Jaattela, M., Wissing, D., Kokholm, K., Kallunki, T., and Egeblad, M. (1998) Hsp70 exerts its anti-apoptotic function downstream of caspase-3-like proteases. *EMBO J.* **17,** 6124–6134.

113. Volbracht, C., Leist, M., Kolb, S. A., and Nicotera, P. (2001) Apoptosis in caspase-inhibited neurons. *Mol. Med.* **7,** 36–48.

114. Huppertz, B., Frank, H. G., and Kaufmann, P. (1999) The apoptosis cascade—morphological and immunohistochemical methods for its visualization. *Anat. Embryol.* **200,** 1–18.

115. Willingham, M. C. (1999) Cytochemical methods for the detection of apoptosis. *J. Histochem. Cytochem.* **47,** 1101–1110.

116. Gavrieli, Y., Sherman, Y., and Ben-Sasson, S. A. (1992) Identification of programmed cell death in situ via specific labeling of nuclear DNA fragmentation. *J. Cell Biol.* **119,** 483–501.

117. Gold, R., Schmied, M., Giegerich, G., Breitschopf, H., Hartung, H. P., Toyka, K., and Lassmann, H. (1994) Differentiation between cellular apoptosis and necrosis by combined use of in situ tailing and nick translation techniques. *Lab. Invest.* **71,** 219-225.

118. Grasl-Kraupp, B., Ruttkay-Nedecky, B., Kondelka, H., Bukowska, K., Bursch, W., and Schulte-Hermann, R. (1995) In situ detection of fragmented DNA (TUNEL assay) fails to discriminate among apoptosis, necrosis and autolytic cell death: a cautionary note. *Hepatology* **21,** 1465–1468.

119. Tatton, N. A., Maclean-Fraser, A., Tatton, W. G., Perl, D. P., and Olanow, C. W. (1998) A fluorescent double-labeling method to detect and confirm apoptotic nuclei in Parkinson's disease. *Ann. Neurol.* **44(Suppl. 1),** S142–S148.

120. Lassmann, H., Bancher, C., Breitschopf, H., Wegiel, J., Bobinski, M., Jellinger, K., and Wisniewski, H. M. (1995) Cell death in Alzheimer's disease evaluated by DNA fragmentation in situ. *Acta Neuropathol.* **89,** 35–41.

121. Anderson, A. J., Su, J. H., and Cotman, C. W. (1996) DNA damage and apoptosis in Alzheimer's disease: colocalization with c-Jun immunoreactivity, relationship to brain area and effect of post mortem delay. *J. Neurosci.* **16,** 1710–1719.

122. Kingsbury, A. E., Fester, O. J. F., Nisbet, A. P., et al. (1995) Tissue pH as an indicator of mRNA preservation in post-mortem brain. *Mol. Brain. Res.* **28,** 311–318.

123. Dragunow, M., Faull, R., Lawlor, P., Beilharz, E. J., Singleton, K., Walker, E. B., and Mee, E. (1995) In situ evidence for DNA fragmentation in Huntington's disease striatum and Alzheimer's disease temporal lobes. *Neuroreport* **6,** 1053–1057.

124. Kingsbury, A. E., Mardsen, C. D., and Foster, O. J. (1998) DNA fragmentation in human substantia nigra: apoptosis or perimortem effect? *Mov. Disord.* **13,** 877–884.

125. Tompkins, M. M., Basgall, E. J., Zamrini, E., and Hill, W. D. (1997) Apoptotic-like changes in Lewy-body-associated disorders and normal aging in substantia nigral neurons. *Am. J. Pathol.* **150,** 119–131.

126. Kockx, M. M., Muhring, J., Knaapen, M. W., and de Meyer, G. R. (1998) RNA synthesis and splicing interferes with DNA in situ end labeling techniques used to detect apoptosis. *Am. J. Pathol.* **152,** 885–888.

127. Srinivasan, A., Roth, K. A., Sayers, R. O., Shindler, K. S., Wong, A. M., Fritz, L. C., and Tomaselli, K. (1998) In situ immunodetection of activated caspase-3 in apoptotic neurons in the developing nervous system. *Cell Death Differ.* **5,** 1004–1016.

128. Stadelmann, C., Deckwerth, T. L., Srinivasan, A., Bancher, C., Brück, W., Jellinger, K., and Lassmann, H. (1999) Activation of caspase-3 in single apopototic neurons and granules of granulovacuolar degeneration in Alzheimer disease and Down's syndrome: a role for autophagy as antiapoptotic counterregulatory mechanism? *Am. J. Pathol.* **155,** 1459–1466.

129. Guo, Q., Sebastian, L., Sopher, B. L., Miller, M. W., Ware, C. B., Martin, G. M., and Mattson, M. P. (1999) Increased vunerability of hippocampal neurons from presenilin-1 mutant knock-in mice to amyloid β-peptide toxicity: central roles of superoxide production and caspase activation. *J. Neurochem.* **72,** 1019–1029.

130. Chui, D. H., Tanahashi, H., Ozawa, K., Ikeda, S., Checler, F., Ueda, O., et al. (1999) Transgenic mice with Alzheimer presenilin 1 mutations show accelerated neurodegeneration without amyloid plaque formation. *Nature Med.* **5,** 560–564.

131. Lucassen, P. J. (2000) Presenilins and cellular damage: a link through amyloid? *J. Alz. Dis.* **2,** 61–67.

132. Yo, Y. H. and Fortini, M. E. (1999) Apopototic activities of wild-type and Alzheimer's disease related mutant presenilins in Drosophila melanogaster. *J. Cell Biol.* **146,** 1351–1364.

133. Guo, Q., Fu, W., Sopher, B. L., Miller, M. W., Ware, C. B., Martin, G. M., and Mattson, M. P. (1999) Increased vulnerability of hippocampal neurons to excitotoxic necrosis in presenilin-1 mutant knock-in mice. *Nature Med.* **5,** 101–106.

133a. Zhou, Y., Zhang, W., Easton, R., Ray, J. W., Lampe, P., Jiang, Z., et al. (2002) Presenilin-1 protects against neuronal apoptosis caused by its interacting protein PAG. *Neurobiol. Dis.* **9,** 126–138.

133b. Terro, F., Czech, C., Esclaire, F., Elyaman, W., Yardin, C., Baclet, M. C., et al. (2002) Neurons overexpressing mutant presenilin-1 are more sensitive to apoptosis induced by endoplasmic reticulum-Golgi stress. *J. Neurosci. Res.* **69,** 530–539.

133c. Mori, M., Nakagami, H., Morishita, R., Mitsuda, N., Yamamoto, K., Yoshimura S., et al. (2002) N141I mutant presenilin-2 gene enhances neuronal cell death and decreases bcl-2 expression. *Life Sci.* **70,** 2567–2580.

134. Kwok, J. B., Li, Q. X., Hallupp, M., Whyte, S., Ames, D., Beyreuther, K., et al. (2000) Novel Leu723Pro amyloid precursor protein mutation increases amyloid β42(43) peptide levels and induces apoptosis. *Ann. Neurol.* **47,** 249–253.

135. Uetsuki, T., Takemoto, K., Nishimura, I., Okamoto, M., Niinobe, M., Momoi, T., et al. (1999) Activation of neuronal caspase-3 by intracellular accumulation of wild-type Alzheimer amyloid precursor protein. *J. Neurosci.* **19,** 6955–6964.

135a. Nishimura, I., Uetsuki, T., Kuwako, K., Hara, T., Kawakami, T., Aimoto, S., and Yoshikawa, K. (2002) Cell death induced by a caspase-cleaved transmembrane fragment of the Alzheimer amyloid precursor protein. *Cell Death Differ.* **9,** 199–208.

136. Ohyagi, Y., Yamada, T., Nishioka, K., Clarke, N. J., Tomlinson, A. J., Naylor, S., et al. (2000) Selective increase in cellular A beta 42 is related to apoptosis but not necrosis. *Neuroreport* **11,** 167–171.

136a. Borghi, R., Pellegrini, L., Lacana, E., Diaspro, A., Pronzato, M. A., Vitali, A., et al. (2002) Neuronal apoptosis is accompanied by amyloid beta-protein accumulation in the endoplasmic reticulum. *J. Alzheimer's Dis.* **4,** 31–37.

137. Ramsden, M., Shukla, C., Pearson, H. A., and Bridges, L. R. (2000) Neurodegeneration induced by aggregated Aβ in central neurons may be apoptotic. *Neuropathol. Appl. Neurobiol.* **26,** 205–206.

138. Ditaranto, K., Tekirian, T. L., and Yang, A. J. (2001) Lysosomal membrane damage in soluble Aβ-mediated cell death in Alzheimer's disease. *Neurobiol. Dis.* **8,** 19–31.

139. Kaltschmidt, B., Uherek, M., Wellmann, H., Volk, B., and Kaltschmidt, C. (1999) Inhibition of NF-κB potentiates amyloid β-mediated neuronal apoptosis. *Proc. Natl. Acad. Sci. USA* **96,** 9409–9414.

140. Lu, D. C., Rabizadeh, S., Chandra, S., Shayya, R. F., Ellerby, L. M., Ye, X., et al. (2000) A second cytotoxic proteolytic peptide derived from amyloid β-protein precursor. *Nature Med.* **6,** 397–404.

141. Milligan, C. E. (2000) Caspase cleavage of APP results in a cytotoxic proteolytic peptid. *Nature Med.* **6,** 385–386.

142. Saez-Valero, J., Angeretti, N., and Forloni, G. (2000) Caspase-3 activation by β-amyloid and prion protein peptides is independent from their neurotoxic effect. *Neurosci. Lett.* **293,** 207–210.

143. Selznick, L. A., Zheng, T. S., Flavell, R. A., Rakic, P., and Roth, K. A. (2000) Amyloid β-induced neuronal death is Bax-dependent but caspase-independent. *J. Neuropathol. Exp. Neurol.* **59,** 271–279.

144. Troy, C. M., Rabacchi, S. A., Xu, Z. H., Maroney, A. C., Connors, T. J., Shelanski, M. L., and Greene, L. A. (2001) β-amyloid-induced neuronal apoptosis requires c-Jun N-terminal kinase activation. *J. Neurochem.* **77,** 157–164.

145. Saito, T., Kijima, H., Kiuchi, Y., Isobe, Y., and Fukushima, K. (2001) β-amyloid induces caspase-dependent early neurotoxic change in PC12 cells: correlation with H2O2 neurotoxicity. *Neurosci. Lett.* **305,** 61–64.

146. White, A. R., Guirguis, R., Brazier, M. W., Jobling, M. F., Hill, A. F., Beyreuther, K., et al. (2001) Sublethal concentrations of prion peptide PrP106-126 or the amyloid β peptide of Alzheimer's disease activates expression of proapoptotic markers in primary cortical neurons. *Neurobiol. Dis.* **8,** 299–316.

147. Giovanni, A., Keramaris, E., Morris, E. J., Hou, S. T., O'Hare, M., Dyson, N., et al. (2000) E2F1 mediates death of β-amyloid-treated cortical neurons in a manner independent of p53 and dependent on Bax and caspase 3. *J. Biol. Chem.* **275,** 11,553–11,560.

148. Troy, C. M., Rabacchi, S. A., Friedman, W. J., Frappier, T. F., Brown, K., and Shelanski, M. L. (2000) Caspase-2 mediates neuronal cell death induced by β-amyloid. *J. Neurosci.* **20,** 1386–1392.

149. Bozyczko-Coyne, D., O'Kane, T. M., Wu, Z. L., Dobrzanski, P., Murthy, S., Vaught, J. L., and Scott, R. W. (2001) CEP-1347/KT-7515, an inhibitor of SAPK/JNK pathway activation, promotes survival and blocks multiple events associated with Aβ-induced cortical neuron apoptosis. *J. Neurochem.* **77,** 849–863.

150. Xu, J., Chen, S. W., Ku, G., Ahmed, S. H., Xu, M., Chen, H., and Hsu, C. Y. (2001) Amyloid-β peptide-induced cerebral endothelial cell death involves mitochondrial dysfunction and caspase activation. *J. Cereb. Blood Flow Metab.* **21,** 702–710.

151. Kajkowski, E. M., Lo, C. F., Ning, X. P., Walker, S., Sofia, H. J., Wang, W., et al. (2001) β-amyloid peptide-induced apoptosis regulated by a novel protein containing a G protein activation module. *J. Biol. Chem.* **276,** 18,748–18,756.

152. Luetjens, C. M., Lankiewicz, S., Bui, N. T., Krohn, A. J., Poppe, M., and Prehn, J. H. (2001) Up-regulation of Bcl-xL in response to subtoxic beta-amyloid: role in neuronal resistance against apoptotic and oxidative injury. *Neuroscience* **102,** 139–150.

153. Passer, B. J., Pellegrini, L., Vito, P., Ganjei, J. K., and D'Adamio, L. (1999) Interaction of Alzheimer's presenilin-1 and presenilin-2 with Bcl-X-L—a potential role in modulating the threshold of cell death. *J. Biol. Chem.* **274,** 2400–2414.

154. Marcon, G., Atzori, C., Srinivasan, A. N., Okazawa, H., Ghetti, B., and Migheli, A. (2000) Caspase-3 is activated in Alzheimer disease but not in frontotemporal dementia. (Abstr.) *Neurobiol. Aging* **21,** 81.

155. Litersky, J. M. and Johnson, G. V. (1995) Phosphorylation of tau in situ: inhibition of calcium-dependent proteolysis. *J. Neurochem.* **65,** 903–911.

156. Chung, C. W., Song, Y. H., Kim, I. K., Yoon, W. J., Ryu, B. R., Jo, D. G., et al. (2001) Proapoptotic effects of tau cleavage product generated by caspase-3. *Neurobiol. Dis.* **8,** 162–172.

157. Eckert, A., Steiner, B., Marques, C., Leutz, S., Romig, H., Haass, C., and Muller, W. E. (2001) Elevated vulnerability to oxidative stress-induced cell death and activation of caspase-3 by the Swedish amyloid precursor protein mutation. *J. Neurosci. Res.* **64,** 183–192.

158. Lorio, G., Avila, J., and Diaz-Nido, J. (2001) Modifications of tau protein during neuronal cell death. *J. Alzheimer's Dis.* **3,** 563–575.

158a. Atzori, C., Ghetti, B., Piva, R., Srinivasan, A. N., Zolo, P., Delisle, M. B., et al. (2001) Activation of the JNK/p38 pathway occurs in diseases characterized by tau protein pathology and is related to tau phosphorylation but not to apoptosis. *J. Neuropathol. Exp. Neurol.* **60,** 1190–1197.

159. Ekinci, F. J., Linsley, M. D., and Shea, T. B. (2000) β-Amyloid-induced calcium influx induces apoptosis in culture by oxidative stress rather than tau phosphorylation. *Mol. Brain Res.* **76,** 389–395.

160. Bhat, N. R. and Zhang, P. (1999) Hydrogen peroxide activation of multiple mitogen-activated protein kinases in an oligodendrocyte cell line: role of extracellular signal-regulated kinase in hydrogen peroxide-induced cell death. *J. Neurochem.* **72,** 112–119.

161. Oh-Hashi, K., Maruyama, W., Yi, H., Takahashi, T., Naoi, M., and Isobe, K. (1999) Mitogen-activated protein kinase pathway mediates peroxynitrite-induced apoptosis in human dopaminergic neuroblastoma SH-SY5Y cells. *Biochem. Biophys. Res. Commun.* **263,** 504–509.

162. Dineley, K. T., Westerman, M., Bui, D., Bell, K., Ashe, K. H., and Sweatt, J. D. (2001) β-amyloid activates the mitogen-activated protein kinase cascade via hippocampal β7 nicotinic acetylcholine receptors: *in vitro* and *in vivo* mechanisms related to Alzheimer's disease. *J. Neurosci.* **21,** 4125–4133.

163. Hetman, M., Kanning, K., Cavanaugh, J. E., and Xia, Z. (1999) Neuroprotection by brain-derived neurotrophic factor is mediated by extracellular signal-regulated kinase and phosphatidylinositol 3-kinase. *J. Biol. Chem.* **274,** 22,569–22,580.

164. Owada, K., Sanjo, N., Kobayashi, T., Mizusawa, H., Muramatsu, H., Muramatsu, T., and Michikawa, M. (1999) Midkine inhibits caspase-dependent apoptosis via the activation of mitogen-activated protein kinase and phosphatidylinositol 3-kinase in cultured neurons. *J. Neurochem.* **73,** 2084–2092.

165. Mielke, K., Brecht, S., Dorst, A., and Herdegen, T. (1999) Activity and expression of JNK1, p38 and ERK kinases, c-Jun N-terminal phosphorylation, and c-Jun promoter binding in the adult rat brain following kainate-induced seizures. *Neuroscience* **91,** 471–483.
166. Zhu, X., Rottkamp, C. A., Boux, H., Takeda, A., Perry, G., and Smith, M. A. (2000) Activation of p38 kinase links tau phosphorylation, oxidative stress, and cell cycle-related events in Alzheimer disease. *J. Neuropathol. Exp. Neurol.* **59,** 880–888.
167. Ferrer, I., Blanco, R., Carmona, M., Ribera, R., Goutan, E., Puig, B., et al. (2001) Phosphorylated MAP kinase (ERK1, ERK2) expression is associated with early tau deposition in neurones and glial cells, but not with increased nuclear DNA vulnerability and cell death, in Alzheimer disease, Pick's disease, progressive supranuclear palsy and corticobasal degeneration. *Brain Pathol.* **11,** 144–158.
168. Hartley, A., Stone, J. M., Heron, C., et al. (1994) Complex I inhibitors induce dose-dependent apoptosis in PC12 cells: relevance to Parkinson's disease. *J. Neurochem.* **63,** 1987–1990.
169. Mochizuki, H., Nakamura, N., Nishi, K., and Mizuno, Y. (1994) Apoptosis is induced by 1-methyl-4-phenylpyridinium ion (MPP+) in ventral mesencephalic striatal co-culture in rat. *Neurosci. Lett.* **170,** 191–194.
170. Sheehan, J. P., Palmter, P. E., Helm, G. A., and Tuttle, J. B. (1997) MPP+ induced apoptotic cell death in SH-SY5Y neuroblastoma cells. An electron microscopic study. *J. Neurosci. Res.* **48,** 226–237.
171. Blum, D., Torch, S., Lambeng, N., Nissou, M., Benabid, A., Sadoul, R., and Verna, J. (2001) Molecular pathways involved in the neurotoxicity of 6-OHDA, dopamine and MPTP: contribution to the apoptotic theory in Parkinson's disease. *Prog. Neurobiol.* **65,** 135–172.
172. Dodel, R. C., Du, Y., Bales, K. R., Ling, Z. D., Carvey, P. M., and Paul, S. M. (1998) Peptide inhibitors of caspase-3-like proteases attenuate 1-methyl-4-phenyl-pyridium-induced toxicity of cultured fetal rat mesencephalic dopamine neurons. *Neuroscience* **86,** 701–707.
173. Oo, T. F. and Burke, R. E. (1997) The time course of developmental cell death in phenotypically defined dopaminergic neurons of the substantia nigra. *Dev. Brain Res.* **98,** 191–196.
174. Tepper, J. M., Damlama, M., and Trent, F. (1994) Postnatal changes in the distribution and morphology of rat substantia nigra dopaminergic neurons. *Neuroscience* **60,** 469–477.
175. Kelly, W. J. and Burke, R. E. (1996) Apoptotic neuron death in rat substantia nigra induced by striatal excitotoxic injury is developmentally dependent. *Neurosci. Lett.* **220,** 85–88.
176. Marti, M. J., James, C. J., Oo, T. F., Kelly, W. J., and Burke, R. E. (1997) Early developmental destruction of terminals in the striatal target induces apoptosis in dopamine neurons of the substantia nigra. *J. Neurosci.* **17,** 2030–2039.
177. Seaton, T. A., Cooper, J. M., and Schapira, A. H. V. (1997) Free radical scavengers protect dopaminergic cell lines from apoptosis induced by complex I inhibitors. *Brain Res.* **777,** 110–118.
178. Ziv, I., Melamed, E., Nardi, N., Luria, D., Achiron, A., Offen, D., and Barzilai, A. (1994) Dopamine induces apoptosis-like cell death in cultured chick sympathetic neurons—a possible novel pathogenetic mechanism in Parkinson's disease. *Neurosci. Lett.* **170,** 136–140.
179. Ziv, I., Offen, D., Barzilai, A., Haviv, R., Stein, R., Zilkhafalb, R., Shirvan, A., and Melamed, E. (1997) Modulation of control mechanisms of dopamine-induced apoptosis—a future approach to the treatment of Parkinson's disease. *J. Neural Transm.* **49(Suppl.),** 195–202.
180. Ziv, I., Offen, D., Haviv, R., Stein, R., Panet, H., Zilkhavan, R., et al. (1997) The proto-oncogene Bcl-2 inhibits cellular toxicity of dopamine—possible implications for Parkinson's disease. *Apoptosis* **2,** 149–155.
181. Ziv, I., Zilkhavan, R., Offen, D., Shirvan, A., Barzilai, A., and Melamed, E. (1997) Levodopa induces apoptosis in cultured neuronal cells—a possible accelerator of nigrostriatal degeneration in Parkinson's disease. *Mov. Disord.* **12,** 17–23.

182. Offen, D., Ziv, I., Panet, H., Wasserman, L., Stein, R., Melamed, E., and Barzilai, A. (1997) Dopamine-induced apoptosis is inhibited in PC12 cells expressing Bcl-2. *Cell. Mol. Neurobiol.* **17,** 289–304.

183. Zhang, J., Price, J. O., Graham, D. G., and Montine, T. J. (1998) Secondary excitotoxicity contributes to dopamine-induced apoptosis of dopaminergic neuronal cultures. *Biochem. Biophys. Res. Commun.* **248,** 812–816.

184. Offen, D., Ziv, I., Barzilai, A., Grordin, S., Glater, E., Hochman, A., and Melamed, E. (1997) Dopamine-melanin induces apoptosis in PC12 cells—possible implications for the etiology of Parkinson's disease. *Neurochem. Int.* **31,** 207–216.

185. Velezpardo, C., Delrio, M. J., Vershueren, H., Ebinger, G., and Vauquelin, G. (1997) Dopamine and iron induce apoptosis in PC12 cells. *Pharmacol. Toxicol.* **80,** 76–84.

186. Ruberg, M., Brugg, B., Prigent, A., Hirsch, E., Brice, A., and Agid, Y. (1997) Is differential regulation of mitochondrial transcription in Parkinson's disease related to apoptosis? *J. Neurochem.* **68,** 2098–2110.

187. Tatton, N. A. and Kish, S. J. (1997) In situ detection of apoptotic nuclei in the substantia nigra compacta of 1-methyl-4-phenyl-1,2,3,6-tetrahydropyridine-treated mice using terminal deoxynucleotidyl transferase labelling and acridine orange staining. *Neuroscience* **77,** 1037–1048.

188. Waters, C. M. and Walkinshaw, G. (1995) Induction of apoptosis in catecholaminergic PC12 cells by L-dopa. Implications for the treatment of Parkinson's disease. *J. Clin. Invest.* **95,** 2458–2464.

189. Chen, S. D., Guo, M., Liu, Z. G., and Chen, H. Z. (1998) The possible role of apoptosis in the pathogenesis of Parkinson's disease (abstr.). *Mov. Disord.* **13(Suppl. 2),** 194.

190. Jackson-Lewis, V. and Przedborski, S. (1998) MPTP-induced necrosis or apoptosis is a dose-dependent phenomenon. (abstr.) *Mov. Disord.* **13(Suppl. 2),** 237.

191. Maruyama, W., Takahashi, T., and Naoi, M. (1998) (-)-deprenyl protects human dopaminergic neuroblastoma SH-SY5Y cells from apoptosis induced by peroxynitrite and nitric oxide. *J. Neurochem.* **70,** 2510–2515.

192. Li, F., Srinivasan, A., Wang, Y., Armstrong, R. C., Tomaselli, K. J., and Fritz, L. C. (1997) Cell specific induction of apoptosis by microinjection of cytochrome c. *J. Biol. Chem.* **272,** 30,299–30,305.

193. Yang, W. and Sun, A. Y. (1998) Paraquat-induced free radical reaction in mouse brain microsomes. *Neurochem. Res.* **23,** 47–53.

194. Burke, R. E. and Kholodilov, N. G. (1998) Programmed cell death: does it play a role in Parkinson's disease? *Ann. Neurol.* **44(Suppl. 1),** S126–S133.

195. Olanow, C. W. and Tatton, W. G. (1999) Etiology and pathogenesis of Parkinson's disease. *Ann. Rev. Neurosci.* **22,** 123–144.

196. Hirsch, E. C., Hunot, S., and Hartmann, A. (2000) Mechanism of cell death in experimental models of Parkinson's disease. *Funct. Neurol.* **15,** 229–237.

197. Fukudo, T., Takahashi, J., Xiang, F., and Tanaka, J. (2000) Apoptosis in 1-methyl-4-phenyl-1,2,3,6-tetrahydropyridine-treated mice (abstr.). *Brain Pathol.* **10,** 787.

198. Spooren, W. P., Gentsch, C., and Wiessner, C. (1998) TUNEL-positive cells in the substantia nigra of C57BL/6 mice after a single bolus of 1-methyl-4-phenyl-1,2,3,6-tetrahydropyridine. *Neuroscience* **85,** 649–651.

199. Hartmann, A., Hunot, S., Michel, P. P., Muriel, M. P., Vyas, S., Faucheux, B. A., et al. (2000) Caspase-3. A vulnerability factor and final effector in apoptotitc death of dopaminergic neurons in Parkinson's disease. *Proc. Natl. Acad. Sci. USA* **97,** 2875–2880.

200. Turmel, H., Hartmann, A., Parain, K., Douhou, A., Srinivasan, A., Agid, Y., and Hirsch, E. C. (2001) Caspase-3 activation in 1-methyl-4-phenyl-1,2,3,6-tetrahydropyridine (MPTP)-treated mice. *Mov. Disord.* **16,** 185–189.

201. Hartmann, A., Troadec, J. D., Hunot, S., Kikly, K., Faucheux, B. A., Mouatt-Prigent, A., et al. (2001) Caspase-8 is an effector in apoptotic death of dopaminergic neurons in

Parkinson's disease, but pathway inhibition results in neuronal necrosis. *J. Neurosci.* **21,** 2247–2255.

202. Vila, M., Jackson-Lewis, V., Vukosavic, S., Djaldetti, R., Liberatore, G., Offen, D., et al. (2001) Bax ablation prevents dopaminergic neurodegeneration in the 1-methyl-4-phenyl-1,2,3,6-tetrahydropyridine mouse model of Parkinson's disease. *Proc. Natl. Acad. Sci. USA* **98,** 2837–2842.

203. Offen, D., Beart, P. M., Cheung, N. S., Pascoe, C. J., Hochman, A., Gorodin, S., et al. (1998) Transgenic mice expressing human Bcl-2 in their neurons are resistant to 6-hydroxydopamine and 1-methyl-4-phenyl-1,2,3,6-tetrahydropyridine neurotoxicity. *Proc. Natl. Acad. Sci. USA* **95,** 5789–5794.

204. Oh, Y. J., Wong, S. C., Moffat, M., and O'Malley, K. L. (1995) Overexpression of Bcl-2 attenuates MPP+, but not 6-ODHA, induced cell death in a dopaminergic neuronal cell line. *Neurobiol. Dis.* **2,** 157–167.

205. Hartmann, A., Michel, P. P., Troadec, J. D., Mouatt-Prigent, A., Faucheux, B. A., Ruberg, M., et al. (2001) Is Bax a mitochondrial mediator in apoptotic death of dopaminergic neurons in Parkinson's disease? *J. Neurochem.* **76,** 1785–1793.

206. Neystat, M., Rzhetskaya, M., Oo, T. F., Kholodilov, N., Yarygina, O., Wilson, A., et al. (2001) Expression of cyclin-dependent kinase 5 and its activator p35 in models of induced apoptotic death in neurons of the substantia nigra *in vivo. J. Neurochem.* **77,** 1611–1625.

207. Hattori, A., Luo, Y. Q., Umegaki, H., Munoz, J., and Roth, G. S. (1998) Intrastriatal injection of dopamine results in DNA damage and apoptosis in rats. *Neuroreport* **9,** 2569–2572.

208. Jeon, B. S., Jackson-Lewis, V., and Burke, R. E. (1995) 6-Hydroxydopamine lesion of the rat substantia nigra: time course and morphology of cell death. *Neurodegeneration* **4,** 131–137.

209. Oo, T. F., Henchcliffe, C., Harrison, et al. (1995) Neuronal death in substantia nigra in murine Weaver mutation is nonapoptotic. *Mov. Disord.* **10,** 693–694.

210. Choi, W. S., Yoon, S. Y., Oh, T. H., Choi, E. J., O'Malley, K. L., and Oh, Y. J. (1999) Two distinct mechanisms are involved in 6-hydroxydopamine- and MPP+-induced dopaminergic neuronal cell death: role of caspases, ROS, and JNK. *J. Neurosci. Res.* **57,** 86–94.

211. Duan, W., Zhang, Z., Gash, D. M., and Mattson, M. P. (1999) Participation of prostate apoptosis response-4 in degeneration of dopaminergic neurons in models of Parkinson's disease. *Ann. Neurol.* **46,** 587–597.

212. Yu, H. Z., Rong, J., Chen, S. D., Li, B., and Liu, Z. G. (2001) 6-hydroxydopamine-induced apoptosis and its probable molecular mechanisms (abstr.). *Parkinsonism Rel. Disord.* **7,** S113.

213. Ochu, E. E., Rothwell, N. J., and Waters, C. M. (1998) Caspases mediate 6-hydroxydopamine-induced apoptosis but not necrosis in PC12 cells. *J. Neurochem.* **70,** 2637–2640.

214. Takai, N., Nakanishi, H., Tanabe, K., Nishioku, T., Sugiyama, T., Fujiwara, M., and Yamamoto, K. (1998) Involvement of caspase-like proteinases in apoptosis of neuronal PC12 cells and primary cultured microglia induced by 6-hydroxydopamine. *J. Neurosci. Res.* **54,** 214–222.

215. Dodel, R. C., Du, Y., Bales, K. R., Ling, Z., Carvey, P. M., and Paul, S. M. (1999) Caspase-3-like proteases and 6-hydroxydopamine induced neuronal cell death. *Brain Res. Mol. Brain. Res.* **64,** 141–148.

216. Blum, D., Wu, Y., Nissou, M. F., Arnaud, S., Benabid, A. L., and Verna, J. M. (1997) P53 amd Bax activation in 6-hydroxydopamine-induced apoptosis in PC12 cells. *Brain Res.* **751,** 139–142.

217. Li, X. and Sun, A. Y. (1999) Paraquat induced activation of transcription factor AP-1 and apoptosis in PC12 cells. *J. Neural Transm.* **106,** 1–21.

218. Hou, S. T., Cowan, E., Dostanic, S., Rasquinha, I., Comas, T., Morley, P., and MacManus, J. P. (2001) Increased expression of the transcription factor E2F1 during dopamine-evoked, caspase-3-mediated apoptosis in rat cortical neurons. *Neurosci. Lett.* **306,** 153–156.

219. Hassouna, I., Wickert, H., Zimmermann, M., and Gollardon, F. (1996) Increase in Bax expression in substantia-nigra following 1-methyl-4-phenyl-1,2,3,6-tetrahydropyridine (MPTP) treatment of mice. *Neurosci. Lett.* **204,** 85–88.

220. Nishi, K. (1997) Expression of c-jun in dopaminergic neurons of the substantia nigra in 1-methyl-4-phenyl-1,2,3,6-tetrahydropyridine (MPTP)-trated mice. *Brain Res.* **771,** 133–141.

221. Takai, N., Nakanishi, H., Tanabe, K., Nishioku, T., Sugiyama, T., Fujiwara, M., and Yamamoto, K. (1998) Involvement of caspase-like proteinases in apoptosis of neuronal PC12 cells and primery cultured microglia induced by 6-hydroxydopamine. *J. Neurosci. Res.* **54,** 214–222.

222. Gash, D. M., Zhang, Z., Ovadia, A., Cass, W. A., Yi, A., Simmermann, L., et al. (1996) Functional recovery in parkinsonian monkeys treated with GDNF. *Nature* **380,** 252–255.

223. Gash, D. M., Zhang, Z., and Gerhardt, G. (1998) Neuroprotective and neurorestorative properties of GDNF. *Ann. Neurol.* **44(Suppl. 1),** S121–S125.

224. Burke, R. E., Antonelli, M., and Sulzer, D. (1998) Glial cell line-derived neurotrophic growth factor inhibits apoptotic death of postnatal substantia nigra dopamine neurons in primary culture. *J. Neurochem.* **71,** 517–525.

225. Offen, D., Ziv, I., Sternin, H., Melamed, E., and Hochman, A. (1996) Prevention of dopamine-induced cell death by thiol antioxidants—possible implications for treatment of Parkinson's disease. *Exp. Neurol.* **141,** 32–39.

226. Sawada, H., Ibi, M., Kihara, T., Urushitani, M., Akaike, A., and Shimohama, S. (1998) Estradiol protects mesencephalic dopaminergic neurons from oxidative stress-induced neuronal death. *J. Neurosci. Res.* **54,** 707–719.

227. Ziv, I., Barzilai, A., Daily, D., Shirvan, A., Offen, D., and Melamed, E. (1998) The role of p53 in dopamine induced apoptosis: possible implications for the neuronal loss in Parkinson's disease. (abstr.) *Mov. Disord.* **13(Suppl. 2),** 22.

228. Mayo, J. C., Sainz, R. M., Uria, H., Antolin, I., Esteban, M. M., and Rodriguez, C. (1998) Melatonin prevents Apoptosis induced by 6-hydroxydopamine in neuronal cells—implications for Parkinson's disease. *J. Pineal. Res.* **24,** 179–192.

229. Lotharius, J., Dugan, L. L., and O'Malley, K. L. (1999) Distinct mechanisms underlie neurotoxin-mediated cell death in cultured dopaminergic neurons. *J. Neurosci.* **19,** 1284–1293.

229a. Lambeng, N., Hourez, R., Torch, S., Verna, J. M., and Blum, D. (2002) Biochemical and molecular mechanisms of neuronal cell death in the experimental neurotoxic models of Parkinson's disease. *M S-Med. Sci.* **18,** 457–466.

230. Oh, J. H., Choi, W.-S., Kim, J.-E., Seo, J.-W., O'Mally, K. L., and Oh, Y. J. (1998) Overexpression of HA-Bax but not Bcl-2 or Bcl-XL attenuates 6-hydroxydopamine-induced neuronal apoptosis. *Exp. Neurol.* **154,** 193–198.

231. von Coelln, R., Kugler, S., Bahr, M., Weller, M., Dichgans, J., and Schulz, J. B. (2001) Rescue from death but not from functional impairment: caspase inhibition protects dopaminergic cells against 6-hydroxydopamine-induced apoptosis but not against the loss of their terminals. *J. Neurochem.* **77,** 263–273.

232. Natsume, A., Mata, M., Goss, J., Huang, S. H., Wolfe, D., Oligino, T., et al. (2001) Bcl-2 and GDNF delivered by HSV-mediated gene transfer act additively to protect dopaminergic neurons from 6-OHDA-induced degeneration. *Exp. Neurol.* **169,** 231–238.

233. Jha, N., Jurma, O., Lalli, G., Liu, Y., Pettus, E. H., Greenamyre, J. T., et al. (2000) Glutathione depletion in PC12 results in selective inhibition of mitochondrial complex I activity. Implications for Parkinson's disease. *J. Biol. Chem.* **275,** 26,096–26,101.

234. Tamatani, M., Ogawa, S., Niitsu, Y., and Tohyama, M. (1998) Involvement of Bcl-2 family and caspase-3-like protease in NO-mediated neuronal apoptosis. *J. Neurochem.* **71,** 1588–1596.

235. Chun, H. S., Gibson, G. E., DeGiorgio, L. A., Zhang, H., Kidd, V. J., and Son, J. H. (2001) Dopaminergic cell death induced by MPP(+), oxidant and specific neurotoxicants shares the common molecular mechanism. *J. Neurochem.* **76,** 1010–1021.

236. Fukuhara, Y., Takeshima, T., Kashiwaya, Y., Shimoda, K., Ishitani, R., and Nakashima, K. (2001) GAPDH knockdown rescues mesencephalic dopaminergic neurons from MPP+-induced apoptosis. *Neuroreport* **12,** 2049–2052.

237. Jeon, B. S., Kholodilov, N. G., Oo, T. F., Kim, S. Y., Tomaselli, K. J., Srinivasan, A., et al. (1999) Activation of caspase-3 in developmental models of programmed cell death in neurons of the substantia nigra. *J. Neurochem.* **73,** 322–333.

238. Saporito, M. S., Thomas, B. A., and Scott, R. W. (2000) MPTP activates c-Jun NH(2)-terminal kinase (JNK) and its upstream regulatory kinase MKK4 in nigrostriatal neurons *in vivo*. *J. Neurochem.* **75,** 1200–1208.

238a. Zhou, W., Schaack, J., Zawada. W. M., and Freed, C. R. (2002) Overexpression of human alpha-synuclein causes dopamine neuron death in primary human mesencephalic culture. *Brain Res.* **926,** 41–50.

238b. Stefanova, N., Klimaschewski, L., Poewe, W., Wenning, G. K., and Reindl, M. (2001) Glial cell death induced by overexpression of α-synuclein. *J. Neurosci. Res.* **65,** 432–438.

239. Kakimura, J., Kitamura, Y., Takata, K., Kohno, Y., Nomura, Y., and Taniguchi, T. (2001) Release and aggregation of cytochrome c and α-synuclein are inhibited by the antiparkinsonian drugs, talipexole and pramipexole. *Eur. J. Pharmacol.* **417,** 59–67.

240. Greenamyre, J. T., MacKenzie, G., Peng, T. I., and Stephans, S. E. (1999) Mitochondrial dysfunction in Parkinson's disease. *Biochem. Soc. Symp.* **66,** 85–97.

241. Reichmann, H. and Janetzky, B. (2000) Mitochondrial dysfunction—a pathogenetic factor in Parkinson's disease. *J. Neurol.* **247(Suppl. 2),** II63–II68.

242. Raha, S. and Robinson, B. H. (2001) Mitochondria, oxygen free radicals, and apoptosis. *Am. J. Med. Genet.* **106,** 62–70.

243. Bernardi, P. (1999) Mitochondrial transport of cations: channels, exchangers, and permeability transition. *Physiol. Rev.* **79,** 1127–1155.

244. Nicole, A., Santiard-Baron, D., and Ceballos-Picot, I. (1998) Direct evidence for glutathione as mediator of apoptosis in neuronal cells. *Biomed. Pharmacother.* **52,** 349–355.

245. Merad-Boudia, M., Nicole, A., Santiardbaron, D., Saille, C., and Ceballospicot, I. (1998) Mitochondrial impairment as an early event in the process of apoptosis induced by glutathione depletion in neuronal cells—relevance to Parkinson's disease. *Biochem. Pharmacol.* **56,** 645–655.

246. He, Y., Lee, T., and Leong, S. K. (2000) 6-Hydroxydopamine induced apoptosis of dopaminergic cells in the rat substantia nigra. *Brain Res.* **858,** 163–166.

247. Turmaine, M., Raza, A., Mahal, A., Mangiarini, L., Bates, G. P., and Davies, S. W. (2000) Nonapoptotic neurodegeneration in a transgenic mouse model of Huntington's disease. *Proc. Natl. Acad. Sci. USA* **97,** 8093–8097.

248. Jackson, G. R., Salecker, I., Dong, X., Yao, X., Arnheim, N., Faber, P. W., et al. (1998) Polyglutamine-expanded human huntingtin transgenes induce degeneration of Drosophila photoreceptor neurons. *Neuron* **21,** 633–642.

249. Reddy, P. H., Williams, M., Charles, V., Garrett, L., Pike-Buchanan, L., Whetsell, W. O. Jr., et al. (1998) Behavioural abnormalities and selective neuronal loss in HD transgenic mice expressing mutated full-length HD cDNA. *Nat. Genet.* **20,** 198–202.

250. Wellington, C. L., Singaraja, R., Ellerby, L., Savill, J., Roy, S., Leavitt, B., et al. (2000) Inhibiting caspase cleavage of huntingtin reduces toxicity and aggregate formation in neuronal and nonneuronal cells. *J. Biol. Chem.* **275,** 19,831–19,838.

251. Duan, W., Guo, Z., and Mattson, M. P. (2000) Participation of par-4 in the degeneration of striatal neurons induced by metabolic compromise with 3-nitropropionic acid. *Exp. Neurol.* **165,** 1–11.

252. Vis, J. C., Verbeek, M. M., De Waal, R. M., Ten Donkelaar, H. J., and Kremer, B. (2001) The mitochondrial toxin 3-nitropropionic acid induces differential expression patterns of apoptosis-related markers in rat striatum. *Neuropathol. Appl. Neurobiol.* **27,** 68–76.

253. Petersen, A., Larsen, K. E., Behr, G. G., Romero, N., Przedborski, S., Brundin, P., and Sulzer, D. (2001) Expanded CAG repeats in exon 1 of the Huntington's disease gene stimulate dopamine-mediated striatal neuron autophagy and degeneration. *Hum. Mol. Genet.* **10,** 1243–1254.

253a. Kiechle, T., Dedeoglu, A., Kubilus, J,, Kowall, N. W., Beal, M. F., Friedlander, R., et al. (2002) Cytochrome C and caspase-9 expression in Huntington's disease. *Neuromolecular Med.* **1,** 183–195.

253b. Kim, Y. J., Yi, Y., Sapp, E., Wang, Y., Cuiffo, B., Kegel, K. B., et al. (2001) Caspase-3-cleaved N-terminal fragments of wild-type and mutant huntingtin are present in normal and Huntington's disease brains, associate with membranes, and undergo calpain-dependent proteolysis. *Proc. Natl. Acad. Sci. USA* **48,** 12,784–12,789.

254. Koliatsos, V. F., Portera-Cailliau, C., Schilling, G., Borechlt, D. B., Becher, M. W., and Ross, C. (2001) Mechanisms of neuronal death in Huntington's disease, in *Pathogenesis of Neurodegenerative Disorders* (Mattson, P., ed.), Humana Press, Totowa, NJ, pp. 93–111.

255. Migheli, A., Piva, R., Atzori, C., Troost, D., and Schiffer, D. (1997) c-Jun, JNK/SAPK kinases and transcription factor NF-kappa B are selectively activated in astrocytes, but not motor neurons, in amyotrophic lateral sclerosis. *J. Neuropathol. Exp. Neurol.* **56,** 1314–1322.

256. Dal Canto, M. C. and Gurney, M. E. (1994) Development of central nervous system pathology in a murine transgenic model of human amyotrophic lateral sclerosis. *Am. J. Pathol.* **145,** 1271–1279.

257. Martin, L. J. (1999) Neuronal death in amyotrophic lateral sclerosis is apoptosis: possible contribution of a programmed cell death mechanism. *J. Neuropathol. Exp. Neurol.* **58,** 459–471.

258. He, B. P. and Strong, M. J. (2000) Motor neuronal death in sporadic amyotrophic lateral sclerosis (ALS) is not apoptotic. A comparative study of ALS and chronic aluminium chloride neurotoxicity in New Zealand white rabbits. *Neuropathol. Appl. Neurobiol.* **26,** 150–160.

259. Anderson, A., Stoltzner, S., Lai, F., Su, J., and Nixon, R. A. (2000) Morphological and biochemical assessment of DNA damage and apoptosis in Down syndrome and Alzheimer disease, and effects of postmortem tissue archival on TUNEL. *Neurobiol. Aging* **21,** 511–524.

260. Kitamura, Y., Shimohama, S., Kamoshima, W., Ota, T., Matsuoka, Y., Nomura, Y., et al. (1998) Alteration of proteins regulating apoptosis, Bcl-2, Bcl-x, Bax, Bak, Bad, ICH-1 and CPP32, in Alzheimer's disease. *Brain Res.* **780,** 260–269.

261. MacGibbon, G. A., Lawlor, P. A., Walton, M., Sirimanne, E., Faull, R. L. M., Synek, B., et al. (1997) Expression of Fos, Jun, and Krox family proteins in Alzheimer's disease. *Exp. Neurol.* **147,** 316–332.

262. Ferrer, I., Blanco, R., Cutillas, B., and Ambrosio, S. (2000) Fas and Fas-L expression in Huntington's disease and Parkinson's disease. *Neuropathol. Appl. Neurobiol.* **26,** 424–433.

263. Mogi, M., Togari, A., Kondo, T., Mizuno, Y., Komure, O., Kuno, S., et al. (2000) Caspase activities and tumor necrosis factor receptor R1 (p55) level are elevated in the substantia nigra from Parkinsonian brains. *J. Neural Transm.* **107,** 335–341.

264. Simic, G., Seso-Simic, D., Lucassen, P. L., Islam Arsnik, Z., Cviko, A., Jelasic, D., et al. (2000) Ultrastructural analysis and TUNEL demonstrate motor neuron apoptosis in Werdnig-Hoffmann disease. *J. Neuropathol. Exp. Neurol.* **59,** 398–407.

265. Su, J. H., Deng, G. M., and Cotman, C. W. (1997) Bax protein expression is increased in Alzheimers brain: correlations with DNA damage, Bcl-2 expression, and brain pathology. *J. Neuropathol. Exp. Neurol.* **56,** 86–93.

266. Su, J. H., Satou, T., Anderson, A. J., and Cotman, C. W. (1996) Up-regulation of Bcl-2 is associated with neuronal DNA damage in Alzheimer's disease. *Neuroreport* **7,** 437–440.
267. Mochizuki, H., Mori, H., and Mizuno, Y. (1997) Apoptosis in neurodegenerative disorders. *J. Neural Transm.* **50(Suppl.),** 125–140.
268. Banati, R. B., Daniel, S. E., Path, M. R. C., and Blunt, S. B. (1998) Glial pathology but absence of apoptotic nigral neurons in long-standing Parkinson's disease. *Mov. Disord.* **13,** 221–227.
269. Graeber, M. B., Grasbon-Frodl, E., Abell-Aleff, P., and Kösel, S. (1999) Nigral neurons are likely to die of a mechanism other than classical apoptosis in Parkinson's disease. *Parkinsonism Relat. Disord.* **5,** 187–192.
270. Wüllner, U., Kornhuber, J., Weller, M., Schulz, J. B., Löschmann, P. A., and Riederer, P. (1999) Cell death and apoptosis regulating proteins in Parkinson's disease—a cautionary note. *Acta Neuropathol.* **97,** 408–412.
271. Kösel, S., Egensperger, R., von Eitzen, U., Mehraein, P., and Graeber, M. (1997) On the question of apoptosis in the parkinsonian substantia nigra. *Acta Neuropathol.* **93,** 105–108.
272. Probst-Cousin, S., Rickert, C. H., Schmid, K. W., and Gullotta, F. (1998) Cell mechanisms in multiple system atrophy. *J. Neuropathol. Exp. Neurol.* **57,** 814–821.
273. Selznick, L. A., Holtzman, D. M., Han, B. H., Gokder, M., Srinivasan, A. N., Johnson, M. J., and Roth, K. A. (1999) In situ immunodetection of neuronal caspase-3 activation in Alzheimer disease. *J. Neuropathol. Exp. Neurol.* **58,** 1020–1026.
274. Embacher, N., Kaufmann, W. A., Beer, R., Maier, H., Jellinger, K. A., Poewe, W., and Ransmayr, G. (2001) Apoptosis signals in sporadic amyotrophic lateral sclerosis. An immunocytochemical study. *Acta Neuropathol.* **102,** 426–434.
275. Adamec, E., Vonsattel, J. P., and Nixon, R. A. (1999) DNA strand breaks in Alzheimer's disease. *Brain Res.* **849,** 67–77.
276. Mullaart, E., Boerrigter, M. E., Ravid, R., Swaab, D. F., and Vijg, J. (1990) Increased levels of DNA breaks in cerebral cortex of Alzheimer's disease patients. *Neurobiol. Aging* **11,** 169–173.
277. Smale, J. G., Nichols, N. R., Brady, D. R., Finch, C. E., and Horton, W. E. Jr. (1995) Evidence for apoptotic cell death in Alzheimer's disease. *Exp. Neurol.* **133,** 225–230.
278. Sheng, J. G., Mrak, R. E., and Griffin, W. S. T. (1998) Progressive neuronal DNA damage associated with neurofibrillary tangle formation in Alzheimer's disease. *J. Neuropathol. Exp. Neurol.* **57,** 323–328.
279. de la Monte, S. M., Sohn, Y. K., and Wands, J. R. (1997) Correlates of p53- and Fas (CD95)-mediated apoptosis in Alzheimer's disease. *J. Neurol. Sci.* **152,** 73–83.
280. Chui, D.H., Dobo, E., Makifuchi, T., et al. (2001) Apoptotic neurons in Alzheimer's disease frequently show intracellular Aβ42 labeling. *J. Alzh. Dis.* **3,** 231–239.
281. Wirths, O., Multhaup, G., Czech, C., Blanchard, V., Moussaoui, S., Tremp, G., et al. (2001) Intraneuronal Aβ accumulation precedes plaque formation in β-amyloid precursor protein and presenilin-1 double-transgenic mice. *Neurosci. Lett.* **306,** 116–120.
282. Lucassen, P. J., Chung, W. C. J., Kamphorst, W., and Swaab, D. F. (1997) DNA damage distribution in the human brain as shown by in situ end labelling. Area-specific differences in aging and Alzheimer's disease in the absence of apoptotic morphology. *J. Neuropathol. Exp. Neurol.* **56,** 887–900.
283. Troncoso, J. C., Sukhov, R. R., Kawas, C. H., and Koliatsos, V. E. (1996) In situ labeling of dying cortical neurons in normal aging and in Alzheimer's disease. Correlations with senile plaques and disease progression. *J. Neuropathol. Exp. Neurol.* **55,** 1134–1142.
283a. Kobayashi, K., Hayashi, M., Nakano, H., Fukutani, Y., Sasaki, K., Shimazaki, M., and Koshino, Y. (2002) Apoptosis of astrocytes with enhanced lysosomal activity and oligodendrocytes in white matter lesions in Alzheimer's disease. *Neuropathol. Appl. Neurobiol.* **28,** 238–251.
284. Braak, H. and Braak, E. (1991) Neuropathological staging of Alzheimer-related changes. *Acta Neuropathol.* **82,** 239–259.

285. Ferrer, I., Segui, J., and Planas, A. M. (1996) Amyloid deposition is associated with c-Jun expression in Alzheimer's disease and amyloid angiopathy. *Neuropathol. Appl. Neurobiol.* **22,** 521–526.
286. Nishimura, T., Akiyama, H., Yonehara, S., Kondo, H., Ikeda, K., Kato, M., et al. (1995) Fas antigen expression in brains of patients with Alzheimer-type dementia. *Brain Res.* **695,** 137–145.
287. Guo, Q., Fu, W., Xie, J., Luo, H., Sells, S. F., Geddes, J. W., et al. (1998) Par-4 is a mediator of neuronal degeneration associated with the pathogenesis of Alzheimer disease. *Nature Med.* **4,** 957–962.
288. Marcus, D. L., Strafaci, J. A., Miller, D. C., Masia, S., Thomas, C. G., Rosman, J., et al. (1998) Quantitative neuronal c-Fos and c-Jun expression in Alzheimer's disease. *Neurobiol. Aging* **19,** 393–400.
289. Tortosa, A., López, W., and Ferrer, I. (1998) Bcl-2 and Bax protein expression in Alzheimer's disease. *Acta Neuropathol.* **95,** 407–412.
290. Torp, R., Su, J. H., Deng, G., and Cotman, C. W. (1998) GADD45 is induced in Alzheimer's disease, and protects against apoptosis in vitro. *Neurobiol. Dis.* **5,** 245–252.
291. Engidawork, E., Gulesserian, T., Yoo, B. C., Cairns, N., and Lubec, G. (2001) Alteration of caspases and apoptosis-related proteins in brains of patients with Alzheimer's disease. *Biochem. Biophys. Res. Commun.* **281,** 84–93.
292. Nagy, Z. S. and Esiri, M. M. (1997) Apoptosis-related protein expression in the hippocampus of Alzheimer's disease. *Neurobiol. Aging* **18,** 655–671.
293. Copani, A., Uberti, D., Sortino, M. A., Bruno, V., Nicoletti, F., and Memo, M. (2001) Activation of cell cycle-associated proteins in neuronal death: a mandatory or dispensable path? *Trends Neurosci.* **24,** 25–31.
294. Yang, Y., Geldmacher, D. S., and Herrup, K. (2001) DNA replication precedes neuronal cell death in Alzheimer's disease. *J. Neurosci.* **21,** 2661–2668.
295. Smith, M. A., Zhu, X., Sun, Z., and Perry, G. (2001) Activation of c-jun in Alzheimer disease (abstr.). *J. Neuropathol. Exp. Neurol.* **60,** 546.
296. Vickers, J. C., Dickson, T. C., Adlard, P. A., Saunders, H. L., King, C. E., and McCormack, G. (1999) The cause of neuronal degeneration in Alzheimer's disease. *Prog. Neurobiol.* **60,** 1–27.
297. Giannakopoulos, P., Kövari, E., Savioz, A., De Bilabao, F., Dubois-Dauphin, M., Hof, P. R., and Bouras, C. (1999) Differential distribution of presenilin-1, Bax, and Bcl-X in Alzheimer's disease and frontotemporal dementia. *Acta Neuropathol.* **98,** 141–149.
298. Stadelmann, C., Brück, W., Bancher, C., Jellinger, K., and Lassmann, H. (1998) Alzheimer disease: DNA fragmentation indicates increased neuronal vulnerability but not apoptosis. *J. Neuropathol. Exp. Neurol.* **57,** 456–464.
298a. Su, J. H., Kesslak, J. P., Head, E., and Cotman, C. W. (2002) Caspase-cleaved amyloid precursor protein and activated caspase-3 are co-localized in the granules of granulovacuolar degeneration in Alzheimer's disease and Down' syndrome brain. *Acta Neuropathol.* **104,** 1–6.
299. Masliah, E., Mallory, M., Alford, M., Tanaka, S., and Hansen, L. A. (1998) Caspase dependent DNA fragmentation might be associated with excitotoxicity in Alzheimer's disease. *J. Neuropathol. Exp. Neurol.* **57,** 1041–1052.
300. Bancher, C., Brunner, C., Lassmann, H., Budka, H., Jellinger, K., Wiche, O., et al. (1989) Accumulation of abnormally phosphorylated τ precedes the formation of neurofibrillary tangles in Alzheimer's disease. *Brain Res.* **477,** 90–99.
301. Su, J. H., Zhao, M., Anderson, A. J., Srinivasan, A., and Cotman, C. W. (2001) Activated caspase-3 expression in Alzheimer's and aged control brain: correlation with Alzheimer pathology. *Brain Res.* **898,** 350–357.
301a. Ayala-Grosso, C., Ng, G., Roy, S., and Robertson, G. S. (2002) Caspase-cleaved amyloid precursor protein in Alzheimer's disease. *Brain Pathol.* **12,** 430–441.

302. Ghoshal, N., Smiley, J. F., DeMaggio, A. J., Hoekstra, M. F., Cochran, E. J., Binder, L. I., and Kuret, J. (1999) A new molecular link between the fibrillar and granulovacuolar lesions of Alzheimer's disease. *Am. J. Pathol.* **155,** 1163–1172.

303. Perry, G, Numomura, A., Lucassen, P. J., Lassmann, H., and Smith, M. A. (1998) Apoptosis and Alzheimer's disease. *Science* **282,** 1265.

304. Ferrer, I., Puig, B., Krupinski, J., Marmona, M., and Blanco, R. (2001) Fas and Fas ligand expression in Alzheimer's disease. *Acta Neuropathol.* **102,** 121–131.

305. de la Monte, S. M., Luong, T., Neely, T. R., Robinson, D., and Wands, J. R. (2000) Mitochondrial DNA damage as a mechanism of cell loss in Alzheimer's disease. *Lab. Invest.* **80,** 1323–1335.

306. Pei, J.-J., Braak, E., Braak, H., Grundke-Iqbal, I., Winblad, K., Winblad, B., and Cowburn, E. R. (2001) Localization of active forms of c-Jun kinase (JNK) and p38 kinase in Alzheimer's disease brains at different stages of neurofibrillary degeneration. *J. Alz. Dis.* **3,** 41–48.

307. Galvin, J. E., Lee, V. M., and Trojanowski, J. Q. (2001) Synucleinopathies: clinical and pathological implications. *Arch. Neurol.* **58,** 186–190.

308. Mochizuki, H., Goto, K., Mori, H., and Mizuno, Y. (1996) Histochemical detection of apoptosis in Parkinson's disease. *J. Neurol. Sci.* **137,** 120–123.

309. Marshall, K. A., Daniel, S. E., Cairns, N., Jenner, P., and Halliwell, B. (1997) Upregulation of the anti-apoptotic protein Bcl-2 may be early event in neurodegeneration: studies on Parkinson's incidental Lewy body disease. *Biochem. Biophys. Res. Commun.* **240,** 84–87.

310. Vyas-Boissiere, F., Hibner, U., and Agid, Y. (1997) Expression of Bcl 2 in adult human brain regions with special reference to neurodegenerative disorders. *J. Neurochem.* **69,** 223–231.

310a. Hartmann, A., Mouatt-Prigent, A., Vila, M., Abbas, N., Perier, C., Faucheux, B. A., et al. (2002) Increased expression and redistribution of the antiapoptotic molecule Bcl-xL in Parkinson's disease. *Neurobiol. Dis.* **10,** 28–32.

311. Tortosa, A., Lopez, E., and Ferrer, I. (199) Bcl-2 and Bax proteins in Lewy bodies from patients with Parkinson's disease and diffuse Lewy body disease. *Neurosci. Lett.* **238,** 78–80.

312. Gonzalez-Garcia, M., Garcia, I., Ding, L., O'Shea, S., Boise, L. H., Thompson, C. B., and Nunoz, G. (1995) Bcl-x is expressed in embryonic and postnatal neural tissues and functions to prevent neuronal cell death. *Proc. Natl. Acad. Sci. USA* **92,** 4304–4308.

313. Fearnley, J. M. and Lees, A. J. (1994) Pathology of Parkinson's disease, in *Neurodegenerative Diseases (19)* (Calne, D. B., ed.), Saunders, Philadelphia, pp. 545–554.

314. Tatton, W. G., Chalmers-Redman, R. M., Elstner, M., Leesch, W., Jagodzinski, F. B., Stupak, D. P., et al. (2000) Glyceraldehyde-3-phosphate dehydrogenase in neurodegeneration and apoptosis signaling. *J. Neural Transm.* **60(Suppl.),** 77–100.

315. Anderson, J. K. (2001) Does neuronal loss in Parkinson's disease involve programmed cell death? *Bio Essays* **23,** 640–647.

316. Tompkins, M. M. and Hill, W. D. (1997) Contribution of somal Lewy bodies to neuronal death. *Brain Res.* **775,** 24–29.

317. Ferrer, I., Blanco, R., Marmona, M., Puig, B., Barrachina, M., Gomec, C., and Ambrosio, S. (2001) Active, phosphorylation-dependent mitogen-activated protein kinase (MAPK/ERK), stress-activated protein kinase/c-jun, N-terminal kinase (SAPK/JNK) and p38 kinase expression in Parkinson's disease and dementia with Lewy bodies. *J. Neural Transm.* **108,** 1383–1396.

318. Langston, J. W., Forno, L. S., Tetrud, J., Reeves, A. G., Kaplan, J. A., and Karluk, D. (1999) Evidence of active nerve cell degeneration in the substantia nigra of humans years after 1-methyl-4-phenyl-1,2,3,6-tetrahydropyridine exposure. *Ann. Neurol.* **46,** 598–605.

319. Liu, B., Wang, K., Gao, H. M., Mandavilli, B., Wang, J. Y., and Hong, J. S. (2001) Molecular consequences of activated microglia in the brain: overactivation induces apoptosis. *J. Neurochem.* **77,** 182–189.

320. Kingham, P. J. and Pocock, J. M. (2001) Microglial secreted cathepsin β induces neuronal apoptosis. *J. Neurochem.* **76,** 1475–1484.
321. McGeer, P. L., Pagani, S., Boyes, B. E., and McGeer, E. G. (1988) Reactive microglia are posiitve for HLA-DR in the substantia nigra of Parkinson's and Alzheimer's disease brains. *Neurology* **38,** 1385–1391.
322. McGeer, P. L., Yasojima, K., and McGeer, E. G. (2001) Inflammation in Parkinson's disease. *Adv. Neurol.* **86,** 83–89.
323. Jellinger, K., Paulus, W., Grundke-Iqbal, I., Riederer, P., and Youdim, M. B. (1990) Brain iron and ferritin in Parkinson's and Alzheimer's diseases. *J. Neural Transm. Park. Dis. Dement. Sect.* **2,** 327–340.
324. Shoham, S. and Youdim, M. B. (2000) Iron involvement in neural damage and microgliosis in models of neurodegenerative diseases. *Cell. Mol. Biol.* **46,** 743–760.
325. Linert, W. and Jellinger, K. A. (2001) Cell death mechanisms and the role of iron in neurodegeneration, in *Mechanisms of Degeneration and Protection of the Dopaminergic System* (Segura-Aquilar, J., ed.), FP Graham Publishing, Johnson City, TN, pp. 21–65.
326. Orth, M. and Schapira, A. H. V. (2001) Mitochondria and degenerative disorders. *Am. J. Med. Genet.* **106,** 27–36.
327. Schapira, A. H. V. (1999) Mitochondrial DNA. *Adv. Neurol.* **80,** 233–237.
328. Schapira, A. H. V. (2001) Causes of neuronal death in Parkinson's disease. *Adv. Neurol.* **86,** 155–161.
329. Lantos, P. L. (1999) The definition of multiple system atrophy: A review of recent developments. *J. Neuropathol. Exp. Neurol.* **57,** 1099–1111.
330. Charles, P. D., Robertson, D., Kerr, L. D., Lonce, S., Austin, M. T., Gelbman, B. D., et al. (1997) Evidence of apoptotic cell death in multiple-system atrophy. (abstr.) *Ann. Neurol.* **42,** 408–409.
331. Hauw, J.-J., Daniel, S. E., Dickson, D., Horoupian, D. S., Jellinger, K., Lantos, P. L., et al. (1994) Preliminary NINDS neuropathologic criteria for Steele-Richardson-Olszewski syndrome (progressive supranuclear palsy). *Neurology* **44,** 2015–2019.
332. Komori, T. (1999) Tau-positive glial inclusions in progressive supranuclear palsy, corticobasal degeneration and Pick's disease. *Brain Pathol.* **9,** 663–679.
333. Dickson, D. W. (1999) Neuropathologic differentiation of progressive supranuclear palsy and corticobasal degeneration. *J. Neurol.* **246(Suppl. 2),** II6-II15.
334. Litvan, I., Hauw, J. J., Bartko, J. J., Lantos, P. L., Daniel, S. E., Horoupian, D. S., et al. (1996) Validity and reliability of the preliminary NINDS neuropathologic criteria for progressive supranuclear palsy and related disorders. *J. Neuropathol. Exp. Neurol.* **55,** 97–105.
335. Litvan, I., Dickson, D. W., Buttner-Ennever, J. A., Delacourte, A., Hutton, M., Dubois, B., et al. (2000) Research goals in progressive supranuclear palsy. First International Brainstorming Conference on PSP. *Mov. Disord.* **15,** 446–458.
336. Tortosa, A., Blanco, R., and Ferrer, I. (1998) Bcl-2 and Bax protein expression in neurofibrillary tangles in progressive supranuclear palsy. *NeuroReport* **9,** 1049–1052.
337. Albers, D. S., Swerdlow, R. H., Manfredi, G., Gajewski, C., Yang, L., Parker, W. D. Jr., and Beal, M. F. (1999) Further evidence for mitochondrial dysfunction in progressive supranuclear palsy. *Exp. Neurol.* **168,** 196–198.
338. Litvan, I., Grimes, D. A., Lang, A. E., Jankovic, J., McKee, A., Verny, M., Jellinger, K., et al. (1999) Clinical features differentiating patients with postmortem confirmed progressive supranuclear palsy and corticobasal degeneration. *J. Neurol.* **246(Suppl. 2),** II1–II5.
339. Litvan, I., Goetz, C. G., and Lang, A. E. (2000) Corticobasal degeneration and related disorders. *Adv. Neurol.* **82.**
340. Dickson, D. W., Bergeron, C., Chin, S. S., Duyckaerts, C., Horoupian, D., Ikeda, K., et al. (2002) Neuropathologic criteria for the diagnosis of corticobasal degeneration. *J. Neuropathol. Exp. Neurol.* **62,** in press.

341. Su, J. H., Nichol, K. E., Sitch, T., Sheu, P., Chubb, C., Miller, B. L., et al. (2000) DNA damage and activated caspase-3 expression in neurons and astrocytes: evidence for apoptosis in frontotemporal dementia. *Exp. Neurol.* **163,** 9–19.

342. Martinac, J. A., Craft, D. K., Su, J. H., Kim, R. C., and Cotman, C. W. (2001) Astrocytes degenerate in frontotemporal dementia: possible relation to hypoperfusion. *Neurobiol. Aging* **22,** 195–207.

343. Thomas, L. B., Gates, D. J., Richfield, E. K., Tf, O. B., Schweitzer, J. B., and Steindler, D. A. (1995) DNA end labeling (TUNEL) in Huntington's disease and other neuropathological conditions. *Exp. Neurol.* **133,** 265–272.

344. Portera-Cailliau, C., Hedreen, J. C., Price, D. L., and Koliatsos, V. E. (1995) Evidence for apoptotic cell death in Huntington's disease and excitotoxic animal models. *J. Neurosci.* **15,** 3775–3787.

345. Butterworth, N. J., Williams, L., Bullock, J. Y., Love, D. R., Faull, R. L. M., and Dragunow, M. (1998) Trinucleotide (CAG) repeat length is positively correlated with the degree of DNA fragmentation in Huntington's disease striatum. *Neuroscience* **87,** 49–53.

346. Zeron, M. M., Chen, N., Moshaver, A., Ting-Chun Lee, A., Wellington, C. L., Hayden, M. R., and Raymond, L. A. (2001) Mutant huntingtin enhances excitotoxic cell death. *Mol. Cell. Neurosci.* **17,** 41–53.

347. Sapp, E., Kegel, K. B., Aronin, N., Hashikawa, T., Uchiyama, Y., Tohyama, K., et al. (2001) Early and progressive accumulation of reactive microglia in the Huntington disease brain. *J. Neuropathol. Exp. Neurol.* **60,** 161–172.

347a. Mantamadiotis, T., Lemberger, T., Bleckmann, S. C., Kern, H., Kretz, O., Martin Villalba, A., et al. (2002) Disruption of CREB fiinction in brain leads to neurodegeneration. *Nat. Genet.* **31,** 47–54.

348. Troost, D., Aten, J., Morsink, F., and de Jong, J. M. B. V. (1995) Apoptosis in amyotrophic lateral sclerosis is not restricted to motor neurons: Bcl-2 expression is increased in unaffected post-central gyrus. *Neuropathol. Appl. Neurobiol.* **21,** 498–504.

349. Mu, X., He, J., Anderson, D. W., Trojanowski, J. Q., and Springer, J. E. (1996) Altered expression of Bcl-2 and Bax mRNA in amyotrophic lateral sclerosis spinal cord motor neurons. *Ann. Neurol.* **40,** 379–386.

350. Fitzmaurice, P. S., Shaw, I. C., Kleiner, H. E., Miller, R. T., Monks, T. J., Lau, S. S., et al. (1996) Evidence for DNA damage in amyotrophic lateral sclerosis. *Muscle Nerve* **19,** 797–798.

351. Martin, L. J. (1999) Neuronal death in amyotrophic lateral sclerosis is apoptosis: possible contribution of a programmed cell death mechanism. *J. Neuropathol. Exp. Neurol.* **58,** 459–471.

352. Ekegren, T., Grundstrom, E., Lindholm, D., and Aquilonius, S. M. (1999) Upregulation of Bax protein and increased DNA degradation in ALS spinal cord motor neurons. *Acta Neurol. Scand.* **100,** 317–321.

352a. Sathasivam, S., Ince, P. G., and Shaw, P. J. (2001) Apoptosis in amyotrophic lateral sclerosis: a review of the evidence. *Neuropathol. Appl. Neurobiol.* **27,** 257–274.

353. Kihira, T., Yoshida, S., Hironishi, M., Wakayama, I., and Yase, Y. (1998) Neuronal degeneration in amyotrophic lateral sclerosis is TI mediated by a possible mechanism different from classical apoptosis. *Neuropathology* **18,** 301–308.

354. Martin, L. J. (2000) p53 is abnormally elevated and active in the CNS of patients with amyotrophic lateral sclerosis. *Neurobiol. Dis.* **7,** 613–622.

355. Vukosavic, S., Stefanis, L., Jackson-Lewis, V., Guegan, C., Romero, N., Chen, C., et al. (2000) Delaying caspase activation by Bcl-2, A clue to disease retardation in a transgenic mouse model of amyotrophic lateral sclerosis. *J. Neurosci.* **20,** 9119–9125.

356. Bar-Peled, O., Knudson, M., Korsmeyer, S. J., and Rothstein, J. D. (1999) Motor neuron degeneration is attenuated in bax-deficient neurons *in vitro. J. Neurosci. Res.* **55,** 542–556.

357. Lucassen, P. J., Muller, M. B., Holsboer, F., Bauer, J., Holtrop, A., Wouda, J., et al. (2001) Hippocampal apoptosis in major depression is a minor event and absent from sub-areas at risk for glucocorticoid overexposure. *Am. J. Pathol.* **158,** 453–468.

358. Perry, G., Nunomura, A., and Smith, M. A. (1998) A suicide note from Alzheimer disease neurons? *Nature Med.* **4,** 897–898.

359. Smith, M. A., Raina, A. K., Nunomura, A., Hochman, A., Takeda, A., and Perry, G. (2000) Apoptosis in Alzheimer disease: fact or fiction. *Brain Pathol.* **10,** 797.

360. Trucco, C., Oliver, F. J., de Murcia, G., and Menissier-de Murcia, J. (1998) DNA repair defect in poly(ADP-ribose) polymerase-deficient cell lines. *Nucleic Acids Res.* **26,** 2644–2649.

361. Raina, A. K., Hochman, A., Zhu, X., Rottkamp, C. A., Nunomura, A., Siedlak, S. L., et al. (2001) Abortive apoptosis in Alzheimer's disease. *Acta Neuropathol.* **101,** 305–310.

362. Morsch, R., Simon, W., and Coleman, P. D. (1999) Neurons may live for decades with neurofibrillary tangles. *J. Neuropathol. Appl. Neurol.* **58,** 188–197.

363. Schwab, C., Schulzer, M., Steele, J. C., and McGeer, P. L. (1999) On the survival time of a tangled neuron in the hippocampal CA4 region in parkinsonism dementia complex of Guam. *Neurobiol. Aging* **20,** 57–63.

363a. Nunomura, A. and Chiba, S. (2002) Avoidance of apoptosis in Alzheimer's disease. *J. Alzheimer's Dis.* **2,** 59–60.

364. Clarke, G., Collins, R. A., Leavitt, B. R., Andrews, D. F., Hayden, M. R., Lumsden, C. J., and McInnes, R. R. (2000) A one-hit model of cell death in inherited neuronal degenerations. *Nature* **406,** 195–199.

365. Overmyer, M., Kraszpulki, M., Seppo, H., Hilkka, S., and Alafuzzoff, I. (2001) DNA-fragmentation, gliosis and histological hallmarks of Alzheimer's disease. *Acta Neuropathol.* **100,** 681–687.

366. Broe, M., Shephard, C. E., Milward, E. A., and Halliday, G. M. (2001) Relationship between DNA-fragmentation, morphological changes and dementia with Lewy bodies. *Acta Neuropathol.* **101,** 616–624.

367. Anderson, A. J., Ruehl, W. W., Fleischmann, L. K., Stenstrom, K., Entriken, T. L., and Cummings, B. J. (2000) DNA damage and apoptosis in the aged canine brain: relationship to Aβ deposition in the absence of neuritic pathology. *Prog. Neuro-Psychopharmacol. Biol. Psychiat.* **24,** 787–799.

368. Rohn, T. T., Head, E., Su, J. H., Anderson, A. J., Bahr, B. A., Cotman, C. W., and Cribbs, D. H. (2001) Correlation between caspase activation and neurofibrillary tangle formation in Alzheimer's disease. *Am. J. Pathol.* **158,** 189–198.

369. Beal, M. F. (199) Mitochondria, free radicals, and neurodegeneration. *Curr. Opin. Neurobiol.* **6,** 661–666.

370. Simonian, N. A. and Coyle, J. T. (1996) Oxidative stress in neurodegenerative diseases. *Ann. Rev. Pharmacol. Toxicol.* **36,** 83–106.

371. Multhaup, G., Masters, C. L., and Beyreuther, K. (1998) Oxidative stress in Alzheimer's disease. *Alzheimer's Rep.* **1,** 147–154.

372. Markesberry, W. R. and Carney, J. M. (1999) Oxidative alterations in Alzheimer's disease (Review). *Brain Pathol.* **9,** 133–146.

373. Markesbery, W. R., Montine, D. J., and Lovell, M. A. (2001) Oxidative alterations in neurodegenerative diseases, in *Pathogenesis of Neurodegenerative Disorders* (Mattson, M. P., ed.), Humana Press, Totowa, NJ, pp. 21–51.

374. Smith, M. A., Rottkamp, C. A., Nunomura, A., Raina, A. K., and Perry, G. (2000) Oxidative stress in Alzheimer's disease. *Biochim. Biophys. Acta* **1502,** 139–144.

375. de la Monte, S. M., Neely, T. R., Cannon, J., and Wands, J. R. (2000) Oxidative stress and hypoxia-like injury cause Alzheimer-type molecular abnormalities in central nervous system neurons. *Cell. Mol. Life Sci.* **57,** 1471–1481.

376. de la Monte, S. M., Ganju, N., Feroz, N., et al. (2000) Oxygen free radical injury is sufficient to cause some Alzheimer-type molecular abnormalities in human CNS neuronal cells.
J. Alz. Dis. **2,** 1–21.

377. Aksenov, M. Y., Aksenova, M. V., Butterfield, D. A., Geddes, J. W., and Markesbery,W. R.
(2001) Protein oxidation in the brain in Alzheimer's disease. *Neuroscience* **103,** 373–383.

378. Butterfield, D. A. and Kanski, J. (2001) Brain protein oxidation in age-related neurodegenerative disorders that are associated with aggregated proteins. *Mech. Ageing Dev.*
122, 945–962.

379. de la Monte, S. M., Luong, T., Neely, T. R., Robinson, D., and Wands, J. R. (2000) Mitochondrial DNA damage as a mechanism of cell loss in Alzheimer's disease. *Lab. Invest.*
80, 1323–1335.

380. de la Monte, S. M. and Wands, J. R. (2001) Alzheimer-associated neuronal thread pro-
tein-induced apoptosis and impaired mitochondrial function in human central nervous
system-derived neuronal cells. *J. Neuropathol. Exp. Neurol.* **60,** 195–207.

381. Pollanen, M. S., Dickson, D. W., and Bergeron, C. (1993) Pathology and biology of the
Lewy body. *J. Neuropathol. Exp. Neurol.* **52,** 183–191.

382. Dale, G. E., Probst, A., Luthert, P., Martin, J., Anderton, B. H., and Leigh, P. N. (1992)
Relationships between Lewy bodies and pale bodies in Parkinson's disease. *Acta Neuropathol.* **83,** 525–529.

383. Head, M. W. and Goldman, J. E. (2000) Small heat shock proteins, the cytoskeleton, and
inclusion body formation. *Neuropathol. Appl. Neurobiol.* **26,** 304–312.

384. Baba, M., Nakajo, S., Tu, P. H., Tomita, T., Nakaya, K., Lee, V. M., Trojanowski, J. Q.,
and Iwatsubo, T. (1998) Aggregation of α-synuclein in Lewy bodies of sporadic Parkinson's disease and dementia with Lewy bodies. *Am. J. Pathol.* **152,** 879–884.

385. Spillantini, M. G., Crowther, R. A., Jakes, R., Hasegawa, M., and Goedert, M. (1998)
α-synuclein in filamentous inclusions of Lewy bodies from Parkinson's disease and dementia with Lewy bodies. *Proc. Natl. Acad. Sci. USA* **95,** 6469–6473.

386. Irizarry, M. C., Growdon, W., Gomez-Isla, T., Newell, K., George, J. M., Clayton, D. F.,
and Hyman, B. T. (1998) Nigral and cortical Lewy bodies and dystrophic nigral neurites
in Parkinson's disease and cortical Lewy body disease contain α-synuclein immunoreactivity. *J. Neuropathol. Exp. Neurol.* **57,** 334–337.

387. Trojanowski, J. Q., Goedert, M., Iwatsubo, T., and Lee, V. M. Y. (1998) Fatal attractions
—abnormal protein aggregation and neuron death in Parkinson's disease and Lewy body-
dementia. *Cell Death Different.* **5,** 832–837.

388. Hashimoto, M. and Masliah, E. (1999) α-synuclein in Lewy body disease and Alzheimer's
disease. *Brain Pathol.* **9,** 707–720.

389. Galvin, J. F., Lee, V. M. Y., Schmidt, L., Tu, P.-H., Iwatsubo, T., and Trojanowski, J. Q.
(1999) Pathobiology of the Lewy body. *Adv. Neurol.* **80,** 313–324.

390. Wakabayashi, K., Engelender, S., Yoshimoto, M., Tsuji, S., Ross, C. A., and Takahashi,
H. (2000) Synphilin-1 is present in Lewy bodies in Parkinson's disease. *Ann. Neurol.* **47,**
521–523.

391. Gai, W. P., Yuan, H. X., Li, X. Q., Power, J. T., Blumbergs, P. C., and Jensen, P. H.
(2000) In situ and *in vitro* study of colocalization and segregation of α-synuclein, ubiquitin,
and lipids in Lewy bodies. *Exp. Neurol.* **166,** 324–333.

392. Castellani, R. J., Siedlak, S. L., Perry, G., and Smith, M. A. (2000) Sequestration of iron
by Lewy bodies in Parkinson's disease. *Acta Neuropathol.* **100,** 111–114.

393. Nakamura, S., Kawamoto, Y., Nakano, S., Akiguchi, I., and Kimura, J. (1998) Cyclin–
dependent kinase 5 and mitogen-activated protein kinase in glial cytoplasmic inclusions
in multiple system atrophy. *J. Neuropathol. Exp. Neurol.* **57,** 690–698.

394. Braak, E., Sandmann-Keil, D., Rüb, U., Gai, W. P., de Vos, R. A. I., Jansen Steur, E. N. H.,
Arai, K., and Braak, H. (2001) α-Synuclein immunopositive Parkinson's disease-related
inclusion bodies in lower brain stem nuclei. *Acta Neuropathol.* **101,** 195–202.

395. Narhi, L., Wood, S. J., Stevenson, S., Jiang, Y., Wu, G. M., Anafi, D., et al. (1999) Both familial Parkinson's disease mutations accelerate α-synuclein aggregation. *J. Biol. Chem.* **273,** 9843–9846.

396. Farrer, M., Gwinn-Hardy, K., Hutton, M., and Hardy, J. (1999) The genetics of disorders with synuclein pathology and parkinsonism. *Hum. Mol. Genet.* **8,** 1901–1905.

397. Volles, M. J., Lee, S. J., Rochet, J. C., Shtilerman, M. D., Ding, T. T., Kessler, J. C., and Lansbury, P. T. Jr. (2001) Vesicle permeabilization by protofibrillar α-synuclein: implications for the pathogenesis and treatment of Parkinson's disease. *Biochemistry* **40,** 7812–7819.

398. Davidson, W. S., Jonas, A., Clayton, D. F., and George, J. M. (1998) Stabilization of α-synuclein secondary structure upon binding to synthetic membranes. *J. Biol. Chem.* **273,** 9443–9449.

399. Jellinger, K. A. (1999) The role of iron in neurodegeneration. Prospects for pharmacotherapy of Parkinson's disease. *Drugs Aging* **14,** 115–140.

400. Riederer, P., Reichmann, H., Janetzky, B., Sian, J., Lesch, K.-P., Lange, K. W., et al. (2001) Neural degeneration in Parkinson's disease. *Adv. Neurol.* **86,** 125–136.

401. Hashimoto, M., Hsu, L. J., Xia, Y., Takeda, A., Sisk, A., Sundsmo, M., and Masliah, E. (1999) Oxidative stress induces amyloid-like aggregate formation of NACP/α-synuclein *in vitro. Neuroreport* **10,** 717–721.

402. Olesen, O. F., Mikkelsen, J. D., Gerdes, C., and Jensen, P. H. (1997) Isoform-specific binding of human apolipoprotein E to the non-amyloid beta component of Alzheimer's disease amyloid. *Mol. Brain. Res.* **44,** 105–112.

403. Krüger, R., Vieira-Saecker, A. M., Kuhn, W., Berg, D., Muller, T., Kuhnl, N., et al. (1999) Increased susceptibility to sporadic Parkinson's disease by a certain combined α-synuclein/apolipoprotein E genotype. *Ann. Neurol.* **45,** 611–617.

404. Zhou, W., Hurlbert, M. S., Schaak, J., Prasad, K. N., and Freed, C. R. (1999) Overexpression of mutant human α-synuclein (A53T) causes dopamine neuron death in rat primary culture and in rat mesencephalon-derived cell line (1RB3AN27) (Abs.). *Soc. Neurosci.* **25,** 27. 25.

405. Masliah, E., Rockenstein, E., Veinbergs, I., Mallory, M., Hashimoto, M., Takeda, A., et al. (2000) Dopaminergic loss and inclusion body formation in α-synuclein mice: implications for neurodegenerative disorders. *Science* **287,** 1265–1269.

406. Feany, M. B. and Bender, W. W. (2000) A Drosophila model of Parkinson's disease. *Nature* **404,** 394–398.

407. Lee, M., Hyun, D., Halliwell, B., and Jenner, P. (2001) Effect of the overexpression of wild-type or mutant alpha-synuclein on cell susceptibility to insult. *J. Neurochem.* **76,** 998–1009.

408. Kholodilov, N. G., Oo, T. F., and Burke, R. E. (1999) Synuclein expression is decreased in rat substantia nigra following induction of apoptosis by intrastriatal 6-hydroxydopamine. *Neurosci. Lett.* **275,** 105–108.

409. Neystat, M., Lynch, T., Przedborski, S., Kholodilov, N., Rzhatskaya, M., and Burke, R. E. (1999) α-synuclein expression in substantia nigra and cortex in Parkinson's disease. *Mov. Disord.* **14,** 417–422.

410. Itoh, K., Weis, S., Mehraein, P., and Müller-Höcker, J. (1997) Defects of cytochrome c oxidase in the substantia nigra of Parkinson's disease: an immunohictochemical and morphometric study. *Mov. Disord.* **12,** 9–16.

411. Hashimoto, M., Takeda, A., Hsu, L. J., Takenouchi, T., and Masliah, E. (1999) Role of cytochrome c as a stimulator of α-synuclein aggregation in Lewy body disease. *J. Biol. Chem.* **274,** 28,849–28,852.

412. Hsu, L. J., Sagara, Y., Arroyo, A., Rockenstein, E., Sisk, A., Mallory, M., et al. (2000) α-synuclein promotes mitochondrial deficit and oxidative stress. *Am. J. Pathol.,* **157,** 401–440.

413. Saha, A. R., Ninkina, N. N., Hanger, D. P., Anderton, B. H., Davies, A. M., and Buchman, V. L. (2000) Induction of neuronal death by α-synuclein. *Eur. J. Neurosci.* **12,** 3073–3077.

414. Hansen, L. and Masliah, E. (2001) Neurobiology of disorders with Lewy bodies, in *Functional Neurobiology of Aging* (Hof, R. R. and Mobbs, L. C. K., eds.), Academic Press, San Diego, CA, pp. 173–182.

415. Munch, G., Luth, H. J., Wong, A., Arendt, T., Hirsch, E., Ravid, R., and Riederer, P. (2000) Crosslinking of α-synuclein by advanced glycation endproducts—an early pathophysiological step in Lewy body formation? *J. Chem. Neuroanat.* **20**, 253–257.

416. Kingsbury, A. E., Marsden, C. D., and Foster, O. J. F. (1999) The vulnerability of nigral neurons to Parkinson's disease is unrelated to their intrinsic capacity for dopamine synthesis: an in situ hybridisation study. *Mov. Disord.* **14**, 206–219.

417. Calingasan, N. Y., Park, L. C., Calo, L. L., Trifiletti, R. R., Gandy, S. E., and Gibson, G. E. (1998) Induction of nitric oxide synthase and microglial responses precede selective cell death induced by chronic impairment of oxidative metabolism. *Am. J. Pathol.* **153**, 599–610.

418. McRae, A., Dahlstrom, A., and Ling, E. A. (1997) Microglial in neurodegenerative disorders: emphasis on Alzheimer's disease. *Gerontology* **43**, 95–108.

419. McGeer, P. L. and McGeer, E. G. (2000) Autotoxicity and Alzheimer disease. *Arch. Neurol.* **57**, 789–790.

420. Hirsch, E. C. (2000) Glial cells and Parkinson's disease. *J. Neurol.* **247(Suppl. 2)**, II58–II62.

421. Tu, P.-H., Galvin, J. E., Baba, M., Giasson, B., Tomita, T., Leight, S., Nakajo, S., et al. (1998) Glial cytoplasmic inclusions in white matter oligodendrocytes of multiple system atrophy brains contain insoluble α-synuclein. *Ann. Neurol.* **44**, 415–422.

422. Gai, W. P., Power, J. H., Blumbergs, P. C., Culvenor, J. G., and Jensen, P. H. (1999) α-Synuclein immunoisolation of glial inclusions from multiple system atrophy brain tissue reveals multiprotein components. *J. Neurochem.* **73**, 2093–2100.

423. Dickson, D. W., Liu, W., Hardy, J., Farrer, M., Mehta, N., et al. (1999) Widespread alterations of α-synuclein in multiple system atrophy. *Am. J. Pathol.* **155**, 1241–1251.

424. Spillantini, M. G., Crowther, R. A., Jakes, R., Cairns, N. J., Lantos, P. L., and Goedert, M. (1998) Filamentous α-synuclein inclusions link multiple system atrophy with Parkinson's disease and dementia with Lewy bodies. *Neurosci. Lett.* **251**, 205–208.

425. Arima, K., Ueda, K., Sunohara, N., Arakawa, K., Hirai, S., Nakamura, M., et al. (1998) NACP/α-synuclein immunoreactivity in fibrillary components of neuronal and oligodendroglial cytoplasmic inclusions in the pontine nuclei in multiple system atrophy. *Acta Neuropathol.* **96**, 439–444.

426. Nakamura, S., Kawamoto, Y., Nakano, S., Akiguchi, I., and Kimura, J. (1998) Cyclin-dependent kinase 5 and mitogen-activated protein kinase in glial cytoplasmic inclusions in multiple system atrophy. *J. Neuropathol. Exp. Neurol.* **57**, 690–698.

427. Campbell, B. C., McLean, C. A., Culvenor, J. G., Gai, W. P., Blumbergs, P. C., Jakala, P., et al. (2001) Solubility of α-synuclein differs between multiple system atrophy and dementia with Lewy bodies. *J. Neurochem.* **76**, 87–96.

428. Morrison, B. M., Hof, P. R., and Morrison, J. H. (1998) Determinants of neuronal vulnerability in neurodegenerative diseases. *Ann. Neurol.* **44(Suppl. 1)**, S32–S44.

429. Arendt, T., Holzer, M., Fruth, R., Brückner, M. K., and Gärtner, U. (1998) Phosphorylation of tau, Aβ-formation, and apoptosis after *in vivo* inhibition of PP-1 and PP-2A. *Neurobiol. Aging* **19**, 3–13.

430. Ferreira, S. T. and De Felice, F. G. (2001) Protein dynamics, folding and misfolding: from basic physical chemistry to human conformational diseases. *FEBS Lett.* **498**, 129–134.

3

Role(s) of Mitogen and Stress-Activated Kinases in Neurodegeneration

Christopher C. J. Miller, Steven Ackerley, Janet Brownlees, Andrew J. Grierson, and Paul Thornhill

1. THE MAPK/SAPK FAMILY

Neurones respond to extracellular stimuli via a variety of intracellular signaling pathways. Some of the most intensely studied of these are those involving the mitogen-activated protein kinases (MAPKs). MAPKs are serine/threonine protein kinases and their substrates include both cytoplasmic and nuclear targets. The founder members of the MAPK family are the p42/p44 MAPKs or extracellular-regulated kinases (ERKs). In non-neuronal cells, these kinases are mainly activated by mitogenic stimuli, but in neurones, such signals include those emanating from neurotrophic and neurotransmitter receptors.

The remaining members of the MAPK family comprise the stress-activated protein kinases (SAPKs), which are mainly activated in response to cellular stresses and proinflammatory cytokines. Well-characterized members of the SAPKs include the SAPK1 members (also known as Jun-N-terminal kinases [JNKs]) and p38 (SAPK2) kinases. SAPK1 members (SAPK1a, SAPK1b, and SAPK1c) are encoded by distinct genes, but these undergo alternative splicing such that there are 4 different SAPK1a/SAPK1b/SAPK1c isoforms, giving rise to a total of at least 12 different SAPK1 family members.

Activation of MAPKs and SAPKs is regulated by signal transduction cascades of regulatory kinases and different nomenclatures exist for both the MAPK/SAPK family and their regulatory kinases. To facilitate a proper understanding of this review, these nomenclatures are listed in Tables 1 and 2. For a more detailed review of the MAPK/SAPK family nomenclature, *see* ref. *1*.

2. REGULATION OF MAPK/SAPKs

Both p42 and p44 MAPK activities are controlled by a cascade of regulatory kinases and signaling molecules that emanate from the cell surface. Both are activated by phosphorylation on threonine and tyrosine residues by the dual-specificity kinases MKK1 and MKK2. These are activated by the Raf kinases, which, in turn, are regulated by the GTP-binding protein Ras. In the classical MAPk cascade, Ras activity is coupled to stimulation

From: *Neuroinflammation, 2nd Edition: Mechanisms and Management*
Edited by: P. L. Wood © Humana Press Inc., Totowa, NJ

Table 1
The Mitogen- and Stress-Activated Protein Kinase Family

Kinase	Other names
p42 MAPK	Erk2; MAPK1
p44 MAPK	Erk1; MAPK2
SAPK1a	JNK2; SAPKα
SAPK1b	JNK3; SAPKβ
SAPK1c	JNK1; SAPKγ
SAPK2a	p38; CSBP; Mxi2; Mpk2; RK
SAPK2b	p38β
SAPK3	Erk6; p38γ
SAPK4	
SAPK5	Erk5; BMK1

Table 2
Regulatory Kinases for the Mitogen- and Stress-Activated Protein Kinase Family

Kinase	Other names	Substrates
MKK1	MAPKK1; MEK1	p42/p44 MAPK
MKK2	MAPKK2; MEK2	p42/p44 MAPK
MKK3	SAPKK2; SKK2	SAPK2
MKK4	SKK1; SEK1; JNKK; SAPKK1	SAPK1; SAPK2
MKK6	SKK3; MEK6; SAPKK3	SAPK2; SAPK3; SAPK4
MKK7	SKK4; SAPKK4; SAPKK5	SAPK1
MEK5	SKK5	SAPK5

of receptor tyrosine kinases at the plasma membrane by src homology 2 (SH2) domain containing proteins such as the Grb2–Sos complex. Alternative routes to activation of p42/p44 MAPKs include those involving increased intracellular Ca^{2+} and cAMP concentrations. Stimulation of glutamate receptors induces elevation of Ca^{2+} and protein kinase C (PKC)-mediated Ras/Raf activation. Stimulation of receptors such as β-adrenergic receptors leads to PKA-mediated Rap1 and B-Raf activation of MEK1/2 (Fig. 1).

Similarly, the SAPKs are regulated via protein kinase cascades although these are less well-characterised than those of the MAPKs (Fig. 2). Perhaps the best studies are those involving the SAPK1 and SAPK2 family members (JNK and p38 kinases). SAPK1 members are activated by phosphorylation by MKK4 and MKK7; SAPK2 members are activated by phosphorylation by MKK3 and MKK6. These, in turn, are activated by MAPKKK family members. A diverse array of cellular stresses leads to SAPK1/2 activation and these include oxidant and osmotic stresses and ultraviolet radiation. In addition, inflammatory cytokines can also lead to activation of SAPKs with tumor necrosis factor-α (TNF-α) and interleukin-1 stimulating both SAPK1 and SAPK2 isoforms (for reviews, *see* refs. *2–4*).

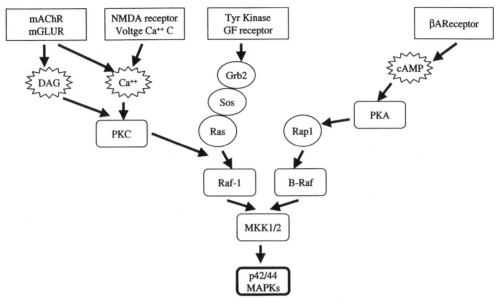

Fig. 1. Signaling pathways for MAPKs.

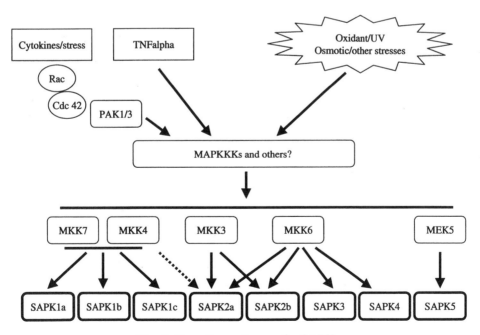

Fig. 2. Signaling pathways for SAPKs.

3. MAPKs AND SAPKs AND NEURODEGENERATIVE DISEASE

Both p42 and p44 MAPKs and SAPK1/2 isoforms are abundantly expressed in neurones of the mature central and peripheral nervous system of mammals. Substrates for MAPK/SAPK isoforms include a number of proteins involved in human neurodegenerative diseases, and as such, much effort has focused on the pharmacological manipulation of their activities as a therapeutic strategy.

3.1. SAPK1, c-Jun and Neuronal Survival and Apoptosis

The SAPK1 pathway has been implicated in both cell survival and cell death by apoptosis (see refs. 5 and 6). One known substrate for SAPK1 of direct relevance to apoptosis is the transcription factor c-Jun, which is a component of the activator protein (AP-1) transcription factor complex. Phosphorylation of c-Jun by SAPK1 on the c-Jun amino-terminal regulatory residues serine-63 and serine-73 causes increased transcriptional activity of AP-1 *(7,8)*.

Activation of SAPK1 members and phosphorylation of c-Jun is known to be involved in some forms of apoptosis, including those involving neurones (see, e.g., refs. *2, 4, 9,* and *10*). However, the effectors of apoptosis downstream of c-Jun are not properly characterized. c-Jun is essential for normal mouse development because homozygous knockout c-Jun mice die in mid-gestation *(11)*. However, replacing endogenous c-Jun with a mutant isoform in which the SAPK1 targeted regulatory residues, serine-63 and serine-73 are replaced with alanine to preclude phosphorylation leads to viable mice *(12)*. Most interestingly, these mice are resistant to epileptic seizures induced by the glutamate receptor agonist kainate. Furthermore, primary neurones obtained from these animals are also protected from glutamate-induced apoptosis although SAPK1 levels are normal *(12)*.

Studies of mice in which the SAPK1a, 1b, and 1c genes have been ablated either alone or in combination have reinforced the role of SAPK1s and c-Jun phosphorylation in glutamate-induced neuronal cell death and apoptosis. Individual SAPK1a, 1b, and 1c knockout mice are all viable and carry no overt phenotype *(13,14)*. Likewise, double knockouts involving SAPK1a and 1b and SAPK1b and 1c are also viable. In contrast, SAPK1a and 1c double knockouts are embryonic lethal with a phenotype involving dysregulation of apoptosis in the brain *(14)*. SAPK1b is a major SAPK in the adult brain and studies of the SAPK1b knockout mice have revealed that these mice are resistant to excitotoxic insults involving kainate. The animals have reduced seizures following administration of kainate, and kainate-induced apoptosis of hippocampal neurones is also reduced *(13)*. Thus, these studies of c-Jun phosphorylation mutant mice and SAPK1 knock-out mice together provide compelling evidence that SAPK1 (and, in particular, SAPK1b-mediated phosphorylation of c-Jun) plays a major role in glutamate-induced excitotoxic damage of neurones.

3.2. SAPK Inhibitors and Ischemia

Glutamate is known to activate both SAPK1 and SAPK2 isoforms in neurones *(15–17)* and the development of specific SAPK inhibitors has enabled their therapeutic value to be determined following ischemic insults. Recently, SB239063, a second generation SAPK2 inhibitor, has been shown to be neuroprotective in rodent models of cerebral focal ischemia *(18)*. Interestingly, SB239063 also reduced stroke-induced expression of interleukin (IL)-1β and TNF-α, which are cytokines that contribute to stroke-induced brain injury *(19)*. These studies demonstrate the potential of SAPK inhibitors as therapeutics for neurodegenerative disease.

3.3. MAPKs/SAPKs and the Neuronal Cytoskeleton in Neurodegenerative Diseases

Abnormal accumulations of the cytoskeleton are seen in several neurodegenerative diseases; these include Alzheimer's disease, amyotrophic lateral sclerosis (ALS), and Parkinson's disease. The constituent proteins in these accumulations are often characterized by being abnormally or inappropriately phosphorylated. Thus, in Alzheimer's disease, accumulations of hyperphosphorylated tau are the major component of the paired helical filaments (PHF) seen in the neurofibrillary tangle (for reviews, see refs. *20* and *21*). Likewise, neurofilament accumulations found in degenerating motor neurones in ALS contain phosphorylated neurofilament-middle- and -heavy-chain (NF-M/NF-H) side arms *(22)*. Neurofilaments containing phosphorylated NF-M/NF-H side-arms are also a component of the Lewy bodies in Parkinson's disease and Alzheimer's disease *(23,24)*.

Many of the sites that are phosphorylated in the tau and neurofilament accumulations are serine and threonine residues that precede a proline (SP/TP motifs). These are consensus sites for phosphorylation by members of the MAPK/SAPK family. Although the best candidate kinases for phosphorylating tau on the SP/TP motifs are glycogen synthase kinase-3 (GSK-3) and cyclin-dependent kinase-5 (cdk5/p35) *(25,26)*, there is evidence that p42/p44 MAPKs, SAPK1, and SAPK2 may also phosphorylate these sites *(27,28)*. Indeed, MAPKs have been shown to be associated with components of the cytoskeleton, including tau *(29)*. Thus, aberrant activation of MAPK/SAPKs may be part of the pathogenic process that leads to hyperphosphorylation of tau in Alzheimer's disease.

More robust data, however, links MAPK/SAPKs with phosphorylation of neurofilament side arms. Several laboratories have now shown that NF-M and NF-H side arms are substrates for both p42/p44 MAPKs and SAPK1 members *(17,30–34)*.

Neurofilaments are synthesized in cell bodies and then transported into and through axons, and their accumulation in disease states indicates that this transport might be perturbed. Indeed, defective transport of the cytoskeleton is an early pathogenic feature in transgenic mice expressing mutant forms of Cu/Zn superoxide dismutase-1 (SOD1) *(35, 36)*. Mutations in SOD1 are the cause of some familial forms of ALS and mice expressing mutant SOD1 are a model of this disease.

The mechanisms that regulate the speed of transport of neurofilaments down the axon are not properly understood, but much correlative evidence links slowing of neurofilament transport with increased phosphorylation of NF-M/NF-H side arms (see, e.g., refs. *32* and *37–43*). Thus, one possibility is that aberrant activation of MAPK/SAPKs occurs in these disease states, leading to increased NF-M/NF-H side-arm phosphorylation, slowing of neurofilament transport, and the formation of neurofilament accumulations.

3.4. SAPKs and Familial Forms of ALS and Alzheimer's Disease: Mutant SOD1, the Amyloid Precursor Protein, and Presenilins

Some forms of ALS and Alzheimer's disease are familial; mutations in the SOD1 gene can cause ALS, and mutations in the amyloid precursor protein (APP) and presenilin genes can cause Alzheimer's disease. The precise mechanisms by which these mutations induce disease are not properly understood. However, for mutant SOD1, some of its proposed pathogenic properties are predicted to lead to activation of MAPK/ SAPKs. One suggestion is that mutant SOD1 acts as a peroxidase to catalyze the formation of damaging hydroxyl radicals *(44)*. A second possibility is that the mutations cause the enzyme to catalyze the conversion of oxygen to superoxide (effectively running the normal reaction backward); superoxide would then combine with nitric oxide to produce peroxynitrite *(45,46)*. Both of these agents are known to activate MAPK/SAPKs; hydroxyl radicals can activate SAPK1 isoforms *(47)* and peroxynitrite can activate p42/p44 MAPKs and SAPK2 isoforms *(48)*. Thus, some of the pathogenic events downstream of mutant SOD1 may be mediated by members of the MAPK/SAPK family.

Increased production and deposition of the 39–43 amino acid Aβ peptide derived from APP is believed to be mechanistic in Alzheimer's disease. The mutations in APP and presenilin-1 that cause Alzheimer's disease are known to modulate APP processing and Aβ production. APP is a phosphoprotein and phosphorylation of threonine 668 within its cytoplasmic domain may regulate APP processing and Aβ production *(49)*. Recently, we have shown that this site is phosphorylated by SAPK1b *(50)* and so activation of SAPKs may lead to increased Aβ production. Mutant presenilin has been shown to increase vulnerability of neurones to excitotoxic damage by glutamate *(51)*. Mutant presenilin may therefore lead indirectly to activation of SAPKs. Thus, as with ALS, activation of SAPKs may also play a part in neuronal cell death in Alzheimer's disease.

REFERENCES

1. Cohen, P. (1997) The search for physiological substrates of MAP and SAP kinases in mammalian cells. *Trends Cell Biol.* **7,** 353–361.
2. Tibbles, L. A. and Woodgett, J. R. (1999) The stress-activated protein kinase pathways. *Cell. Mol. Life Sci.* **55,** 1230–1254.
3. Ichijo, H. (1999) From receptors to stress-activated MAP kinases. *Oncogene* **18,** 6087–6093.
4. Davis, R. J. (2000) Signal transduction by the JNK group of MAP kinases. *Cell* **103,** 239–252.
5. Ip, Y. T. and Davis, R. J. (1998) Signal transduction by the c-Jun N-terminal kinase (JNK) —from inflammation to development. *Curr. Opin. Cell Biol.* **10,** 205–219.
6. Mielke, K. and Herdegen, T. (2001) JNK and p38 stresskinases—degenerative effectors of signal-transduction-cascades in the nervous system. *Prog. Neurobiol.* **61,** 45–60.
7. Pulverer, B. J., Kyriakis, J. M., Avruch, J. M., Nikolakaki, E., and Woodgett, J. R. (1991) Phosphorylation of c-jun mediated by MAP kinases. *Nature* **353,** 670–674.
8. Smeal, T., Binetruy, B., Mercola, D. A., Birrer, M., and Karin, M. (1991) Oncogenic and transcriptional cooperation with Ha-Ras requires phosphorylation of c-jun on serines 63 and 73. *Nat. Neurosci.* **354,** 494–496.
9. Ham, J., Babij, C., Whitfield, J., Pfarr, C. M., Lallemand, D., Yaniv, M., et al. (1995) A c-Jun dominant negative mutant protects sympathetic neurons against programmed cell death. *Neuron* **14,** 927–939.
10. Ham, J., Eilers, A., Whitfield, J., Neame, S. J., and Shah, B. (2000) c-Jun and the transcriptional control of neuronal apoptosis. *Biochem. Pharmacol.* **60,** 1015–1021.
11. Hilberg, F., Aguzzi, A., Howells, N., and Wagner, E. F. (1993) c-jun is essential for normal mouse development and hepatogenesis. *Nature* **365,** 179–181.
12. Behrens, A., Sibilia, M., and Wagner, E. F. (1999) Amino-terminal phosphorylation of c-Jun regulates stress-induced apoptosis and cellular proliferation. *Nat. Genet.* **21,** 326–329.

13. Yang, D. D., Kuan, C.-Y., Whitmarsh, A. J., Rincon, M., Zheng, T. S., Davis, R. J., et al. (1997) Absence of excitotoxicity-induced apoptosis in the hippocampus of mice lacking the Jnk3 gene. *Nature* **389**, 865–870.
14. Kuan, C. Y., Yang, D. D., Roy, D. R. S., Davis, R. J., Rakic, P., and Flavell, R. A. (1999) The Jnk1 and Jnk2 protein kinases are required for regional specific apoptosis during early brain development. *Neuron* **22**, 667–676.
15. Kawasaki, H., Morooka, T., Shimohama, S., Kimura, J., Hirano, T., Gotoh, Y., et al. (1997) Activation and involvement of p38 mitogen-activated protein kinase in glutamate-induced apoptosis in rat cerebellar granule cells. *J. Biol. Chem.* **272**, 18,518–18,521.
16. Schwarzschild, M. A., Cole, R. L., and Hyman, S. E. (1997) Glutamate, but not dopamine, stimulates stress-activated protein kinase and AP-1-mediated transcription in striatal neurons. *J. Neurosci.* **17**, 3455–3466.
17. Brownlees, J., Yates, A., Bajaj, N. P., Davis, D., Anderton, B. H., Leigh, P. N., et al. (2000) Phosphorylation of neurofilament heavy chain side-arms by stress activated protein kinase-1b/Jun N-terminal kinase-3. *J. Cell Sci.* **113**, 401–407.
18. Barone, F. C., Irving, E. A., May, A. M., Lee, J. C., Kassis, S., Kumar, S., et al. (2001) SB 239063, a second-generation p38 mitogen-activated protein kinase inhibitor, reduces brain injury and neurological deficits in cerebral focal ischemia. *J. Pharmacol. Exp. Ther.* **296**, 312–321.
19. Barone, F. C., Irving, E. A., Ray, A. M., Lee, J. C., Kassis, S., Kumar, S., et al. (2001) Inhibition of p38 mitogen-activated protein kinase provides neuroprotection in cerebral focal ischemia. *Med. Res. Rev.* **21**, 129–145.
20. Goedert, M. (1993) Tau protein and the neurofibrillary pathology of Alzheimer's disease. *Trends Neurosci.* **16**, 460–465.
21. Trojanowski, J. Q. and Lee, V. M.-Y. (1994) Phosphorylation of neuronal cytoskeletal proteins in Alzheimer's disease and Lewy body dementias. *Ann. NY Acad. Sci.* **747**, 92–109.
22. Hirano, A. (1991) Cytopathology of amyotrophic lateral sclerosis, in *Amyotrophic Lateral Sclerosis and Other Motor Neuron Diseases* (Rowland, L. P., ed.), Advances in Neurology, Vol. 56, Raven, New York, pp. 91–101.
23. Trojanowski, J. Q., Schmidt, M. L., Shin, R.-W., Bramblett, G. T., Rao, D., and Lee, V. M.-Y. (1993) Altered *Tau* and neurofilament proteins in neurodegenerative diseases: diagnostic implications for Alzheimer's disease and Lewy body dementias. *Brain Pathol.* **3**, 45–54.
24. Schmidt, M. L., Martin, J. A., Lee, V. M. Y., and Trojanowski, J. Q. (1996) Convergence of Lewy bodies and neurofibrillary tangles in amygdala neurons of Alzheimer's disease and Lewy body disorders. *Acta Neuropathol. (Berl.)* **91**, 475–481.
25. Lovestone, S., Reynolds, C. H., Latimer, D., Davis, D. R., Anderton, B. H., Gallo, J.-M., et al. (1994) Alzheimer's disease-like phosphorylation of the microtubule-associated protein tau by glycogen synthase kinase-3 in transfected mammalian cells. *Curr. Biol.* **4**, 1077–1086.
26. Patrick, G., Zukerberg, L., Nikolic, M., de la Monte, S., Dikkes, P., and Tsai, L.-H. (1999) Conversion of p35 to p25 deregulates Cdk5 activity and promotes neurodegeneration. *Nature* **402**, 615–622.
27. Drewes, G., Lichtenberg-Kraag, B., Doring, F., Mandelkow, E.-M., Biernat, J., Goris, J., et al. (1992) Mitogen activated protein (MAP) kinase transforms tau protein into an Alzheimer-like state. *EMBO J.* **11**, 2131–2138.
28. Reynolds, C. H., Betts, J. C., Blackstock, W. P., Nebreda, A. R., and Anderton, B. H. (2000) Phosphorylation sites on tau identified by nanoelectrospray mass spectrometry: differences in vitro between the mitogen-activated protein kinases ERK2, c-Jun N-terminal kinase and P38, and glycogen synthase kinase-3beta. *J. Neurochem.* **74**, 1587–1595.
29. Veeranna, G. J., Shetty, K. T., Takahashi, M., Grant, P., and Pant, H. C. (2000) Cdk5 and MAPK are associated with complexes of cytoskeletal proteins in rat brain. *Mol. Brain Res.* **76**, 229–236.

30. Giasson, B. I. and Mushynski, W. E. (1996) Aberrant stress-induced phosphorylation of perikaryal neurofilaments. *J. Biol. Chem.* **271**, 30,404–30,409.

31. Giasson, B. I. and Mushynski, W. E. (1997) Study of proline-directed kinases involved in phosphorylation of the heavy neurofilament subunit. *J. Neurosci.* **17**, 9466–9472.

32. Ackerley, S., Grierson, A. J., Brownlees, J., Thornhill, P., Anderton, B. H., Leigh, P. N., et al. (2000) Glutamate slows axonal transport of neurofilaments in transfected neurons. *J. Cell Biol.* **150**, 165–175.

33. Veeranna, Amin, N. D., Ahn, N. G., Jaffe, H., Winters, C. A., Grant, P., et al. (1998) Mitogen-activated protein kinases (Erk1,2) phosphorylate Lys-Ser-Pro (KSP) repeats in neurofilament proteins NF-H and NF-M. *J. Neurosci.* **18**, 4008–4021.

34. Li, B. S., Veeranna, Gu, J. G., Grant, P., and Pant, H. C. (1999) Activation of mitogen-activated protein kinases (Erk1 and Erk2) cascade results in phosphorylation of NF-M tail domains in transfected NIH 3T3 cells. *Eur. J. Biochem.* **262**, 211–217.

35. Zhang, P., Tu, P.-H., Abtahian, F., Trojanowski, J. Q., and Lee, V. M.-Y. (1997) Neurofilaments and orthograde transport are reduced in ventral root axons of transgenic mice that express human SOD1 with a G93A mutation. *J. Cell Biol.* **139**, 1307–1315.

36. Williamson, T. L. and Cleveland, D. W. (1999) Slowing of axonal transport is a very early event in the toxicity of ALS-linked SOD1 mutants to motor neurons. *Nat. Neurosci.* **2**, 50–56.

37. Watson, D. F., Hoffmann, P. N., Fittro, K. P., and Griffin, J. W. (1989) Neurofilament and tubulin transport slows along the course of mature motor axons. *Brain Res.* **477**, 225–232.

38. Watson, D. F., Fittro, K. P., Hoffman, P. N., and Griffin, J. W. (1991) Phosphorylation-related immunoreactivity and the rate of transport of neurofilaments in chronic 2,5-hexanedione intoxication. *Brain Res.* **539**, 103–109.

39. Archer, D. R., Watson, D. F., and Griffin, J. W. (1994) Phosphorylation-dependent immunoreactivity of neurofilaments and the rate of slow axonal transport in the central and peripheral axons of the rat dorsal root ganglia. *J. Neurochem.* **62**, 1119–1125.

40. Nixon, R. A., Paskevich, P. A., Sihag, R., and Thayer, C. (1994) Phosphorylation on carboxy terminus domains of neurofilament proteins in retinal ganglion cell neurons in vivo: influences on regional neurofilament spacing, and axon caliber. *J. Cell Biol.* **126**, 1031–1046.

41. Nixon, R. A., Lewis, S. E., Mercken, M., and Sihag, R. K. (1994) Orthophosphate and methionine label separate pools of neurofilaments with markedly different axonal transport kinetics. *Neurochem. Res.* **19**, 1445–1453.

42. Jung, C. W. and Shea, T. B. (1999) Regulation of neurofilament axonal transport by phosphorylation in optic axons in situ. *Cell Motil. Cytoskeleton* **42**, 230–240.

43. Sanchez, I., Hassinger, L., Sihag, R. K., Cleveland, D. W., Mohan, P., and Nixon, R. A. (2000) Local control of neurofilament accumulation during radial growth of myelinating axons in vivo. Selective role of site-specific phosphorylation. *J. Cell Biol.* **151**, 1013–1024.

44. Wiedau-Pazos, M., Goto, J. J., Rabizadeh, S., Gralla, E. D., Roe, J. A., Valentine, J. S., et al. (1996) Altered reactivity of superoxide dismutase in familial amyotrophic lateral sclerosis. *Science* **271**, 515–518.

45. Beckman, J. S., Carson, M., Smith, C. D., and Koppenol, W. H. (1993) ALS, SOD and peroxynitrite. *Nature* **364**, 584.

46. Estevez, A. G., Crow, J. P., Sampson, J. C., Reiter, C., Zhuand, Y., Richardson, G. J., et al. (1999) Induction of nitric oxide-dependent apoptosis in motor neurons by zinc-deficient superoxide dismutase. *Science* **286**, 2498–2500.

47. Verheij, M., Bose, R., Lin, X. H., Yao, B., Jarvis, W. D., Grant, S., et al. (1996) Requirement for ceramide-initiated SAPK/JNK signalling in stress-induced apoptosis. *Nature* **380**, 75–79.

48. Jope, R. S., Zhang, L., and Song, L. (2000) Peroxynitrite modulates the activation of p38 and extracellular regulated kinases in PC12 cells. *Arch. Biochem. Biophys.* **376**, 365–370.

49. Suzuki, T., Oishi, M., Marshak, D. R., Czernik, A. J., Nairn, A. C., and Greengard, P. (1994) Cell cycle-dependent regulation of the phosphorylation and metabolism of the Alzheimer amyloid presursor protein. *EMBO J.* **13,** 1114–1122.
50. Standen, C. L., Brownlees, J., Grierson, A. J., Kesavapany, S., Lau, K. F., McLoughlin, D. M., et al. (2001) Phosphorylation of thr^{668} in the cytoplasmic domain of the Alzheimer's disease amyloid precursor protein by stress-activated protein kinase 1b (Jun N-terminal kinase-3). *J. Neurochem.* **76,** 316–320.
51. Guo, Q., Fu, W. M., Sopher, B. L., Miller, M. W., Ware, C. B., Martin, G. M., and Mattson, M. P. (1999) Increased vulnerability of hippocampal neurons to excitotoxic necrosis in presenilin-1 mutant knock-in mice. *Nat. Med.* **5,** 101–106.

4

Roles of Chemokines and Their Receptors in Neuroinflammation

Andrzej R. Glabinski and Richard M. Ransohoff

1. CHEMOKINES: OVERVIEW

Chemokines are chemoattractant cytokines that stimulate directional migration of inflammatory cell in vitro and in vivo. Because of this, chemokines can be included into a large group of cytokines involved in the pathogenesis of inflammatory processes. All chemokines were identified within the last 20 yr and our knowledge about their roles in biology is rapidly growing. At present, an enormous amount of literature about chemokines and chemokine receptors is published each year.

There are now more than 50 different chemokines described in the literature. They are virtually all 8- to 10-kDa proteins with 20–70% homology in amino acid sequence. Chemokines are divided according to their structure into four main subfamilies: XCL, CCL, CXCL, and CX3CL. The criterion is the presence or absence of intervening amino acids between the first two cysteines near the N-terminus. If these cysteines are adjacent, the chemokine belongs to the CCL subfamily. The presence of one or three separating amino acids assigns the chemokine to the CXCL or CX3CL subfamily. The XCL subfamily possesses only one cysteine at the N-terminus. The CX3CL subfamily comprises only one chemokine: fractalkine. The XCL subfamily includes two chemokines: XCL1 (lymphotactin) and XCL2 (SCM1). The two other chemokine subfamilies, CXCL and CCL, are much larger and can be further subdivided. The CXCL family consists of at least 16 members, CCL is even larger—at least 28 members identified to date. The criterion for further division of the CXCL subfamily is the presence of the ELR motif (glutamate–leucine–arginine) near the N-terminus. This subdivision also has functional significance. Chemokines with the ELR motif attract neutrophils, whereas non-ELR CXC chemokines attract predominantly mononuclear inflammatory cells: monocytes and lymphocytes. CC chemokines can also be subdivided further into monocyte chemoattractant proteins (MCP-1–5) and others (1).

Originally described as chemoattractant factors, chemokines turned out to be involved also in a large diversity of other physiological and pathological processes. They can not only guide leukocytes to inflammatory sites but also activate target cells at sites of injury. ELR-positive CXC chemokines (interleukin [IL]-8, GRO possess angiogenic activity, whereas ELR-negative interferon-inducible protein [IP-10], Mig) are angiostatic. Several chemokines may also induce smooth-muscle proliferation and induce cytokine production in lymphocytes (2).

From: *Neuroinflammation, 2nd Edition: Mechanisms and Management*
Edited by: P. L. Wood © Humana Press Inc., Totowa, NJ

2. CHEMOKINE RECEPTORS

Chemokines influence their target cells through chemokine receptors, which belong to seven transmembrane domain receptors signaling through the G-protein system. The homology between chemokine receptors is between 25% and 80% *(3)*. There are 11 chemokine receptors described so far for CC chemokines (CCR1-11), 6 for CXC chemokines (CXCR1-6), and 1 for each CX3C and C chemokine (XCR1 and CX3CR1, respectively). The interactions between chemokines and their receptors are complex and significant redundancy in this system is observed. It is frequently observed that multiple chemokines can bind to single chemokine receptor and that several chemokine receptors can respond to an individual chemokine ligand. Several chemokine receptors provide exceptions to this rule and are "monogamous." In general, chemokine receptors do not respond to ligands from distinct subfamilies. These properties resulted in the division of chemokine receptors into four main functional groups: shared, specific, promiscuous, and viral *(4)*. The examples of specific receptors are CXCR5 (ligand BCA-1), CCR6 (ligand MIP-3a), and CCR9 (ligand TECK). The cardinal example of a promiscuous chemokine receptor is Duffy antigen receptor for chemokines (DARC), which is expressed mainly on erythrocytes and postcapillary venules. Because it does not signal and is present abundantly on circulating erythrocytes, it was suggested that this receptor may serve as a "sink for chemokines," eliminating excess and promoting maintenance of chemokine gradient *(5)*. The last group of chemokine receptors are virus-encoded receptors. Their biological role(s) is not known. The most plausible hypothesis is that viruses "pirated" chemokine receptors to degrade host defenses during infection. In addition to virus-encoded chemokine receptors, some viruses also encode chemokine homologs and chemokine-binding proteins, presumably to achieve the same goals during infection *(6)*.

Chemokine receptors are present mainly on blood inflammatory cells (leukocytes). Some of them may be constitutively expressed like CCR2 on monocytes; others (like CCR5 on lymphocytes) have to be upregulated by inflammatory stimuli (e.g., IL-2) *(7)*. It was also recently shown that the division of acivated T-helper lymphocytes (Th) cells into pro-inflammatory Th1 and anti-inflammatory Th2 cells is also reflected by the expression of a different spectrum of chemokine receptors. Th1 cells express mainly CCR1, CCR5, and CXCR3 receptors, whereas Th2 predominantly CCR3, CCR4, and CCR8 *(8)*. Our knowledge about chemokine receptors increased significantly after the discovery that the human immunodeficiency virus (HIV) uses CCR5, CXCR4, and many other chemokine receptors as coreceptors for invasion of T-cells and monocytes.

Chemokine binding to chemokine receptors initiates upregulation of inositol triphosphate and intracellular calcium flux. Moreover, the activation of Ras and Rho families is induced. The Rho family plays a role in the formation of pseudopods involved in directional migration of inflammatory cells *(9)*.

3. CHEMOKINES AND LEUKOCYTE EXTRAVASATION

Chemokines are produced at tissue sites of inflammation by parenchymal cells that thereby induce the migration of inflammatory cells from the blood. Moreover, chemokines are produced by migrating leukocytes, thus augmenting the inflammatory process. The extravasation of leukocytes and their accumulation in an inflamed region is a complicated and multistep process. It is initiated by adhesion molecules (i.e., selectins expressed on endothelium and leukocytes that interact with their carbohydrate receptors). This interaction causes leukocyte "rolling" on endothelium. At this stage, another group of adhesion molecules, integrins, initiate firm attachement of leukocytes to endothelium and their exposure to chemokine gradient. This signal stimulates transmigration of inflammatory cells from the vessel lumen to inflamed tissue. Chemokines play an important role during the extravasation step of this process, but they are also required to activate integrins and initiate leukocyte arrest and, in this way, accelerate the process of transmigration. Expression of specific sets of chemokines in an inflammatory region is responsible for the cellular composition of inflammatory foci.

The inflammatory process in the central nervous system (CNS) has unique features not seen in the periphery. The most important difference is the presence of the blood-brain barrier (BBB), which is composed of the nonfenestrated cerebrovascular endothelium sealed by tight junctions. Inflammatory cells migrating to the CNS must first penetrate the BBB and accumulate in perivascular/subarachnoid space *(10)*. Under physiological conditions, only activated T-cells penetrate this barrier during the process of immunological surveillance of the CNS *(11)*. Inflammation of immunological origin starts when patrolling T-cells encounter cognate antigen within the CNS parivascular space *(12)*. As a result, proinflammatory cytokines are produced by both T-cells and perivascular macrophages, stimulating CNS parenchymal cells to express chemokines (*see* Section 4). Astrocyte-derived chemokines may influence the BBB endothelium and attract antigen-nonspecific inflammatory cells to the nascent site of inflammation *(13)*. Inflammatory responses in the CNS also result from diverse other types of injury, including infection and mechanical, physical, chemical, and ischemic damage. Regardless of its origin, this response is usually characterized by chemokine overexpression (*see* Section 5).

4. EXPRESSION OF CHEMOKINES BY CNS CELLS IN VITRO

Although inflammatory leukocytes are the principal producers of chemokines and bearers of their receptors, cells of neural origin are also able to express chemokines and chemokine receptors. Initial studies showed that human glioma cell lines produce MCP-1 and IL-8 *(14,15)*. Cultured astrocytes stimulated with tumor necrosis factor-α (TNF-α) and transforming growth factor (TGF-β) express MCP-1 at both mRNA and protein levels *(16)* and astrocytoma cells stimulated with interferon-γ (IFN-γ) produce MCP-1 *(17)*. Stimulated astrocytes are also able to express monocyte inflammatory protein (MIP)-1α, MIP-1β, RANTES, and IP-10 *(18–20)*. Infection of cultured astrocytes with paramyxovirus NDV stimulates expression of IP-10 and RANTES *(21)*, and HIV-1 infection stimulates expression of IL-8 and IP-10 in affected astrocytes *(22)*. Infection of cultured human astrocytes with neurotropic coronavirus OC43 leads to increased expression of cytokines IL-6 and TNF-α, as well as chemokine MCP-1 *(23)*.

Microglial cells (especially after stimulation) have been also shown to be potent sources of some chemokines. After stimulation with IL-6 and colony-stimulating factor-1 (CSF-1) brain macrophages express MCP-1 *(24)*. Other inflammatory cytokines like TNF-α and IL-1β and lipopolysaccharide (LPS) may stimulate cultured microglia to produce MCP-1, MIP-1α, and MIP-1β *(25)*, IL-8 *(26)*, and RANTES *(27)*. It has been also shown that some infectious agents may stimulate overexpression of chemokines by microglia. For example, the simian immunodeficiency (SIV) virus *(28)* and cryptococcal polysaccharide *(29)* can induce expression of IL-8 in cultured microglia.

Cultured brain endothelial cells can express MCP-1 spontaneously and this expression increases after stimulation with TNF-α *(30)*. It has been recently shown that cultured human cerebromicrovascular endothelial cells are able to express genes for MCP-1 and IL-8 when stimulated by hypoxic astrocytes, mediated by IL-1β *(31)*. Brain microvascular endothelial cells may express CXCR2 as well *(32)*. Parasitic infection of cultured brain endothelium stimulates expression of IL-8 *(33)*. Mixed human brain cell cultures stimulated with TNF-α expressed RANTES and MIP-1β *(34)*.

Chemokine receptors have also been found to be expressed on CNS cells in vitro and in vivo. Numerous studies showed that cultured astrocytes can express CXCR4, CX3CR1, CCR1, CCR10, and CCR11 *(35–37)*. In our studies, TGF-β1, but not IFN-γ, TNF-α, and LPS was able to stimulate primary mouse astrocytes to upregulate selectively CCR1 in vitro *(38)*. In the same model, TNF-α was a potent supressor of CXCR4 expression at the mRNA and protein levels *(39)*.

Microglia can express CCR3, CCR5, CXCR4, and CX3CR1 in vitro *(36,40,41)*. Cultured human neurons were shown to express CCR1, CCR5, CXCR2, and CXCR4 *(42)*. Another group showed that cultured human fetal neurons and the human neuronal cell line NT2.N can express CCR2, CXCR2, CXCR3 and CXCR4 at the mRNA and protein levels. Additionally, it has been shown in those studies that NT2.N neurons may produce chemokine MCP-1 *(43)*.

5. GENETICALLY PROGRAMMED OVEREXPRESSION OF CHEMOKINES IN CNS IN VIVO

Studies on chemokine expression under the control of CNS-specific promoters showed accumulation of appropriate subsets of leukocytes in this organ. These observations indicate that chemokines are potent inducers of selective recruitment of leukocyte subpopulations from the blood to the CNS in vivo. Transgenic mice expressing MCP-1 under control of the oligodendrocyte-specific MBP promoter exhibited selective monocyte accumulation in CNS perivascular spaces *(44)*. Intraperitoneal injection with LPS augmented this accumulation. Despite massive inflammatory infiltrates, transgenic mice did not show any evident neurological and behavioral deficits *(44)*. In another study, transgenic mice expressing chemokine KC in oligodendrocytes massive accumulation of neutrophils was found. The peak of this expression was observed between 2 and 3 wk of age in perivascular, meningeal, and parenchymal sites of CNS tissue *(45)*. Those mice developed delayed (beginning from 40 d of age) neurological syndrome of postural instability and rigidity. Neuropathological analysis showed BBB disruption and microglial activation *(45)*.

Those observations suggested that chemokines may selectively recruit specific subpopulations of inflammatory cells to the CNS and that this chemokine-driven recruitment is not invariably linked to leukocyte activation.

6. CHEMOKINES IN INFECTIOUS NEUROINFLAMMATION

Chemokines are important players in the formation of the inflammatory response of an organism to infectious challenge. Although neuroinflammation is characterized by certain unique features (when compared to inflammation in other organs), other essential characteristics of inflammatory responses are conserved. One conserved mechanism is localized production of chemoattractant agents at sites of inflammation. In an early study analyzing chemokine involvement in the pathogenesis of experimental pneumococcal meningitis, Saukkonen and co-workers observed that intracisternal administration of antibodies blocking MIP-1α, MIP-1β, and MIP-2 during induction of the diseases delayed the onset of inflammation *(46)*. In brains from mice with encephalomyelitis induced by *Listeria*, expression of genes for MIP-1α, MIP-1β, and MIP-2 was also detected *(47)*. This expression was mainly localized in neutrophils accumulating in lateral and third ventricles starting from 12 h after disease induction. The highest level of MIP-1α and MIP-2 in corresponding cerebrospinal fluid (CSF) was found by enzyme-linked immunosorbent assay (ELISA) at 48–72 h postinfection *(47)*.

Many studies analyzed the level of chemokines in the CSF from patients with meningitis of different origins. It has been reported that there is a correlation between IL-8 and GRO-α levels and granulocyte counts in the CSF from patients with bacterial meningitis and between MCP-1 CSF levels and mononuclear cell counts in CSF from patients with nonbacterial meningitis *(48)*. Others observed a correlation between IL-8 concentration in CSF and neutrophil counts in patients with nonpyogenic meningitis *(49)*. In patients with pneumococcal, meningococcal, and *Haemophilus influenzae* bacterial meningitis, MCP-1, IL-8, and GRO-α, as well as also levels of MIP-1α and MIP-1β were elevated in the CSF *(50)*. This observation was extended by studies showing diminished migration of neutrophils after addition to the CSF of anti-IL-8 and anti-GRO-α antibodies. Migration of mononuclear cells was reduced in the same system by anti-MIP-1α, anti-MIP-1β, and anti-MCP antibodies *(50)*. A recent study analyzing the development of experimental brain abscess after embolization of *Staphalococcus aureus* beads showed involvement of chemokines in that process. Increased expression of neutrophil chemoattractant KC was detected 24 h after infection, whereas MCP-1 and MIP-1α were overexpressed in the brain within 24 h after bacterial exposure *(51)*.

Viral infections of the CNS parenchyma were also shown to be connected with increased chemokine levels in the CSF. In encephalitis caused by SIV, a primate model of human AIDS encephalitis, increased expression of MIP-1α, MIP-1β, MCP-1, MCP-3, RANTES, and IP-10 was detected *(52)*. Lymphocytic choriomeningitis infection led to increased expression of genes for MCP-1, MIP-1β, RANTES, IP-10, and MCP-3 in the brains of infected mice by 3 d after infection. A later increased expression of C-10, MIP-2, MIP-1α and lymphotactin was observed *(53)*. In encephalomyelitis caused by mouse hepatitis virus (MHV), increased expression of MIP-1α, MIP-2, IP-10, MCP-1, and RANTES was detected in the infected brain and spinal cord. Astrocytes expressed IP-10 in that model *(54)*. It has been proposed recently that Mig contributes significantly to the clearance of MHV CNS infection, as mice treated with anti-Mig antisera had much more severe disease. In the treated group, accumulation of CD4+ and CD8+ T-cells and expression of IFN-γ in the brain were significantly decreased *(55)*.

Increased expression of MIP-1α, MIP-1β, IP-10, MCP-1, and RANTES was present in murine brain during fatal hemorrhagic encephalopathy induced by infection with mouse adenovirus type-1 (MAV-1) *(56)*. The same infection caused increased upregulation of chemokine receptors CCR1-5 in BALB/c and C57BL/6 mice *(57)*. In dogs infected by canine distemper virus (CDV), an increased level of IL-8 was observed in the CSF *(58)*. During meningoencephalitis induced by infection with the Borna disease virus (BDV), astrocytes were shown to express mRNA for IP-10 *(59)*. Theiler's virus model of multiple sclerosis (MS) was characterized by biphasic overexpression of chemokines IP-10, MCP-1, and RANTES, first during the acute inflammatory stage of the disease and, later, during demyelinating stage of infection *(60)*.

In brains from patients with HIV encephalitis, MCP-1, MIP-1α, and RANTES were detected on macrophages *(61)*. Moreover, in brains from patients with HIV-associated dementia, in which monocytic infiltration of the CNS is present, increased levels of MCP-1 were detected in brains and CSF *(62)*. The human T-cell lymphotrophic virus type-1 (HTLV-1) virus causes chronic progressive myelopathy with neurological and pathological features similar to progressive MS. Expression of MCP-1 has been observed in spinal cord lesions present in that disease *(63)*. Moreover, HTLV-1 -specific T-cell clones from patients with this myelopathy may express chemokines MIP-1α and MIP-1β *(64)*.

Additional information showing involvement of chemokines in infectious neuro-inflammation was obtained by studying mice with disrupted chemokine and chemokine receptor genes. MIP-1α knockout mice infected with neurotropic fungus *Cryptococcus neoformans* had decreased leukocyte recruitment to the brain and impaired cryptococcal clearance from the brain *(65)*. In the CSF from patients with *Cryptococcal* meningitis, an increased level of IL-8 was found *(29)*.

7. CHEMOKINES IN IMMUNE-MEDIATED NEUROINFLAMMATION

The best characterized experimental neuroinflammation of immunological origin is experimental autoimmune encephalomyelitis (EAE). This disease can be induced in susceptible strains of laboratory animals (mice, rats, guinea pigs, monkeys) by immunization with CNS myelin protein antigens like myelin basic protein (MBP), proteolipid protein (PLP), and myelin oligodendrocyte glycoprotein (MOG). EAE is an example of autoimmune inflammation of the CNS and is characterized by the presence of disseminated inflammatory "cuffs" around microvessels of the brain and spinal cord. EAE is considered to be a useful animal model of certain aspects of the human demyelinating disease MS. Both diseases share similar pathological features, although the autoimmune origin of MS has not been rigorously proven yet.

Initial descriptive reports showed that expression of some chemokines (MCP-1, IP-10) occurs during early stages of EAE *(66,67)* and those chemokines are expressed by astrocytes in the vicinity of inflammatory lesions *(67)*. Later, the expression of additional chemokines (MIP-1α, MIP-1β, RANTES, KC, MCP-3, TCA-3, fractalkine, MCP-5) was detected during acute EAE *(68–70)*. Our results showed that chemokine expression by parenchymal astrocytes during EAE do not initiate, but amplify, the ongoing CNS inflammatory process *(13)*. A correlation among MIP-1α, RANTES, and GRO-α expression and intensity of CNS inflammation was also reported *(71)*. A functional study found that the blockade of MIP-1α expression by antibody prevented the appearance of passively transferred acute EAE *(72)*. However, mice with knockout of MIP-1α and its receptor CCR5 were fully susceptible to MOG-induced EAE, showing the complexity of this subject *(73)*. Lately, in a variant of EAE in BALB/c mice characterized by pronounced neutrophil accumulation, increased expression of MIP-2 (chemokine-attracting neutrophils) as well as MIP-1α and MCP-1 was reported. Astrocytes were the main cellular source of MIP-2 and MIP-1α; infiltrating neutrophils expressed MIP-1α and MCP-1 *(74)*.

In chronic relapsing EAE (ChREAE), characterized by the spontaneous appearance of clinical relapses of the disease, we observed overexpression of chemokines MCP-1, IP10, GRO-α, MIP-1α, and RANTES concomittant with relapse. Astrocytes were the producers of MCP-1, IP-10, and GRO-α, whereas infiltrating inflammatory cells expressed RANTES and MIP-1α *(75)*. Anti-MCP-1, but not anti-MIP-1α, antibody was shown to significantly reduce the severity of relapses in ChREAE *(76)*. Recently, mice lacking the MCP-1 gene were shown to be markedly resistant to EAE induction and showed impaired recruitment of macrophages to the CNS *(77)*. The expression of chemokines within the CNS parenchyma during EAE is probably driven by inflammatory cytokines produced by migrating inflammatory cells as supported by observations in IFN-γ knockout mice (GKO) with EAE. In that model, we observed selectively diminished IP-10 expression, whereas non-IFN-γ-dependent chemokines, MCP-1 and GRO-α, were overexpressed *(78)*.

In addition to chemokines, CNS expression of chemokine receptors CXCR2, CXCR3, CXCR4, CX3CR1, CCR2, CCR5, and CCR8 was increased during EAE *(41,79)*. In ChREAE, the expression of CXCR2 and CXCR4 correlated with the appearance of new relapses *(80)*. Protection from EAE induced with altered peptide ligand (APL) was shown to reduce levels of several chemokines and also CXCR2, CXCR3, CCR1, CCR5, and CCR8 *(79)*. CCR6 and its ligand MIP-3a, a potent attractant of dendritic cells (DCs), were both upregulated in the CNS during EAE. In those studies, a prominent infiltration of mature DCs in the spinal cord of mice with acute and chronic EAE was described *(81)*. Two recent publications showed that CCR2 plays a necessary role in the pathogenesis of EAE. Mice with CCR2 knockout did not develop clinical EAE and failed to accumulate mononuclear inflammatory cells in the CNS. Moreover, they failed to upregulate RANTES, MCP-1, IP-10, CCR1, CCR2, and CCR5 during EAE *(82,83)*.

Many reports have been published lately addressing chemokine expression in MS. Reassuringly, these results resemble those obtained earlier in EAE. Hvas and co-workers detected RANTES by *in situ* hybridization in perivascular inflammatory cells *(84)*. In another study, MCP-1 was localized in astrocytes and inflammatory cells and MIP-1α and MIP-1β in inflammatory cells in active MS plaques *(85)*. Reactive astrocytes and inflammatory cells were also shown to express MCP-1, MCP-2, and MCP-3 in active MS lesions by others *(86)*. Recently, our group reported increased levels of IP-10, Mig, and RANTES in the CSF from patients with MS relapse *(87)*. The receptor for IP-10 and Mig (CXCR3) was detected on CSF cells, as well as lymphocytes in perivascular inflammatory cuffs; the receptor for RANTES (CCR5) was present on lymphocytes, macrophages, and microglial cells in active MS lesions *(87)*. Compatible results were published by others *(88)*. Analysis of chemotactic activity of T-cells from MS patients showed increased migratory rate toward chemokines RANTES and MIP-1α. This aberrant migration could be diminished by anti-CCR5 antibodies *(89)*. Moreover CCR2 and, to a smaller extent, CCR3 were detected in MS brains on CNS-infiltrating lymphocytes as well as on macrophages and microglia. Ligands for those receptors, MCP-1 and MCP-3, were localized around inflammatory foci *(85)*. The same group reported Mig, IP-10, and CXCR3 expression in actively demyelinating lesions by macrophages and reactive astrocytes in periplaque CNS tissue *(90)*. Treatment of remitting–relapsing MS with IFN-β reduced RANTES production in sera- and blood-adherent mononuclear cells both in relapse and in remission *(91)*.

8. CHEMOKINES IN TRAUMA TO THE NERVOUS SYSTEM

Several types of physical CNS injury have been shown to be followed by increased expression of chemokines. MCP-1 overexpression was observed after mechanical penetrating injury to the brain *(92,93)*. Astrocytes in the vicinity of the injury site expressed MCP-1 as early as a few hours after trauma *(93)*. Expression of other chemokines studied in that model was not increased. We observed a strict correlation between MCP-1 expression and the intensity of the inflammatory reaction in the brain. Out of four different injury models studied, the paradigm with the lowest intensity of inflammatory reaction (neonatal stab injury model) was typified by the lowest MCP-1 expression *(93)*. Other investigators showed increased expression of RANTES and MIP-1β 24 h after stab injury to the brain *(94)*. Immunohistochemistry localized MIP-1β in reactive astrocytes and macrophages at the site of injury, whereas RANTES was diffusely expressed in surrounding necrotic tissue *(94)*. Augmentation of mechanical cortical injury with LPS led to increased expression of several chemokines: MCP-1, MIP-1α, MIP-1β, RANTES, IP-10, and KC *(95)*. Antisense oligodeoxynucleotides that suppress MCP-1 protein expression diminished the accumulation of macrophages at the site of stab injury to the rat brain *(96)*. In a model of mechanical injury to the spinal cord, the expression of MIP-1α and MIP-1β was observed 1 d after injury diffusely in gray matter, later being present in inflammatory cells at the site of injury *(97)*. In a precisely calibrated contusion injury to the spinal cord, increased expression of other chemokines was also reported. MCP-1 >>> MCP-5 = GRO-α = IP-10 = MIP-3α were expressed within hours after injury and pre-ceded influx of inflammatory cells to the site of injury *(98)*.

Another type of injury, cryolesion of the cerebral cortex, induced increased expression of MCP-1, with a peak at 6 h after trauma. Another chemokine analyzed in that model, IP-10, was not overexpressed *(99)*. During chemical injury to the CNS induced by triethyltin (TET) overexpression of MIP-1α was detected *(100)*. It has been reported recently that MCP-1 may be an important mediator of acute excitotoxic injury induced by *N*-methyl-D-aspartate in the neonatal rat brain *(101)*. In the CSF from patients with severe brain trauma, the IL-8 concentration was significantly elevated. There was a clear correlation between the CSF IL-8 level and BBB disruption measured by the CSF/serum albumin ratio *(102)*.

9. CHEMOKINES IN ISCHEMIC INJURY TO THE NERVOUS SYSTEM

Several studies have demonstrated that one consequence of CNS ischemia is increased expression of chemokines. Early experiments reported increased expression of MCP-1 and MIP-1α 6 h after onset of brain ischemia *(103)*. In that study, endothelial cells and macrophages expressed MCP-1, whereas MIP-1α was described in astrocytic cells *(103)*. In another study, experimental middle cerebral artery occlusion (MCAO) also induced increased expression of MCP-1 beginning 6 h after injury *(104)*. Astrocytes provided the main cellular source of MCP-1 after MCAO; later (after 4 d) MCP-1 was expressed predominantly by macrophages and microglia at the ischemic area *(105)*. Overexpression of MIP-1α was detected in microglia localized in injured brain region after 4–6 h of ischemia *(106)*. In the rat model of MCAO, CXC chemokine cytokine-induced neutrophil chemoattractant (CINC) was overexpressed after 12 h of ischemia *(107)*. In the same model, increased expression of CXCR3 was observed and correlated with leukocyte accumulation after focal brain ischemia *(108)*. In a neonatal model of brain hypoxia–ischemia, the peak of MCP-1 expression was detected at 8–24 h after the onset of ischemia and this expression returned to basal levels by 48 h *(109)*.

Pronounced reperfusion after focal brain ischemia may lead to additional brain damage and is usually a result of the accumulation of neutrophils. Chemokines attracting neutrophils like IL-8 and CINC were shown to be upregulated after brain reperfusion *(110)*. Blocking IL-8 with the antibody significantly reduced the size of infarcted brain tissue *(110)*. In a rat forebrain reperfusion injury model, MCP-1 expression was detected at the transcript level as early as 1 h after reperfusion *(111)*.

In patients with subarachnoid hemorrhage, MCP-1 and IL-8 levels in the CSF were significantly increased compared with patients with unruptured aneurysms *(112)*.

10. CHEMOKINES IN NEURODEGENERATION

In the course of neurodegenerative disorders, the BBB remains intact and migration of inflammatory cells from the blood to the CNS is not observed. Therefore, inflammatory cells detected during chronic neurodegenerative pathologies are CNS macrophages/ microglia. Chemokine and chemokine receptor expression was extensively studied in Alzheimer's disease (AD), the most common nerodegenerative disease causing dementia. It has been shown that CXCR2 immunostaining is present in senile plaques and correlates with APP expression *(113,114)*. Additionally, increased expression of CCR3 and CCR5 in reactive microglia of AD brains was reported. In that study, MIP-1β was detected in AD predominantly in reactive astrocytes *(115)*. The same group reported overexpression of IP-10 and its receptor CXCR3 in astrocytes in AD brains *(116)*. Other authors observed increased expression of MCP-1 in mature senile plaques and reactive microglia from patients with AD *(117)*. Additional information from in vitro studies confirms the possible involvement of chemokines and their receptors in AD pathogenesis. β-Amyloid was shown to stimulate cultured brain macrophages to produce MCP-1 *(118)* and astrocytoma cells for production of IL-8 *(119)*. Interestingly, RANTES was shown to be neuroprotective when added to neuronal cultures exposed to toxic fragment of β-amyloid peptide *(120)*.

In an experimental model of thalamic retrograde neurodegeneration induced by damage to the cerebral cortex, rapid overexpression of MCP-1 in a thalamus ipsilateral to injury was observed. This expression was localized by *in situ* hybridization to glial cells of the lateral geniculate nucleus *(121)*.

11. CHEMOKINES IN PERIPHERAL NERVOUS SYSTEM PATHOLOGY

After axotomy in the peripheral nervous system, macrophages accumulate at the site of nerve transection. It has been hypothesized that this inflammatory reaction is the principal factor promoting regeneration of injured periperal nerve. It has been reported that expression of MCP-1 preceded recruitment of macrophages to the injury region and was localized by *in situ* hybridization in Schwann cells *(122)*. In a recent study analyzing chemokine expression in the experimental lesion of facial and hypoglossal nerves, MCP-1 was expressed by damaged neurons. Expression of RANTES and IP-10 as well as the MCP-1 receptor CCR2 was not elevated *(123)*. In a model of peripheral Wallerian degeneration induced in CCR2 knockout mice, a macrophage invasion after sciatic nerve transection was significantly impaired. In sciatic axotomy of CCR5-deficient mice, this finding was not observed *(124)*.

An animal model of human Guillain–Barre polyneuropathy, experimental autoimmune neuritis (EAN) is characterized by the presence of mononuclear inflammatory cells (lymphocytes and macrophages) in affected nerves. It has been reported that MCP-1 expression increases shortly before clinical signs of this disease *(125)*. In another study in that model, the peak of MIP-1α and MIP-1β expression preceded maximum disease severity, whereas the maximum expression of MCP-1, RANTES, and IP-10 was present at the time of peak of the disease. RANTES expression was localized within invading lymphocytes, and IP-10 was detected mainly in perineurial endothelium *(126)*. In trigeminal ganglia from mice infected at least 5 d earlier with herpes simplex virus type I, increased expression of RANTES was found *(127)*.

12. CONCLUSIONS

The data described in this review show that chemokines play important roles in diverse nervous system pathologies. Originally described as chemotactic agents, chemokines have recently been found to be crucial factors in damage to nervous tissue, as key mediators of inflammatory responses. Chemokine involvement has been reported in neuroinfections, autoimmune pathologies, neurotrauma after mechanical, physical, or chemical injury, and ischemia. Rapidly accumulating information about the extra-inflammatory properties of chemokines has also impacted the neurosciences. Currently, chemokines and their receptors are considered to be important factors in neurodegenerative processes, nervous tissue development, and neuron–glia communication. One may expect that, in the near future, our knowledge in this area will expand further. However, even at present, chemokines can be considered important targets for new therapies, especially in neuroinflammatory conditions.

REFERENCES

1. Rossi, D. and Zlotnik, A. (2000) The biology of chemokines and their receptors. *Ann. Rev. Immunol.* **18,** 217–242.
2. Lukacs, N. W., Hogaboam, C., Campbell, E., and Kunkel, S. L. (1999) Chemokines: function, regulation and alteration of inflammatory responses. *Chem. Immunol.* **72,** 102–120.
3. Kelvin, D. J., Michiel, D. F., Johnston, J. A., Lloyd, A. R., Sprenger, H., Oppenheim, J. J., et al. (1993) Chemokines and serpentines: the molecular biology of chemokine receptors. *J. Leukocyte Biol.* **54,** 604–612.
4. Premack, B. A. and Schall, T. J. (1996) Chemokine receptors: gateways to inflammation and infection. *Nat. Med.* **2,** 1174–1178.
5. Murphy, P. M., Baggiolini, M., Charo, I. F., Hebert, C. A., Horuk, R., Matsushima, K., et al. (2000) International union of pharmacology. XXII. Nomenclature for chemokine receptors. *Pharmacol. Rev.* **52,** 145–176.
6. Lalani, A. S., Barrett, J. W., and McFadden, G. (2000) Modulating chemokines: more lessons from viruses. *Immunol. Today* **21,** 100–106.
7. Alkhatib, G., Combadiere, C., Broder, C. C., Feng, Y., Kennedy, P. E., Murphy, P. M., et al. (1996) CC CKR5: a RANTES, MIP-1alpha, MIP-1beta receptor as a fusion cofactor for macrophage-tropic HIV-1. *Science* **272,** 1955–1958.
8. Sallusto, F., Lanzavecchia, A., and Mackay, C. R. (1998) Chemokines and chemokine receptors in T-cell priming and Th1/Th2-mediated responses. *Immunol. Today* **19,** 568–574.

9. Laudanna, C., Campbell, J. J., and Butcher, E. C. (1996) Role of Rho in chemoattractant-activated leukocyte adhesion through integrins. *Science* **271,** 981–983.

10. Glabinski, A. R. and Ransohoff, R. M. (1999) Sentries at the gate: chemokines and the blood-brain barrier. *J. Neurovirol.* **5,** 623–634.

11. Wekerle, H., Engelhardt, B., Risau, W., and Meyermann, R. (1991) Interaction of T lymphocytes with cerebral endothelial cells in vitro. *Brain Pathol.* **1,** 107–114.

12. Hickey, W. F. (1991) Migration of hematogenous cells through the blood-brain barrier and the initiation of CNS inflammation. *Brain Pathol.* **1,** 97–105.

13. Glabinski, A. R., Tani, M., Tuohy, V. K., Tuthill, R. J., and Ransohoff, R. M. (1995) Central nervous system chemokine mRNA accumulation follows initial leukocyte entry at the onset of acute murine experimental autoimmune encephalomyelitis. *Brain Behav. Immun.* **9,** 315–330.

14. Yoshimura, T., Robinson, E. A., Tanaka, S., Appella, E., Kuratsu, J., and Leonard, E. J. (1989) Purification and amino acid analysis of two human glioma-derived monocyte chemoattractants. *J. Exp. Med.* **169,** 1449–1459.

15. Morita, M., Kasahara, T., Mukaida, N., Matsushima, K., Nagashima, T., Nishizawa, M., et al. (1993) Induction and regulation of IL-8 and MCAF production in human brain tumor cell lines and brain tumor tissues. *Eur. Cytokine Network* **4,** 351–358.

16. Hurwitz, A. A., Lyman, W. D., and Berman, J. W. (1995) Tumor necrosis factor alpha and transforming growth factor beta upregulate astrocyte expression of monocyte chemoattractant protein-1. *J. Neuroimmunol.* **57,** 193–198.

17. Zhou, Z. H., Chaturvedi, P., Han, Y. L., Aras, S., Li, Y. S., Kolattukudy, P. E., et al. (1998) IFN-gamma induction of the human monocyte chemoattractant protein (hMCP)-1 gene in astrocytoma cells: functional interaction between an IFN-gamma-activated site and a GC-rich element. *J. Immunol.* **160,** 3908–3916.

18. Barna, B. P., Pettay, J., Barnett, G. H., Zhou, P., Iwasaki, K., and Estes, M. L. (1994) Regulation of monocyte chemoattractant protein-1 expression in adult human non-neoplastic astrocytes is sensitive to tumor necrosis factor (TNF) or antibody to the 55-kDa TNF receptor. *J. Neuroimmunol.* **50,** 101–107.

19. Majumder, S., Zhou, L. Z., and Ransohoff, R. M. (1996) Transcriptional regulation of chemokine gene expression in astrocytes. *J. Neurosci. Res.* **45,** 758–769.

20. Peterson, P. K., Hu, S., Salak-Johnson, J., Molitor, T. W., and Chao, C. C. (1997) Differential production of and migratory response to beta chemokines by human microglia and astrocytes. *J. Infect. Dis.* **175,** 478–481.

21. Fisher, S. N., Vanguri, P., Shin, H. S., and Shin, M. L. (1995) Regulatory mechanisms of MuRantes and CRG-2 chemokine gene induction in central nervous system glial cells by virus. *Brain Behav. Immun.* **9,** 331–344.

22. Kutsch, O., Oh, J., Nath, A., and Benveniste, E. N. (2000) Induction of the chemokines interleukin-8 and IP-10 by human immunodeficiency virus type 1 tat in astrocytes. *J. Virol.* **74,** 9214–9221.

23. Edwards, J. A., Denis, F., and Talbot, P. J. (2000) Activation of glial cells by human coronavirus OC43 infection. *J. Neuroimmunol.* **108,** 73–81.

24. Calvo, C. F., Yoshimura, T., Gelman, M., and Mallat, M. (1996) Production of monocyte chemotactic protein-1 by rat brain macrophages. *Eur. J. Neurosci.* **8,** 1725–1734.

25. McManus, C. M., Brosnan, C. F., and Berman, J. W. (1998) Cytokine induction of MIP-1 alpha and MIP-1 beta in human fetal microglia. *J. Immunol.* **160,** 1449–1455.

26. Ehrlich, L. C., Hu, S., Sheng, W. S., Sutton, R. L., Rockswold, G. L., Peterson, P. K., et al. (1998) Cytokine regulation of human microglial cell IL-8 production. *J. Immunol.* **160,** 1944–1948.

27. Kitai, R., Zhao, M., Zhang, N., Hua, L. L., and Lee, S. C. (2000) Role of MIP-1beta and RANTES in HIV-1 infection of microglia: inhibition of infection and induction by IFNbeta. *J. Neuroimmunol.* **110,** 230–239.

28. Sopper, S., Demuth, M., Stahl-Hennig, C., Hunsmann, G., Plesker, R., Coulibaly, C., et al. (1996) The effect of simian immunodeficiency virus infection in vitro and in vivo on the cytokine production of isolated microglia and peripheral macrophages from rhesus monkey. *Virology* **220,** 320–329.

29. Lipovsky, M. M., Gekker, G., Hu, S., Ehrlich, L. C., Hoepelman, A. I., and Peterson, P. K. (1998) Cryptococcal glucuronoxylomannan induces interleukin (IL)-8 production by human microglia but inhibits neutrophil migration toward IL-8. *J. Infect. Dis.* **177,** 260–263.

30. Zach, O., Bauer, H. C., Richter, K., Webersinke, G., Tontsch, S., and Bauer, H. (1997) Expression of a chemotactic cytokine (MCP-1) in cerebral capillary endothelial cells in vitro. *Endothelium* **5,** 143–153.

31. Zhang, W., Smith, C., Howlett, C., and Stanimirovic, D. (2000) Inflammatory activation of human brain endothelial cells by hypoxic astrocytes in vitro is mediated by IL-1beta. *J. Cereb. Blood Flow Metab.* **20,** 967–978.

32. Otto, V. I., Heinzel-Pleines, U. E., Gloor, S. M., Trentz, O., Kossmann, T., and Morganti-Kossmann, M. C. (2000) sICAM-1 and TNF-alpha induce MIP-2 with distinct kinetics in astrocytes and brain microvascular endothelial cells. *J. Neurosci. Res.* **60,** 733–742.

33. Bourdoulous, S., Bensaid, A., Martinez, D., Sheikboudou, C., Trap, I., Strosberg, A. D., et al. (1995) Infection of bovine brain microvessel endothelial cells with Cowdria ruminantium elicits IL-1 beta, -6, and -8 mRNA production and expression of an unusual MHC class II DQ alpha transcript. *J. Immunol.* **154,** 4032–4038.

34. Lokensgard, J. R., Gekker, G., Ehrlich, L. C., Hu, S., Chao, C. C., and Peterson, P. K. (1997) Proinflammatory cytokines inhibit HIV-1 (SF162) expression in acutely infected human brain cell cultures. *J. Immunol.* **158,** 2449–2455.

35. Heesen, M., Tanabe, S., Berman, M. A., Yoshizawa, I., Luo, Y., Kim, R. J., et al. (1996) Mouse astrocytes respond to the chemokines MCP-1 and KC, but reverse transcriptase–polymerase chain reaction does not detect mRNA for the KC or new MCP-1 receptor. *J. Neurosci. Res.* **45,** 382–391.

36. Tanabe, S., Heesen, M., Berman, M. A., Fischer, M. B., Yoshizawa, I., Luo, Y., et al. (1997) Murine astrocytes express a functional chemokine receptor. *J. Neurosci.* **17,** 6522–6528.

37. Dorf, M. E., Berman, M. A., Tanabe, S., Heesen, M., and Luo, Y. (2000) Astrocytes express functional chemokine receptors. *J. Neuroimmunol.* **111,** 109–121.

38. Han, Y., Wang, J., Zhou, Z., and Ransohoff, R. M. (2000) TGFbeta1 selectively up-regulates CCR1 expression in primary murine astrocytes. *Glia* **30,** 1–10.

39. Han, Y., Wang, J., He, T., and Ransohoff, R. M. (2001) TNF-alpha down-regulates CXCR4 expression in primary murine astrocytes. *Brain Res.* **888,** 1–10.

40. He, J., Chen, Y., Farzan, M., Choe, H., Ohagen, A., Gartner, S., et al. (1997) CCR3 and CCR5 are co-receptors for HIV-1 infection of microglia. *Nature* **385,** 645–649.

41. Jiang, Y., Salafranca, M. N., Adhikari, S., Xia, Y., Feng, L., Sonntag, M. K., et al. (1998) Chemokine receptor expression in cultured glia and rat experimental allergic encephalomyelitis. *J. Neuroimmunol.* **86,** 1–12.

42. Hesselgesser, J., Halks-Miller, M., DelVecchio, V., Peiper, S. C., Hoxie, J., Kolson, D. L., et al. (1997) CD4-independent association between HIV-1 gp120 and CXCR4: functional chemokine receptors are expressed in human neurons. *Curr. Biol.* **7,** 112–121.

43. Coughlan, C. M., McManus, C. M., Sharron, M., Gao, Z., Murphy, D., Jaffer, S., et al. (2000) Expression of multiple functional chemokine receptors and monocyte chemoattractant protein-1 in human neurons. *Neuroscience* **97,** 591–600.

44. Fuentes, M. E., Durham, S. K., Swerdel, M. R., Lewin, A. C., Barton, D. S., Megill, J. R., et al. (1995) Controlled recruitment of monocytes and macrophages to specific organs through transgenic expression of monocyte chemoattractant protein-1. *J. Immunol.* **155,** 5769–5776.

45. Tani, M., Fuentes, M. E., Peterson, J. W., Trapp, B. D., Durham, S. K., Loy, J. K., et al. (1996) Neutrophil infiltration, glial reaction, and neurological disease in transgenic mice expressing the chemokine N51/KC in oligodendrocytes. *J. Clin. Invest.* **98,** 529–539.

46. Saukkonen, K., Sande, S., Cioffe, C., Wolpe, S., Sherry, B., Cerami, A., et al. (1990) The role of cytokines in the generation of inflammation and tissue damage in experimental gram-positive meningitis. *J. Exp. Med.* **171,** 439–448.

47. Seebach, J., Bartholdi, D., Frei, K., Spanaus, K. S., Ferrero, E., Widmer, U., et al. (1995) Experimental *Listeria* meningoencephalitis. Macrophage inflammatory protein-1 alpha and -2 are produced intrathecally and mediate chemotactic activity in cerebrospinal fluid of infected mice. *J. Immunol.* **155,** 4367–4375.

48. Sprenger, H., Rosler, A., Tonn, P., Braune, H. J., Huffmann, G., and Gemsa, D. (1996) Chemokines in the cerebrospinal fluid of patients with meningitis. *Clin. Immunol. Immunopathol.* **80,** 155–161.

49. Lopez-Cortes, L. F., Cruz-Ruiz, M., Gomez-Mateos, J., Viciana-Fernandez, P., Martinez-Marcos, F. J., and Pachon, J. (1995) Interleukin-8 in cerebrospinal fluid from patients with meningitis of different etiologies: its possible role as neutrophil chemotactic factor. *J. Infect. Dis.* **172,** 581–584.

50. Spanaus, K. S., Nadal, D., Pfister, H. W., Seebach, J., Widmer, U., Frei, K., et al. (1997) C-X-C and C-C chemokines are expressed in the cerebrospinal fluid in bacterial meningitis and mediate chemotactic activity on peripheral blood-derived polymorphonuclear and mononuclear cells in vitro. *J. Immunol.* **158,** 1956–1964.

51. Kielian, T. and Hickey, W. F. (2000) Proinflammatory cytokine, chemokine, and cellular adhesion molecule expression during the acute phase of experimental brain abscess development. *Am. J. Pathol.* **157,** 647–658.

52. Sasseville, V. G., Smith, M. M., Mackay, C. R., Pauley, D. R., Mansfield, K. G., Ringler, D. J., et al. (1996) Chemokine expression in simian immunodeficiency virus-induced AIDS encephalitis. *Am. J. Pathol.* **149,** 1459–1467.

53. Asensio, V. C. and Campbell, I. L. (1997) Chemokine gene expression in the brains of mice with lymphocytic choriomeningitis. *J. Virol.* **71,** 7832–7840.

54. Lane, T. E., Asensio, V. C., Yu, N., Paoletti, A. D., Campbell, I. L., and Buchmeier, M. J. (1998) Dynamic regulation of alpha- and beta-chemokine expression in the central nervous system during mouse hepatitis virus-induced demyelinating disease. *J. Immunol.* **160,** 970–978.

55. Liu, M. T., Armstrong, D., Hamilton, T. A., and Lane, T. E. (2001) Expression of Mig (monokine induced by interferon-gamma) is important in T lymphocyte recruitment and host defense following viral infection of the central nervous system. *J. Immunol.* **166,** 1790–1795.

56. Charles, P. C., Weber, K. S., Cipriani, B., and Brosnan, C. F. (1999) Cytokine, chemokine and chemokine receptor mRNA expression in different strains of normal mice: implications for establishment of a Th1/Th2 bias. *J. Neuroimmunol.* **100,** 64–73.

57. Charles, P. C., Chen, X., Horwitz, M. S., and Brosnan, C. F. (1999) Differential chemokine induction by the mouse adenovirus type-1 in the central nervous system of susceptible and resistant strains of mice. *J. Neurovirol.* **5,** 55–64.

58. Tipold, A., Moore, P., Zurbriggen, A., Burgener, I., Barben, G., and Vandevelde, M. (1999) Early T cell response in the central nervous system in canine distemper virus infection. *Acta Neuropathol. (Berl.)* **97,** 45–56.

59. Sauder, C., Hallensleben, W., Pagenstecher, A., Schneckenburger, S., Biro, L., Pertlik, D., et al. (2000) Chemokine gene expression in astrocytes of Borna disease virus-infected rats and mice in the absence of inflammation. *J. Virol.* **74,** 9267–9280.

60. Murray, P. D., Krivacic, K., Chernosky, A., Wei, T., Ransohoff, R. M., and Rodriguez, M. (2000) Biphasic and regionally-restricted chemokine expression in the central nervous system in the Theiler's virus model of multiple sclerosis. *J. Neurovirol.* **6(Suppl. 1),** S44–S52.

61. Sanders, V. J., Pittman, C. A., White, M. G., Wang, G., Wiley, C. A., and Achim, C. L. (1998) Chemokines and receptors in HIV encephalitis. *AIDS* **12,** 1021–1026.

62. Conant, K., Garzino-Demo, A., Nath, A., McArthur, J. C., Halliday, W., Power, C., et al. (1998) Induction of monocyte chemoattractant protein-1 in HIV-1 Tat-stimulated astrocytes and elevation in AIDS dementia. *Proc. Natl. Acad. Sci. USA* **95,** 3117–3121.

63. Umehara, F., Izumo, S., Takeya, M., Takahashi, K., Sato, E., and Osame, M. (1996) Expression of adhesion molecules and monocyte chemoattractant protein-1 (MCP-1) in the spinal cord lesions in HTLV-I-associated myelopathy. *Acta Neuropathol.* **91,** 343–350.

64. Biddison, W. E., Kubota, R., Kawanishi, T., Taub, D. D., Cruikshank, W. W., Center, D. M., et al. (1997) Human T cell leukemia virus type I (HTLV-I)-specific CD8+ CTL clones from patients with HTLV-I-associated neurologic disease secrete proinflammatory cytokines, chemokines, and matrix metalloproteinase. *J. Immunol.* **159,** 2018–2025.

65. Huffnagle, G. B. and McNeil, L. K. (1999) Dissemination of C. neoformans to the central nervous system: role of chemokines, Th1 immunity and leukocyte recruitment. *J. Neurovirol.* **5,** 76–81.

66. Hulkower, K., Brosnan, C. F., Aquino, D. A., Cammer, W., Kulshrestha, S., Guida, M. P., et al. (1993) Expression of CSF-1, c-fms, and MCP-1 in the central nervous system of rats with experimental allergic encephalomyelitis. *J. Immunol.* **150,** 2525–2533.

67. Ransohoff, R. M., Hamilton, T. A., Tani, M., Stoler, M. H., Shick, H. E., Major, J. A., et al. (1993) Astrocyte expression of mRNA encoding cytokines IP-10 and JE/MCP-1 in experimental autoimmune encephalomyelitis. *FASEB J.* **7,** 592–600.

68. Godiska, R., Chantry, D., Dietsch, G. N., and Gray, P. W. (1995) Chemokine expression in murine experimental allergic encephalomyelitis. *J. Neuroimmunol.* **58,** 167–176.

69. Pan, Y., Lloyd, C., Zhou, H., Dolich, S., Deeds, J., Gonzalo, J. A., et al. (1997) Neurotactin, a membrane-anchored chemokine upregulated in brain inflammation. *Nature* **387,** 611–617.

70. Sun, D., Tani, M., Newman, T. A., Krivacic, K., Phillips, M., Chernosky, A., et al. (2000) Role of chemokines, neuronal projections, and the blood-brain barrier in the enhancement of cerebral EAE following focal brain damage. *J. Neuropathol. Exp. Neurol.* **59,** 1031–1043.

71. Glabinski, A. R., Tuohy, V. K., and Ransohoff, R. M. (1998) Expression of chemokines RANTES, MIP-1alpha and GRO-alpha correlates with inflammation in acute experimental autoimmune encephalomyelitis. *Neuroimmunomodulation* **5,** 166–171.

72. Karpus, W. J., Lukacs, N. W., McRae, B. L., Strieter, R. M., Kunkel, S. L., and Miller, S. D. (1995) An important role for the chemokine macrophage inflammatory protein-1 alpha in the pathogenesis of the T cell-mediated autoimmune disease, experimental autoimmune encephalomyelitis. *J. Immunol.* **155,** 5003–5010.

73. Tran, E. H., Kuziel, W. A., and Owens, T. (2000) Induction of experimental autoimmune encephalomyelitis in C57BL/6 mice deficient in either the chemokine macrophage inflammatory protein-1alpha or its CCR5 receptor. *Eur. J. Immunol.* **30,** 1410–1415.

74. Nygardas, P. T., Maatta, J. A., and Hinkkanen, A. E. (2000) Chemokine expression by central nervous system resident cells and infiltrating neutrophils during experimental autoimmune encephalomyelitis in the BALB/c mouse. *Eur. J. Immunol.* **30,** 1911–1918.

75. Glabinski, A. R., Tani, M., Strieter, R. M., Tuohy, V. K., and Ransohoff, R. M. (1997) Synchronous synthesis of alpha- and beta-chemokines by cells of diverse lineage in the central nervous system of mice with relapses of chronic experimental autoimmune encephalomyelitis. *Am. J. Pathol.* **150,** 617–630.

76. Karpus, W. J. and Kennedy, K. J. (1997) MIP-1alpha and MCP-1 differentially regulate acute and relapsing autoimmune encephalomyelitis as well as Th1/Th2 lymphocyte differentiation. *J. Leukocyte Biol.* **62,** 681–687.

77. Huang, D., Wang, J., Kivisakk, P., Rollins, B. J., and Ransohoff, R. M. (2001) Absence of monocyte chemoattractant protein 1 in mice leads to decreased local macrophage recruitment and antigen-specific T helper cell type 1 immune response in experimental autoimmune encephalomyelitis. *J. Exp. Med.* **193,** 713–725.

78. Glabinski, A. R., Krakowski, M., Han, Y., Owens, T., and Ransohoff, R. M. (1999) Chemokine expression in GKO mice (lacking interferon-gamma) with experimental auto-immune encephalomyelitis. *J. Neurovirol.* **5,** 95–101.

79. Fischer, F. R., Santambrogio, L., Luo, Y., Berman, M. A., Hancock, W. W., and Dorf, M. E. (2000) Modulation of experimental autoimmune encephalomyelitis: effect of altered peptide ligand on chemokine and chemokine receptor expression. *J. Neuroimmunol.* **110,** 195–208.

80. Glabinski, A. R., O'Bryant, S., Selmaj, K., and Ransohoff, R. M. (2000) CXC chemokine receptor expression during chronic relapsing experimental autoimmune encephalomyelitis. *Ann. NY Acad. Sci.* **917,** 135–144.

81. Serafini, B., Columba-Cabezas, S., Di Rosa, F., and Aloisi, F. (2000) Intracerebral recruitment and maturation of dendritic cells in the onset and progression of experimental auto-immune encephalomyelitis. *Am. J. Pathol.* **157,** 1991–2002.

82. Fife, B. T., Huffnagle, G. B., Kuziel, W. A., and Karpus, W. J. (2000) CC chemokine receptor 2 is critical for induction of experimental autoimmune encephalomyelitis. *J. Exp. Med.* **192,** 899–906.

83. Izikson, L., Klein, R. S., Charo, I. F., Weiner, H. L., and Luster, A. D. (2000) Resistance to experimental autoimmune encephalomyelitis in mice lacking the CC chemokine receptor (CCR)2. *J. Exp. Med.* **192,** 1075–1080.

84. Hvas, J., McLean, C., Justesen, J., Kannourakis, G., Steinman, L., Oksenberg, J. R., et al. (1997) Perivascular T cells express the pro-inflammatory chemokine RANTES mRNA in multiple sclerosis lesions. *Scand. J. Immunol.* **46,** 195–203.

85. Simpson, J. E., Newcombe, J., Cuzner, M. L., and Woodroofe, M. N. (1998) Expression of monocyte chemoattractant protein-1 and other beta-chemokines by resident glia and inflammatory cells in multiple sclerosis lesions. *J. Neuroimmunol.* **84,** 238–249.

86. McManus, C., Berman, J. W., Brett, F. M., Staunton, H., Farrell, M., and Brosnan, C. F. (1998) MCP-1, MCP-2 and MCP-3 expression in multiple sclerosis lesions: an immuno-histochemical and in situ hybridization study. *J. Neuroimmunol.* **86,** 20–29.

87. Sorensen, T. L., Tani, M., Jensen, J., Pierce, V., Lucchinetti, C., Folcik, V. A., et al. (1999) Expression of specific chemokines and chemokine receptors in the central nervous system of multiple sclerosis patients. *J. Clin. Invest.* **103,** 807–815.

88. Balashov, K. E., Rottman, J. B., Weiner, H. L., and Hancock, W. W. (1999) CCR5(+) and CXCR3(+) T cells are increased in multiple sclerosis and their ligands MIP-1alpha and IP-10 are expressed in demyelinating brain lesions. *Proc. Natl. Acad. Sci. USA* **96,** 6873–6878.

89. Zang, Y. C., Samanta, A. K., Halder, J. B., Hong, J., Tejada-Simon, M. V., Rivera, V. M., et al. (2000) Aberrant T cell migration toward RANTES and MIP-1 alpha in patients with multiple sclerosis. Overexpression of chemokine receptor CCR5. *Brain* **123,** 1874–1882.

90. Simpson, J. E., Newcombe, J., Cuzner, M. L., and Woodroofe, M. N. (2000) Expression of the interferon-gamma-inducible chemokines IP-10 and Mig and their receptor, CXCR3, in multiple sclerosis lesions. *Neuropathol. Appl. Neurobiol.* **26,** 133–142.

91. Iarlori, C., Reale, M., Lugaresi, A., De Luca, G., Bonanni, L., Di Iorio, A., et al. (2000) RANTES production and expression is reduced in relapsing-remitting multiple sclerosis patients treated with interferon-beta-1b. *J. Neuroimmunol.* **107,** 100–107.

92. Berman, J. W., Guida, M. P., Warren, J., Amat, J., and Brosnan, C. F. (1996) Localization of monocyte chemoattractant peptide-1 expression in the central nervous system in experimental autoimmune encephalomyelitis and trauma in the rat. *J. Immunol.* **156,** 3017–3023.

93. Glabinski, A. R., Balasingam, V., Tani, M., Kunkel, S. L., Strieter, R. M., Yong, V. W., et al. (1996) Chemokine monocyte chemoattractant protein-1 is expressed by astrocytes after mechanical injury to the brain. *J. Immunol.* **156,** 4363–4368.

94. Ghirnikar, R. S., Lee, Y. L., He, T. R., and Eng, L. F. (1996) Chemokine expression in rat stab wound brain injury. *J. Neurosci. Res.* **46,** 727–733.

95. Hausmann, E. H., Berman, N. E., Wang, Y. Y., Meara, J. B., Wood, G. W., and Klein, R. M. (1998) Selective chemokine mRNA expression following brain injury. *Brain Res.* **788,** 49–59.

96. Ghirnikar, R. S., Lee, Y. L., Li, J. D., and Eng, L. F. (1998) Chemokine inhibition in rat stab wound brain injury using antisense oligodeoxynucleotides. *Neurosci. Lett.* **247,** 21–24.

97. Bartholdi, D. and Schwab, M. E. (1997) Expression of pro-inflammatory cytokine and chemokine mRNA upon experimental spinal cord injury in mouse: an in situ hybridization study. *Eur. J. Neurosci.* **9,** 1422–1438.

98. McTigue, D. M., Tani, M., Krivacic, K., Chernosky, A., Kelner, G. S., Maciejewski, D., et al. (1998) Selective chemokine mRNA accumulation in the rat spinal cord after contusion injury. *J. Neurosci. Res.* **53,** 368–376.

99. Grzybicki, D., Moore, S. A., Schelper, R., Glabinski, A. R., Ransohoff, R. M., and Murphy, S. (1998) Expression of monocyte chemoattractant protein (MCP-1) and nitric oxide synthase-2 following cerebral trauma. *Acta Neuropathol. (Berl.)* **95,** 98–103.

100. Mehta, P. S., Bruccoleri, A., Brown, H. W., and Harry, G. J. (1998) Increase in brain stem cytokine mRNA levels as an early response to chemical-induced myelin edema. *J. Neuroimmunol.* **88,** 154–164.

101. Galasso, J. M., Liu, Y., Szaflarski, J., Warren, J. S., and Silverstein, F. S. (2000) Monocyte chemoattractant protein-1 is a mediator of acute excitotoxic injury in neonatal rat brain. *Neuroscience* **101,** 737–744.

102. Kossmann, T., Stahel, P. F., Lenzlinger, P. M., Redl, H., Dubs, R. W., Trentz, O., et al. (1997) Interleukin-8 released into the cerebrospinal fluid after brain injury is associated with blood-brain barrier dysfunction and nerve growth factor production. *J. Cereb. Blood Flow Metab.* **17,** 280–289.

103. Kim, J. S. (1996) Cytokines and adhesion molecules in stroke and related diseases. *J. Neurol. Sci.* **137,** 69–78.

104. Wang, X., Yue, T. L., Barone, F. C., and Feuerstein, G. Z. (1995) Monocyte chemoattractant protein-1 messenger RNA expression in rat ischemic cortex. *Stroke* **26,** 661–665; discussion 665–666.

105. Gourmala, N. G., Buttini, M., Limonta, S., Sauter, A., and Boddeke, H. W. (1997) Differential and time-dependent expression of monocyte chemoattractant protein-1 mRNA by astrocytes and macrophages in rat brain: effects of ischemia and peripheral lipopolysaccharide administration. *J. Neuroimmunol.* **74,** 35–44.

106. Takami, S., Nishikawa, H., Minami, M., Nishiyori, A., Sato, M., Akaike, A., et al. (1997) Induction of macrophage inflammatory protein MIP-1alpha mRNA on glial cells after focal cerebral ischemia in the rat. *Neurosci. Lett.* **227,** 173–176.

107. Liu, T., Young, P. R., McDonnell, P. C., White, R. F., Barone, F. C., and Feuerstein, G. Z. (1993) Cytokine-induced neutrophil chemoattractant mRNA expressed in cerebral ischemia. *Neurosci. Lett.* **164,** 125–128.

108. Wang, X., Li, X., Schmidt, D. B., Foley, J. J., Barone, F. C., Ames, R. S., et al. (2000) Identification and molecular characterization of rat CXCR3: receptor expression and interferon-inducible protein-10 binding are increased in focal stroke. *Mol. Pharmacol.* **57,** 1190–1198.

109. Ivacko, J., Szaflarski, J., Malinak, C., Flory, C., Warren, J. S., and Silverstein, F. S. (1997) Hypoxic–ischemic injury induces monocyte chemoattractant protein-1 expression in neonatal rat brain. *J. Cereb. Blood Flow Metab.* **17,** 759–770.

110. Matsumoto, T., Yokoi, K., Mukaida, N., Harada, A., Yamashita, J., Watanabe, Y., et al. (1997) Pivotal role of interleukin-8 in the acute respiratory distress syndrome and cerebral reperfusion injury. *J. Leukocyte Biol.* **62,** 581–587.

111. Yoshimoto, T., Houkin, K., Tada, M., and Abe, H. (1997) Induction of cytokines, chemokines and adhesion molecule mRNA in a rat forebrain reperfusion model. *Acta Neuropathol. (Berl.)* **93,** 154–158.

112. Gaetani, P., Tartara, F., Pignatti, P., Tancioni, F., Rodriguez y Baena, R., and De Benedetti, F. (1998) Cisternal CSF levels of cytokines after subarachnoid hemorrhage. *Neurol. Res.* **20,** 337–342.

113. Xia, M., Qin, S., McNamara, M., Mackay, C., and Hyman, B. T. (1997) Interleukin-8 receptor B immunoreactivity in brain and neuritic plaques of Alzheimer's disease. *Am. J. Pathol.* **150,** 1267–1274.

114. Horuk, R., Martin, A. W., Wang, Z., Schweitzer, L., Gerassimides, A., Guo, H., et al. (1997) Expression of chemokine receptors by subsets of neurons in the central nervous system. *J. Immunol.* **158,** 2882–2890.

115. Xia, M. Q., Qin, S. X., Wu, L. J., Mackay, C. R., and Hyman, B. T. (1998) Immunohistochemical study of the beta-chemokine receptors CCR3 and CCR5 and their ligands in normal and Alzheimer's disease brains. *Am. J. Pathol.* **153,** 31–37.

116. Xia, M. Q., Bacskai, B. J., Knowles, R. B., Qin, S. X., and Hyman, B. T. (2000) Expression of the chemokine receptor CXCR3 on neurons and the elevated expression of its ligand IP-10 in reactive astrocytes: in vitro ERK1/2 activation and role in Alzheimer's disease. *J. Neuroimmunol.* **108,** 227–235.

117. Ishizuka, K., Kimura, T., Igata-yi, R., Katsuragi, S., Takamatsu, J., and Miyakawa, T. (1997) Identification of monocyte chemoattractant protein-1 in senile plaques and reactive microglia of Alzheimer's disease. *Psychiatry Clin. Neurosci.* **51,** 135–138.

118. Meda, L., Baron, P., Prat, E., Scarpini, E., Scarlato, G., Cassatella, M. A., et al. (1999) Proinflammatory profile of cytokine production by human monocytes and murine microglia stimulated with beta-amyloid[25–35]. *J. Neuroimmunol.* **93,** 45–52.

119. Gitter, B. D., Cox, L. M., Rydel, R. E., and May, P. C. (1995) Amyloid beta peptide potentiates cytokine secretion by interleukin-1 beta-activated human astrocytoma cells. *Proc. Natl. Acad. Sci. USA* **92,** 10,738–10,741.

120. Bruno, V., Copani, A., Besong, G., Scoto, G., and Nicoletti, F. (2000) Neuroprotective activity of chemokines against *N*-methyl-D-aspartate or beta-amyloid-induced toxicity in culture. *Eur. J. Pharmacol.* **399,** 117–121.

121. Muessel, M. J., Berman, N. E., and Klein, R. M. (2000) Early and specific expression of monocyte chemoattractant protein-1 in the thalamus induced by cortical injury. *Brain Res.* **870,** 211–221.

122. Ransohoff, R. M. (1997) Chemokines in neurological disease models: correlation between chemokine expression patterns and inflammatory pathology. *J. Leukocyte Biol.* **62,** 645–652.

123. Flugel, A., Hager, G., Horvat, A., Spitzer, C., Singer, G. M., Graeber, M. B., et al. (2001) Neuronal MCP-1 expression in response to remote nerve injury. *J. Cereb. Blood Flow Metab.* **21,** 69–76.

124. Siebert, H., Sachse, A., Kuziel, W. A., Maeda, N., and Bruck, W. (2000) The chemokine receptor CCR2 is involved in macrophage recruitment to the injured peripheral nervous system. *J. Neuroimmunol.* **110,** 177–185.

125. Fujioka, T., Kolson, D. L., and Rostami, A. M. (1999) Chemokines and peripheral nerve demyelination. *J. Neurovirol.* **5,** 27–31.

126. Kieseier, B. C., Krivacic, K., Jung, S., Pischel, H., Toyka, K. V., Ransohoff, R. M., et al. (2000) Sequential expression of chemokines in experimental autoimmune neuritis. *J. Neuroimmunol.* **110,** 121–129.

127. Halford, W. P., Gebhardt, B. M., and Carr, D. J. (1996) Persistent cytokine expression in trigeminal ganglion latently infected with herpes simplex virus type 1. *J. Immunol.* **157,** 3542–3549.

5

Neurotoxic Mechanisms of Nitric Oxide

Kathleen M. K. Boje

1. INTRODUCTION

Nitric oxide (nitrogen monoxide; NO) is a simple molecular specie that has diverse, yet versatile biological functions. As a biological emissary, NO and its redox derivatives participate in necessary physiological functions or undesired pathological disturbances *(1)*. NO is a ubiquitous second messenger *(2)*, exerting physiological roles in many organ systems (e.g., musculature, pulmonary, gastrointestinal, immune, renal, endocrinological, and reproductive systems) *(3–8)*, in addition to modulation of regional and systemic circulation *(9–11)*. NO is also involved in the developmental aspects of neurosynaptic plasticity *(12,13)*.

The scientific literature is replete with numerous and diverse contemporary reviews of NO and nitric oxide synthase (NOS) *(14)*. For a succinct global overview with a historical perspective, the reader may wish to consult the review by Moncada and Furchgott *(15)*, who each are established leaders for his seminal contribution to the discovery and identification of NO.

Nitric oxide and related species exert intricate physiological and pathophysiological effects in the central nervous system (CNS). Under the "right" conditions, NO and its derivatives can initiate and mediate toxicity either exclusively or synergistically with other effectors. As such, NO neurotoxicity represents only one mechanism of neurotoxicity; other mechanisms are operative, depending on the brain region, cell type, and initial cellular insult. The focus of this chapter is restricted to a discussion of the neurotoxicity of NO and its derivatives.

2. BIOCHEMISTRY OF NITRIC OXIDE

2.1. Nitric Oxide Synthase Isoforms: Biochemical Characteristics

Nitric oxide is synthesized by one of three isoforms of NOS, each of which has been characterized and cloned from many tissues from diverse animal species, including humans *(15)*. The naming conventions for the isoforms are based on the isoform's original expression characteristics (tissue source, constituitive or inducible) and, as such, the currently accepted nomenclature still uses such designations with the implicit understanding that the designation is not to be construed literally. For example, the inducible isoform (iNOS) is constituitively expressed in some tissues (including fetal tissue); the

From: *Neuroinflammation, 2nd Edition: Mechanisms and Management*
Edited by: P. L. Wood © Humana Press Inc., Totowa, NJ

constitutive isoforms can be upregulated under certain conditions; and the neuronal isoform (nNOS) is expressed in skeletal muscle *(15,16)*. It is apparent that the expression (constitutive versus inducible) and functional activity (transient versus sustained synthesis) of the NOS isoforms are influenced by cell/tissue type *(14)*.

Various CNS cell types express all three isoforms. Discrete neuronal populations constituitively express nNOS with a subset that also coexpresses the endothelial isoform (eNOS). However, nNOS is dynamically regulated during CNS development, plasticity, and injury *(17)*. Cerebrovascular endothelial cells express eNOS constituitively, whereas glial and cerebrovascular endothelial cells express iNOS following exposure to immunostimulatory agents.

In broad strokes, eNOS and nNOS typically function in cell signaling processes, with transient picomolar production of NO governed by elevations of intracellular calcium and subsequent NOS enzyme association with calcium/calmodulin. This is in general contrast to the gene-mediated regulation of iNOS, where sustained micromolar synthesis of NO occurs in an apparent calcium-independent manner because of a tightly coupled association between iNOS and calcium/calmodulin *(18)*. eNOS is primarily a membrane-associated enzyme, whereas nNOS and iNOS are predominately cytosolic enzymes *(15,16)*.

2.2. Biochemical Synthesis and Metabolism of Nitric Oxide

The biosynthesis of NO is highly regulated, requiring either elevations in intracellular calcium (for constituitively expressed eNOS and nNOS) or immunostimulatory agents (for *de novo* expression of iNOS) *(19)*. Synthesis of NO requires L-arginine, O_2, NADPH, flavin adenine dinucleotide, flavin mononucleotide, and tetrahydrobiopterin in a five-electron oxidation of the guanidino moiety of arginine *(20–22)*.

Nitric oxide is typically produced upon demand; by itself, it is not stored in cellular pools for future use. However, a fraction of endogenously synthesized NO may be captured as NO–thiol protein adducts (i.e., *S*-nitrosothiols), thereby creating a "NO reservoir" for cell signaling, modulation of protein function, or antimicrobial defense. For example, intracellular *S*-nitrosothiol stores, particularly *S*-nitrosoglutathione, may attain micromolar concentrations *(23)*. The seminal work of Stamler et al. postulated that S-nitrosylated hemogloblin serves as a local oxygen biosensor *(24–26)* and as a circulating "delivery agent and bank" of bioactive NO *(27,28)*. Moreover, exogenous, nonprotein *S*-nitrosothiols are in drug development for potential therapeutic use *(29)*. Thus, the biological effects of NO should also be considered from the *S*-nitrosothiol perspective. As the biological chemistry of *S*-nitrosothiols is distinctive from NO, and the roles of *S*-nitrosothiols in the CNS are intriguingly diverse, the reader is urged to consult excellent contemporary reviews in these areas *(23,30–32)*.

The biological chemistry of NO and related species is amazingly elegant, yet bewilderingly complex, as lucidly presented in several excellent reviews *(33–36)*. NO is a paramagnetic free radical that possesses an odd number of electrons. Owing to its relative stability (a biological half-life of several seconds under ideal conditions), a net ionic charge of zero and a high 1-octanol/water partition coefficient, NO can diffuse across several lipid membranes to exert effects distal from its biosynthetic source *(33–35)*.

A number of factors influence the biological fate of NO: its local concentration, physiological milieu (redox environment, pO_2, pH, CO_2 concentration), and local concentrations of other bioreactants (metalloproteins, thiols, reactive oxygen species) *(35,37,38)*. It is generally accepted that NO is a very modest effector of toxicity *(35–38)*. Oxidation, reduction, or adduction of NO (Fig. 1) produces to a variety of nitrogen oxide species (e.g., NO^-, NO^+, $ONOO^-$, $ONOOCO_2^-$, NO_2, NO_2^-, NO_2^+, and N_2O_3), collectively referred to as reactive nitrogen oxide species (RNOS). The initial reaction of NO with superoxide ($O_2^{\bullet-}$) forms peroxynitrite ($ONOO^-$) (Fig. 1). Similarly, NO can react with O_2 to produce NO_2, which then reacts with NO to produce N_2O_3. Although $ONOO^-$ is considered the primary and most injurious effector of nitric oxide toxicity, NO_2 and N_2O_3 may exert substantial biological toxicity as well *(33,36)*. The RNOS are capable of damaging a number of molecular targets (Fig. 1), thereby creating cellular distress. When the number of insults sustained by critical molecular targets reaches an upper limit, cell death ensues via apoptotic or necrotic mechanisms *(39,40)*. Cells resistant to NO neurotoxicity (e.g., neurons with nNOS), survive because of the $O_2^{\bullet-}$ detoxifying effects of manganese superoxide dismutase *(41)*.

2.3. Biological Effects of Nitric Oxide and Reactive Nitrogen Oxide Species

The biological effects of NO and its corresponding RNOS are broad and diverse, encompassing regulatory, protective and deleterious effects *(33,36)*. The type of biological effect is highly dependent on the conditions of NO synthesis (e.g., molecular events triggering biosynthesis, cell synthesis source, duration of synthesis, localized NO concentration, and biological target [cell type and protein]) *(36)*. Regulatory effects of NO are likely to occur under conditions of low synthesis and to promote physiological homeostasis that is commonly mediated through a variety of signal transduction mechanisms, including interactions with guanylate cyclase and other hemoproteins *(2,23,33, 42)*. For instance, NO and corresponding *S*-nitrosothiols are involved in the maintenance of cardiovascular tone, renal function, and bronchodilation *(33)*, as well as synaptic neurotransmission and neuroplasticity *(31,43,44)*. The protective effects of NO are exerted through inactivation of oxidants and reactive oxygen species known to evoke oxidant stress. For example, under certain conditions, NO can protect cells against oxidant stress through a direct reaction with alkoxyl ($RO^{\bullet}$) and alkyl hydroperoxyl ($ROO^{\bullet}$) radicals, lipid peroxidation through chain termination of the lipid free-radical reaction *(33,36,40,45,46)* and apoptosis through multiple mechanisms *(40)*. The deleterious effects of NO are manifest as cytotoxic insults that contribute to the underlying pathology of various disease states *(47)*.

3. MECHANISMS OF TOXICITY

Nitric oxide is chemically reactive toward O_2 or $O_2^{\bullet-}$, depending on the local concentrations of NO and oxygen tension, NO scavenging agents (e.g., glutathione or other thiols), and $O_2^{\bullet-}$ scavenging agents (e.g., superoxide dismutase [SOD]) *(48)*. Typically, high concentrations of NO produced by upregulated NOS isoforms will indiscriminately attack critical cellular targets via the $NO/O_2^{\bullet-}$ and NO/O_2 reaction pathways, as illustrated in Fig. 1 *(36,49)*. RNOS can oxidize, nitrate (covalent bond formation with NO_2), nitrosate (addition of NO^+), or nitrosylate (covalent bond formation with NO) at critical residues of biomolecules. The contribution of the RNOS to cytotoxic mechanisms will be discussed in this section.

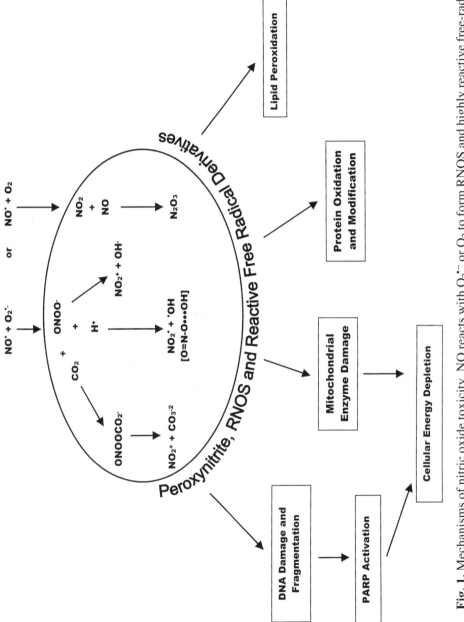

Fig. 1. Mechanisms of nitric oxide toxicity. NO reacts with $O_2^{\bullet-}$ or O_2 to form RNOS and highly reactive free-radical derivatives. These potent oxidizing compounds attack critical cellular targets. For illustrative purposes, NO is depicted with an unpaired electron.

3.1. Universal Molecular Mechanisms of Cytotoxicity

3.1.1. NO and Its Redox Metabolites NO^- and NO^+

Compared to $ONOO^-$, NO is a relatively weak oxidant. NO itself can directly react with the transition metals (e.g., iron, copper, zinc) of metalloproteins, thereby altering protein structure/function through nitrosylation reactions. NO nitrosylates the iron centers of heme proteins to either enhance (e.g., guanylate cyclase) or inhibit (e.g., cytochrome P450s, hemoglobin and cyclooxygenase–2) protein functional activity *(2,36, 50)*. NO also reacts with copper proteins, although the biological significance is an area of future research *(51)*. Quite possibly, NO nitrosylation of proteins may accentuate cellular distress under conditions favoring formation of $ONOO^-$ and RNOS.

The nitroxyl anion (NO^-) is a one-electron reduction product of NO and is postulated to be physiologically synthesized by NOS, SOD, and/or *S*-nitrosothiols *(52–55)*. In the presence of hydrogen peroxide or transition metals, NO^- forms strong oxidants capable of evoking cytotoxity *(56)*. NO^- also reacts rapidly with NO to form reactive $N_2O_2^{\bullet-}$ or with O_2 to form $ONOO^-$ *(34)*.

The metabolite NO^+ is postulated to derive from a redox reaction involving NO. However, based on the reduction potentials of NO^+/NO, it is not physiologically possible to form NO^+ as a unique specie, no matter how fleeting its existence *(15,34)*.

3.1.2. $ONOO^-$ and Its Metabolites NO_2 and NO_2^+

The major molecular mechanism of NO toxicity is the formation of $ONOO^-$ via reaction of NO with $O_2^{\bullet-}$ (Fig. 1). Under normal physiological conditions, intracellular concentrations of NO and $O_2^{\bullet-}$ are relatively low because NO is utilized for regulatory functions and $O_2^{\bullet-}$ is effectively scavenged by SOD. However, under conditions of prodigious NO synthesis, NO effectively competes with SOD for reaction with $O_2^{\bullet-}$, forming $ONOO^-$ at cytotoxic concentrations *(48,57,58)*. $ONOO^-$ is sufficiently stable to diffuse across several cell membranes before reacting with critical cellular targets in nitrative, nitrosative, and/or oxidative reactions *(23,36)* (*see* Section 3.2).

At physiological pH, $ONOO^-$ also exists in equilibrium as its conjugate acid ONOOH (pK_a 6.6–6.8), with the concentration of $ONOO^-$ four to six times greater than ONOOH, as estimated by the Henderson–Hasselbalch equation. Depending on the cellular environment, ONOOH may decompose to cytotoxic species (i.e., hydroxyl radical [HO^-] and nitrogen dioxide [NO_2]), or to the relatively inert nitrates/nitrates (NO_2^-, NO_3^-) via nitric acid (e.g., ONOOH $\rightarrow HNO_3$) *(33–35,57)*.

3.1.3. $ONOOCO_2^-$ and Its Metabolites NO_2^+ and CO_3^{2-}

Given the high concentration (approx 1 m*M*) and ubiquitous physiological distribution of CO_2, $ONOO^-$ rapidly and readily reacts with CO_2 to form $ONOOCO_2^-$, a far more potent nitrating specie than $ONOO^-$ *(37)*. The $ONOO^-/CO_2$ reaction is pH independent and occurs at a rate faster than the spontaneous decomposition of $ONOO^-$. In fact, this reaction is now believed to be the major pathway of metabolic biotransformation for $ONOO^-$, and the $ONOOCO_2^-$ specie likely serves as the major intermediate in the nitration of critical tyrosine, tryptophan, and nucleic acid residues *(37,38)*.

3.1.4. NO_2 and N_2O_3

Another mechanism of NO toxicity can occur with the reaction of NO with O_2 to yield cytotoxic NO_2. Subsequent NO auto-oxidation with NO_2 produces yet another toxic specie, N_2O_3 (Fig. 1). Given the relative lipophilicity of NO and O_2, it is postulated that the NO/O_2 auto-oxidation reactions are preferentially promoted in the hydrophobic environment of lipid membranes. Compared to HO^- and $ONOO^-$, NO_2 and N_2O_3 are relatively mild oxidants, yet the extent of their damage can be significant. NO_2 and N_2O_3 effectively nitrosate amines and nitrosylate thiols, forming carcinogenic nitrosoamines and *S*-nitrosothiols, respectively *(33,36)*.

3.2. Universal Cell Biological Mechanisms of RNOS Cytotoxicity

Compared to NO or $O_2^{\cdot-}$, $ONOO^-$, and subsequent RNOS derivatives are a much more potent nitrative, nitrosative, and oxidative stressors, possessing substantially greater reactivity toward critical cellular components. Targets of $ONOO^-$ and RNOS toxicity include amines, thiols, tyrosine and tryptophan residues, nucleic acids, iron–sulfur centers, and metalloproteins.

3.2.1. DNA Damage and Fragmentation, PARP Activation, and Cellular Energy Depletion

Reactive nitric oxide synthase can damage DNA by three mechanisms *(36)*: (1) direct interaction with DNA, resulting in mutagenic insults and DNA strand breakage; (2) inhibition of enzymes involved in DNA repair processes; and (3) chemical reaction with endogenous molecules to form DNA alkylating carcinogens, such as the nitrosoamines. RNOS promote mutagenic effects by nitrosative deamination of cytosine, adenine, and guanine bases *(36,59)*. RNOS inhibits DNA ligase, thereby abolishing the repair of RNOS-mediated DNA strand breaks *(60–62)*. RNOS-mediated DNA damage activates a DNA repair enzyme, poly(ADP-ribose) polymerase (PARP). In an ironic set of circumstances, PARP undergoes RNOS nitrosylation, which further enhances PARP activity. This leads to excessive consumption and subsequent depletion of cellular energy reserves in compromised neurons *(63,64)*.

3.2.2. Mitochondrial Enzyme Damage and Cellular Energy Depletion

Disruption of cellular energetics through mitochondrial dysfunction may be an important mechanism for numerous neurodegenerative diseases *(65)*. RNOS can reversibly and irreversibly depress mitochondrial respiration by three mechanisms *(66)*: (1) reversible inhibition of cytochrome-*c* oxidase (complex IV) by low concentrations of NO; (2) irreversible inhibition of mitochondrial complexes I–IV (e.g., *cis*-aconitase, NADH: ubiquinone oxidoreductase, and succinate:ubiquinone oxidoreductase) by RNOS nitrosylation of the iron–sulfate centers *(67,68)* and (3) induction of the mitochondrial permeability transition (PT) pore. RNOS inhibition of key respiratory enzymes promotes an intracellular mobilization and loss of iron. Moreover, RNOS nitrosylates the thiol residues of glyceraldehyde-3-phosphate dehydrogenase (GAPDH), thereby inhibiting glycolysis *(66,69)*. RNOS inhibition of mitochondrial complexes coupled with impairment of GAPDH activity creates a synergistic effect of crippling cellular ATP synthesis. The last RNOS mechanism of mitochondrial damage, the PT pore mechanism, is controversial, as data from one group contend that this last mechanism is an artifact of the experimental conditions *(47)*.

3.2.3. Protein Oxidation and Modification

Reactive nitric oxide synthase can nitrosate and nitrate key amino acids in proteins, thereby altering function through alteration of active sites, conformation, or blockade of phosphorylation sites. S-Nitrosylation of cysteinyl or methionyl sulhydryls and nitration of tyrosine residues may represent another new class of cellular signaling mechanisms, not unlike that which occurs with serine, threonine, and tyrosine phosphorylation *(23)*. Alternatively, these reactions may be destructive under conditions of cytotoxicity.

Reactive nitric oxide synthase-mediated S-nitrosylation of thiol groups may alter protein structure/function or promote oxidation of vicinal thiols or histidine residues *(23, 38,70)*. The activities of G-proteins, ion channels, and kinases are altered by S-nitrosylation *(23,66)*. For example, S-nitrosylation will promote a structural activation of tissue plasminogen activator *(23)*. A more complicated example is observed with RNOS effects on the ryanodine-sensitive calcium release channel. S-Nitrosylation will reversibly activate the ryanodine-sensitive calcium channel, with an irreversible activation occurring with RNOS-mediated oxidation of vicinal thiols *(71)*. RNOS influences gene regulation through nitrosylation of cysteine residues of transcriptional regulators *(66, 72)*, as exemplified by NO-mediated functional alterations of the transcription factors NF-κB and AP-1.

Depending on the conditions, RNOS-mediated nitration of proteins may be necessary or superfluous. Nitration of tyrosine residues is essentially irreversible and may represent an alternative mechanism of tyrosine phosphorylation regulation. Nitrated proteins are typically targeted for degradation, which suggests a physiological regulation of protein turnover. RNOS tyrosine nitration, without exception, diminishes protein function, as observed with SOD and glutamine synthetase *(49)*. RNOS can also nitrate tryptophan residues *(73)*, although the significance of this reaction is unclear.

3.2.4. Lipid Peroxidation

Although NO itself acts to quench lipid alkoxyl (LO·) or lipid hydroperoxyl (LOO·) radicals, RNOS can initiate and propagate lipid peroxidation *(33,45,49)*. The reaction mechanisms are unclear but may include a direct oxidant effect on fatty acids and/or oxidation of endogenous antioxidants, such as α-tocopherol *(45)*.

3.3. RNOS Cytotoxic Cell Biological Mechanisms Unique to the CNS

3.3.1. N-Methyl-D-Aspartate Receptor-Mediated Excitotoxicity

The *N*-methyl-D-aspartate (NMDA) receptor is a ligand-gated ion channel that permits the influx of calcium upon ligand binding by the coneurotransmitters glutamate and glycine. Under conditions of excessive synaptic glutamate, the subsequent chronic influx of calcium via the NMDA receptor complex promotes excitotoxicity through prolonged nNOS synthesis of NO and mitochondrial production of $O_2^{\cdot-}$. NMDA receptor activation in non-nNOS neurons stimulates the mitochondrial synthesis of $O_2^{\cdot-}$. These products (NO and $O_2^{\cdot-}$) diffuse from their sources to form $ONOO^-$, which is cytoxic to neighboring neurons *(74,75)*. nNOS neurons are uniquely resistant to NMDA toxicity because of overexpression of MnSOD *(41)*. This hypothesis of NO-mediated NMDA excitotoxicity is not universally accepted, as the phenomenon has been inconsistently replicated by various laboratory groups using in vitro cell cultures *(76)*.

S-Nitrosylation of a redox modulatory site of the NMDA receptor complex causes a downregulation of the frequency of NMDA receptor channel opening *(31)*. Specifically, S-nitrosylation of the Cys-399 residue of the NR2A subunit represents a feedback mechanism to curtail not only excessive receptor activation but also provide neuroprotection *(77)*. However, not all laboratory groups are convinced that NO acts exclusively at the redox modulatory site, as other evidence suggests that NO may act at multiple sites of the NMDA receptor complex *(76)*.

3.3.2. Inhibition or Stimulation of Synaptic Neurotransmitter Reuptake Transporters

Excessive glutamate in the synaptic cleft is an initiating event for excitotoxcity. NO was observed to not only reduce the uptake of glutamate from synaptosomes or cultured astrocytes *(78–81)* by S-nitrosylation *(82)* but also to augment the release of glutamate *(83,84)*, which may promote a vicious cycle of glutamate neurotoxicity via activation of glutamate receptor subtypes.

Although a direct neurotoxic effect is not evident, NO can mediate the uptake and release of the neurotransmitters. Specifically, S-nitrosylation is thought to underlie the inhibition of norepinephrine reuptake *(85)*; NO also mediates norepinephrine and L-glutamate release following NMDA receptor stimulation *(84)*. It is believed that NO-mediated alterations in calcium and sodium flux are operative mechanisms for a NO-dependent enhancement of GABA release *(86)*. Interestingly, NO can enhance the reuptake of seratonin *(87)*. Divergent results were reported for the effects of NO on dopamine reuptake *(81,85,88)*.

4. NITRIC OXIDE NEUROTOXICITY IN CENTRAL NERVOUS SYSTEM DISEASES: AN OVERVIEW

4.1. NO: CNS Cell Sources and Targets

Nitric oxide synthase activity is implicated in many human neurological diseases and animal disease models. The sources and targets of NO toxicity are CNS cells themselves (Fig. 2). Subsets of neuronal populations can express two or three NOS isoforms. For example, cultured cerebellar granule neurons express all three isoforms *(64,89)* and hippocampal pyramidal CA1–CA3 neurons express eNOS *(64)*. Astrocytes can express all three isoforms *(72)*, microglial cells clearly express iNOS *(20,90–96)*, and cerebrovascular endothelial cells express eNOS and iNOS. nNOS neurons make up <5% of the neuronal population *(97)* and show remarkable resistance to RNOS toxicity because of their exceptional expression and activity of manganese SOD *(41)*. However, given the diffusivity of NO, neighboring cells are susceptible to RNOS toxicity. Neurons *(75,93)*, oligodendrocytes *(98)*, choroid epithelium of the blood–cerebrospinal fluid barrier *(99)*, and endothelial cells of the blood-brain barrier *(100–102)* are susceptible to RNOS toxicity. Neurons and oligodendrocytes, in contrast to astrocytes, are exquisitely vulnerable to RNOS toxicity owing to an inability to sustain a high level of glycolysis and lower reserves of intracellular antioxidants, such as glutathione and α-tocopherol (vitamin E) *(67)*.

4.2. RNOS Evidence in CNS Diseases

The following subsections present an overview of literature evidence that implicates RNOS participation in a variety of CNS diseases. Although the pathological causes of

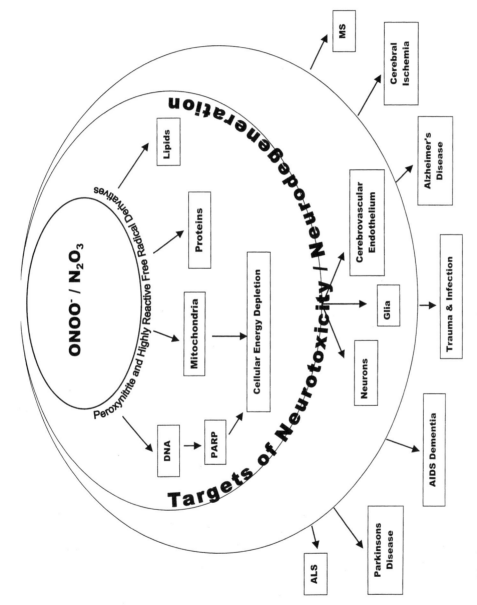

Fig. 2. Targets of nitric oxide-mediated neurotoxicity and neurodegeneration. Neurons, glia, and the cerebrovascular endothelium incur toxic insults, which, depending on the brain region and extent of toxic injury, may initiate or promote a variety of neurological afflictions.

many neurodegenerative diseases remain elusive, basic and clinical research has identified a participatory role of RNOS in the neurodegenerative disease process (Fig. 2).

4.2.1. Infection, Inflammation, and Blood-Brain Barrier Disruption

Alterations in blood-brain barrier function are evident during CNS infection or inflammatory processes *(103)* [e.g., multiple sclerosis *(104,105)*, HIV-1 dementia *(106)*, cerebral ischemia *(107)*, brain tumors *(108)*, and meningitis *(109)*]. Given that excessive NO synthesis may be a pathological process involved in a number of these disease states (*see* following subsections), it is tempting to speculate that NO might be involved in permeability changes of the blood-brain barrier.

Morphologically, the blood-brain barrier consists of astrocytic processes enveloping cerebral endothelial capillaries. Support for the hypothesis that excessive production of NO mediates blood-brain-barrier disruption is derived from in vitro studies identifying iNOS induction in human astrocytes *(110,111)*, fibroblasts *(112)*, and endothelial cells *(113,114)*. In rats, intracisternal administration of lipopolysaccharide (LPS) provoked blood-brain- and blood–cerebrospinal fluid (CSF)-barrier disruption, accompanied by meningeal inflammation and NO synthesis *(99,100,115,116)*. Treatment with a specific iNOS inhibitor, aminoguanidine, during meningeal inflammation significantly diminished meningeal NO production and preserved normal blood-brain- and blood–CSF-barrier integrity *(99,100)*. Finally, pharmacological provision of NO (as NO, NO^- and NO^+ redox forms) enhanced the blood-brain-barrier permeability in normal rodents *(102,117)*. NO itself provoked only modest elevations in permeability; the greatest enhancement of barrier occurred with NO donors that released the NO^- and NO^+ redox forms *(102)*.

4.2.2. Cerebral Ischemia

The roles of NO and RNOS during cerebral ischemia appear to be contradictory *(118)*. However, reasonable amounts of NO, particularly that produced by eNOS, aid in promoting regional cerebral blood flow during ischemia, thereby minimizing the extent of the infarct. During cerebral ischemia, excessive glutamate release and persistence in the synaptic cleft promotes excitotoxicity via activation of NMDA receptors and nNOS. $ONOO^-$ and RNOS are produced most likely during the reperfusion phase following ischemia *(119)* and contribute to neuronal cell death *(118)*. One to two days following the ischemic event, microglia and invading phagocytes supply NO (via iNOS) and $O_2^{\bullet-}$ to generate additional $ONOO^-$ *(120)*. 3-Nitrotyrosinated proteins, a fingerprint for $ONOO^-$, are found in the infarcted region, along with evidence for nNOS activity throughout the peri-infarct and infarct regions *(121)*.

4.2.3. Trauma

Traumatic injury of the brain or spinal cord triggers the expression of iNOS in rodent models *(122–124)*. The traumatic event triggers the elevation of iNOS induction signals (i.e., inflammatory cytokines, followed by iNOS is expression by injured astrocytes and microglia). Infiltrating phagocytes provides a second source of iNOS. Whether or not CNS trauma-induced RNOS toxicity occurs in humans is an issue that requires critical evaluation.

4.2.4. Amyotrophic Lateral Sclerosis

Amyotrophic lateral sclerosis (ALS) is a chronic and progressive paralytic disorder characterized by neurodegeneration of the large motor neurons in the brain and spinal cord. In the early 1990s, Beckman hypothesized that SOD mutations observed in ALS patients may contribute to an increased nitration of proteins (especially neurofilaments) and impaired phosphorylation *(58,125)* because of ineffective scavenging of $O_2^{\bullet-}$ and subsequent formation of RNOS. In support of this hypothesis, motor neuron death occurs in a transgenic mouse model of ALS, with features of mutated SOD and upregulation of iNOS and no change in nNOS *(126–128)*. Moreover, patients with sporadic ALS have elevated CSF concentrations of NO_2^- and NO_3^- *(129,130)* and evidence of iNOS expression and nitrotyrosinated proteins, as revealed by immunostaining *(131)*.

4.2.5. Multiple Sclerosis

Multiple sclerosis (MS) is an inflammatory demyelinating neurological disorder with a variable clinical presentation. Evidence for a participatory role of RNOS in MS is based on patient data and animal models of MS (i.e., experimental allergic encephalomyelitis [EAE]) *(132)*. iNOS and nitrotyrosinated proteins are prominent features found in the CNS lesions of patients and EAE animals *(38,67,132,133)*. MS patients with active disease also present with increased CSF NO_2^- and NO_3^- *(134–136)*. Of interest is the finding that interferon-β, an approved therapy for MS, inhibits iNOS expression in human astroglial cell cultures *(137)*.

4.2.6. Parkinson's Disease

Parkinson's disease is a chronic and progressive neurologic disorder involving "shaking palsy" (paralysis agitans) with four cardinal features of tremor, rigidity, bradykinesia, and postural instability. A pathological marker of Parkinson's disease is a significant loss of dopaminergic neurons in the substantia nigra. The current thinking is that mitochondrial dysfunction and oxidative stress may be major contributing mechanisms in the development of Parkinson's disease *(138)*, although the precise identities of the oxidative free-radical species awaits further characterization. RNOS are postulated to contribute to the Parkinsonian neurodegenerative processes *(139,140)*. A destructive role for nNOS was observed in animal models of MPTP-induced parkinsonism *(141,142)*. In addition, parkinsonian patients have increased CSF concentrations of NO_2^- and NO_3^- *(143)*, iNOS expression in glial cells of the substantia nigra *(140)*, and 3-nitrotyrosinated proteins *(144)*.

4.2.7. Alzheimer's Disease

Alzheimer's disease is a progressive degenerative dementia, characterized by a decline in memory and cognitive functions. Although Alzheimer's is likely to be an etiologically heterogeneous syndrome, neuroinflammation and oxidative stress are likely to contribute to the underlying pathology. It is suspected that β-amyloid plaques contribute to or are a consequence of the underlying pathology. Although some studies demonstrated that β-amyloid is directly neurotoxic in cell cultures *(145,146)*, other reports identified an indirect mechanism, involving β-amyloid induction of iNOS *(147,148)* and activation of NADPH oxidase to synthesize $O_2^{\bullet-}$ *(149)*. In addition, it is hypothesized that aberrant expression of cerebrovascular eNOS also promotes the formation of neurotoxic RNOS

(150–152). Finally, elevations of nitrotyrosinated proteins in Alzheimer's patients provides data that RNOS contribute to the pathology *(153,154)*.

4.2.8. AIDS Dementia

Acquired immunodeficiency syndrome (AIDS) dementia is a neurologic complication of human immunodeficiency virus type 1 (HIV-1) infection. A current, though not universally accepted hypothesis is that RNOS neurotoxicity contributes to AIDS dementia *(155–159)*. The HIV-1 coat protein, gp41, triggers neurotoxicity through induction of iNOS *(160)*. Postmortem examination of patients with AIDS dementia typically reveal iNOS immunoreactivity that is colocalized with gp41 *(155–158)* and accompanied by an increased in nitrotyrosinated brain proteins *(158)*.

REFERENCES

1. Liaudet, L., Soriano, F. G., and Szabo, C. (2000) Biology of nitric oxide signaling. *Crit. Care Med.* **28,** N37–N52.
2. Denninger, J. W. and Marletta, M. A. (1999) Guanylate cyclase and the NO/cGMP signaling pathway. *Biochim. Biophys. Acta* **1411,** 334–350.
3. Blake, D. J. and Kroger, S. (2000) The neurobiology of duchenne muscular dystrophy: learning lessons from muscle? *Trends Neurosci.* **23,** 92–99.
4. Bredt, D. S. (1999) Endogenous nitric oxide synthesis: biological functions and pathophysiology. *Free Radical Res.* **31,** 577–596.
5. McCann, S. M., Mastronardi, C., Walczewska, A., Karanth, S., Rettori, V., and Yu, W. H. (1999) The role of nitric oxide in reproduction. *Braz. J. Med. Biol. Res.* **32,** 1367–1379.
6. Noris, M. and Remuzzi, G. (1999) Physiology and pathophysiology of nitric oxide in chronic renal disease. *Proc. Assoc. Am. Physicians* **111,** 602–610.
7. Boger, R. H. (1999) Nitric oxide and the mediation of the hemodynamic effects of growth hormone in humans. *J. Endocrinol. Invest.* **22,** 75–81.
8. Gerlach, M., Blum-Degen, D., Lan, J., and Riederer, P. (1999) Nitric oxide in the pathogenesis of Parkinson's disease. *Adv. Neurol.* **80,** 239–245.
9. Bigatello, L. M. (2000) Nitric oxide: modulation of the pulmonary circulation. *Minerva Anestesiol.* **66,** 307–313.
10. Matheson, P. J., Wilson, M. A., and Garrison, R. N. (2000) Regulation of intestinal blood flow. *J. Surg. Res.* **93,** 182–196.
11. Ignarro, L. J. (1999) Nitric oxide: a unique endogenous signaling molecule in vascular biology. *Biosci. Rep.* **19,** 51–71.
12. Contestabile, A. (2000) Roles of NMDA receptor activity and nitric oxide production in brain development. *Brain Res. Rev.* **32,** 476–509.
13. Agerman, K., Canlon, B., Duan, M., and Ernfors, P. (1999) Neurotrophins, NMDA receptors, and nitric oxide in development and protection of the auditory system. *Ann. NY Acad. Sci.* **884,** 131–142.
14. Michel, T. and Feron, O. (1997) Nitric oxide synthases: which, where, how, and why? *J. Clin. Invest.* **100,** 2146–2152.
15. Moncada, S., Higgs, A., and Furchgott, R. (1997) International Union of Pharmacology nomenclature in nitric oxide research. *Pharmacol. Rev.* **49,** 137–142.
16. Boje, K. M. K. (1997) The role of glial nitric oxide in neurotoxicity and central nervous system diseases, in *Neuroinflammation* (Wood, P. L., ed.), Humana, Totowa, NJ.
17. Dawson, T. M., Sasaki, M., Gonzalez-Zulueta, M., and Dawson, V. L. (1998) Regulation of neuronal nitric oxide synthase and identification of novel nitric oxide signaling pathways. *Prog. Brain Res.* **118,** 3–11.

18. Cho, H. J., Xie, Q. W., Calaycay, J., Mumford, R. A., Swiderek, K. M., Lee, T. D., et al. (1992) Calmodulin is a subunit of nitric oxide synthase from macrophages. *J. Exp. Med.* **176,** 599–604.

19. Nathan, C. and Xie, Q. W. (1994) Regulation of biosynthesis of nitric oxide. *J. Biol. Chem.* **269,** 13,725–13,728.

20. Galea, E., Feinstein, D. L., and Reis, D. J. (1992) Induction of calcium-independent nitric oxide synthase activity in primary rat glial cultures. *Proc. Natl. Acad. Sci. USA* **89,** 10,945–10,949.

21. Nathan, C. and Xie, Q. W. (1994) Nitric oxide synthases: roles, tolls, and controls. *Cell* **78,** 915–918.

22. Förstermann, U., Gath, I., Schwarz, P., Closs, E. I., and Kleinert, H. (1995) Isoforms of nitric oxide synthase: properties, cellular distribution and expressional control. *Biochem. Pharmacol.* **50,** 1321–1332.

23. Gaston, B. (1999) Nitric oxide and thiol groups. *Biochim. Biophys. Acta* **1411,** 323–333.

24. Jia, L., Bonaventura, C., Bonaventura, J., and Stamler, J. S. (1996) *S*-Nitrosohaemoglobin: a dynamic activity of blood involved in vascular control. *Nature* **380,** 221–226.

25. Gow, A. J. and Stamler, J. S. (1998) Reactions between nitric oxide and haemoglobin under physiological conditions. *Nature* **391,** 169–173.

26. Stamler, J. S., Jia, L., Eu, J. P., McMahon, T. J., Demchenko, I. T., Bonaventura, J., et al. (1997) Blood flow regulation by *S*-nitrosohemoglobin in the physiological oxygen gradient. *Science* **276,** 2034–2037.

27. Gross, S. S. and Lane, P. (1999) Physiological reactions of nitric oxide and hemoglobin: a radical rethink. *Proc. Natl. Acad. Sci. USA* **96,** 9967–9969.

28. Gow, A. J., Luchsinger, B. P., Pawloski, J. R., Singel, D. J., and Stamler, J. S. (1999) The oxyhemoglobin reaction of nitric oxide. *Proc. Natl. Acad. Sci. USA* **96,** 9027–9032.

29. Wang, K., Zhang, W., Xian, M., Hou, Y. C., Chen, X. C., Cheng, J. P., et al. (2000) New chemical and biological aspects of S-nitrosothiols. *Curr. Med. Chem.* **7,** 821–834.

30. Hogg, N. (2000) Biological chemistry and clinical potential of *S*-nitrosothiols. *Free Radical Biol. Med.* **28,** 1478–1486.

31. Lipton, S. A., Rayudu, P. V., Choi, Y. B., Sucher, N. J., and Chen, H. S. (1998) Redox modulation of the NMDA receptor by NO-related species. *Prog. Brain Res.* **118,** 73–82.

32. Chiueh, C. C. and Rauhala, P. (1999) The redox pathway of *S*-nitrosoglutathione, glutathione and nitric oxide in cell to neuron communications. *Free Radical Res.* **31,** 641–650.

33. Grisham, M. B., Jourdheuil, D., and Wink, D. A. (1999) Nitric oxide I. Physiological chemistry of nitric oxide and its metabolites: implications in inflammation. *Am. J. Physiol.* **276,** G315–G321.

34. Koppenol, W. H. (1998) The basic chemistry of nitrogen monoxide and peroxynitrite. *Free Radical Biol. Med.* **25,** 385–391.

35. Kelm, M. (1999) Nitric oxide metabolism and breakdown. *Biochim. Biophys. Acta* **1411,** 273–289.

36. Wink, D. and Mitchell, J. B. (1998) Chemical biology of nitric oxide: insights into regulatory, cytotoxic and cytoprotective mechanisms of nitric oxide. *Free Radical Biol. Med.* **25,** 434–456.

37. Squadrito, G. L. and Pryor, W. A. (1998) Oxidative chemistry of nitric oxide: the roles of superoxide, peroxynitrite, and carbon dioxide. *Free Radical Biol. Med.* **25,** 392–403.

38. Torreilles, F., Salman-Tabcheh, S., Guerin, M.-C., and Torreilles, J. (1999) Neurodegenerative disorders: the role of peroxynitrite. *Brain Res. Rev.* **30,** 153–163.

39. Leist, M. and Nicotera, P. (1998) Apoptosis, excitotoxicity and neuropathology. *Exp. Cell Res.* **239,** 183–201.

40. Kim, Y.-M., Bombeck, C. A., and Billiar, T. R. (1999) Nitric oxide as a bifunctional regulator of apoptosis. *Circ. Res.* **84,** 253–256.

41. Gonzalez-Zulueta, M., Ensz, L. M., Mukhina, G., Lebovitz, R. M., Zwacka, R. M., Engelhardt, J. F., et al. (1998) Manganese superoxide dismutase protects nNOS neurons from NMDA and nitric oxide-mediated neurotoxicity. *J. Neurosci.* **18,** 2040–2055.
42. Colasanti, M. and Suzuki, H. (2000) The dual personality of NO. *Trends Pharmacol. Sci.* **21,** 249–252.
43. Hawkins, R. D., Son, H., and Arancio, O. (1998) Nitric oxide as a retrograde messenger during long-term potentiation in hippocampus. *Prog. Brain Res.* **118,** 155–172.
44. Kara, P. and Friedlander, M. J. (1998) Dynamic modulation of cerebral cortex synaptic function by nitric oxide. *Prog. Brain Res.* **118,** 183–198.
45. Hogg, N. and Kalyanaraman, B. (1999) Nitric oxide and lipid peroxidation. *Biochim. Biophys. Acta* **1411,** 378–384.
46. Chiueh, C. C. (1999) Neuroprotective properties of nitric oxide. *Ann. NY Acad. Sci.* **890,** 301–311.
47. Hensley, K., Tabatabaie, T., Stewart, C. A., Pye, Q., and Floyd, R. A. (1997) Nitric oxide and derived species as toxic agents in stroke, AIDS dementia, and chronic neurodegenerative disorders. *Chem. Res. Toxicol.* **10,** 527–532.
48. Crow, J. P. and Beckman , J. S. (1995) The role of peroxynitrite in nitric oxide-mediated toxicity. *Curr. Topics Microbiol. Immunol.* **196,** 57–73.
49. Patel, R. P., McAndrew, J., Sellak, H., White, C. R., Ho, H., Freeman, B. A., et al. (1999) Biological aspects of reactive nitrogen species. *Biochim. Biophys. Acta* **1411,** 385–400.
50. Cooper, C. E. (1999) Nitric oxide and iron proteins. *Biochim. Biophys. Acta* **1411,** 290–309.
51. Torres, J. and Wilson, M. T. (1999) The reactions of copper proteins with nitric oxide. *Biochim. Biophys. Acta* **1411,** 310–322.
52. Schmidt, H. H. H. W., Hofmann, H., Schindler, U., Shutenko, Z. S., Cunningham, D. D., and Feelish, M. (1996) No NO• from NO synthase. *Proc. Natl. Acad. Sci. USA* **93,** 14,492–14,497.
53. Rusche, K. M., Spiering, M. M., and Marletta, M. A. (1998) Reactions catalyzed by tetrahydrobiopterin-free nitric oxide synthase. *Biochemistry* **37,** 15,503–15,512.
54. Arnelle, D. R. and Stamler, J. S. (1995) NO+, NO, and NO− donation by *S*-nitrosothiols: implications for regulation of physiological functions by S-nitrosylation and acceleration of disulfide formation. *Arch. Biochem. Biophys.* **318,** 279–285.
55. Murphy, M. E. and Sies, H. (1991) Reversible conversion of nitroxyl anion to nitric oxide by superoxide dismutase. *Proc. Natl. Acad. Sci. USA* **88,** 10,860–10,864.
56. Chazotte-Aubert, L., Oikawa, S., Gilibert, I., Bianchini, F., Kawanishi, S., and Ohshima, H. (1999) Cytotoxicity and site-specific DNA damage induced by nitroxyl anion (NO(−)) in the presence of hydrogen peroxide. Implications for various pathophysiological conditions. *J. Biol. Chem.* **274,** 20,909–20,915.
57. Beckman, J. S., Chen, J., Crow, J. P., and Ye, Y. Z. (1994) Reactions of nitric oxide, superoxide and peroxynitrite with superoxide dismutase in neurodegeneration. *Prog. Brain Res.* **103,** 371–380.
58. Beckman, J. S. (1994) Peroxynitrite versus hydroxyl radical: the role of nitric oxide in superoxide-dependent cerebral injury. *Ann. NY Acad. Sci.* **738,** 69–75.
59. Wink, D. A., Kasprzak, K. S., Maragos, C. M., Elespuru, R. K., Misra, M., Dunams, T. M., et al. (1991) DNA deaminating ability and genotoxicity of nitric oxide and its progenitors. *Science* **254,** 1001–1003.
60. Derojaswalker, T., Tamir, S., Ji, H., Wishnok, J. S., and Tannenbaum, S. R. (1995) Nitric oxide induces oxidative damage in addition to deamination in macrophage DNA. *Chem. Res. Toxicol.* **8,** 473–477.
61. Graziewicz, M., Wink, D. A., and Laval, F. (1996) Nitric oxide inhibits DNA ligase activity: potential mechanisms for NO-mediated DNA damage. *Carcinogenesis* **17,** 2501–2505.
62. Kwon, N. S., Stuehr, D. J., and Nathan, C. (1991) Inhibition of tumor cell ribonucleotide reductase by macrophage-derived nitric oxide. *J. Exp. Med.* **174,** 761–768.

63. Zhang, J., Dawson, V. L., Dawson, T. M., and Snyder, S. H. (1994) Nitric oxide activation of poly(ADP-ribose) synthetase in neurotoxicity. *Science* **263,** 687–689.
64. Dawson, V. L. and Dawson, T. (1998) Nitric oxide in neurodegeneration. *Prog. Brain Res.* **118,** 215–229.
65. Beal, M. F. (1999) Mitochondria, NO and neurodegeneration. *Biochem. Soc. Symp.* **66,** 43–54.
66. Murphy, M. P. (1999) Nitric oxide and cell death. *Biochim. Biophys. Acta* **1411,** 401–414.
67. Heales, J. R., Bolanos, J. P., Stewart, V. C., Brookes, P. S., Land, J. M., and Clark, J. B. (1999) Nitric oxide, mitochondria and neurological disease. *Biochim. Biophys. Acta* **1410,** 215–228.
68. Brown, G. C. (1999) Nitric oxide and mitochondrial respiration. *Biochim. Biophys. Acta* **1411,** 351–369.
69. Dimmeler, S., Lottspeich, F., and Brune, B. (1992) Nitric oxide causes ADP-ribosylation and inhibition of glyceraldehyde-3-phosphate dehydrogenase. *J. Biol. Chem.* **267,** 16,771–16,774.
70. Brolliet, M.-C. (1999) S-Nitrosylation of proteins. *Cell. Mol. Life Sci.* **55,** 1036–1042.
71. Xu, L., Eu, G., Meissner, G., and Stamler, J. S. (1988) Activation of the cardiac calcium release channel (Ryanodine receptor) by poly-S-nitrosylation. *Science* **279,** 234–237.
72. Murphy, S. (2000) Production of nitric oxide by glial cells: regulation and potential roles in the CNS. *Glia* **29,** 1–14.
73. Alvares, B., Rubbo, H., Kirk, M., Barnes, S., Freeman, B. A., and Radi, R. (1996) Peroxynitrite-dependent tryptophan nitration. *Chem. Res. Toxicol.* **9,** 390–396.
74. Dawson, V. L., Dawson, T. M., London, E. D., Bredt, D. S., and Snyder, S. H. (1991) Nitric oxide mediates glutamate neurotoxicity in primary cortical cultures. *Proc. Natl. Acad. Sci. USA* **88,** 6368–6371.
75. Dawson, V. L., Brahmbhatt, H. P., Mong, J. A., and Dawson, T. M. (1994) Expression of inducible nitric oxide synthase causes delayed neurotoxicity in primary mixed neuronal–glial cortical cultures. *Neuropharmacology* **33,** 1425–1430.
76. Aizenman, E., Brimecombe, J. C., Potthoff, W. K., and Rosenberg, P. A. (1998) Why is the role of nitric oxide in NMDA receptor function and dysfunction so controversial? *Prog. Brain Res.* **118,** 53–71.
77. Kim, W. K., Choi, Y. B., Rayudu, P. V., Das, P, Asaad, W., Arnelle, D. R., et al. (1999) Attenuation of NMDA receptor activity and neurotoxicity by nitroxyl anion, NO–. *Neuron* **24,** 461–469.
78. Pogun, S., Dawson, V., and Kuhar, M. J. (1994) Nitric oxide inhibits 3H-glutamate transport in synaptosomes. *Synapse* **18,** 21–26.
79. Sorg, O., Horn, T. F., Yu, N., Gruol, D. L., and Bloom, F. E. (1997) Inhibition of astrocyte glutamate uptake by reactive oxygen species: role of antioxidant enzymes. *Mol. Med.* **2,** 431–440.
80. Hu, S., Sheng, W. S., Ehrlich, L. C., Peterson, P. K., and Chao, C. C. (2000) Cytokine effects on glutamate uptake by human astrocytes. *Neuroimmunomodulation* **7,** 153–159.
81. Lonart, G. and Johnson, K. M. (1994) Inhibitory effects of nitric oxide on the uptake of [3H]dopamine and [3H]glutamate by striatal synaptosomes. *J. Neurochem.* **63,** 2108–2117.
82. Wolosker, H., Reis, M., Assreuy, J., and de Meis, L. (1996) Inhibition of glutamate uptake and proton pumping in synaptic vesicles by S-nitrosylation. *J. Neurochem.* **66,** 1943–1948.
83. Ye, Z.-C. and Sontheimer, H. (1998) Glial glutamate transport as target for nitric oxide: consequences for neurotoxicity. *Prog. Brain Res.* **118,** 241–251.
84. Montague, P. R., Gancayco, C. D., Winn, M. J., Marchase, R. B., and Friedlander, M. J. (1994) Role of NO production in NMDA receptor-mediated neurotransmitter release in cerebral cortex. *Science* **263,** 973–977.
85. Kaye, D. M., Gruskin, S., Smith, A. I., and Esler, M. D. (2000) Nitric oxide mediated modulation of norepinephrine transport: identification of a potential target for S-nitrosylation. *Br. J. Pharmacol.* **130,** 1060–1064.

86. Ohkuma, S., Katsura, M., Chen, D. Z., Narihara, H., and Kuriyama, K. (1996) Nitric oxide-evoked [3H] gamma-aminobutyric acid release is mediated by two distinct release mechanisms. *Mol. Brain Res.* **36,** 137–144.

87. Miller, K. J. and Hoffman, B. J. (1994) Adenosine A3 receptors regulate serotonin transport via nitric oxide and cGMP. *J. Biol. Chem.* **269,** 27,351–27,356.

88. Pogun, S., Baumann, M. H., and Kuhar, M. J. (1994) Nitric oxide inhibits [3H]dopamine uptake. *Brain Res.* **641,** 83–91.

89. Hemeka, M. T., Feinstein, D. L., Galea, E., Gleichmann, M., Wullner, U., and Klockgether, T. (1999) Peroxisome proliferator-activated receptor gamma agonists protect cerebellar granule cells from cytokine-induced apoptotic cell deathe by inhibition of inducible nitric oxide synthase. *J. Neuroimmunol.* **100,** 156–168.

90. Agullo, L., Baltrons, M. A., and Garcia, A. (1995) Calcium-dependent nitric oxide formation in glial cells. *Brain Res.* **686,** 160–168.

91. Egberongbe, Y. I., Gentleman, S. M., Falkai, P., Bogerts, B., Polak, J. M., and Roberts, G. W. (1994) The distribution of nitric oxide synthase immunoreactivity in the human brain. *Neuroscience* **59,** 561–578.

92. Kugler, P. and Drenckhahn, D. (1996) Astrocytes and Bergmann glia as an important site of nitric oxide synthase I. *Glia* **16,** 165–173.

93. Boje, K. M. and Arora, P. K. (1992) Microglial-produced nitric oxide and reactive nitrogen oxides mediate neuronal cell death. *Brain Res.* **587,** 250–256.

94. Chao, C. C., Hu, S., Molitor, T. W., Shaskan, E. G., and Peterson, P. K. (1992) Activated microglia mediate neuronal cell injury via a nitric oxide mechanism. *J. Immunol.* **149,** 2736–2741.

95. Simmons, M. L. and Murphy, S. (1992) Induction of nitric oxide synthase in glial cells. *J. Neurochem.* **59,** 897–905.

96. Zielasek, J., Tausch, M., Toyka, K. V., and Hartung, H.-P. (1992) Production of nitrite by neonatal rat microglial cells/brain macrophages. *Cell Immunol.* **141,** 111–120.

97. Dawson, T. M., Bredt, D. S., Fotuhi, M., Hwang, P. M., and Snyder, S. H. (1991) Nitric oxide synthase and neuronal NADPH diaphorase are identical in brain and peripheral tissues. *Proc. Natl. Acad. Sci. USA* **88,** 7797–7801.

98. Merrill, J. E., Ignarro, L. J., Sherman, M. P., Melinek, J., and Lane, T. E. (1993) Microglial cell cytotoxicity of oligodendrocytes is mediated through nitric oxide. *J. Immunol.* **151,** 2132–2141.

99. Boje, K. M. K. (1995) Inhibition of nitric oxide synthase partially attenuates alterations in the blood–cerebrospinal fluid barrier during experimental meningitis in the rat. *Eur. J. Pharmacol.* **272,** 297–300.

100. Boje, K. M. K. (1996) Inhibition of nitric oxide synthase attenuates blood-brain barrier disruption during experimental meningitis. *Brain Res.* **720,** 75–83.

101. Borgerding, R. A. and Murphy, S. (1995) Expression of inducible nitric oxide synthase in cerebral endothelial cells is regulated by cytokine-activated astrocytes. *J. Neurochem.* **65,** 1342–1347.

102. Boje, K. M. K. and Lakhman, S. S. (2000) Direct cerebrovascular infusion of nitric oxide prodrugs elicit disruption of the blood-brain barrier. *J. Pharmacol. Exp. Ther.* **293,** 545–550.

103. de Vries, H. E., Kuiper, J., de Boer, A. G., Van Berkel, T. J. C., and Breimer, D. D. (1997) The blood-brain barrier in neuroinflammatory diseases. *Pharmacol. Rev.* **49,** 143–156.

104. Poser, C. M. (1994) Notes on the pathogenesis of multiple sclerosis. *Clin. Neurosci.* **2,** 258–265.

105. Stone, L. A., Smith, M. E., Albert, P. S., Bash, C. N., Maloni, H., Frank, J. A., et al. (1995) Blood-brain barrier disruption on contrast-enhanced MRI in patients with mild relapsing-remitting multiple sclerosis: relationship to course, gender, and age. *Neurology* **45,** 1122–1126.

106. Power, C. and Johnson, R. T. (1995) HIV-1 associated dementia: clinical features and pathogenesis. *Can. J. Neurol. Sci.* **22,** 92–100.

107. Zhang, J., Benveniste, H., Klitzman, B., and Piantadosi, C. A. (1995) Nitric oxide synthase inhibition and extracellular glutamate concentration after cerebral ischemia/reperfusion. *Stroke* **26,** 298–304.

108. Stewart, D. J. (1994) A critique of the role of the blood-brain barrier in the chemotherapy of human brain tumors. *J. Neuro–Oncol.* **20,** 121–139.

109. Tunkel, A. R. and Scheld, W. M. (1993) Pathogenesis and pathophysiology of bacterial–meningitis. *Ann. Rev. Med.* **44,** 103–120.

110. Lee, S. C., Dickson, D. W., Liu, W., and Brosnan, C. F. (1993) Induction of nitric oxide synthase activity in human astrocytes by interleukin-1 beta and interferon-gamma. *J. Neuroimmunol.* **46,** 19–24.

111. Lee, S. C., Dickson, D. W., Brosnan, C. F., and Casadevall, A. (1994) Human astrocytes inhibit *Cryptococcus* neoformans growth by a nitric oxide-mediated mechanism. *J. Exp. Med.* **180,** 365–369.

112. Skaper, S. D., Facci, L., and Leon, A. (1995) Inflammatory mediator stimulation of astrocytes and meningeal fibroblasts induces neuronal degeneration via the nitridergic pathway. *J. Neurochem.* **64,** 266–276.

113. Kilbourn, R. G. and Belloni, P. (1990) Endothelial cell production of nitrogen oxides in response to interferon gamma in combination with tumor necrosis factor, interleukin-1, or endotoxin. *J. Natl. Cancer Inst.* **82,** 772–776.

114. Rosenkranz-Weis, P., Sessa, W. C., Milstien, S., Kaufman, S., Watson, C. A., and Pober, J. S. (1994) Regulation of nitric oxide synthesis by proinflammatory cytokines in human umbilical vein endothelial cells. Elevations in tetrahydrobiopterin levels enhance endothelial nitric oxide synthase specific activity. *J. Clin. Invest.* **93,** 2236–2243.

115. Boje, K. M. K. (1995) Cerebrovascular permeability changes during experimental meningitis in the rat. *J. Pharmacol. Exp. Ther.* **274,** 1199–1203.

116. Korytko, P. J. and Boje, K. M. K. (1996) Pharmacological characterization of nitric oxide production in a rat model of meningitis. *Neuropharmacology* **35,** 231–237.

117. Mayhan, W. G. (2000) Nitric oxide donor-induced increase in permeability of the blood-brain barrier. *Brain Res.* **866,** 101–108.

118. Dalkara, T., Endres, M., and Moskowitz, M. A. (1998) Mechanisms of NO neurotoxicity. *Prog. Brain Res.* **118,** 231–239.

119. Gursoy-Ozdemir, Y., Bolay, H., Saribas, O., and Dalkara, T. (2000) Role of endothelial nitric oxide generation and peroxynitrite formation in reperfusion injury after focal cerebral ischemia. *Stroke* **31,** 1974–1980.

120. Iadecola, C., Xu, X., Zhang, F., el-Fakahany, E. E., and Ross, M. E. (1995) Marked induction of calcium-independent nitric oxide synthase activity after focal cerebral ischemia. *J. Cereb. Blood Flow Metab.* **15,** 52–59.

121. Eliasson, M. J., Huang, Z., Ferrante, R. J., Sasamata, M., Molliver, M. E., Snyder, S. H., et al. (1999) Neuronal nitric oxide synthase activation and peroxynitrite formation in ischemic stroke linked to neural damage. *J. Neurosci.* **19,** 5910–5918.

122. Bethea, J. R., Castro, M., Keane, R. W., Lee, T. T., Dietrich, W. D., and Yezierski, R. P. (1998) Traumatic spinal cord injury induces nuclear factor-kappaB activation. *J. Neurosci.* **18,** 3251–3260.

123. Wada, K., Chatzipanteli, K., Kraydieh, S., Busto, R., and Dietrich, W. D. (1998) Inducible nitric oxide synthase expression after traumatic brain injury and neuroprotection with aminoguanidine treatment in rats. *Neurosurgery* **43,** 1427–1436.

124. Chabrier, P.-E., Demerle-Pallardy, C., and Auguet, M. (1999) Nitric oxide synthases: targets for therapeutic strategies in neurological diseases. *Cell. Mol. Life Sci.* **55,** 1029–1035.

125. Beckman, J. S., Carson, M., Smith, C. D., and Koppenol, W. H. (1993) ALS, SOD and peroxynitrite. *Nature* **364,** 584.

126. Estevez, A. G., Crow, J. P., Sampson, J. B., Reiter, C., Zhuang, Y., Richardson, G. J., et al. (1999) Induction of nitric oxide-dependent apoptosis in motor neurons by zinc-deficient superoxide dismutase. *Science* **286,** 2498–2500.

127. Facchinetti, F., Sasaki, M., Cutting, F. B., Zhai, P., MacDonald, J. E., Reif, D., et al. (1999) Lack of involvement of neuronal nitric oxide synthase in the pathogenesis of a transgenic mouse model of familial amyotrophic lateral sclerosis. *Neurosci.* **90,** 1483–1492.

128. Almer, G., Vukosavic, S., Romero, N., and Przedborski, S. (1999) Inducible nitric oxide synthase up-regulation in a transgenic mouse model of familial amyotrophic lateral sclerosis. *J. Neurochem.* **72,** 2415–2425.

129. Taskiran, D., Sagduyu, A., Yuceyar, N., Kutay, F. Z., and Pogun, S. (2000) Increased cerebrospinal fluid and serum nitrite and nitrate levels in amyotrophic lateral sclerosis. *Int. J. Neurosci.* **101,** 65–72.

130. Tohgi, H., Abe, T., Yamazaki, K., Murata, T., Ishizaki, E., and Isobe, C. (1999) Increase in oxidized NO products and reduction in oxidized glutathione in cerebrospinal fluid from patients with sporadic form of amyotrophic lateral sclerosis. *Neurosci. Lett.* **260,** 204–206.

131. Sasaki, S., Shibata, N., Komori, T., and Iwata, M. (2000) iNOS and nitrotyrosine immunoreactivity in amyotrophic lateral sclerosis. *Neurosci. Lett.* **291,** 44–48.

132. Parkinson, J. F., Mitrovic, B., and Merril, J. E. (1997) The role of nitric oxide in multiple sclerosis. *J. Mol. Med.* **75,** 174–186.

133. Cross, A. H., Manning, P. T., Keeling, R. M., Schmidt, R. E., and Misko, T. P. (1998) Peroxynitrite formation within the central nervous system in active multiple sclerosis. *J. Neuroimmunol.* **88,** 45–56.

134. Giovannoni, G., Heales, S. J., Silver, N. C., O'Riordan, J., Miller, R. F., Land, J. M., et al. (1997) Raised serum nitrate and nitrite levels in patients with multiple sclerosis. *J. Neurol. Sci.* **145,** 77–81.

135. Svenningsson, A., Petersson, A. S., Andersen, O., and Hansson, G. K. (1999) Nitric oxide metabolites in CSF of patients with MS are related to clinical disease course. *Neurology* **53,** 1880–1882.

136. Brundin, L., Morcos, E., Olsson, T., Wiklund, N. P., and Andersson, M. (1999) Increased intrathecal nitric oxide formation in multiple sclerosis: cerebrospinal fluid nitrite as activity marker. *Eur. J. Neurol.* **6,** 585–590.

137. Hua, L. L., Liu, J. S., Brosnan, C. F., and Lee, S. C. (1998) Selective inhibition of human glial inducible nitric oxide synthase by interferon-beta: implications for multiple sclerosis. *Ann. Neurol.* **43,** 384–387.

138. Zhang, Y., Dawson, V. L., and Dawson, T. M. (2000) Oxidative stress and genetics in the pathogenesis of Parkinson's disease. *Neurobiol. Dis.* **7,** 240–250.

139. Hunot, S., Dugas, N., Faucheux, B., Hartmann, A., Tardieu, M., Debre, P., et al. (1999) Fc epsilonRII/CD23 is expressed in Parkinson's disease and induces, in vitro, production of nitric oxide and tumor necrosis factor-alpha in glial cells. *J. Neurosci.* **19,** 3440–3447.

140. Hunot, S., Boissiere, F., Faucheux, B., Brugg, B., Mouatt-Prigent, A., Agid, Y., et al. (1996) Nitric oxide synthase and neuronal vulnerability in Parkinson's disease. *Neuroscience* **72,** 355–363.

141. Hantraye, P., Brouillet, E., Ferrante, R., Palfi, S., Dolan, R., Matthews, R. T., et al. (1996) Inhibition of neuronal nitric oxide synthase prevents MPTP-induced parkinsonism in baboons. *Nat. Med.* **2,** 1017–1021.

142. Przedborski, S., Jackson-Lewis, V., Yokoyama, R., Shibata, T., Dawson, V. L., and Dawson, T. M. (1996) Role of neuronal nitric oxide in 1-methyl-4-phenyl-1,2,3,6-tetrahydropyridine (MPTP)-induced dopaminergic neurotoxicity. *Proc. Natl. Acad. Sci. USA* **93,** 4565–4571.

143. Qureshi, G. A., Baig, S., Bednar, I., Sodersten, P., Forsberg, G., and Siden, A. (1995) Increased cerebrospinal fluid concentration of nitrite in Parkinson's disease. *Neuroreport* **6,** 1642–1644.

144. Good, P. F., Hsu, A., Werner, P., Perl, D. P., and Olanow, C. W. (1998) Protein nitration in Parkinson's disease. *Neuropathol. Exp. Neurol.* **57,** 338–342.

145. Kowall, N. W. (1994) Beta amyloid neurotoxicity and neuronal degeneration in Alzheimer's disease. *Neurobiol. Aging* **15,** 257–258.

146. Cotman, C. W. and Anderson, A. J. (1995) A potential role for apoptosis in neurodegeneration and Alzheimer's disease. *Mol. Neurobiol.* **10,** 19–45.

147. Meda, L., Cassatella, M. A., Szendrei, G. I., Otvos, L. Jr., Baron, P., Villalba, M., et al. (1995) Activation of microglial cells by beta-amyloid protein and interferon-gamma. *Nature* **374,** 647–650.

148. Goodwin, J. L., Uemura, E., and Cunnick, J. E. (1995) Microglial release of nitric oxide by the synergistic action of beta-amyloid and IFN-gamma. *Brain Res.* **692,** 207–214.

149. Bianca, V. D., Dusi, S., Bianchini, E., Dal Pra, I., and Rossi, F. (1999) Beta-amyloid activates the O-2 forming NADPH oxidase in microglia, monocytes, and neutrophils. A possible inflammatory mechanism of neuronal damage in Alzheimer's disease. *J. Biol. Chem.* **274,** 15,493–15,499.

150. de la Monte, S. M., Lu, B. X., Sohn, Y. K., Etienne, D., Kraft, J., Ganju, N., et al. (2000) Aberrant expression of nitric oxide synthase III in Alzheimer's disease: relevance to cerebral vasculopathy and neurodegeneration. *Neurobiol. Aging* **21,** 309–319.

151. de la Monte, S. M., Sohn, Y. K., Etienne, D., Kraft, J., and Wands, J. R. (2000) Role of aberrant nitric oxide synthase-3 expression in cerebrovascular degeneration and vascular-mediated injury in Alzheimer's disease. *Ann. NY Acad. Sci.* **903,** 61–71.

152. Sohn, Y. K., Ganju, N., Bloch, K. D., Wands, J. R., and de la Monte, S. M. (1999) Neuritic sprouting with aberrant expression of the nitric oxide synthase III gene in neurodegenerative diseases. *J. Neurol. Sci.* **162,** 133–151.

153. Tohgi, H., Abe, T., Yamazaki, K., Murata, T., Ishizaki, E., and Isobe, C. (1999) Alterations of 3-nitrotyrosine concentration in the cerebrospinal fluid during aging and in patients with Alzheimer's disease. *Neurosci. Lett.* **269,** 52–54.

154. Hensley, K., Maidt, M. L., Yu, Z., Sang, H., Markesbery, W. R., and Floyd, R. A. (1998) Electrochemical analysis of protein nitrotyrosine and dityrosine in the Alzheimer brain indicates region-specific accumulation. *J. Neurosci.* **18,** 8126–8132.

155. Adamson, D. C., McArthur, J. C., Dawson, T. M., and Dawson, V. L. (1999) Rate and severity of HIV-associated dementia (HAD): correlations with Gp41 and iNOS. *Mol. Med.* **5,** 98–109.

156. Rostasy, K., Monti, L., Yiannoutsos, C., Kneissl, M., Bell, J., Kemper, T. L., et al. (1999) Human immunodeficiency virus infection, inducible nitric oxide synthase expression, and microglial activation: pathogenetic relationship to the acquired immunodeficiency syndrome dementia complex. *Ann. Neurol.* **46,** 207–216.

157. Vincent, V. A., De Groot, C. J., Lucassen, P. J., Portegies, P., Troost, D., Tilders, F. J., et al. (1999) Nitric oxide synthase expression and apoptotic cell death in brains of AIDS and AIDS dementia patients. *AIDS* **13,** 317–326.

158. Boven, L. A., Gomes, L., Hery, C., Gray, F., Verhoef, J., Portegies, P., et al. (1999) Increased peroxynitrite activity in AIDS dementia complex: implications for the neuropathogenesis of HIV-1 infection. *J. Immunol.* **162,** 4319–4327.

159. Bagasra, O., Bobroski, L., Sarker, A., Bagasra, A., Saikumari, P., and Pomerantz, R. J. (1997) Absence of the inducible form of nitric oxide synthase in the brains of patients with the acquired immunodeficiency syndrome. *J. Neurovirol.* **3,** 153–167.

160. Adamson, D. C., Kopnisky, K. L., Dawson, T. M., and Dawson, V. L. (1999) Mechanisms and structural determinants of HIV-1 coat protein, gp41-induced neurotoxicity. *J. Neurosci.* **19,** 64–71.

6
Chronic Intracerebral LPS as a Model of Neuroinflammation

Gary L. Wenk and Beatrice Hauss-Wegrzyniak

1. INTRODUCTION

This chapter will describe our recent progress in studying the effects of chronic neuroinflammation in order to better understand the contribution of this condition to neurodegeneration. The animal model to be discussed is directed at a better understanding of the role of chronic neuroinflammation in the progression of Alzheimer's disease (AD). The most important point to make at the onset of this review is that we do not believe that neuroinflammation causes AD. Rather, this chapter will outline the details of an animal model that has provided compelling evidence that inflammation develops in response to existing genetically determined conditions within brains of AD patients.

Animal models of neurodegenerative diseases are useful because they offer the opportunity to investigate the effectiveness of potential pharmacotherapies that can be administered to affected individuals. An animal model of a neurodegenerative disease is useful and valid if it reproduces important aspects of the pathology seen in humans and its ability to predict an effective therapy. This chapter will discuss the features of an animal model of chronic neuroinflammation in rats that meets many aspects of these criteria.

1.1. Significance of the Model

A major problem in AD research today is that none of the hypothesized mechanisms either previously or currently in vogue have been able to explain the cellular and regional distribution patterns that characterize the neuropathology of AD. None of the current animal models of AD, particularly those involving transgenic animals [1,2], being investigated are able to explain why certain neural systems (e.g., cholinergic) or specific brain regions (e.g., temporal lobe) degenerate to such a great extent in the AD brain.

In the course of studying the effects of chronic neuroinflammation, we became increasingly persuaded that this process qualifies as a promising candidate mechanism to help explain the pattern of the pathophysiology of AD. Our model allows the investigation and comparison of important parallels between consequences of the neuroinflammation-induced neurodegeneration in young and old rats and the neurodegeneration seen in AD patients. These similarities include the pattern of neurodegeneration within the brain, the types of neurons affected, the abnormal expression of specific proteins, and the chronic nature of the neurodegenerative process. We will show that the disseminated pattern of the

From: *Neuroinflammation, 2nd Edition: Mechanisms and Management*
Edited by: P. L. Wood © Humana Press Inc., Totowa, NJ

neuronal degeneration induced by our animal model of chronic neuroinflammation bears a close resemblance to the pattern of selected neurodegeneration, neurofibrillary changes, and senile plaque distribution seen in the brains of AD patients. It is important to note that no other animal model of AD has been able to explain as closely the pattern of neurodegeneration associated with specific neurotransmitter systems or the regional pattern of neuropathology that characterize AD.

The chronic inflammatory condition in the brains of AD patients most likely develops in response to prolonged activation by deposits of soluble β-amyloid *(3)*. The extent and degree of the brain's inflammatory response to the presence of β-amyloid varies in different regions. Brain regions that develop the greatest and most prolonged inflammatory response are also those that show the greatest degree of neuropathology. The consequences of neuroinflammation may also underlie the loss of specific neural systems. For example, we have shown that chronic exposure to inflammatory proteins leads to the selective degeneration of basal forebrain cholinergic neurons *(4,5)*. Furthermore, our recent findings *(6,7)* suggest that inflammation plays an important role during the initial phases of neurodegeneration (i.e., the pattern of highest degree of inflammation in the brain parallels the pattern of changes glucose utilization that has been documented by positron emission tomography (PET) studies in humans during the early stages of dementia and neurodegeneration) *(8,9)*. The results of our studies also offer a plausible explanation for why people who consume high doses of anti-inflammatory drugs have a reduced incidence of these disorders, particularly AD *(10–12)*. Given the unique interactions of inflammatory processes with the pathological hallmarks associated with AD *(13–15)*, it is not surprising that conventional therapy using nonsteroidal anti-inflammatory drugs (NSAIDs) can slow the progress, or delay the onset, of AD *(11,12,16,17)*.

Recently, we introduced a novel animal model that uses chronic infusion of the inflammogen lipopolysaccharide (LPS) into the brain *(6,18–21)*. Two different approaches using LPS infusions have been investigated. The first involves the chronic infusion of LPS directly into the fourth ventricle in order to produce widespread inflammation throughout the brain. The fourth ventricle was chosen in order to avoid mechanical injury to forebrain structures that might impair the behavioral performance of the animals. The second approach involves the infusion of LPS directly into the basal forebrain region. We have found that rats infused with very low levels of LPS do not develop fevers or seizures nor do they show any overt signs of illness. In contrast, chronic exposure to elevated levels of tumor necrosis factor-α (TNF-α) in transgenic mice *(22)* or to high doses of LPS *(23)* produced extensive inflammation that was associated with ataxia, fever, and seizures.

2. STUDIES OF WIDESPREAD BRAIN INFLAMMATION

We have used chronic infusion of LPS into the fourth ventricle of young rats in order to investigate the time-dependent evolution of the distribution and density of activated microglia. Rats were sacrificed at various times after the infusion began. The brains of these rats were then prepared for immunohistochemical analysis to determine the location and density of activated microglia. After 2 d of infusion, activated microglia could be seen scattered evenly throughout the brain. This finding was consistent with the hypothesis that the LPS had spread throughout the cerebral hemispheres within a relatively short period of time. In contrast to our findings 2 d after the infusion began, 21 d later we found that the greatest inflammatory response was now concentrated within the hippocampus, particularly the dentate gyrus, and within the entorhinal and piriform cortex within the temporal lobe as well as throughout the cingulate gyrus *(7)*. Surprisingly, even though the LPS was still being infused throughout the brain, these results demonstrated that temporal lobe regions responded differently to the continued presence of an inflammogen such as LPS. Given these results, we speculate that it is possible that the continued activation of microglia by an inflammogenic molecule could lead to the long-term exposure of temporal lobe neurons to potentially toxic levels of cytokines and complement proteins. A similar condition of chronic neuroinflammation is thought to exist in the brains of AD patients, possibly in response to the presence of elevated levels of soluble β-amyloid *(3)*. These data might explain the apparent concentration of AD-related pathology (i.e., senile plaques and tangles) within the temporal regions and the hippocampus. Recent studies using PET have confirmed that these brain regions that demonstrate the greatest inflammatory response also show the earliest detectable declines in glucose utilization *(8)*, particularly in those patients who are vulnerable to AD because of their APOE status *(9)*.

Chronic LPS infusion into the fourth ventricle reproduced many of the pathological components of AD. The brains of these young rats had increased activation of microglia and astrogliosis. Many regions had elevated levels of various inflammatory proteins, such as interleukin-1 (IL-1α and IL-1β) and TNF-α, as well as elevated levels of β-amyloid precursor proteins (APP). The glial cell activation and presence of elevated inflammatory proteins was associated with significant temporal lobe pathology (i.e., pyramidal with cell loss) *(6,7)*. Furthermore, and most importantly, these rats also demonstrated impairments in tasks that depend on an intact hippocampus (e.g., performance in the Morris water maze and a spatial memory task on a T maze). In contrast, these chronically inflamed rats were not impaired in the performance of tasks that do not require an intact hippocampus (e.g., object recognition) *(6,18,19)*. The working memory impairment demonstrated in the Morris water maze task correlated with the degree of inflammation within the hippocampal region and could be made worse if the LPS infusion was increased to 74 d *(20)*. A much shorter infusion of IL-2 into the lateral ventricles (i.e., for only 14 d) also impaired performance in the water maze task *(24)*. Similarly, transgenic mice that chronically overexpress IL-6 in their resident astrocytes, under the control of a GFAP promoter, exhibited dose- and age-related learning and memory deficits that were correlated with microglial activation, loss of synapses, and the degeneration of calbindin-containing neurons *(25)*. Finally, the memory impairment produced by LPS infusion could also be attenuated by treatment with an interleukin-1 receptor antagonist *(26)*.

Overall, the chronic infusion of LPS was well tolerated by all of the rats. Initially after surgery, all of the LPS-treated rats lost a few grams of weight. Within a few days, however, all of the rats had regained the lost weight and continued to grow normally for the duration of the study. The chronic infusion of LPS did not induce seizures or fever and our preliminary studies have found that the blood-brain barrier remains intact. Although LPS is a lipophilic molecule, it does not appear to leave the brain in significant amounts when infused centrally *(27)*. The effects of the LPS infusion in our model appeared to be limited to the central nervous system inasmuch as we have not found detectable levels of inflammatory proteins in the blood of LPS-treated rats. Therefore, a general malaise did not cause the impaired behavioral performance observed in these studies. It is also conceivable that the pathological effects of the chronic LPS infusion are the result of nonspecific cytotoxic effects of LPS upon the brain. This is highly unlikely inasmuch as the cytotoxicity of LPS can be completely attenuated by coadministration of either cyclo-oxygenase inhibitors or glutamate receptor antagonists (*see* Section 3.2.).

2.1. MRI Studies of the Effects of Chronic LPS Infusion

The chronic infusion of LPS into the fourth ventricle was associated with a loss of hippocampal pyramidal cells *(6)*. Recently, we have used magnetic resonance imaging (MRI) to study the consequences of the cell loss. These MRI studies have identified enlarged lateral ventricles with shrinkage of the temporal lobe regions in rats chronically exposed to the effects of low-level neuroinflammation *(28)*. We speculate that this ventricular enlargement and atrophy of temporal lobe regions, which is very similar to that seen in patients with AD in the early phases of the disease *(29–34)*, is related to the cell loss that we have observed within temporal lobe structures *(6)* and probably also underlies the spatial working memory impairment.

2.2. Studies Using Anti-Inflammatory Drugs

Taken together, these findings predict that the effects of inflammation upon working memory could be reversed by treatment with an anti-inflammatory drug. We tested this prediction using a novel cyclo-oxygenase (COX) inhibitor, nitroflurbiprofen (NFP, NicOx, France). NFP (2-fluoro-a-methyl [1,1'-biphenyl]-4-acetic acid, 4-(nitrooxy) butyl ester) is a novel NSAID drug that shows a significant attenuation of the gastrointestinal side effects *(35,36)*. NFP is a derivative of flurbiprofen, an inhibitor of COX enzymatic activity (IC_{50} type $1/IC_{50}$ type $2 = 10.27$). NFP was produced by the incorporation of a nitric oxide (NO) moiety through an ester linkage to the carboxyl group of flurbiprofen. NO plays an important role in gastric mucosal defense; drugs that generate NO reduce the severity of gastric mucosal injury in vivo and ex vivo *(37,38)*. The usefulness of NFP in the current model is most likely the result of its ability to cross the blood-brain barrier and inhibit both isoforms of the COX enzyme leading to the suppression of prostaglandin synthesis *(39)*. The behavioral effects of chronic NFP treatment that are described in detail (*see* Section 3.1.) are probably the result of the inhibition of prostanoid synthesis within activated glia and the subsequent decreased production and release of cytokines and complement proteins. Because AD is a disease that is closely associated with aging, we also investigated whether aged rats could benefit from the chronic anti-inflammatory therapy.

Young (3 mo), adult (10 mo), and aged (24 mo) rats were infused with LPS for 42 d. The aged control (those infused with artificial cerebrospinal fluid [CSF]) rats demonstrated poor performance on the Morris water maze, as compared to young or adult control rats. The performance of adult rats was significantly worse than young rats, but significantly better than old rats. When the brains of young rats were chronically infused with LPS, their performance became as impaired as that seen in adult and old rats infused with only CSF. Also, in contrast to young rats, the brains of control adult and aged rats had significantly more activated microglia. As expected, aged rats had the greatest number of activated microglia. Probably for this reason, the young and adult rats demonstrated a much greater inflammatory response following the chronic LPS infusion, as compared to the aged rats. We speculate that the level of inflammation reaches a maximum level with normal aging and that the chronic LPS infusion was not able to increase the level of inflammation further. Overall, the results of these studies are consistent with the hypothesis that performance in the Morris water maze can be impaired by the presence of a chronic inflammatory condition, particularly within the hippocampus. Furthermore, these data are consistent with the hypothesis that the increase in inflammation that accompanies normal aging *(40–42)* might underlie the age-associated impairment in performance in this and other tasks that depend on normal temporal lobe function.

The degree of inflammation and the level of inflammatory proteins as well as the impairment in performance in the Morris water maze task were all significantly attenuated in young rats given chronic NFP therapy *(18,19)*. In contrast to the benefits of chronic NFP therapy that was seen in young rats, NFP therapy did not improve the working memory ability of either adult or old rats infused with LPS *(19)*. The chronic NFP therapy was also ineffective against the level of activated microglia in the older rats *(19)* (i.e., the therapy did not decrease the number of activated microglia within the temporal lobe regions).

The results of these studies suggest that future investigations of NSAID therapies that are designed to slow the onset and development of diseases associated with chronic neuroinflammation should be initiated in genetically predisposed adults before age-associated inflammatory processes have a chance to develop. This finding and the fact that many common NSAIDs do not cross the blood brain barrier may explain why recent attempts to alter the course of progression of the dementia in elderly humans with AD using NSAID therapy has not been successful. The results of our studies suggest that NSAIDs can provide neuroprotection but not therapy.

2.3. Ultrastructural Changes Induced by Chronic LPS Infusion

Electron microscopic studies revealed numerous changes in the cellular components involved in protein synthesis within the hippocampal neurons of young rats chronically infused with LPS into the fourth ventricle for 37 d *(43)*. These ultrastructural changes probably underlie the loss of neurons that was responsible for the significant learning and memory impairments reported earlier. The cytoplasm of the effected hippocampal neurons contained only a few polyribosomes, a few cisternae of rough endoplasmic reticulum (RER), and a small Golgi apparatus, all suggestive of either impaired or reduced protein synthesis. In addition, the presence of myelinic figures within the nucleus and the high concentration of ribosomes sequestered within the nuclear infoldings are consistent with nuclear dysfunction. However, the presence of deep invaginations of the nuclear envelope would result in an increase in the exchange area between the nuclear matrix and the cytoplasm. The nucleolus was also located in an eccentric position; this usually occurs in cells actively engaged in protein synthesis. These two ultrastructural changes would represent a possible attempt by the neurons to compensate for impaired protein synthesis within the cytoplasm. Because this initial study *(43)* only examined for changes at a single time-point, it is not possible to know whether the observed changes reflect either progressing or recovering changes in neuronal integrity.

The most interesting and unusual electron microscopic finding was that almost all of the RER cisternae were paired between themselves and also with the perinuclear cisterna. The meaning of this pairing is unknown, but certainly at the regions of the paired tracts, the ribosomes would be prevented from interacting with the RER surface; this would be consistent with a further impairing in protein synthesis and normal cellular function. In contrast to these changes seen in the nucleus, no problems were associated with the organelles of cellular respiration (mitochondria) or catalytic processes (lysosomes, pigments). The cytoskeleton also appeared normal; these findings would be consistent with normal axonal transport. These ultrastructural findings are consistent with the hypothesis that long-term inflammation within the brain may lead to degenerative changes within hippocampal pyramidal neurons.

3. STUDIES OF LOCALIZED INFLAMMATION WITHIN THE BASAL FOREBRAIN

We have used chronic infusion of LPS into the nucleus basalis magnocellularis within the basal forebrain to selectively destroy cholinergic cells in a time-dependent, but not dose-dependent, manner *(21,44)*. Similarly, the chronic infusion of interleukin (IL)-1β or TNF-α can also destroy the cholinergic projection to the neocortex that originates within the nucleus basalis *(5)*. Medial septal cholinergic neurons that innervate the hippocampus may also be selectively vulnerable to immune-mediated processes *(45)*. In addition, the level of cholinergic enzyme activity was significantly reduced within the septal cholinergic neurons of transgenic mice that express elevated levels of TNF-α *(46)*. Therefore, the entire forebrain cholinergic system may be vulnerable to elevated levels of inflammatory proteins, particularly to TNF-α. Stimulation of TNF-α receptors may induce cell death by "silencing of survival signals" via the inhibition of insulin-like growth factor-1-mediated signaling within neurons *(47)*. TNF-α can also inhibit glutamate reuptake into astrocytes and may potentiate glutamate-receptor-mediated toxicity *(48–51)* within the basal forebrain, a region that is vulnerable to excess glutamatergic function *(52)*.

Alzheimer's disease is characterized by a forebrain deficiency of acetylcholine *(53, 54)* that may underlie aspects of the cognitive impairments associated with AD *(55,56)*. The mechanism underlying the degeneration of basal forebrain cholinergic cells is unknown. The results of these studies using LPS infusions have led us to speculate that the long-term exposure to elevated inflammatory proteins may lead to the selective degeneration of basal forebrain cholinergic neurons in AD *(5,21)*. A potential role for neuroinflammation and the specificity of its effects upon cholinergic neurons were initially suggested by a study that isolated antibodies from the sera of AD patients that selectively recognized and destroyed basal forebrain cholinergic cells when injected into a rat brain *(57)*. In addition, head trauma in humans is a significant risk factor for AD *(58)* and is associated with increased levels of inflammatory proteins *(59)* and a decline in the number of basal forebrain cholinergic neurons *(60)*. In vitro studies also indicate that brain inflammation may selectively destroy basal forebrain cholinergic neurons *(61)*.

3.1. Neuroprotection Studies Using Anti-Inflammatory Drugs

Basal forebrain cholinergic neurons can be rescued from the cytotoxic effects of chronic LPS infusion by treatment with anti-inflammatory drugs. We have investigated two different NSAIDs that lack gastrointestinal toxicity (i.e., NFP [discussed earlier] and NCX 2216 [both supplied by NicOx S.A., Nice, France]). NFP and NCX 2216 {*trans*-3-4[4-[2-fluoro-α-methyl-1(1,1'-biphenyl)-4-acetyloxy]-3-methoxyphyenyl]-2-propenoic acid 4-(nitrooxy)butyl ester} are derivatives of flurbiprofen; NCX2216 also possesses significant antioxidant proclivities. Previous experiments have demonstrated that these novel NSAIDs have good anti-inflammatory efficacy and gastrointestinal tolerability in rats *(35,36)*. We have also determined previously that daily peripheral administration of NFP can significantly reduce the number of activated microglial cells in young rats *(6,18)*. Therefore, we felt confident that these drugs would significantly attenuate the inflammation induced by the chronic LPS infusion into the basal forebrain.

The infusion of LPS into the basal forebrain of young rats produced extensive inflammation and a significant loss of cholinergic cells that was significantly attenuated by cotreatment with either NFP or NCX2216 *(4)*. These results suggest that exposure to LPS can activate glia and that this activation can lead to the release of cytokines and inflammatory proteins that can be detrimental to cellular function *(26,62)*. Our histological and biochemical analyses have confirmed that the cytotoxic effects of chronic neuroinflammation within the basal forebrain were selective for cholinergic neurons *(21)*. For example, radioimmunoassays found no significant LPS-induced decline in the endogenous levels of galanin, leu-enkephalin, neurokinin B, neurotensin, somatostatin, or substance P within the basal forebrain region. However, the histological analyses found a dense distribution of silver-stained cells within the basal forebrain that had a general morphology and size that was consistent with cholinergic magnocellular neurons. When considered together with the decrease in neocortical ChAT activity and the reduced number of ChAT-immunoreactive cells, the results strongly suggests that the inflammatory processes reduced cholinergic cell number and not simply cholinergic cellular function. The medial septal cholinergic neurons that innervate the hippocampus may also be selectively vulnerable to immune-mediated processes *(45)*. The level of ChAT activity was also significantly reduced within the septal cholinergic neurons of transgenic mice that express elevated levels of the inflammatory protein TNF-α *(46)*. Whether TNF-α acts directly upon cholinergic cells, or via stimulation of the release of other cytokines, or via its influence on nerve growth factor *(46)* remains to be determined. Finally, when mixed neuronal/glial cell cultures were exposed to LPS, only cholinergic neurons died; this cell loss was significantly attenuated by coincubation with the *N*-methyl-D-aspartate (NMDA) open-channel antagonist MK-801 *(61)*. During inflammatory processes in the brain that are associated with disease, a marked rise in the synthesis of prostaglandins can also be seen *(63–65)*. Taken together, the results of these studies predict that the long-term inhibition of prostaglandin synthesis by chronic NSAID therapy might attenuate the loss of basal forebrain cholinergic neurons in the AD brain and reduce the symptoms of the dementia associated with AD.

3.2. The Role of NMDA Receptors in the Death of Basal Forebrain Cholinergic Neurons

A fundamental question that has challenged previous hypotheses of neurodegeneration in AD is why specific neuronal populations are affected more than others. The results of the studies discussed earlier are consistent with the hypothesis that basal forebrain cholinergic neurons are vulnerable to the consequences of elevated levels of cytokines and prostaglandins produced by the condition of chronic neuroinflammation that characterizes the brain of AD patients *(66)*. Although the mechanism of this toxicity is unknown, we have recently demonstrated an important role for NMDA-sensitive glutamate receptors *(7,21)* in the toxicity of LPS. We have hypothesized *(7)* that the following cascade of events occurs in the basal forebrain in response to the chronic infusion of LPS that leads to the degeneration of cholinergic cells:

$$LPS \rightarrow IL\text{-}1, TNF\alpha \rightarrow Prostaglandins \rightarrow \uparrow[Glutamate]_{ext} \rightarrow NMDA_{recpt}$$
$$\rightarrow Calcium\ Influx \rightarrow \uparrow[Nitric\ Oxide] \rightarrow Cell\ Death$$

Lipopolysaccharide infusions stimulate the endogenous production of inflammatory cytokines by activated astrocytes and microglia *(26,62)*; cytokines, in turn, stimulate the production of other inflammatory mediators such as prostaglandins *(67)*; prostaglandins induce the release of glutamate from astrocytes *(68)*, leading to increased levels of extracellular glutamate and the stimulation of glutamate receptors, the depolarization-dependent unblocking of NMDA receptors by Mg^{2+}, and the entry of toxic amounts of Ca^{2+} into neurons and the subsequent generation of toxic levels of nitric oxide and initiate a cascade of reactive oxygen intermediates *(69,70)*. Prostaglandins and various cytokines may also indirectly elevate the extracellular concentration of glutamate by inhibiting its reuptake by astrocytes *(71,72)*. Neurons in astrocyte-poor cultures are more vulnerable to glutamate excitotoxicity, and blocking astrocytic uptake of glutamate results in neurodegeneration *(71,72)*. Recent evidence suggests that elevated levels of inflammatory proteins may selectively target cholinergic basal forebrain neurons *(5,21,44,61)*. The selective vulnerability of cholinergic neurons within the basal forebrain may also be related to the fact that they receive a dense glutamate projection from the pedunculopontine tegmentum *(58)* and are sensitive to excess stimulation of glutamate NMDA receptors *(52)*. Consistent with this hypothesized process is the recent finding that the cytokine IL-6 can increase NMDA-mediated calcium influx and potentiate its neurotoxicity *(73)*.

We hypothesize that this model of inflammation-induced lesions of the basal forebrain cholinergic neurons mimics the human condition associated with chronic glial activation and may thereby replicate the initial steps of the neurotoxic cascade that may lead to cell death in the AD brain. Taken together with the results of previous studies *(6,7,19,21,44,74–76)*, the following scenario can be suggested that depends on the ultimate role of glutamate as endogenous neurotoxin acting within the basal forebrain. The interaction of microglia within senile plaques containing β-amyloid protein results in a chronic activation of these cells and the release of various cytokines and complement proteins. Elevated levels of cytokines and β-amyloid proteins may then advance glutamate excitotoxicity within the AD brain *(77)* by releasing glutamate to overstimulate NMDA receptors on glia as well as impairing glutamate uptake mechanisms and detoxification processes vital to neuronal survival *(78)*. Stimulated glia would also release cytokines that would then potentiate the toxicity of glutamate *(79)*. The long-term exposure of cholinergic cells to elevated levels of extracellular glutamate may ultimately lead to their gradual degeneration as the behavioral symptoms progress, particularly during the advanced stages of the disease *(54,80)*. Taken together, the results of these studies are consistent with a role for both prostaglandins and NMDA receptors in the cascade of biochemical processes that lead to the degeneration of cholinergic cells within the basal forebrain region of AD patients. Recent evidence suggests that β-amyloid acts to enhance the susceptibility of neurons to glutamate toxicity and that this toxicity can be attenuated by treatment with NMDA receptor channel antagonists *(81)*. Our findings predict that combination therapy using a safe and effective NSAID, such as NFP, and a NMDA receptor channel antagonist, such as memantine *(52)*, would attenuate the loss of forebrain cholinergic neurons and delay the onset of the cognitive impairments associated with their degeneration.

3.3. The Role of Caspase in the Death of Basal Forebrain Cholinergic Neurons

Chronic infusion of LPS into the basal forebrain region was associated with a significant increase in the level of caspase activity (i.e., caspases-3, -8, and -9) *(7)*. In order to investigate whether the generation of caspase activity contributes to the degeneration of basal forebrain cholinergic cells, some rats that were also infused with a potent and irreversible pan-caspase synthesis inhibitor, z-Val-Ala-Asp(OMe)-fluoromethyl ketone (zVAD) *(82)*. Treatment with zVAD significantly reduced endogenous caspase levels; however, the inhibition of caspase synthesis inhibition did not provide neuroprotection for cholinergic cells. These findings suggested that the increase in endogenous levels of caspase produced in response to the LPS infusion was not responsible for the degeneration of basal forebrain cholinergic neurons. Elevated caspase activity is associated with the process of apoptosis *(83)*. We have used the terminal transferase-mediated deoxyuridine-digoxigenin nick end labeling (TUNEL) method to examine for the presence of DNA fragmentation as a marker for detection of apoptosis at the cellular level. Our investigations have found no TUNEL-positive cells within the basal forebrain region of LPS-infused rats. Taken together, these findings are consistent with the hypothesis that LPS-induced death of cholinergic neurons is not via apoptotic processes. Recent evidence suggests that β-amyloid induces necrosis rather than apoptosis *(84)*.

4. CONCLUSIONS

At the beginning of this chapter, we stated that an animal model of a neurodegenerative disease is useful and valid if it reproduces important aspects of the pathology seen in humans and if it predicts an effective therapy. We have presented evidence that the chronic infusion of LPS into the brain, either via the fourth ventricle to produce global neuroinflammation or via the basal forebrain to produce local effects, reproduces important aspects of the pathobiology of AD. In addition, this model suggests a unique mechanism to explain the cellular and regional distribution pattern of the neuropathology that characterizes AD. The results of our studies suggest that the condition of chronic neuroinflammation underlies the degeneration of specific neural systems (e.g., cholinergic) or selected brain regions (e.g., temporal lobe) in the AD brain. No other animal model of AD has been able to explain as closely the pattern of neurodegeneration associated with specific neurotransmitter systems or the regional pattern of neuropathology that characterizes AD. These aspects of our model make it useful for testing potential pharmacotherapies for the prevention of AD.

ACKNOWLEDGMENT

This work was supported by a National Institute of Aging grant (ROI AG10546).

REFERENCES

1. Guenette, S. Y. and Tanzi, R. E. (1999) Progress toward valid transgenic mouse models for Alzheimer's disease. *Neurobiol. Aging* **20**, 201–211.
2. Holcombe, L. A., Gordon, M. N., Jantzen, P., Hsiao, K., Duff, K., and Morgan, D. (1999) Behavioral changes in transgenic mice expressing both amyloid precursor protein and presenilin-1 mutations: lack of association with amyloid deposits. *Behav. Genet.* **29**, 177–185.
3. Cotman, C. W., Tenner, A. J., and Cummings, B. J. (1996) Beta-amyloid converts an acute phase injury response to chronic injury responses. *Neurobiol. Aging* **17**, 723–731.
4. Wenk, G. L., McGann, K., Fiorucci, S., Mencarelli, A., Hauss-Wegrzyniak, B., and Del Soldato, P. (2000) Mechanisms to prevent the toxicity of chronic neuroinflammation on forebrain cholinergic neurons. *Eur. J. Pharmacol.* **402**, 77–85.
5. Wenk, G. L. and Willard, L. B. (1999) The neural mechanisms underlying cholinergic cell death within the basal forebrain. *Int. J. Dev. Neurosci.* **16**, 729–735.
6. Hauss-Wegrzyniak, B., Dobrzanski, P., Stoehr, J. D., and Wenk, G. L. (1998) Chronic neuroinflammation in rats reproduces components of the neurobiology of Alzheimer's disease. *Brain Res.* **780**, 294–303.
7. Wenk, G. L., Hauss-Wegrzyniak, B., and Willard, L. B. (2000) Pathological and biochemical studies of chronic neuroinflammation may lead to therapies for Alzheimer's Disease, in *Research and Perspectives in Neurosciences: Neuro-Immune Neurodegenerative and Psychiatric Disorders and Neural Injury* (Patterson, P., Kordon, C., and Christen, Y., eds.), Springer-Verlag, Heidelberg, pp. 73–77.
8. Friedland, R. P., Jagust, W. J., Huesman, R. H., Koss, E., Knittel, B., Mathis, C. A., et al. (1989) Regional cerebral glucose transport and utilization in Alzheimer's disease. *Neurology* **39**, 1427–1434.
9. Reiman, E. M., Caselli, R. J., Yun, L. S., Chen, K., Bandy, D., Minoshima, S., et al. (1996) Preclinical evidence of Alzheimer's disease in persons homozygous for the e4 allele for apolipoprotein E. *N. Engl. J. Med.* **334**, 752–758.
10. Stewart, W. F., Kawas, C., Corrada, M., and Metter, J. (1997) Risk of Alzheimer's disease and duration of NSAID use. *Neurology* **48**, 626–632.
11. Andersen, K., Launer, L. J., Ott, A., Hoes, A. W., Breteler, M. M. B., and Hoffman, A. (1995) Do nonsteroidal anti-inflammatory drugs decrease the risk of Alzheimer's disease? *Neurology* **45**, 1441–1445.
12. Breitner, J. C. S., Gau, B. A., Welsh, K. A., Plassman, B. L., McDonald, W. M., Helmas, M. J., et al. (1994) Inverse association of anti-inflammatory treatments and Alzheimer's disease. *Neurology* **44**, 227–232.
13. Eikelenboom, P., Roxemuller, J. M., and van Muiswinkel, F. L. (1998) Inflammation and Alzheimer's disease: relationships between pathogenic mechanisms and clinical expression. *Exp. Neurol.* **154**, 89–98.
14. Eikelenboom, P., Zhan, S. S., van Gool, A., and Allsop, D. (1994) Inflammatory mechanisms in Alzheimer's disease. *TIPS* **15**, 447–450.
15. Mrak, R. E., Sheng, J. G., and Griffin, W. S. T. (1995) Glial cytokines in Alzheimer's disease: review and pathogenic implications. *Hum. Pathol.* **26**, 816–823.
16. McGeer, P. L., McGeer, E. G., Rogers, J., and Sibley, J. (1990) Anti-inflammatory drugs and Alzheimer's disease. *Lancet* **335**, 1037.
17. Rogers, J., Kirby, L. C., Hempelman, S. R., Berry, D. L., McGeer, P. L., Kaszniak, A. W., et al. (1993) Clinical trial of indomethacin in Alzheimer's disease. *Neurology* **43**, 1609–1611.
18. Hauss-Wegrzyniak, B., Willard, L. B, Pepeu, G., Del Soldato, P., and Wenk, G. L. (1999) Peripheral administration of novel anti-inflammatories can attenuate the effects of chronic inflammation within the CNS. *Brain Res.* **815**, 36–43.
19. Hauss-Wegrzyniak, B., Vraniak, P., and Wenk, G. L. (1999) The effects of a novel NSAID upon chronic neuroinflammation are age dependent. *Neurobiol. Aging* **20**, 305–313.

20. Hauss-Wegrzyniak, B., Vraniak, P., and Wenk, G. L. (2000) LPS-induced neuroinflammatory effects do not recover with time. *NeuroReport* **11,** 1759–1763.
21. Willard, L. B., Hauss-Wegrzyniak, B., Danysz, W., and Wenk, G. L. (2000) The cytotoxicity of chronic neuroinflammation upon basal forebrain cholinergic neurons of rats can be attenuated by glutamatergic antagonism or cyclooxygenage-2 inhibition. *Exp. Brain Res.* **134,** 58–65.
22. Probert, L., Akassaglou, K., Kassiotis, G., Pasparakis, M., Alexopoulou, L., and Kollias, G. (1997) TNFα transgenic and knockout models of CNS inflammation and degeneration. *J. Neuroimmunol.* **72,** 137–141.
23. Mouihate, A. and Pittman, Q. J. (1998) Lipopolysaccharide-induced fever is dissociated from apoptotic cell death in the rat brain. *Brain Res.* **805,** 95–103.
24. Hanisch, U.-K., Neuhaus, J., Rowe, W., Van Rossum, D., Möller, M., Kettenmann, H., et al. (1997) Neurotoxic consequences of central long-term administration of interleukin-2 in rats. *Neuroscience* **79,** 799–818.
25. Heyser, C. J., Masliah, E., Samimi, A., Campbell, I. L., and Gold, L. H. (1997) Progressive decline in avoidance learning paralleled by inflammatory neurodegeneration in transgenic mice overexpressing interleukin 6 in the brain. *Proc. Natl. Acad. Sci. USA* **94,** 1500–1505.
26. Bluthe, R.-M., Dantzer, R., and Kelley, K. W. (1992) Effects of interleukin-1 receptor antagonist on the behavioral effects of lipopolysaccharide in rat. *Brain Res.* **573,** 318–320.
27. De Simoni, M. G., Del Bo, R., De Luigi, A., Simard, S., and Forloni, G. (1995) Central endotoxin induces different patterns of interleukin (IL)-1beta and IL-6 messenger ribonucleic acid expression and IL-6 secretion in the brain and periphery. *Endocrinology* **136,** 897–902.
28. Hauss-Wegrzyniak, B., Galons, J. P., and Wenk, G. L. (2000) Quantitative volumetric analysis of brain magnetic resonance imaging from rat with chronic neuroinflammation and correlation with histology. *Exp. Neurol.* **165,** 347–354.
29. Cuenod, C.-A., Denys, A., Michot, J.-L., Jehenson, P., Forette, F., Kaplan, D., et al. (1993) Amygdala atrophy in Alzheimer's disease—an in vivo magnetic resonance imaging study. *Arch. Neurol.* **50,** 941–945.
30. DeToledo-Morrell, L., Sullivan, M. P., Morrell, F., Wilson, R. S., Bennett, D. A., and Spencer, S. (1997) Alzheimer's disease: in vivo detection of differential vulnerability of brain regions. *Neurobiol. Aging* **18,** 463–468.
31. De Leon, M. J., George, A. E., Golomb, J., Tarshish, C., Convit, A., Kluger, A., et al. (1997) Frequency of hippocampus atrophy in normal elderly and Alzheimer's disease patients. *Neurobiol. Aging* **18,** 1–11.
32. Juottonen, K., Laakso, M. P., Insausti, R., Pitkanen, A., Partanen, K., and Soininen, H. (1998) Volumes of the entorhinal and perirhinal cortices in Alzheimer's disease. *Neurobiol. Aging* **19,** 15–22.
33. Killiany, R. J., Moss, M. B., Albert, M. S., Sandor, T., Tieman, J., and Jolesz, F. (1993) Temporal lobe regions on magnetic resonance imaging identify patients with early Alzheimer's disease. *Arch. Neurol.* **50,** 949–954.
34. Laakso, M. P., Soininen, H., Partanen, K., Lehtovirta, M., Hallikainen, M., Hanninen, H., et al. (1998) MRI of the hippocampus in Alzheimer's disease: sensitivity, specificity, and analysis of the incorrectly classified subjects. *Neurobiol. Aging* **19,** 23–31.
35. Fiorucci, S., Antonelli, E., Santucci, L., Morelli, O., Miglietti, M., Federici, B., et al. (1999) Gastrointestinal safety of NO-derived aspirin is related to inhibition of ICE-like cysteine proteases. *Gastroenterology* **116,** 1089–1106.
36. Wallace, J. L., Reuter, B., Cicala, C., McKnight, W., Grisham, M. B., and Cirino, G. (1994) Novel nonsteroidal anti-inflammatory drug derivatives with markedly reduced ulcerogenic properties in the rat. *Gastroenterology* **107,** 173–179.
37. Kitagawa, H., Takeda, F., and Kohei, H. (1990) Effect of endothelium-derived relaxing factor on the gastric lesion induced by HCl in rats. *J. Pharmacol. Exp. Therap.* **253,** 1133–1137.

38. MacNaughton, W. K., Cirino, G., and Wallace, J. L. (1989) Endothelium-derived releasing factor (nitric oxide) has protective actions in the stomach. *Life Sci.* **45,** 1869–1876.
39. Santini, G., Sciulli, M. G., Padovano, R., di Giamberardino, M., Rotondo, M.T., Del Soldato, P., et al. (1996) Effects of flurbiprofen and flurbinitroxybutylester on prostaglandin endoperoxidase synthases. *Eur. J. Pharmacol.* **316,** 65–72.
40. Ogura, K., Ogawa, M., and Yoshida, M. (1998) Effects of ageing on microglia in the normal rat brain: immunohistochemical observations. *NeuroReport* **5,** 1224–1226.
41. Sheffield, L. G. and Berman, N. E. J. (1998) Microglial expression of MHC class II increases in normal aging of nonhuman primates. *Neurobiol. Aging* **19,** 47–55.
42. Sheng, J. G., Mrak, R. E., and Griffin, W. S. T. (1998) Enlarged and phagocytic, but not primed, interleukin-1-alpha-immunoreactive microglia increase with age in normal human brain. *Acta Neuropathol.* **95,** 229–234.
43. Hauss-Wegrzyniak, B., Vannucchi, M. G., and Wenk, G. L. (2000) Behavioral and ultrastructural changes induced by chronic neuroinflammation in young rats. *Brain Res.* **859,** 157–166.
44. Willard, L. B., Hauss-Wegrzyniak, B., and Wenk, G. L. (1999) The pathological and biochemical consequences of acute and chronic neuroinflammation within the basal forebrain of rats. *Neuroscience* **88,** 193–200.
45. Kalman, J., Engelhardt, J. I., Le, W. D., Xie, W., Kovacs, I., Kasa, P., et al. (1997) Experimental immune-mediated damage of septal cholinergic neurons. *J. Neuroimmunol.* **77,** 63–74.
46. Aloe, L., Fiore, M., Probert, L., Turrini, P., and Tirassa, P. (1999) Overexpression of tumor necrosis factor-alpha in the brain of transgenic mice differentially alters nerve growth factor levels and choline acetyltransferase activity. *Cytokine* **11,** 45–54.
47. Venters, H. D., Dantzer, R., and Kelley, K. W. (2000) A new concept in neurodegeneration: TNFα is a silencer of survival signals. *TINS* **23,** 175–180.
48. Soliven, B. and Albert, J. (1992) Tumor necrosis factor modulates Ca^{2+} currents in cultured sympathetic neurons. *J. Neuroscience* **12,** 2665–2671.
49. Chao, C. and Hu, S. (1994) Tumor necrosis factor alpha potentiates glutamate neurotoxicity in human fetal brain cell cultures. *Dev. Neurosci.* **16,** 172–179.
50. Chao, C. C., Hu, S., Ehrlich, L., and Peterson, P. K. (1995) Interleukin-1 and tumor necrosis factor alpha synergistically mediate neurotoxicity: involvement of nitric oxide and *N*-methyl-D-aspartate receptors. *Brain Behav. Immunol.* **9,** 355–365.
51. Kim, W. K. and Ko, K. H. (1998) Potentiation of *N*-methyl-D-aspartate-mediated neurotoxicity by immunostimulated murine microglia. *J. Neurosci. Res.* **54,** 17–26.
52. Wenk, G. L., Danysz, W., and Mobley, S. L. (1995) MK-801, memantine and amantadine show potent neuroprotective activity against NMDA toxicity in the NBM—a dose response study. *Eur. J. Pharmacol.* **293,** 267–270.
53. Whitehouse, P. J., Price, D. L., Clark, A. W., Coyle, J. T., and DeLong, M. R. (1981) Alzheimer disease: evidence for selective loss of cholinergic neurons in the nucleus basalis. *Ann. Neurol.* **10,** 122–126.
54. Davis, K. L., Mohs, R. C., Marin, D., Purohit, D. P., Perl, D. P., Lantz, M., et al. (1999) Cholinergic markers in elderly patients with early signs of Alzheimer disease. *JAMA* **281,** 1401–1406.
55. McGeer, P. L., McGeer, E. G., Suzuki, J., Dolman, C. E., and Nagai, T. (1984) Aging, Alzheimer's disease and the cholinergic system of the basal forebrain. *Neurology* **34,** 741–745.
56. Muir, J. L. (1997) Acetylcholine, aging, and Alzheimer's disease. *Phamacol. Biochem. Behav.* **56,** 687–696.
57. Foley, P., Bradford, H. F., Dochart, M., Fillet, H., Luine, V. N., McEwen, B., et al. (1988) Evidence for the presence of antibodies to cholinergic neurons in the serum of patients with Alzheimer disease. *J. Neurol.* **235,** 466–471.
58. Rasmusson, D. X., Brandt, J., Martin, D. B., and Folstein, M. F. (1995) Head injury as a risk factor in Alzheimer's disease. *Brain Injury* **9,** 213–219.

59. Griffin, D. E., Wesselingh, S. L., and McArthur, J. C. (1994) Elevated central nervous system prostaglandins in human immunodeficiency virus-associated dementia. *Ann. Neurol.* **35,** 592–597.

60. Murdoch, I., Perry, E. K., Court, J. A., Graham, D. I., and Dewar, D. (1998) Cortical cholinergic dysfunction after head injury. *J. Neurotrauma* **15,** 295–305.

61. McMillian, M., Kong, L.-Y., Sawin, S. M., Wilson, B., Das, K., Hudson, P., et al. (1995) Selective killing of cholinergic neurons by microglial activation in basal forebrain mixed neuronal/glial cultures. *Biochem. Biophys. Res. Commun.* **215,** 572–577.

62. Quan, N., Sundar, S. K., and Weiss, J. M. (1994) Induction of interleukin-1 in various brain regions after peripheral and central injections of lipopolysaccharide. *J. Neuroimmunol.* **49,** 125–134.

63. Dubois, R. N., Abramson, S. B., Crofford, L., Gupta, R. A., Simon, L. S., Van De Putte, L. B. A., et al. (1998) Cyclooxygenase in biology and disease. *FASEB J.* **12,** 1063–1073.

64. Griffin, W. S. T., Sheng, J. G., Royston, M. C., Gentleman, S. M., Mckenzie, J. E., Graham, D. I., et al. (1998) Glial–neuronal interactions in Alzheimer's disease: the potential role of a "cytokine cycle" in disease progression. *Brain Pathol.* **8,** 65–72.

65. Prasad, K. N., Hovland, A. R., LaRosa, F., and Hovland, P. G. (1998) Prostaglandins as putative neurotoxins in Alzheimer's disease. *Proc. Soc. Exp. Biol. Med.* **219,** 120–125.

66. Akiyama, H., Barger, S., Barnum, S., Bradt, B., Bauer, J., Cooper, N. R., et al. (2000) Inflammation in Alzheimer's disease. *Neurobiol. Aging* **21,** 383–421.

67. Katsuura, G., Gottschall, P. E., Dahl, R. R., and Airmura, A. (1989) Interleukin-1β increases prostaglandin E$_2$ in rat astrocyte cultures: modulatory effect of neuropeptides. *Endocrinology* **124,** 3125–3127.

68. Bezzi, P., Carmignoto, G., Pasti, L., Vesce, S., Rossi, D., Rizzini, B. L., et al. (1998) Prostaglandins stimulate calcium-dependent glutamate release in astrocytes. *Nature* **391,** 281–285.

69. Goossens, V., Grooten, J., De Vos, K., and Fiers, W. (1995) Direct evidence for tumor necrosis factor-induced mitochondrial reative oxygen intermediates and their involvement in cytotoxicity. *Proc. Natl. Acad. Sci. USA* **92,** 8115–8119.

70. Minghetti, L. and Levi, G. (1995) Induction of prostanoid biosynthesis by bacterial lipopolysaccharide and isoproterenol in rat microglial cultures. *J. Neurochem.* **65,** 2690–2698.

71. Robinson, M. B., Djali, S., and Buchhalter, J. R. (1993) Inhibition of glutamate uptake with L-transpyrrolidine-2,4-dicarboxylate potentiates glutamate neurotoxicity in primary hippocampal cultures. *J. Neurochem.* **61,** 2099–2103.

72. Rothstein, J. D., Jin, L., Dykes-Hoberg, M., and Kuncl, R. W. (1993) Chronic inhibition of glutamate uptake produces a model of slow neurotoxicity. *Proc. Natl. Acad. Sci. USA* **90,** 6591–6595

73. Qiu, Z., Sweeney, D. D., Netzeband, J. G., and Gruol, D. L. (1998) Chronic interleukin-6 alters NMDA receptor-mediated membrane responses and enhances neurotoxicity in developing CNS neurons. *J. Neurosci.* **18,** 10,445–10,456.

74. Gahtan, E. and Overmier, J. B. (1999) Inflammatory pathogenesis in Alzheimer's disease: biological mechanisms and cognitive sequeli. *Neurosci. Biobehav. Rev.* **23,** 615–633.

75. Giulian, D. (1999) Neurogenetics '99: microglia and the immune pathology of Alzheimer disease. *Am. J. Hum. Genet.* **65,** 13–18.

76. McGeer, E. G. and McGeer, P. L. (1998) The importance of inflammatory mechanisms in Alzheimer disease. *Exp. Gerontol.* **33,** 371–378.

77. Abbas, A. K., Lichtman, A. H., and Pober, J. S. (1994) *Cellular and Molecular Immunology*, Saunders, London.

78. Aisen, P. S. and Davis, K. L. (1994) Inflammatory mechanisms in Alzheimer's disease: implications for therapy. *Am. J. Psychiatry* **151,** 1105–1113.

79. Dommergues, M.-A., Patkai, J., Renauld, J.-C., Evrard, P., and Gressens, P. (2000) Proinflammatory cytokines and interleukin-9 exacerbate excitotoxic lesions of the newborn murine neopallium. *Ann. Neurol.* **47,** 54–63.

80. Pappas, B. A., Bayley, P. J., Bui, B. K., Hansen, L. A., and Thal, L. J. (2000) Choline acetyl-transferase activity and cognitive domain scores of Alzheimer's patients. *Neurobiol. Aging* **21,** 11–17.
81. Morimoto, K., Tonohiro, Y. T., Yamada, N., Oda, T., and Kaneko, I. (1998) Co-injection of beta-amyloid with ibotenic acid induces synergistic loss of rat hippocampal neurons. *Neuroscience* **84,** 479–487.
82. Harada, J. and Sugimoto, M. (1998) Inhibitors of interleukin-1beta-converting enzyme-family proteases (caspases) prevent apoptosis without affecting decreased cellular ability to reduce 3-(4,5-dimethylthiazol-2-yl)-2,5-diphenyltetrazolium bromide in cerebellar granule neurons. *Brain Res.* **793,** 231–243.
83. Miller, D. K. (1997) The role of the caspase family of cyteine proteases in apoptosis. *Semin. Immunol.* **9,** 35–49.
84. Behl, C., Davis, J. B., Klier, F. G., and Schubert, D. (1994) Amyloid beta peptide induces necrosis rather than apoptosis. *Brain Res.* **645,** 253–264.

7

Peroxisome Proliferator-Activated Receptor Gamma Agonists

Potential Therapeutic Agents for Neuroinflammation

Gary E. Landreth, Sophia Sundararajan, and Michael T. Heneka

1. NEUROINFLAMMATION AND NEUROLOGICAL DISEASE

Inflammatory processes play a central role in a number of diseases afflicting the nervous system. There is considerable controversy over whether inflammatory mechanisms are the cause or consequence of neurodegenerative changes. Moreover, if inflammatory changes are secondary to more primary neuropathological changes, do they exacerbate neuronal dysfunction and promote cell death? In some neurological diseases, the inflammatory cells are the primary effectors of the pathology; for example, in multiple sclerosis, T cells direct macrophage-mediated loss of myelin. In other diseases such as stroke, peripheral leukocytes are recruited to the lesion site along with the parallel activation of the endogenous microglia. These cells act in concert to mount a robust pro-inflammatory response that greatly expands and exacerbates the primary infarct. Traumatic brain injury is also associated with inflammatory cell infiltration and induction of a local inflammatory response. More recently, human immunodeficiency virus (HIV) and Creutzfeldt–Jakob disease have been shown to have an inflammatory component arising secondary to the primary neuropathological process. The involvement of an inflammatory component in the etiology of Alzheimer's disease (AD) has recently received considerable attention (1). Indeed, the only demonstrated effective therapy for AD patients is long-term treatment with nonsteroidal anti-inflammatory drugs (NSAIDs). The mechanistic basis of the efficacy of NSAIDs in AD remains unclear. However, the recent recognition that NSAIDs can bind to and activate the nuclear receptor peroxisome proliferator-activated receptor gamma (PPARγ) has offered an explanation for the efficacy of these drugs in AD and has opened new therapeutic approaches to this disease. Indeed, the newly appreciated anti-inflammatory actions of PPARγ agonists may allow novel therapies for other central nervous system (CNS) indications with an inflammatory component.

1.1. Inflammatory Mechanisms in Alzheimer's Disease

The pathophysiological relevance of inflammation in AD has been established by multiple lines of converging evidence (1). A substantial body of literature has documented the elevated expression of cytokines, complement proteins, chemokines, and other acute-

From: *Neuroinflammation, 2nd Edition: Mechanisms and Management*
Edited by: P. L. Wood © Humana Press Inc., Totowa, NJ

phase and pro-inflammatory molecules in the AD brain. In the AD brain, the inflammatory response is most evident in those areas that show prominent neurodegenerative changes and abundant amyloid plaques. Inflammation is absent in brain regions not affected by AD pathology, such as the cerebellum. AD inflammation arises from within the brain, with no significant involvement of lymphocytes *(1)*.

Abundant, activated microglia are invariably found to be associated with amyloid deposits in the AD brain *(2,3)* as well as in transgenic mouse models of AD that develop extensive plaque pathology *(4)*. Microglia interact with β-amyloid plaques through cell surface receptors linked to tyrosine–kinase-based pro-inflammatory signal transduction cascades *(5–8)*. The interaction of microglia with the deposited fibrillar forms of β-amyloid leads to the conversion of the microglia into an activated phenotype and results in the synthesis and secretion of cytokines and other acute-phase proteins that are neurotoxic *(9)*.

Nonsteroidal anti-inflammatory drugs have been shown to be remarkably effective in reducing the risk of AD *(10,11)*. Five epidemiological surveys, three population-based studies and one prospective clinical study have investigated the protective effects of NSAIDs in AD *(12,13)*. Sustained treatment with NSAIDs lowers the risk of AD by 55%, delays disease onset, attenuates symptomatic severity, and slows the loss of cognitive abilities. Importantly, several studies indicate that the beneficial effects of NSAIDs are positively correlated with the duration of the medication *(12)*. The principal cellular target of NSAID actions is believed to be the microglia. This view is supported by the finding that the number of activated microglia is reduced by approx 65% in patients receiving anti-inflammatory drugs *(14)*. The canonical targets of NSAID actions are the cyclo-oxygenases (COX), which are effectively inhibited by this class of drugs. However, treatment of AD patients with a COX-2 selective inhibitor was without effect, raising the question of whether the protective effects of NSAIDs might be mediated through other mechanisms *(15)*. It is of particular importance that Lehmann et al. have identified another target of NSAID actions, the ligand-activated nuclear receptor PPARγ *(16)*. NSAIDs directly bind to PPARγ and activate its transcriptional regulatory activities. Thus, it has been argued that the anti-inflammatory actions of NSAIDs may be mediated principally through their ability to activate PPARγ *(17–20)*.

1.2. Inflammatory Mechanisms in Stroke

The recognition that PPARγ agonists act to suppress microglial pro-inflammatory responses led us to consider the utility of these drugs in other CNS indications with an inflammatory component, such as stroke. Stroke is the third leading cause of death and the principal cause of disability in the United States. Inflammation plays a critical role in ischemic injury in the brain. The activation and proliferation of macrophages and microglia are the hallmark of this reaction. Previously, these cells were thought to play a beneficial role by phagocytosing necrotic debris and helping to restore blood supply. More recently, it has been appreciated that the inflammatory response is also detrimental in ischemic injury *(21)*. Manipulation of the immune reaction to stroke is a promising strategy for reducing injury following cerebral ischemia and several lines of evidence support the idea that blocking inflammation is beneficial. The inflammatory reaction begins soon after the onset of ischemia and precedes the onset of apoptotic cell death. Microglial activation is observed as early as 30 min after the onset of cerebral ischemia. Microglia synthesize and secrete the pro-inflammatory cytokines interleukin (IL)-lβ and tumor necrosis factor (TNF)-α, as well as reactive oxygen and nitrogen species. Together, these microglial products trigger neuronal apoptosis. In addition to the inflammatory response of microglia within the brain, a vigorous systemic response also occurs. Soon after the onset of ischemia, neutrophils, followed by monocytes and macrophages, are recruited to the lesion site and infiltrate the brain. The accumulation of these cells at the site of injury restricts blood flow in the microvasculature, thereby exacerbating ischemic injury *(21)*. Like microglia, these cells produce pro-inflammatory molecules that lead to further neuronal damage. Experimental strategies that block cytokine action *(22,23)* or reduce systemic influx of inflammatory cells have been shown to reduce infarct size *(24)*.

1.3. Multiple Sclerosis

Multiple sclerosis is an immune-cell-mediated disease with a complex etiology *(25)*. The most prevalent forms of the disease arise following migration of autoreactive T cells into the nervous system. The infiltrating T cells release pro-inflammatory cytokines, facilitating the development of an inflammatory response and injury to myelin-forming oligodendrocytes. Macrophages are recruited to the site of the lesion and, together with endogenous microglia, undergo phenotypic activation, produce a host of pro-inflammatory molecules, and participate in the degradation of the myelin sheath. The demyelination of axons results in impaired axonal conduction and the appearance of neurological deficits. The disease lesions are characterized by the inflammatory reaction, and therapeutic strategies that target these processes have proven to be efficacious both in humans and animal models of the disease. The ability of PPARγ agonists to inhibit T-cell activation *(26)* and inhibit pro-inflammatory gene expression *(27)* suggests that these agents may be efficacious in treating this disease. This view is supported by recent evidence demonstrating that PPARγ agonists suppress the development of neurological deficits in a mouse experimental autoimmune encephalomyelitis [*(28)*; Feinstein et al., unpublished].

2. ROLE OF PPARs IN INFLAMMATION

2.1. PPAR Family of Nuclear Receptors

The nuclear receptor superfamily of transcription factors represents an important class of regulators of gene expression whose best recognized members include the steroid, thyroid, and retinoid receptors *(29)*. A subclass of this receptor family are the peroxisome proliferator-activated receptors (PPARs), which were so named because of the ability of PPARα to respond to synthetic hypolipidemic drugs by the induction of peroxisome proliferation. There are three PPAR genes, encoding the highly related receptor isoforms α, γ, and β/δ (designated *NR1C1*, *NR1C2* and *NR1C3*, respectively) that share a common structure and mechanism of regulation *(30)*. This nuclear receptor subfamily binds a variety of lipid ligands whose transcriptional regulatory actions are activated upon binding of the ligand to the receptor. These receptors have been the subject of several recent reviews *(27,30–34)*.

Peroxisome proliferator-activated receptor gamma is expressed at highest levels in adipose tissue but is also found in the lymphocytes, vascular smooth muscle, and myeloid cells. PPARγ acts to regulate lipid metabolism and the differentiation of adipocytes.

It has only recently been appreciated that PPARs also regulate pro-inflammatory gene expression *(27,31,32)*. The regulation of genes involved in inflammatory responses are regulated principally through PPARγ. PPARγ acts to positively regulate the gene expression of a number of genes through transcriptional transactivation as well as to inhibit gene expression by transcriptional transrepression. These two distinct consequences of PPARγ activation are mediated through allied but mechanistically different processes. PPARα is expressed principally in brown adipose tissue, liver, kidney, and heart. PPARβ/δ is ubiquitously expressed and very little is known about its biological functions.

Peroxisome proliferator-activated receptors are DNA-binding proteins and are structurally similar to other members of the superfamily with respect to their domain architecture. The crystal structure of PPARγ has been solved and the details of its interactions with DNA and its ligands are reasonably well understood *(35,36)*. The PPARs possess a DNA-binding domain positioned near the N-terminus of the molecule that is separated by a hinge region from the C-terminal ligand-binding domain. The DNA binding domain has two zinc fingers that are highly conserved within this subfamily. PPARα and PPARγ bind to identical consensus DNA sequences, whereas PPARβ/δ binds to a slightly divergent sequence *(37,38)*. A common feature of the nonsteroidal nuclear receptor superfamily is their requirement for heterodimerization with the retinoid receptors (RXRs), which confers high-affinity binding to DNA *(29)*. The PPAR isoforms form obligate heterodimers with RXRs; thus, their binding to DNA requires paired PPAR–RXR recognition elements (termed PPREs) found in the promoters of its target genes. It is noteworthy that it is sufficient to bind the ligand to only one member of the receptor pair to elicit the transcriptional regulatory activity of the receptor.

The ligand-binding domain of the PPARs is characterized by a large binding pocket lined with a rather diverse range of putative interactive residues *(35,36)*. This domain, as might be expected, exhibits lower levels of sequence homology between family members, reflective of the distinctive pattern of ligand specificity exhibited by the individual PPAR isoforms. The binding of the ligand induces a conformational change in the receptor, allowing its interaction with transcriptional coactivators and corepressors.

2.1.1. PPARγ Ligands

The ligands of the PPARs are lipids, with each PPAR isoform exhibiting a distinctive substrate specificity. Importantly, it is now apparent that the individual ligands can produce different conformational changes in the receptor and elicit quantitatively and qualitatively different biological effects *(30)*. There is considerable controversy over the identity of the endogenous ligand for PPARγ. A number of long-chain polyunsaturated fatty acids and eicosinoids bind to PPARγ, including linoleic, eicosapentaenoic, and docosahexanoic acids (DHA) *(39)*. It is not clear whether intracellular levels of these compounds are sufficient to activate PPARγ and, thus, it remains unclear if interactions with the endogenous fatty acids are biologically significant *(30)*. The cyclopentone prostaglandin 15-deoxy–Δ12,14 prostaglandin J2 (PGJ2) binds to and activates PPARγ and there is a substantial literature suggesting that it is the natural ligand for this receptor *(40)*. Indeed, PGJ2 exerts potent anti-inflammatory effects and inhibits cytokine expression. The interpretation of many of these studies on the anti-inflammatory effects of PGJ2 must be re-evaluated in light of the recent reports that PGJ2 is a direct inhibitor of the inhibitory κB (IκB) kinase, IKKα, and results in the covalent modification of the p65 subunit of nuclear factor κB (NF-κB), leading to inhibition of NF-κB action *(41,42)*. Given the central role of NF-κB in pro-inflammatory gene expression, many of the effects attributed to PPARγ may have arisen, at least in part, through the action of this compound on the NF-κB pathway. These findings have forced a re-evaluation of much of the previous literature in which PGJ2 actions were thought to be mediated exclusively through PPARγ.

Recently, additional natural ligands have been identified that are components of oxidized low-density lipoprotein (LDL). The modified oxidized lipids 9-hydroxyoctadecadienoic acid (HODE) and 13-HODE have been shown to bind and activate PPARγ *(43)*. Macrophages take up oxidized LDL, followed by the intracellular release of the oxidized lipid species, resulting in PPARγ activation as a consequence of the binding of these compounds to the receptor.

There has been a substantial effort by the pharmaceutical industry to develop synthetic ligands for PPARγ. The most prominent class of synthetic ligands that act as PPARγ agonists are the thiazolidinediones (TZDs) *(30)*. These compounds were developed as therapeutic agents for the treatment of type II diabetes, because in adipose tissue, PPARγ regulates fat cell differentiation and the expression of a number of enzymes of lipid metabolism. The TZDs act principally in adipose tissue to regulate lipid metabolism, enhance target organ sensitivity to insulin, and regulate blood glucose and lipid metabolite levels *(44,45)*. There are currently three members of the thiazolidinedione class that have been approved by the Food and Drug Administration (FDA) for treatment of type II diabetes. They are pioglitazone (Actos™), rosiglitazone (Advandia™), and troglitazone (Rezulin™). There are currently approx 3.5 million people worldwide who use these drugs for the treatment of diabetes. The thiazolidinediones are generally safe drugs with good oral bioavailability and have no adverse effects in normal individuals. Other chemical classes of PPARγ agonists have been developed and are currently in clinical trials *(30)*.

Lehmann and colleagues found that a number of classical NSAIDs bind to PPARγ and activate its transcriptional activities *(16)*. Specifically, indomethacin, fenoprofen, flufenamic acid, and ibuprofen act as PPARγ agonists. These findings suggest that the anti-inflammatory effects of NSAIDs may not occur exclusively through their inhibition of cyclo-oxygenases, but rather may occur as a consequence of the ability of these drugs to directly activate PPARγ and inhibit pro-inflammatory gene expression. The recognition that NSAIDs are PPARγ agonists has been argued to explain the discrepancy between clinically efficacious doses of NSAIDs that are typically substantially greater than those required for the inhibition of cyclo-oxygenases, but consistent with occupancy of PPARγ *(46)*. It is noteworthy that studies examining the efficacy of NSAIDs in AD have shown that aspirin and acetaminophen are not linked to a reduction in AD risk, although they are very effective inhibitors of cyclo-oxygenases *(12,13)*. Indeed, epidemiological studies have reported efficacy in only that subset of NSAIDs that are PPARγ agonists *(16)*.

The ability of PPARγ to bind and be activated by fatty acids and long-chain polyunsaturated lipids, and particularly the omega-3 fatty acids and their immediate metabolites *(39)*, suggests that these transcription factors may be responsive to dietary regulation *(47)*. The role of the long-chain fatty acid DHA in the activation of PPARγ and its anti-inflammatory actions was initially demonstrated by Ricote *(20)*. Combs et al. reported that DHA efficiency inhibited IL-6 and TNF-α reporter activity and the elaboration of neurotoxic factors by monocytes *(17)*. These findings are of particular interest given the recent observation that DHA is also a ligand for RXR *(48)*.

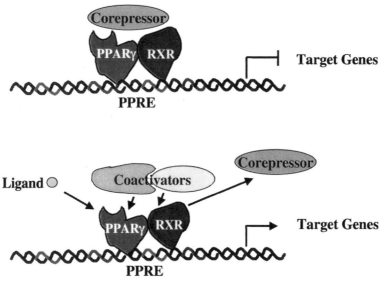

Fig. 1. Mechanism of transcriptional transactivation by PPARγ.

2.1.2. Mechanism of Transcriptional Transactivation by PPARγ

Transcriptional transactivation is mediated through the ligand-stimulated displacement of a transcriptional corepressor that is constitutively associated with the PPAR/RXR heterodimer (Fig. 1). The transcriptionally inactive PPAR complex interacts with any of a number of corepressor molecules, such as N-CoR or SMRT, suppressing its interaction with DNA and coactivators. The specific corepressor employed is likely to be distinct in different cell types. Upon ligand binding, the corepressor is displaced, the receptor complex then associates with coactivator molecules and binds to the PPRE in the promoter of its target genes. The principal coactivators interacting with PPARγ are the ubiquitously expressed SRC-1 and functionally related molecules (e.g., PGC-1, PBP, TIF-2, and p/CIP). A second class of coactivators are the cyclic AMP response element binding protein (CREB)-binding protein (CBP) and its homolog, p300. The current model for formation of a transcriptionally active PPAR complex involves the formation of a multimeric complex of SRC-1 and CBP/p300 with PPARs *(27)*. The coactivators serve both to bridge the receptor complex to the basal transcriptional apparatus and to alter chromatin structure through their intrinsic histone acetylase activities. The assembly of these complexes on the promoter then results in transcription of the target gene.

Of particular significance to neuroinflammation is the ability of PPARγ activation to regulate gene expression in myeloid lineage cells. In general, the genes that are positively regulated by PPARγ are molecules linked to lipid metabolism, reflective of the primary function of this transcription factor in adipose tissue. In monocytes and macrophages, the B-class scavenger receptor CD36 is dramatically upregulated following PPARγ ligation. CD36 is the principal high-density lipoprotein receptor in myeloid cells and also binds oxidized LDL; thus, it has been argued that the enhanced uptake of lipid

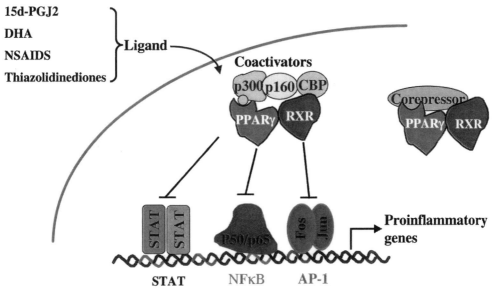

Fig. 2. Mechanism of transcriptional repression by PPARγ.

precursors of natural PPARγ ligands as well as modified lipid ligands leads to increased cellular levels of these species *(43,49)*. In macrophages, PPARγ activation stimulates the differentiation of these cells, a process facilitated by increased uptake and metabolism of lipid precursors of the natural ligands of the receptor. In addition, it has recently been shown that PPARγ agonists induce lipoprotein lipase in macrophages *(50)*. These mechanisms are important in the pathogenesis of atherosclerosis but may have limited relevance to neuroinflammatory phenomena.

2.1.3. Mechanisms of Transcriptional Transrepression by PPARγ

In myeloid lineage cells, the principal result of PPARγ activation is the inhibition of gene expression induced by inflammatory stimuli. The mechanisms of transrepression, which are particularly relevant to the anti-inflammatory actions of PPARγ, have not been fully resolved and, indeed, multiple mechanisms may be in play. The binding of a ligand to PPARγ results in the inhibition of pro-inflammatory gene expression (Fig. 2). PPARγ antagonizes the actions of the positively acting transcription factors AP-1, STAT, and NF-κB. Ricote and colleagues provided direct evidence for this antagonism by examination of reporter constructs containing minimal promoters containing only the binding sites for AP-1 and NF-κB and found that PPARγ agonists effectively inhibited the activation of these reporters *(20)*. Importantly, the transrepressive actions of PPARs do not involve the binding of the receptor to DNA. PPARγ inhibits gene expression through its capacity to bind and sequester coactivator molecules, preventing their association with the positively acting promoter elements, a process termed "squelching" *(51,52)*. In these settings, PPARγ ligand binding results in dissociation of the corepressor from the PPAR–RXR complex and the subsequent association of the coactivator molecules

SRC-1 in concert with CBP/p300. The amount of the coactivators is thought to be rate limiting and thus their sequestration by PPARγ arrests gene expression. It has been postulated that the repressive effects are mediated through coactivator sequestration by the PPAR complex and also by the ability of PPARs to functionally inactive the coactivator by conformationally constraining these molecules and thus inhibiting their interaction with the basal transcriptional apparatus *(51)*. The net consequence of PPARγ agonist binding is the arrest of expression of a number of pro-inflammatory genes. It should be noted that this type of transcription transrepression requires substantially higher levels of receptor occupancy and, thus, higher ligand concentrations, because it involves sequestration of the bulk of the coactivators by the PPARγ–ligand complex. Indeed, the experimental data indicate that transrepression occurs at a higher drug concentration than that required for transactivation *(20)*.

Peroxisome proliferator-activated receptors have been shown to antagonize the action of NF-κB. There is compelling evidence that this can happen at the level of the promoter-associated transcriptional complexes. However, recent evidence has been provided that indicates that the anti-inflammatory actions of the PPARs may also be mediated in part thorough the ability of this receptor class to induce the expression of IκB *(53)*. IκB serves to hold the NF-κB complex in an inactive state in the cytoplasm. PPARα ligands have been shown to induce the elevation of cytoplasmic levels of IκB mRNA and protein and it seems likely that PPARγ agonists may elicit a similar response. The presence of high levels of cytoplasmic IκB would serve to block the translocation of NF-κB to the nucleus and act to inhibit pro-inflammatory gene expression.

The study of the actions of PPARγ on pro-inflammatory gene expression and promoter activity have shown an unexpected complexity in the action of this receptor. In the initial reports of PPARγ-dependent inhibition of pro-inflammatory cytokine expression, it was observed that PPARγ agonists effectively suppressed the expression of these genes when macrophages were stimulated with phorbol ester or okadaic acid, but not the endotoxin lipopolysaccharide (LPS). This finding is supported by Thieringer et al., who also found that LPS-stimulated cytokine production was not affected by treatment of the cells with a large panel of PPARγ agonists *(54)*, although this latter finding is controversial *(55,56)*. Phorbol ester and LPS employ distinct signal transduction cascades to stimulate gene expression through the phosphorylation of transcription factors. It is possible that the LPS stimulation provokes the formation of transcriptional complexes on the promoters of the target genes that are molecularly distinct from those elicited by other stimuli and utilize elements whose activity is not compromised by PPARγ activation.

2.2. The Role of PPARγ in Inflammation

The initial indications that PPARγ may play a role in suppressing inflammatory gene expression came from studies of adipocytes. In diabetic models, PPARγ agonists were shown to suppress the expression of the pro-inflammatory cytokine TNF-α, which acts to inhibit insulin signaling and contributes to insulin resistance *(33,57)*. There is now substantial evidence that PPARγ agonists have potent anti-inflammatory activity *(27, 30–33)*. The demonstration that PPARγ plays a role in myeloid cell biology arose from two seminal studies by Ricote and Jiang and colleagues *(20,46)*. These studies demonstrated that exposure of monocytes or macrophages to both synthetic and natural ligands of PPARγ resulted in the inhibition of expression of the pro-inflammatory cytokines IL-1β, IL-6, and TNF-α following treatment of the cells with phorbol ester *(46)*. These effects were a consequence of the transcriptional inactivation of the promoters of the cytokine genes. PPARγ agonists have also been shown to inhibit the expression of the chemokine MCP-1 gene, whose expression is induced upon activation of macrophages *(58)*, as well as the chemokine-inducted migration of these cells *(59)*. Importantly, activation of PPARγ by both synthetic and natural ligands blocked the phenotypic conversion of the monocytes into reactive macrophages *(17,60)* and the activation of microglia *(17)*.

2.2.1. iNOS

Ricote et al. reported that both natural and synthetic PPARγ agonists inhibited inducible nitric oxide synthase (iNOS) expression in interferon (IFN)γ-stimulated monocytes and macrophages *(20)*. Subsequently, PPARγ-mediated inhibition of iNOS expression has been described in a variety of cells such as macrophages, microglia, astrocytes, and neurons *(18,20,61,62)*. Long-term iNOS-mediated production of nitric oxide (NO) is an important element of the innate immune response that exerts toxic effects on tumor cells and invading micro-organisms *(63)*. It is of particular significance that in the CNS, iNOS expression has been shown to provoke neuronal cell death. Induction of iNOS in astrocytes and microglia leads to NO-mediated neuronal cell death in vitro. Moreover, iNOS was found to be induced in neurons that undergo neurodegenerative changes in the brains of patients suffering from AD. A number of experimental studies in stroke rodent models of multiple sclerosis and neurodegeneration have demonstrated a neuroprotective action of iNOS inhibitors *(64)*.

The induction of iNOS expression in neurons by IFN-γ and LPS was antagonized by the PPARγ agonists troglitazone and PGJ2, both in vitro *(18)* and in vivo *(19)*. In vitro, iNOS induction in response to LPS and various cytokines resulted in NO-dependent activation of caspase-3 activity, subsequent DNA fragmentation, and apoptotic cell death. Exposure of neurons to PPARγ agonists and NSAIDs blocked both iNOS expression and cell death. These studies were extended into an animal model in which LPS and IFN-γ were directly injected into the brain. Neuronal iNOS induction was blocked and neuronal apoptosis was inhibited when the PPARγ agoinsts troglitazone, PGJ2, and the NSAID ibuprofen were coinjected. Consistent with these findings and supporting a protective role in the neuroinflammation associated with AD, Ogawa and colleagues reported that β-amyloid-induced iNOS expression in macrophages was inhibited by indomethacin and ibuprofen, but not aspirin *(65)*. In summary, inhibition of both glial and neuronal iNOS by PPARγ-agonists may exert neuroprotective effects in a variety of inflammatory brain disorders, including AD.

The molecular details of the mechanism of repression of iNOS gene transcription have recently been reported by Li et al. and conform to the scheme shown in Fig. 2 *(51)*. The inhibition of iNOS gene expression is principally the result of the stable interaction of PPARγ with a complex of the coactivators SRC-1 and CBP/p300. The coactivator CBP/p300 is required for the transcriptional activation by the positively acting transcription factors NF-κB and AP-1. Competitive binding of PPARγ to limiting amounts of these coactivators prevented their association with AP-1 and NF-κB elements in the iNOS promoter, thereby blocking promotor activity and the synthesis of iNOS *(51)*.

Although transcriptional transrepression is likely to be the principal mechanism through which iNOS expression is suppressed, additional mechanisms are also involved. There is evidence that PPARγ agonists also act to inhibit NF-κB activation through regulation of its regulatory IκB subunit. Cytoplasmic IκB levels are stabilized through inhibition of proteosome function by HSP70, which stabilizes the cytosolic NF-κB inhibitors IκBα and IκBβ, thereby limiting the nuclear translocation of NF-κB, which has been shown to be essential for iNOS gene transcription *(64)*. HSP70 expression is induced by both troglitazone and the endogenous PPARγ agonist PGJ2 and this may contribute to iNOS inhibition *(66)*. In addition, activation of PPARα results in the induction of IκB gene expression *(53)* and these events may act in concert to elevate cytoplasmic levels of IκB and, thus, inhibit iNOS expression.

2.2.2. COX-2

The cyclooxygenase-2 gene is an immediate early gene that is rapidly induced in response to a variety of stimuli and in many cell types. The induction of COX-2 results in the acute upregulation of prostaglandin synthesis. The principal product of COX-2 that has been postulated to have a role in pro-inflammatory responses is prostaglandin E2 and it has been argued that this pro-inflammatory product participates in the etiology of AD *(67)*. COX-2 levels in the AD cortex were reported to be elevated by 50% over age-matched control patients *(68)*. The argument that COX plays a role in the etiology of AD is derived principally from the epidemiological studies, demonstrating a protective effect of NSAIDs in AD, rather than compelling evidence that prostaglandins have deleterious actions in the CNS *(17)*. Combs et al. have recently demonstrated that β-amyloid-stimulated COX-2 expression in primary microglia and in monocytes was inhibited by the PPARγ agonists troglitazone and PGJ2. In agreement with these studies, Kitamura et al. have reported that the NSAIDs indomethacin and PGJ2 inhibit COX-2 expression in mixed glial cultures *(68)*, a finding similar to that reported in macrophages by Inoue et al. *(69)*.

Subbaramaiah and colleagues have nicely dissected the mechanism by which PPARγ acts to inhibit COX-2 expression *(52)*. The induction of COX-2 in macrophages requires the binding of AP-1 (a c-fos and c-jun heterodimer) to the CRE element in the COX-2 promoter. PPARγ acts to prevent AP-1 binding to the promoter through two independent mechanisms. First, PPARγ ligands block the expression of the c-jun gene, thus preventing the formation of a functional AP-1 complex and resulting in the inhibition of promoters requiring an AP-1 promoter element for gene expression. In addition, the ligand-bound PPARγ associates with both CBP and p300, sequestering these coactivators and preventing their interaction with the promoter-associated AP-1 complex (and presumably other transcription factors) in a fashion directly analogous to that reported for inhibition of iNOS expression *(51)*. These complementary mechanisms serve to silence the iNOS gene.

The discussion of the action of COX-1 and COX-2 and their enzymatic products have been focused principally on the pro-inflammatory actions of these molecules; however, recent evidence suggests that the cyclo-oxygenases are responsible for the production of anti-inflammatory compounds. It has recently been appreciated that COX-2 activation results in the production of the anti-inflammatory cyclopentenone prostaglandins, principally PGJ2. These prostanoids are produced during the later stages of an inflammatory response *(70)*, facilitating the resolution of the inflammatory episode. The cyclopentenone prostaglandins act to inhibit inflammatory gene expression through a direct inhibition of IκB kinases and, thus, inhibition of NF-κB activation, as well as through binding of PGJ2 and activation of PPARγ *(41,42,70)*.

2.2.3. Other Targets

Peroxisome proliferator-activated receptor gamma acts broadly to inhibit other elements of the pro-inflammatory response of myeloid lineage cells. Ricote et al. demonstrated that the synthesis of the metalloproteinase 9 (MMP9, also termed gelatinase B) induced upon macrophage activation was inhibited by PPARγ agonists *(20)*. This effect was shown to be a consequence of PPARγ acting directly on the promoter of the MMP9 gene through antagonism of positively acting AP-1 elements. Similarly, expression of the A-class scavenger receptor, SRA, was blocked by PPARγ agonists. SRA binds to a diverse range of substrates and oxidized LDL and participates in their uptake. SRA expression was stimulated upon phenotypic activation of macrophages, and its induction was blocked by treatment of the cells with PPARγ ligands. The SRA promoter contains AP-1 sites whose ability to stimulate SRA expression was antagonized by activation of PPARγ *(20)*.

2.2.4. Alternative Mechanisms of Action of PPARγ Agonists

A criticism of the studies on the anti-inflammatory actions of PPARγ agonists, and TZDs in particular, has been that high levels of these agents are required to elicit anti-inflammatory effects. The anti-inflammatory actions are typically observed at concentrations that do not correspond well with the binding affinity of the receptor. Whether this is a reflection of the bioavailibility of the drug or that there are other mechanisms of drug action is presently controversial. Moreover, it has been reported that some TZDs exhibit antidiabetic actions while failing to significantly activate PPARγ *(71,72)*. One potential explanation is that the PPARγ agonists, and particularly the TZDs, may act through other mechanisms. Recent studies using macrophages in which the PPARγ gene was knocked out revealed that TZDs exhibited anti-inflammatory activity at high drug concentrations *(55,56)*. These findings demonstrate that the TZDs can operate through non-PPARγ-dependent mechanisms, perhaps through binding of these drugs to the related PPARα isoform. A number of studies using dominant negative and dominant positive PPARγ molecules clearly show the dependence of gene expression on PPARγ function and support the conclusion that these mechanisms are physiologically relevant to the action of TZDs. However, these findings raise the question of whether the drugs have additional cellular targets.

3. ACTIONS OF PPARγ IN THE NERVOUS SYSTEM

There is remarkably little known about the functional roles of PPAR isoforms in the nervous system in general, and PPARγ in particular.

3.1. PPAR Expression in the Brain

There is limited data on the expression of PPARs in the brain. There are only a small number of studies investigating PPAR expression and the conclusions of these studies are frequently conflicting. Significantly, PPARγ was found to be expressed only at very low levels in the brain. This finding is consistent with a previous report that PPARγ mRNA was present in the brain at barely detectable levels by Northern analysis *(73)*. Primary cultures of neurons and astrocytes revealed a similar pattern of PPAR expression;

however, PPARγ was found most prominently in adult (but not neonatal) cortical astrocytes *(74)*. Examination of the developmental regulation of PPARγ expression in the rat brain by *in situ* hybridization revealed that this isoform is transiently expressed at substantial levels in the hindbrain at E13.5, but its expression falls to very low levels by birth. PPARγ expression was also found at early developmental stages in the forebrain, midbrain, and spinal cord *(75)*. Interestingly, PPARγ expression in these early embryos was only observed within the CNS and not elsewhere in the embryo and raises the possibility that this transcription factor may play a role in the ontogeny of the nervous system. There is a single report of PPARγ expression in the human brain. Kitamura and colleagues reported that PPARγ expression could be detected by Western analysis in the temporal cortex of the human brain *(68)*. Significantly, they found an approx 50% increase in the amount of immunoreactive PPARγ protein in the brains of AD patients.

A detailed analysis of PPAR expression in the adult brain using RNAase protection assays found that PPARδ was the predominant isoform expressed in the brain and was expressed in all areas examined. PPARα was present in most brain regions at substantially lower levels. Kainu reported that PPARα was found at highest levels in cerebellar granule neurons and also in oligodendrocytes *(76)*.

3.2. PPARγ and Microglia

Peroxisome proliferative-activated receptor gamma is constitutively expressed in microglial cells and monocytes *(9,77,78)*, although the expression of this nuclear receptor is subject to both positive and negative regulation. The best documented example of the regulated expression of PPARγ is in myeloid lineage cells. PPARγ expression is induced upon differentiation of monocytes into macrophages *(49)* and upon activation of resting macrophages *(20)*. Kitamura and colleagues have also recently reported that the T-cell-derived cytokine IL-4 provoked the induction of PPARγ expression in microglial cultures *(78)*. Treatment of the cells with IL-4 resulted in stimulation of PPARγ expression within 10 h and was sustained for at least 60 h, the longest interval tested. Interestingly, LPS treatment of IL-4-stimulated macrophages inhibited the induction of PPARγ expression. Bernado et al. have reported a similar finding, demonstrating that treatment of microglial cells with LPS resulted in diminished cellular levels of PPARγ, thus facilitating the pro-inflammatory activation of the cells *(77)*.

β-Amyloid treatment of monocytes and microglia results in their pro-inflammatory activation and stimulation of the synthesis and secretion of neurotoxins. Combs et al. have reported that treatment of either microglia or monocytes with PPARγ agonists arrested the secretion of the neurotoxic factors *(17)*. This study also demonstrated that the PPARγ agonists inhibited the β-amyloid-induced stimulation of IL-6 and TNF-α expression. This effect was observed following treatment of the cells with the PPARγ agonists ciglitazone and troglitazone, the NSAIDs ibuprofen and indomethacin, as well as the natural PPARγ ligands PGJ2 and DHA. Significantly, the COX-2-specific inhibitor NS398 was without effect. The phenotypic activation of monocytes by phorbol ester and microglia by β-amyloid was also blocked by PPARγ ligands, as was the expression of the complement receptor CR3/Mac1. Activation of PPARγ also resulted in the greatly reduced expression of β-amyloid-stimulated expression of COX-2 in both monocytes and microglia. This latter finding was also reported by Kitamura and colleagues using mixed glial cultures *(68)*.

Bernardo et al. reported that activation of microglia with LPS or IFN-γ resulted in the induction of iNOS, TNF-α and major histocompatibility complex (MHC) class II expression that was fully suppressed by treatment with PGJ2 *(77)*. Similarly, Van Eldik and colleagues reported that microglial activation resulted in the induction of iNOS expression that was effectively inhibited by PGJ2 *(79)*. These latter authors argued that iNOS expression was not regulated by PPARγ but rather by other targets of PGJ2, as troglitazone was unable to inhibit iNOS activity in LPS-stimulated microglia. Moreover, they were unable to detect any stimulation of a PPRE reporter by troglitazone in these cells. The conclusion that PPARγ is not acting to regulate iNOS conflicts with a number of reports in the literature *(18,19,51)*. Glass and colleagues have now dissected the role of PPARγ at this promoter in considerable detail *(51)*. It seems likely that the failure of Petrova et al. to observe an effect of troglitazone on the PPRE reporter in the microglia may be the result of their use of a concentration of troglitazone that was insufficient to activate PPARγ. Alternatively, PPARγ agonists may be ineffective in suppressing the LPS-driven activation of the iNOS promoter, a situation similar to that reported by Moller and colleagues for cytokine gene expression *(52)*. The author's conclusion that PGJ2 may act through mechanisms independent of PPARγ has subsequently been substantiated by the demonstration that PGJ2 does indeed inhibit NF-κB-stimulated gene expression through direct inhibition of IκB kinases and NF-κB *(41,42)*.

Lim et al. have recently published a provocative study in which ibuprofen was administered to an animal model of AD *(80)*. In APP-overexpressing Tg2576 mice fed ibuprofen for 6 mo, there was a reduction in the amount of IL-1β in the brain and reduced levels of activated microglia. These findings are consistent with observations in the AD patients where NSAIDs usage was correlated with a 65% reduction of activated microglia associated with senile plaques *(14)*. The transgenic animals also exhibited a significant reduction in the amount of soluble β-amyloid in the brain and reduced area occupied by amyloid plaques. It is not clear whether this is the result of the action of ibuprofen on PPARγ or on the cyclo-oxygenases. The high levels of ibuprofen administered to the animals are likely to be sufficient to activate PPARγ. A number of studies have documented the effect of the NSAIDs ibuprofen and indomethacin on pro-inflammatory gene expression by monocytes and microglia *(1)*. Both Combs et al. and Klegeris et al. have shown that these NSAIDs suppress the expression of the pro-inflammatory cytokines and neurotoxicity *(17,81)*. Combs et al. have argued that these effects are likely to be the result of the action of NSAIDs on activation of PPARγ rather than on their canonical targets, the cyclo-oxygenases *(17)*.

3.3. PPARγ Actions in Astrocytes

Astrocytes from rat brain were found to express all three PPAR isoforms *(74)*. However, astrocytes derived from adult rat cortex express PPARγ at higher levels than those found from neonatal brain. Human astrocytes were reported to express only PPARγ and PPARδ, as detected by reverse transcriptase–polymer chain reaction *(82)*.

Chattopadhyay et al. reported that the PPARγ agonists ciglitazone and PGJ2 reduced the viability of primary human astrocyte within 4 h and induced apoptosis. These authors also reported that PPARγ agonists provoked apoptosis of the malignant human astrocytoma cell line T98G *(82)*. These latter findings are consistent with our observation that transformed astrocytic cell lines respond to PPARγ agonists by induction of apoptosis (Heneka et al., unpublished observations). However, we failed to observe this effect with primary rodent astrocytes under similar conditions *(83)*. The susceptibililty of transformed or neoplastic cells to induction of apoptosis by PPARγ activation has been well documented in a number of cell types, although the molecular basis of this effect is presently unclear *(27,31)*.

Kitamura and colleagues have reported that PPARγ agonists inhibit iNOS expression in mixed rodent glial cultures provoked by treatment of the cells with IFN-γ and LPS *(61)*. These cultures are approx 90% astrocytes, so it is likely that the iNOS inhibition by PPARγ activators observed in these studies reflects the effects of the drugs on this cell type. Microglial cells also respond to PPARγ agonists by suppression of iNOS expression and the mixed nature of the cultures does not allow firm conclusions to be drawn as to the relative contribution of each cell type to the observed effects. These authors have also demonstrated that treatment of the mixed glial cultures with PPARγ agonists resulted in the induction of heme oxygenase-1 *(61)*.

3.4. PPARγ Actions in Neurons

The literature on the actions of PPARγ in neurons are contained in two reports by Heneka and colleagues *(18,19)*. Treatment of primary cerebellar granule cell cultures with the pro-inflammatory activators LPS and IFN-γ, resulted in the induction of iNOS expression and subsequent apoptotic death of the neurons arising from the toxic effects of the reactive nitrogen species. However, treatment of the cells with the PPARγ agonists ciglitazone and PGJ2 resulted in the suppression of iNOS expression and enhanced neuronal survival. A similar effect was observed following treatment with the NSAIDs ibuprofen and indomethacin, reinforcing the view that NSAIDs exert their actions through their capacity to bind and activate PPARγ. These studies have been extended into an animal model *(19)*. In this study, LPS and IFN-γ were injected into the cerebellum of rats, resulting in the induction of neuronal iNOS expression and death of neurons at the site of injection. If, however, PPARγ agonists troglitazone, ibuprofen, or PGJ2 were injected simultaneously, the iNOS induction was suppressed and cell death was significantly attenuated.

The TZD troglitazone has recently been reported to promote the survival of rat spinal motor neurons, but not other classes of neurons *(84)*. This effect appears not to be mediated through PPARγ, as other TZDs were without effect. Rohn and colleagues have found that PGJ2 provoked neuronal apoptosis in cultures of primary cortical neurons and in a neuroblastoma cell line *(85)*. These data were interpreted as an effect of PPARγ activation; however, given the ability of PGJ2 to inhibit the NF-κB pathway and the role of NF-κB on neuronal survival, the conclusion that this is an effect on PPARγ is unsubstantiated.

4. CONCLUSIONS

The recent recognition that PPARγ plays important roles in the regulation of pro-inflammatory gene expression has provided new insight and investigative opportunities in the roles this transcription factor plays in the nervous system. It is of particular interest and importance to ascertain if the efficacy of NSAIDs in reducing AD risk is a consequence of the action of these drugs on PPARγ. The discovery of the anti-inflammatory actions of PPARγ agonists argue that these compounds may be of utility in other

CNS inflammatory indications and the early experimental evidence supports this view. The availability of potent, FDA-approved PPARγ agonists will allow the rapid translation of experimental findings into the clinic.

REFERENCES

1. Akiyama, H., Barger, S., Barnum, S., Bradt, B., Bauer, J., Cole, G. M., et al. (2000) Inflammation and Alzheimer's disease. *Neurobiol. Aging* **21,** 383–421.
2. McGeer, P. L., Kawamata, T., Walker, D. G., Akiyama, H., Tooyama, I., and McGeer, E. G. (1993) Microglia in degenerative neurological disease. *Glia* **7,** 84–92.
3. Perlmutter, L. S., Barron, E., and Chui, H. C. (1990) Morphologic association between microglia and senile plaque amyloid in Alzheimer's disease. *Neurosci. Lett.* **119,** 32–36.
4. Bornemann, K. D., Wiederhold, K. H., Pauli, C., Ermini, F., Stalder, M., Schnell, L., et al. (2001) Abeta-induced inflammatory processes in microglia cells of APP23 transgenic mice. *Am. J. Pathol.* **158,** 63–73.
5. Bamberger, M. and Landreth, G. (2001) Microglial Interaction with β-amyloid—implications for the pathogenesis of Alzheimer's disease. *Microsc. Res. Tech.* **54,** 59–70.
6. Combs, C. K., Johnson, D. J., Cannady, S. B., Lehman, T. M., and Landreth, G. E. (1999) Identification of microglial signal transduction pathways mediating a neurotoxic response to amyloidogenic fragments of β-amyloid and prion proteins. *J. Neurosci.* **19,** 928–939.
7. McDonald, D. R., Brunden, K. R., and Landreth, G. E. (1997) Amyloid fibrils activate tyrosine kinase-dependent signaling and superoxide production in microglia. *J. Neurosci.* **17,** 2284–2294.
8. McDonald, D., Bamberger, M., Combs, C., and Landreth, G. (1998) β-Amyloid fibrils activate parallel mitogen-activated protein kinase pathways in microglia and THP-1 monocytes. *J. Neurosci.* **18,** 4451–4460.
9. Combs, C., Bates, P., Karlo, J., and Landreth, G. (2001) Regulation of beta-amyloid stimulated proinflammatory responses by peroxisome proliferator-activated receptor alpha. *Neurochem. Int.* **39,** 449–557.
10. McGeer, P. L. and Rogers, J. (1992) Anti-inflammatory agents as a therapeutic approach to Alzheimer's disease. *Neurology* **42,** 447–449.
11. McGeer, P. L. and McGeer, E. G. (1999) Inflammation of the brain in Alzheimer's disease: implications for therapy. *J. Leukocyte Biol.* **65,** 409–415.
12. Breitner, J. C. (1996) The role of anti-inflammatory drugs in the prevention and treatment of Alzheimer's disease. *Annu. Rev. Med.* **47,** 401–411.
13. Stewart, W. F., Kawas, C., Corrada, M., and Metter, E. J. (1997) Risk of Alzheimer's disease and duration of NSAID use. *Neurology* **48,** 626–632.
14. Mackenzie, I. R. and Munoz, D. G. (1998) Nonsteroidal anti-inflammatory drug use and Alzheimer-type pathology in aging. *Neurology* **50,** 986–990.
15. Sainati, S., Ingram, D., Talwalker, G., and Geis, G. (2000) Results of a double-blind, randomized, placebo-controlled study of celecoxib in the treatment of Alzheimer's disease. Sixth International Stockholm/Springfield Symposium on Advances in Alzheimer Therapy, p. 180.
16. Lehmann, J. M., Lenhard, J. M., Oliver, B. B., Ringold, G. M., and Kliewer, S. A. (1997) Peroxisome proliferator-activated receptors alpha and gamma are activated by indomethacin and other non-steroidal anti-inflammatory drugs. *J. Biol. Chem.* **272,** 3406–3410.
17. Combs, C. K., Johnson, D. E., Karlo, J. C., Cannady, S. B., and Landreth, G. E. (2000) Inflammatory mechanisms in Alzheimer's disease: inhibition of beta-amyloid-stimulated proinflammatory responses and neurotoxicity by PPARgamma agonists. *J. Neurosci.* **20,** 558–567.
18. Heneka, M. T., Feinstein, D. L., Galea, E., Gleichmann, M., Wullner, U., and Klockgether, T. (1999) Peroxisome proliferator-activated receptor gamma agonists protect cerebellar granule cells from cytokine-induced apoptotic cell death by inhibition of inducible nitric oxide synthase. *J. Neuroimmunol.* **100,** 156–168.

19. Heneka, M. T., Klockgether, T., and Feinstein, D. L. (2000) Peroxisome proliferator-activated receptor-gamma ligands reduce neuronal inducible nitric oxide synthase expression and cell death in vivo. *J. Neurosci.* **20,** 6862–6867.

20. Ricote, M., Li, A. C., Willson, T. M., Kelly, C. J., and Glass, C. K. (1998) The peroxisome proliferator-activated receptor-gamma is a negative regulator of macrophage activation. *Nature* **391,** 79–82.

21. Barone, F. C. and Feuerstein, G. Z. (1999) Inflammatory mediators and stroke: new opportunities for novel therapeutics. *J. Cereb. Blood Flow Metab.* **19,** 819–834.

22. Loddick, S. A. and Rothwell, N. J. (1996) Neuroprotective effects of human recombinant interleukin-1 receptor antagonist in focal cerebral ischaemia in the rat. *J. Cereb. Blood Flow Metab.* **16,** 932–940.

23. Matsuo, Y., Onodera, H., Shiga, Y., Nakamura, M., Ninomiya, M., Kihara, T., et al. (1994) Correlation between myeloperoxidase-quantified neutrophil accumulation and ischemic brain injury in the rat. Effects of neutrophil depletion. *Stroke* **25,** 1469–1475.

24. Chopp, M., Li, Y., Jiang, N., Zhang, R. L., and Prostak, J. (1996) Antibodies against adhesion molecules reduce apoptosis after transient middle cerebral artery occlusion in rat brain. *J. Cereb. Blood Flow Metab.* **16,** 578–584.

25. Noseworthy, J. H., Lucchinetti, C., Rodriguez, M., and Weinshenker, B. G. (2000) Multiple sclerosis. *N. Engl. J. Med.* **343,** 938–952.

26. Yang, X. Y., Wang, L. H., Chen, T., Hodge, D. R., Resau, J. H., DaSilva, L., et al. (2000) Activation of human T lymphocytes is inhibited by peroxisome proliferator-activated receptor gamma (PPARgamma) agonists. PPARgamma co-association with transcription factor NFAT. *J. Biol. Chem.* **275,** 4541–4544.

27. Gelman, L., Fruchart, J. C., and Auwerx, J. (1999) An update on the mechanisms of action of the peroxisome proliferator-activated receptors (PPARs) and their roles in inflammation and cancer. *Cell. Mol. Life Sci.* **55,** 932–943.

28. Niino, M., Iwabuchi, K., Kikuchi, S., Ato, M., Morohashi, T., Ogata, A., et al. (2001) Amelioration of experimental autoimmune encephalomyelitis in C57BL/6 mice by an agonist of peroxisome proliferator-activated receptor-gamma. *J. Neuroimmunol.* **116,** 40–48.

29. Blumberg, B. and Evans, R. M. (1998) Orphan nuclear receptors—new ligands and new possibilities. *Genes Dev.* **12,** 3149–3155.

30. Willson, T. M., Brown, P. J., Sternbach, D. D., and Henke, B. R. (2000) The PPARs: from orphan receptors to drug discovery. *J. Med. Chem.* **43,** 527–550.

31. Murphy, G. J. and Holder, J. C. (2000) PPAR-gamma agonists: therapeutic role in diabetes, inflammation and cancer. *Trends Pharmacol. Sci.* **21,** 469–474.

32. Ricote, M., Huang, J. T., Welch, J. S., and Glass, C. K. (1999) The peroxisome proliferator-activated receptor (PPARgamma) as a regulator of monocyte/macrophage function. *J. Leukocyte Biol.* **66,** 733–739.

33. Desvergne, B. and Wahli, W. (1999) Peroxisome proliferator-activated receptors: nuclear control of metabolism. *Endocr. Rev.* **20,** 649–688.

34. Willson, T. M. and Wahli, W. (1997) Peroxisome proliferator-activated receptor agonists. *Curr. Opin. Chem. Biol.* **1,** 235–241.

35. Nolte, R. T., Wisely, G. B., Westin, S., Cobb, J. E., Lambert, M. H., Kurokawa, R., et al. (1998) Ligand binding and co-activator assembly of the peroxisome proliferator-activated receptor-gamma. *Nature* **395,** 137–143.

36. Uppenberg, J., Svensson, C., Jaki, M., Bertilsson, G., Jendeberg, L., and Berkenstam, A. (1998) Crystal structure of the ligand binding domain of the human nuclear receptor PPARgamma. *J. Biol. Chem.* **273,** 31,108–31,112.

37. He, T. C., Chan, T. A., Vogelstein, B., and Kinzler, K. W. (1999) PPARdelta is an APC-regulated target of nonsteroidal anti-inflammatory drugs. *Cell* **99,** 335–345.

38. Lemberger, T., Desvergne, B., and Wahli, W. (1996) Peroxisome proliferator-activated receptors: a nuclear receptor signaling pathway in lipid physiology. *Annu. Rev. Cell. Dev. Biol.* **12,** 335–363.
39. Yu, K., Bayona, W., Kallen, C. B., Harding, H. P., Ravera, C. P., McMahon, G., et al. (1995) Differential activation of peroxisome proliferator-activated receptors by eicosanoids. *J. Biol. Chem.* **270,** 23,975–23,983.
40. Forman, B. M., Tontonoz, P., Chen, J., Brun, R. P., Spiegelman, B. M., and Evans, R. M. (1995) 15-Deoxy-delta 12, 14-prostaglandin J2 is a ligand for the adipocyte determination factor PPAR gamma. *Cell* **83,** 803–812.
41. Rossi, A., Kapahi, P., Natoli, G., Takahashi, T., Chen, Y., Karin, M., et al. (2000) Anti-inflammatory cyclopentenone prostaglandins are direct inhibitors of IkappaB kinase. *Nature* **403,** 103–108.
42. Straus, D. S., Pascual, G., Li, M., Welch, J. S., Ricote, M., Hsiang, C. H., et al. (2000) 15-Deoxy-delta 12,14-prostaglandin J2 inhibits multiple steps in the NF-kappa B signaling pathway. *Proc. Natl. Acad. Sci. USA* **97,** 4844–4849.
43. Nagy, L., Tontonoz, P., Alvarez, J. G., Chen, H., and Evans, R. M. (1998) Oxidized LDL regulates macrophage gene expression through ligand activation of PPARgamma. *Cell* **93,** 229–240.
44. Olefsky, J. M. (2000) Treatment of insulin resistance with peroxisome proliferator-activated receptor gamma agonists. *J. Clin. Invest.* **106,** 467–472.
45. Steppan, C. M., Bailey, S. T., Bhat, S., Brown, E. J., Banerjee, R. R., Wright, C. M., et al. (2001) The hormone resistin links obesity to diabetes. *Nature* **409,** 307–312.
46. Jiang, C., Ting, A. T., and Seed, B. (1998) PPAR-gamma agonists inhibit production of monocyte inflammatory cytokines. *Nature* **391,** 82–86.
47. Jump, D. B. and Clarke, S. D. (1999) Regulation of gene expression by dietary fat. *Annu. Rev. Nutr.* **19,** 63–90.
48. de Urquiza, A. M., Liu, S., Sjoberg, M., Zetterstrom, R. H., Griffiths, W., Sjovall, J., et al. (2000) Docosahexaenoic acid, a ligand for the retinoid X receptor in mouse brain. *Science* **290,** 2140–2144.
49. Tontonoz, P., Nagy, L., Alvarez, J. G., Thomazy, V. A., and Evans, R. M. (1998) PPARgamma promotes monocyte/macrophage differentiation and uptake of oxidized LDL. *Cell* **93,** 241–252.
50. Michaud, S. E. and Renier, G. (2001) Direct regulatory effect of fatty acids on macrophage lipoprotein lipase: potential role of PPARs. *Diabetes* **50,** 660–666.
51. Li, M., Pascual, G., and Glass, C. K. (2000) Peroxisome proliferator-activated receptor gamma-dependent repression of the inducible nitric oxide synthase gene. *Mol. Cell. Biol.* **20,** 4699–4707.
52. Subbaramaiah, K., Lin, D. T., Hart, J. C., and Dannenberg, A. J. (2001) Peroxisome proliferator-activated receptor gamma ligands suppress the transcriptional activation of cyclooxygenase-2. Evidence for involvement of activator protein-1 and CREB-binding protein/p300. *J. Biol. Chem.* **276,** 12,440–12,448.
53. Delerive, P., Gervois, P., Fruchart, J. C., and Staels, B. (2000) Induction of IkappaBalpha expression as a mechanism contributing to the anti-inflammatory activities of peroxisome proliferator-activated receptor-alpha activators. *J. Biol. Chem.* **275,** 36,703–36,707.
54. Thieringer, R., Fenyk-Melody, J. E., Le Grand, C. B., Shelton, B. A., Detmers, P. A., Somers, E. P., et al. (2000) Activation of peroxisome proliferator-activated receptor gamma does not inhibit IL-6 or TNF-alpha responses of macrophages to lipopolysaccharide in vitro or in vivo. *J. Immunol.* **164,** 1046–1054.
55. Chawla, A., Barak, Y., Nagy, L., Liao, D., Tontonoz, P., and Evans, R. M. (2001) PPAR-gamma dependent and independent effects on macrophage-gene expression in lipid metabolism and inflammation. *Nat. Med.* **7,** 48–52.

56. Moore, K. J., Rosen, E. D., Fitzgerald, M. L., Randow, F., Andersson, L. P., Altshuler, D., et al. (2001) The role of PPAR-gamma in macrophage differentiation and cholesterol uptake. *Nat. Med.* **7,** 41–47.
57. Lemberger, T., Braissant, O., Juge-Aubry, C., Keller, H., Saladin, R., Staels, B., et al. (1996) PPAR tissue distribution and interactions with other hormone-signaling pathways. *Ann. NY Acad. Sci.* **804,** 231–251.
58. Murao, K., Ohyama, T., Imachi, H., Ishida, T., Cao, W. M., Namihira, H., et al. (2000) TNF-alpha stimulation of MCP-1 expression is mediated by the Akt/PKB signal transduction pathway in vascular endothelial cells. *Biochem. Biophys. Res. Commun.* **276,** 791–796.
59. Kintscher, U., Goetze, S., Wakino, S., Kim, S., Nagpal, S., Chandraratna, R. A., Graf, K., et al. (2000) Peroxisome proliferator-activated receptor and retinoid X receptor ligands inhibit monocyte chemotactic protein-1-directed migration of monocytes. *Eur. J. Pharmacol.* **401,** 259–270.
60. Ricote, M., Huang, J., Fajas, L., Li, A., Welch, J., Najib, J., et al. (1998) Expression of the peroxisome proliferator-activated receptor gamma (PPARgamma) in human atherosclerosis and regulation in macrophages by colony stimulating factors and oxidized low density lipoprotein. *Proc. Natl. Acad. Sci. USA* **95,** 7614–7619.
61. Kitamura, Y., Kakimura, J., Matsuoka, Y., Nomura, Y., Gebicke-Haerter, P. J., and Taniguchi, T. (1999) Activators of peroxisome proliferator-activated receptor-gamma (PPARgamma) inhibit inducible nitric oxide synthase expression but increase heme oxygenase-1 expression in rat glial cells. *Neurosci. Lett.* **262,** 129–132.
62. Colville-Nash, P. R., Qureshi, S. S., Willis, D., and Willoughby, D. A. (1998) Inhibition of inducible nitric oxide synthase by peroxisome proliferator-activated receptor agonists: correlation with induction of heme oxygenase 1. *J. Immunol.* **161,** 978–984.
63. Wiesinger, H. (2001) Arginine metabolism and the synthesis of nitric oxide in the nervous system. *Prog. Neurobiol.* **64,** 365–391.
64. Heneka, M. T. and Feinstein, D. L. (2001) Expression and function of inducible nitric oxide synthase in neurons. *J. Neuroimmunol.* **114,** 8–18.
65. Ogawa, O., Umegaki, H., Sumi, D., Hayashi, T., Nakamura, A., Thakur, N. K., et al. (2000) Inhibition of inducible nitric oxide synthase gene expression by indomethacin or ibuprofen in beta-amyloid protein-stimulated J774 cells. *Eur. J. Pharmacol.* **408,** 137–141.
66. Maggi, L. B. Jr., Sadeghi, H., Weigand, C., Scarim, A. L., Heitmeier, M. R., and Corbett, J. A. (2000) Anti-inflammatory actions of 15-deoxy-delta 12,14-prostaglandin J2 and troglitazone: evidence for heat shock-dependent and -independent inhibition of cytokine-induced inducible nitric oxide synthase expression. *Diabetes* **49,** 346–355.
67. Pasinetti, G. M. and Aisen, P. S. (1998) Cyclooxygenase-2 expression is increased in frontal cortex of Alzheimer's disease brain. *Neuroscience* **87,** 319–324.
68. Kitamura, Y., Shimohama, S., Koike, H., Kakimura, J., Matsuoka, Y., Nomura, Y., et al. (1999) Increased expression of cyclooxygenases and peroxisome proliferator-activated receptor-gamma in Alzheimer's disease brains. *Biochem. Biophys. Res. Commun.* **254,** 582–586.
69. Inoue, H., Tanabe, T., and Umesono, K. (2000) Feedback control of cyclooxygenase-2 expression through PPARgamma. *J. Biol. Chem.* **275,** 28,028–28,032.
70. Gilroy, D. W., Colville-Nash, P. R., Willis, D., Chivers, J., Paul-Clark, M. J., and Willoughby, D. A. (1999) Inducible cyclooxygenase may have anti-inflammatory properties. *Nat. Med.* **5,** 698–701.
71. Lohray, B. B., Bhushan, V., Reddy, A. A., Rao, P. B., Reddy, N. J., Harikishore, P., et al. (1999) Novel euglycemic and hypolipidemic agents. 4. Pyridyl- and quinolinyl-containing thiazolidinediones. *J. Med. Chem.* **42,** 2569–2581.
72. Reddy, K. A., Lohray, B. B., Bhushan, V., Reddy, A. S., Rao Mamidi, N. V., Reddy, P. P., et al. (1999) Novel antidiabetic and hypolipidemic agents. 5. Hydroxyl versus benzyloxy containing chroman derivatives. *J. Med. Chem.* **42,** 3265–3278.

73. Zhu, Y., Alvares, K., Huang, Q., Rao, M. S., and Reddy, J. K. (1993) Cloning of a new member of the peroxisome proliferator-activated receptor gene family from mouse liver. *J. Biol. Chem.* **268,** 26,817–26,820.
74. Cullingford, T. E., Bhakoo, K., Peuchen, S., Dolphin, C. T., Patel, R., and Clark, J. B. (1998) Distribution of mRNAs encoding the peroxisome proliferator-activated receptor alpha, beta, and gamma and the retinoid X receptor alpha, beta, and gamma in rat central nervous system. *J. Neurochem.* **70,** 1366–1375.
75. Braissant, O. and Wahli, W. (1998) Differential expression of peroxisome proliferator-activated receptor-alpha, -beta, and -gamma during rat embryonic development. *Endocrinology* **139,** 2748–2754.
76. Kainu, T., Wikstrom, A. C., Gustafsson, J. A., and Pelto-Huikko, M. (1994) Localization of the peroxisome proliferator-activated receptor in the brain. *Neuroreport* **5,** 2481–2485.
77. Bernardo, A., Levi, G., and Minghetti, L. (2000) Role of the peroxisome proliferator-activated receptor-gamma (PPAR-gamma) and its natural ligand 15-deoxy-delta12, 14-prostaglandin J2 in the regulation of microglial functions. *Eur. J. Neurosci.* **12,** 2215–2223.
78. Kitamura, Y., Taniguchi, T., Kimura, H., Nomura, Y., and Gebicke-Haerter, P. J. (2000) Interleukin-4-inhibited mRNA expression in mixed rat glial and in isolated microglial cultures. *J. Neuroimmunol.* **106,** 95–104.
79. Petrova, T. V., Akama, K. T., and Van Eldik, L. J. (1999) Cyclopentenone prostaglandins suppress activation of microglia: down-regulation of inducible nitric-oxide synthase by 15-deoxy-Delta12,14-prostaglandin J2. *Proc. Natl. Acad. Sci. USA* **96,** 4668–4673.
80. Lim, G. P., Yang, F., Chu, T., Chen, P., Beech, W., Teter, B., et al. (2000) Ibuprofen suppresses plaque pathology and inflammation in a mouse model for Alzheimer's disease. *J. Neurosci.* **20,** 5709–5714.
81. Klegeris, A., Walker, D. G., and McGeer, P. L. (1997) Interaction of Alzheimer beta-amyloid peptide with the human monocytic cell line THP-1 results in a protein kinase C-dependent secretion of tumor necrosis factor-alpha. *Brain Res.* **747,** 114–121.
82. Chattopadhyay, N., Singh, D. P., Heese, O., Godbole, M. M., Sinohara, T., Black, P. M., et al. (2000) Expression of peroxisome proliferator-activated receptors (PPARS) in human astrocytic cells: PPARgamma agonists as inducers of apoptosis. *J Neurosci. Res.* **61,** 67–74.
83. Fitch, M. T., Doller, C., Combs, C. K., Landreth, G. E., and Silver, J. (1999) Cellular and molecular mechanisms of glial scarring and progressive cavitation: in vivo and in vitro analysis of inflammation-induced secondary injury after CNS trauma. *J. Neurosci.* **19,** 8182–8198.
84. Nishijima, C., Kimoto, K., and Arakawa, Y. (2001) Survival activity of troglitazone in rat motoneurones. *J. Neurochem.* **76,** 383–390.
85. Rohn, T. T., Wong, S. M., Cotman, C. W., and Cribbs, D. H. (2001) 15-deoxy-delta12,14-prostaglandin J2, a specific ligand for peroxisome proliferator-activated receptor-gamma, induces neuronal apoptosis. *Neuroreport* **12,** 839–843.

8

Neuroinflammation-Mediated
Neurotoxin Production in Neurodegenerative Diseases

Potential of Nitrones as Therapeutics

Robert A. Floyd and Kenneth Hensley

1. INTRODUCTION

There is critical need for the development of novel therapeutics for the treatment of neurodegenerative diseases. Of the many risk factors involved in their etiology, age is by far the largest. This is seen in the increased incidences of stroke and dementia with age even in the optimally healthy individual. The reason why age plays such a large role is probably related to the age-associated enhanced susceptibility of brain to insults which cause oxidative damage. An enhanced tendency of brain to have a smoldering neuroinflammatory state reflects this heightened age-associated risk. Exacerbated neuroinflammation is considered the cause of dementia associated with neurodegenerative conditions. Exacerbated neuroinflammation is associated with localized glia activation and results in the enhanced production of toxins that act more selectively on neurons. Nitric oxide and its oxidation products is considered the major neurotoxin produced, however, reactive oxygen species and lipid oxidation products are also produced at high levels in the localized region of exacerbated neuroinflammation. Agents, which quell the exacerbated neuroinflammation state, appear the most promising for future therapeutics. The nitrones are especially promising as novel therapeutics for various neurodegenerative diseases especially stroke and Alzheimer's disease. They are considered in more depth.

2. NEURODEGENERATIVE DISEASES,
THEIR IMPORTANCE, AND NEED FOR NOVEL THERAPEUTICS

Diseases associated with neurodegenerative conditions represent a large array of disorders that exact an enormous cost to society. The total cost of just Alzheimer's disease (AD), stroke, and Parkinson's disease (PD) represent well over 100 billion dollars per year. Their burden could be minimized significantly if effective treatment modalities existed. However, this is not the case; in fact, only minimally effective therapy exists for only a fraction of the subjects having each of these three diseases now. Basic research on both the etiology as well as novel therapeutic approaches have escalated over the last few years. It is likely this will yield successful novel therapies in the future. A major breakthrough in the development of successful novel therapies seems likely based on

From: *Neuroinflammation, 2nd Edition: Mechanisms and Management*
Edited by: P. L. Wood © Humana Press Inc., Totowa, NJ

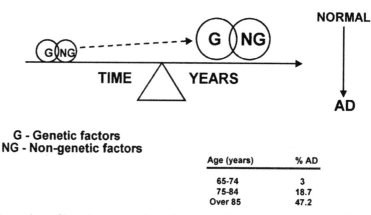

Age (years)	% AD
65-74	3
75-84	18.7
Over 85	47.2

Fig. 1. Illustration of how both genetic and nongenetic (i.e., environmental) factors become more dominant with age, influencing the onset of AD. The percentage of AD in certain age ranges are summaries of the results of community-based study. (From ref. *1*.)

the emerging recognition that a state of exacerbated neuroinflammation exists in specific brain regions of subjects suffering from several neurodegenerative disorders and that this leads to localized production of toxins that causes damage to neurons, leading to their dysfunction or death. This is considered the cause of dementia suffered by many subjects afflicted with several neurodegenerative diseases.

3. AGE ENHANCEMENT OF NEURODEGENERATIVE DISEASES

Age is, by far, the largest risk factor in the development of neurodegenerative diseases. It is not known why advanced age becomes such a dominate factor; however, it is expected that it brings into play several otherwise minor factors so that their additive effect then becomes a collective group that increasingly dominates with advancing age. Figure 1 illustrates not only the influence of age on the number of Alzheimer's cases computed on a community-based study *(1)* but also depicts the notion of the increasingly additive influence of genetic and nongenetic factors with advanced age. Nongenetic factors may relate to environmental or lifestyle factors. Studies on the incidence of Alzheimer's disease have shown that prior head trauma was a positive factor *(2)*. A recent study using a transgenic mouse of Alzheimer's disease has shown that amyloid plaque deposition in this animal is increased by head trauma *(3)*.

A recent study examining the development of optimally healthy very elderly people (85 yr and older) illustrates the enormous importance of increasing age in the development of cognitive impairment in the very old *(4)*. From 4494 elderly individuals in the Portland, Oregon area, 174 who were rated optimally healthy were studied. This group included 100 individuals that were 85 yr or older. At entry into the study, all subjects were free of cognitive impairment, stroke, heart disease, hypertension, cancer, diabetes, and neurological diseases. During the 5.6 ± 0.3 yr of following these people, 38% developed hypertension, 37% developed cognitive impairment, 33% developed heart disease (arrhythmia, myocardial infarction), 30% developed cancer, 23% Alzheimer's disease, 23% had a stroke or transient ischemia attack, 6% developed diabetes, and 2% became blind (Fig. 2). Several individuals had multiple maladies, so the groups are not mutu-

At Entry Time 100 Optimally Healthy 85 Years or Older		5.6 ± 0.3 Yrs. Later
• Cognitive Impairment	0%	37%
• Hypertension	0%	38%
• Heart Disease, Arrhythmia, MI	0%	33%
• Cancer (all types)	0%	30%
• Alzheimer's Disease	0%	23%
• Stroke	0%	23%
• Diabetes	0%	6%
• Blind	0%	2%

Fig. 2. A summary of results of a study by McNeal et al. *(4)* in which 100 optimally healthy 85-yr-old individuals were followed for 5.6 yr. The data illustrate the effect of age on the onset of specific age-associated diseases.

ally exclusive. With respect to brain aging, of the 100 who entered the study after 5.6 yr, 34 had cognitive impairment, and of these 34, 23 had Alzheimer's disease (Fig. 2). This is in comparison to 58 who showed no cognitive impairment (note that 8 of the 100 had indeterminate cognitive status, so they are excluded from these totals).

A very recent report emphasizes the importance of age in the occurrence of stroke *(5)*. Figure 3 presents the data obtained. The exponential rise in the occurrence of stroke after midlife is extremely striking. In our earlier work *(6,7)* with different aged gerbils, we had noted the extreme sensitivity of the older animals when compared to the younger ones. Until now, these findings in animals had not been demonstrated so remarkably in humans.

4. EXACERBATED NEUROINFLAMMATION IN NEURODEGENERATIVE DISEASES

It is now becoming more widely accepted that exacerbated neuroinflammatory processes are important in the events involved in causing damage to neurons in several diseases. Early observations made by the McGeers and colleagues *(8–11)* and Rogers et al. *(12)* provided the first clues to the now increasingly important realization that exacerbated neuroinflammation is important in several neurodegenerative diseases. We have reviewed this area recently *(13)*.

Figure 4 presents a simplified view of the general state of exacerbated neuroinflammation in advanced states of certain neurodegenerative diseases such as Alzheimer's disease. In this figure, we have represented glia cells (astrocytes and microglia) as large circles, emphasizing the generally much larger number of them in relation to neurons (about 10-fold). Glia surround and interact with neurons, generally providing various growth factors and nourishment factors. In Fig. 4, we have shown, under certain conditions, irritants and activators acting upon glia, causing them to be activated. Glia cells are resident inflammatory cells and can be activated to produce various growth factors as well as toxins, some of which are very damaging to neurons. The activators in the case of Alzheimer's disease include β-amyloid plaques *(14)*, which form in specific brain

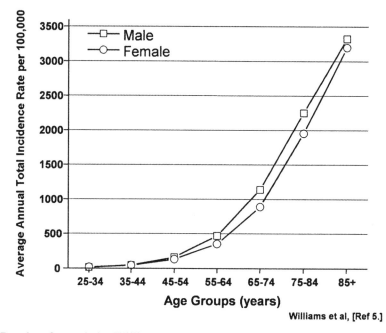

Williams et al, [Ref 5.]

Fig. 3. Results of a study by Williams *(5)*, showing the incidence of stroke as a function of age and gender in the US population.

regions of the affected individuals *(15)*. It is known that activated microglia produce factors that activate astrocytes, which, in turn, activate microglia *(15)*. This escalating cascade of glia activation proceeds, leading to a localized cluster of activated cells that produce many factors and toxins, some of which cause damage to the bystander neurons (Fig. 4). It is this localized exacerbated neuroinflammation that damages neurons, leading to either dysfunction and/or death. The dysfunction or death of neurons under these conditions appears to be the major cause of dementia associated with some neurodegenerative diseases.

Some of the known toxins and other factors that activated glia produce that can cause damage to neurons include reactive oxygen species (ROS) and nitric oxide (NO). Microglia have many properties of macrophages, including the production of superoxide when they are activated *(16)*. When activated by proinflammatory cytokines, astrocytes will not only produce H_2O_2 but, in turn, are also activated by H_2O_2 *(17)*. Nitric oxide production by glia is synthesized by inducible nitric oxide synthase (iNOS). NO production by iNOS is constant and sustained, thus yielding much higher levels than that produced by either neuronal NOS (nNOS) or endothelial (eNOS), both of which are calcium dependent and hence yield only very low levels of NO in short bursts *(18,19)*. The larger levels of NO produced by glia have a damaging effect on the bystander neurons *(see* ref. *20* and references herein). Neurons are much more susceptible to NO and its oxidation products than are the glia that produced it. In the case of Alzheimer's disease, it has been demonstrated that β-amyloid stimulates glia to produce NO *(14,21)*.

The mechanistic basis of the enhanced susceptibility of neurons to toxins produced by the activated glia in coculture is still not entirely resolved; however, it has been

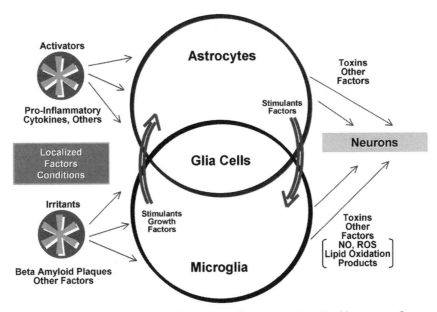

Fig. 4. Scheme illustrating cellular influences and processes involved in a state of exacerbated neuroinflammation. Shown are glia cells (both astrocytes and microglia) that become activated by irritants and or activators. Glial activation sets up an escalating cascade of heightened activation brought on by microglia providing stimulants and growth factors that further activate astrocytes, which, in turn, produce various stimulants and factors that further activate microglia. The end result is that various toxins and other factors such as NO (nitric oxide), ROS (reactive oxygen species), lipid oxidation products, and so forth, are produced that cause damage to neurons that are more susceptible than the glia, that produced the toxins. Damage to neurons result in their death or dysfunction, which is the root cause of dementia associated with neurodegenerative diseases.

shown that neuron respiratory metabolism is very sensitive to NO. Additionally, it has been proposed that NO acts on neurons to mediate glutamate release, which acts via the *N*-methyl-D-aspartate (NMDA) receptor to trigger a large influx of calcium into neurons, leading to apoptotic or necrotic death *(20)*. Many other theories exist to explain why neurons are damaged by activated glia. Nevertheless, most agree that NO is an important mediator of damage to neurons.

There is much supporting evidence for the occurrence of exacerbated neuroinflammatory processes in neurodegenerative human subjects. Some of the more pertinent observations are those showing the enhanced activation of signal transduction processes in cells surrounding β-amyloid plaques. It has been shown that there is a high level of activated p38 in the surrounding cells of neuritic plaques in brains of Alzheimer's subjects when they are compared to age-matched controls *(22)*. It should be noted that p38 activation occurs in signal transduction processes that are involved in the upregulation of iNOS expression. The importance of enhanced production of NO and its reaction products was confirmed by the demonstration, using electrochemical analysis, that protein nitrotyrosine and dityrosine adducts were significantly enhanced in the affected brain region of Alzheimer's subjects when compared to age-matched controls *(23)*. The fact that

- **Prevent build-up of beta-amyloid plaques**

  - • **Beta-amyloid vaccination (trials ceased)**
  - • **Inhibitors of beta-secretase (selectivity?)**

- **Catalytic inhibitors of enzymes that produce toxins**

  - • **COX-II inhibitors (trials ceased)**
  - • **iNOS inhibitors (not specific enough?)**

- **Blockers of reactive oxygen species/oxidative damage**

  - • **Vitamin E (prolonged home stay/no effect on dementia)**
  - • **Lipoic acid (no effect in HIV-dementia)**

- **Quell/Selectively alter enhanced signal transduction cascade**

  - • **Inhibit iNOS induction (MAP kinase inhibitors, nitrones?)**
  - • **Inhibitors of p38/NF-κB pathway (nitrones?)**
  - • **Non-steroidal anti-inflammatory drugs (mechanism of action?)**

Fig. 5. List of possible novel therapeutics approaches to quelling exacerbated neuroinflammation. The examples given relate mostly to the treatment of Alzheimer's disease.

gliosis and microglia clustering were found around degenerating neurons in the substantia nigra area of Parkinson's disease subjects, caused by 1-methyl-4-phenyl-1,2,3,6-tetrahydropyridine (MPTP) ingestion, clearly implicates the importance of neuroinflammation in this neurodegenerative disease also *(24)*.

5. TREATMENTS BASED ON QUELLING EXACERBATED NEUROINFLAMMATION

Many academic laboratories and pharmaceutical companies have active research ongoing with the aim of developing treatments for neurodegenerative diseases. Most attention is directed to Alzheimer's disease. The accumulation of β-amyloid plaques is the most definitive neuropathologic feature of this disease and, hence, measures to prevent buildup of the plaques is one treatment rationale. As noted earlier, β-amyloid plaques are considered irritants that activate the glia; hence, plaque buildup would be expected to contribute to exacerbated neuroinflammation. Exacerbated neuroinflammation is now widely considered of significant importance in the clinical manifestations of the Alzheimer's disease. Novel therapeutics based on the goal of quelling exacerbated neuroinflammation is certainly a rationale approach. Four readily apparent possible major approaches are presented in Fig. 5.

Regarding approaches to *prevent the buildup of β-amyloid plaques*, at least two are now under active development. The first is the design of inhibitors of the enzyme β-secretase, which cleaves the amyloid precursor protein into peptides responsible for the eventual buildup of β-amyloid plaques. The second is the β-amyloid vaccination approach. In the case of the vaccination approach, early work with an Alzheimer's disease transgenic mouse model showed promising results in preventing the buildup of β-amyloid plaques *(25,26)*. Subsequent β-amyloid vaccination clinical trials in humans were started, but according to very recent public press releases, these have been discontinued because of toxicity complications. Regarding the approach of developing inhibitors of β-secretase, it appears that major hurdles have to be overcome. Two of the most obvious include the development of an inhibitor that localizes exclusively to the brain (i.e., β-secretase is an enzyme present in most cells of the body). Additionally, there is no evidence that β-secretase is more expressed in Alzheimer's subjects versus control subjects. Despite the hurdles involved, significant research and development activities are occurring using this approach not only in academic laboratories but in major pharmaceutical companies as well. Another possible approach may be forthcoming based on recent observations. Utilizing a copper chelator, it was shown that amyloid plaque formation was abated in a preclinical mouse model *(27)*.

Regarding *catalytic inhibitors of enzymes that produce toxins*, in recent years extensive clinical trials have been initiated utilizing catalytic inhibitors of cycloxygenase II (COX-II) with the anticipation that this would prevent the enzymatic production of products, which may be of importance in the development of clinical aspects of Alzheimer's disease *(28)*. However, from the best accounts now available, these trials have ceased because of the lack of clinical effects. The results of Yermakova and O'Banion *(29)* had implicated the unlikely possibility of COX-II contributing to end-stage pathology in Alzheimer subjects because the levels of this enzyme decreases during this important time frame. Because iNOS is an important enzyme in the synthesis of the toxin considered responsible for neuron damage, it is reasonable to expect inhibitors of this enzyme to be evaluated. Studies based on this approach in stroke in animal models *(30,31)* have yielded results implicating the importance of iNOS. Presently, there seems to be limitations to this approach because of a lack of specific inhibitors for iNOS. Several inhibitors of iNOS also show activity with eNOS and nNOS. If these latter enzymes are inhibited, it is expected to cause significant complications in blood pressure regulation and possibly cognitive functions or muscle action if eNOS or nNOS, respectively, are inhibited to any extent.

The recent increased activity in nonsteroidal anti-inflammatory drugs (NSAIDs) came from the earlier pioneering research on neuroinflammatory processes and the potential and the highly suggestive effectiveness of indomethacein and ibuprofen *(32,33)*. Increased interest in this area is certain to result from a recent prospective study that involved 6989 subjects, where the association between the use of NSAIDs and Alzheimer's disease and vascular dementia were evaluated *(34)*. The study clearly showed that the relative risk of developing Alzheimer's disease decreased significantly when long-term use of NSAIDs had occurred. Short-term use of NSAIDs showed no effect. NSAID treatment had no effect on the development of vascular dementia. No definitive proof in humans has been given that NSAID treatment decreases exacerbated neuroinflammation in the brains of the Alzheimer's subjects. It is possible that this occurs. A recent study in rats using novel NSAID derivatives clearly implicated this conclusion *(35)*. The novel NSAIDs derivatives used were nitroflurbiprofen or nitro-aspirin. Neuroinflammation was induced in the fourth ventricular space of the rat's brain by continuous direct lipopolysaccharide (LPS) infusion. Daily peripheral (subcutaneous) administration of these novel drugs attenuated brain inflammation, as reflected by a decrease in OX-6-positive microglia within the affected brain regions.

When approaches to *minimize reactive oxygen species and/or oxidative damage* are considered, the immediate reaction is to consider using antioxidants. Although this is certainly rational, antioxidant is a term that connotes different ideas and facts attributed to each individual; unfortunately, its very use as a catch-all term and the mixed results that invariably come from biological experiments with various antioxidants have considerably confused the field. As the knowledge of the biological production as well as the consequences of ROS and oxidative damage has increased, our understanding of the various antioxidants and their highly unique properties and specific and varied biological effects have also increased. Almost invariably, vitamin E is readily considered. In fact, clinical trials with very high levels of vitamin E supplementation have been conducted in the treatment of Alzheimer's disease. The results showed that very high levels of vitamin E had an effect on prolonging the entrance of Alzheimer's subjects into institutionalized living centers, but there was no effect on abating the onset of dementia *(36)*. It should also be noted that there are no data available indicating that the vitamin E supplementation had any effect on the amount of oxidative damage in the brains of the treated subjects. Vitamin E loading of the brain does not readily occur *(37)*; so, as an ideal antioxidant for brain, it seems to be severely lacking. Lipoic acid is another low-molecular-weight antioxidant that may have promise in neurodegenerative diseases because it penetrates the brain and shows protection from stroke in rats if given before the stroke *(38)*. It has been tested in a clinical trial to ameliorate the development of dementia in advanced cases of human immunodeficiency virus (HIV)-infected individuals *(39)*. In this trial, it showed no effectiveness. It should be noted that some promising low-molecular-weight superoxide dismutase mimetics have shown neuroprotective activity in prelclinical models *(40,41)*. It is not known if clinical trials with these compounds are underway.

Regarding the approaches of *inhibiting the induction of iNOS and quelling or selectively altering enhanced signal transduction pathways*, this seems the most attractive approach at the present time. As noted earlier, the NSAID effect could possibly be acting through these mechanisms. Much effort is most likely now being focused on these attractive approaches. However, we are aware of only the effort being made with nitrones and some closely related chemicals. In Section 6, we present a more detailed summary of available data on the nitrones as drug development candidates. It is highly likely that their action involves not only inhibiting the induction of iNOS but also abating and altering enhanced signal transduction pathways but also as a side effect inhibiting the production of reactive oxygen species and oxidative damage to brain. In the following section, we present a more detailed summary of the nitrones as novel potential therapeutics for the treatment of neurodegenerative diseases.

6. NITRONES AS THERAPEUTICS

We have reviewed this subject several times previously *(37,42–48)*, so only a summary will be provided here in the context of this report. Figure 6 presents a general formula for nitrones, and the general nitrone trapping reaction with a free radical • R. Also presented is the formula for α-phenyl-*tert*-butyl nitrone (PBN), the compound that has been most widely used in academic laboratories. Attention to the nitrones arose from their use in analytical chemistry to trap and, hence, stabilize highly reactive free radicals that then made it possible by the use of electron paramagnetic resonance spectroscopy to char-

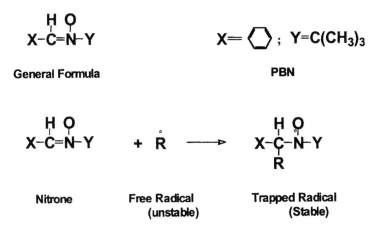

Nitrone Free Radical Trapping Reaction

Fig. 6. Chemical formula illustrating the general nitrone structure and specifically PBN (α-phenyl-*tert*-butyl nitrone) as well as the spin-trapping of a free radical • R by a nitrone to yield a trapped radical stabilized as a nitroxide.

acterize the trapped (spin-trapped) free radicals. This spin-trapping ability of nitrones made it possible to definitely show their presence of free radicals and help characterize them in many chemical reactions *(49,50)*. Their use was then applied more widely [i.e., in biochemical experiments *(51–54)*, and, finally, in experimental animals *(55–59)*].

In addition to the first use of nitrones as analytical agents in chemical and then biochemical systems *(see* Fig. 7), they were shown to have pharmacologic properties. The first observation implicating this possibility was made by Novelli et al. *(60)* in 1985. They showed that PBN protected rats from trauma caused by placing them in a tumbling mill. Although not listed in Fig. 7, the next demonstration of the therapeutic potential of the nitrones occurred shortly thereafter, when it was shown that they were protective in LPS-induced septic shock in rats *(61–63)*. PBN was protective only when given prior to the LPS dosing.

Another major milestone occurred when it was demonstrated that PBN was protective in experimental stroke in gerbils *(6)*. This result has been repeated many times *(64–68)* and extended to several other stroke models and, importantly, PBN (as well as some of its derivatives) is neuroprotective if given within an hour or so after the start of brain reperfusion. The 1,3 disulfonyl phenyl derivative of PBN (referred to as NXY-059) is in commercial development for stroke and has been shown to be much more potent than PBN in a middle cerebral artery occlusion stroke model in rats *(69)* and marmosets *(70)*. Illustrations of the scope of the various neuroprotective indications where PBN has been demonstrated to be effective include a hearing loss model mediated by noise plus carbon monoxide *(71)*, an intense white-light-mediated retinal damage model in rats *(72)*, and kainic acid-mediated epilepsy-type lesions in a rat model *(73)*. In the latter case, neuroprotection was most effective when administered at least 90 min after the subcutaneous injection of kainate.

1957   1969   1975   1985   1990   1998   2001

First Synthesis of PBN (α-phenyl-*tert*-butyl nitrone)[87]

First used to trap free radicals in chemical systems[49]

First used to trap radicals in biological systems[88]

Demonstrated pharmaco-protective activity[60]

Demonstrated neuro-protective activity[6]

Demonstrated anti-aging activity[89]

Demonstrated anti-cancer activity[75,76]

Fig. 7. Illustration of the major historical events in the development of nitrones as therapeutics. The superscripted numbers are reference numbers.

As noted in Fig. 7, the nitrones have been shown to have both antiaging activity, an area we have summarized previously *(47,74)*, as well as anticancer activity, which we discovered recently *(75,76)*. In addition to the indications listed in this chapter, there are many more biological indications where the nitrones have shown pharmacologic activity and are not presented here. Some of the most intriguing observations regarding the pharmacological implications of PBN, and certainly clues to its mechanism of action, came very early after our discovery of its neuroprotective activity. It was shown that older gerbils were much more sensitive to a global stroke than were younger gerbils *(6)*. Additionally, it was shown that oxidized protein in the brains was much higher, approximately twofold, in older gerbils than younger gerbils *(77)*. However, after 14 d of chronic low-level (30 mg/kg/d) treatment of the older gerbils with PBN, the oxidized protein content fell back to the levels observed in younger gerbils, but it then rebounded back to the normal old brain levels within 2 wk after ceasing chronic PBN treatment *(77)*. The susceptibility of older gerbils to stroke was significantly reduced after 14 d of chronic PBN administration and, surprisingly, the protectiveness rendered by PBN remained for at least 5 d after ceasing chronic dosing *(78)*. The half-life of PBN, which readily penetrates the brain within 20 min after dosing *(79)*, is about 132 min in rats *(79)*, so it is highly likely that no PBN remained in the brain of the chronically administered gerbils when they were stroked 1 d, and certainly not at 3 and 5 d, after ceasing dosing them. Our interpretation has been that PBN, because of its chronic presence in the dosed older gerbil brain, alters the brain to be less susceptible to the massive stroke insult. Much evidence has suggested that the older brain is more susceptible to insults. The older brain shows a much enhanced glia activation than the younger brain *(80,81)* and suggests that it has a smoldering, only partially manifest, tendency toward neuroinflammatory processes *(13)*. The data suggested that the increased protein oxidation level in the old brain reflects this increased susceptibility. In this view, the presence of chronic PBN shifts the balance, hence reversing the processes that increase with age and results in an unbalanced highly susceptible state in the older brain.

Clearly, the nitrones have enormous potential as novel therapeutics for several indications and especially in the treatment of neurodegenerative diseases. Their potent activity is not the result of their classical spin-trapping reactions for which they were originally used in analytical chemistry. Several readily apparent facts support this conclusion. Primary is the fact that in traditional solution-phase spin-trapping reactions, a very high concentration of the nitrone is required (i.e., 10–100 mM) in order to be able to trap a significant fraction (10–50%) of the free radicals produced. The requirement for a large amount of nitrone is the result of their low reaction rates with most free radicals (i.e., usually 10^4–10^7 s/M). It should also be noted that the ROS level increases very rapidly after a stroke but then becomes much less within a few minutes after the reperfusion phase has started. However, nitrones are active when given within 1 h or so after the reperfusion phase has started. This reinforces the conclusion that their biological action does not depend on their classical mass action reaction activity. Another important fact is that they are not very effective antioxidants in classical biological lipid peroxidation systems. For instance, about 5 mM of PBN is required to inhibit 50% of the lipid peroxidation rate in rat microsomal systems in comparisons to vitamin E or butylated hydroxy toluene where 1–5 μM will curtail 50% of the lipid peroxidation rate in the same systems *(82)*. It should also be noted that the level of nitrone in the stroked brain is on the order of 100 μM or less (i.e., much less than the 100 mM expected to be needed to trap a major portion of the free radicals in a robust solution-phase free-radical reaction system).

The mechanistic basis of the potent neuroprotective action of the nitrones depends on their ability to quell and selectively alter signal transduction pathways that are exacerbated in many neurodegenerative conditions. Their ability to act potentially in the mitogen-activated protein kinase pathways explains their ability to inhibit the induction of p38 and NF-κB in activated cultured cells *(17,83)* as well as in the brain of experimental animals *(73)*. One important result in this regard is their ability to inhibit the induction iNOS in cultured cells *(84,85)* as well as in the intact brain *(86)*. They have been shown to decrease the increased production of pro-inflammatory cytokines that are normally produced as a result of large insults to the brain *(73)*. In primary astrocytes, hydrogen peroxide is as effective as the pro-inflammatory cytokine interleukin (IL)-1 in the induction of p38 as well as iNOS; in this case, PBN was shown to effectively inhibit the activation of signal transduction pathways triggered by either one *(17)*. Therefore, with the exacerbated neuroinflammation model in mind (Fig. 4), we consider the nitrones to have enormous potential to inhibit the damaging effects caused by exacerbated neuroinflammation. This is because they will act *first* to downregulate glial activation caused by activators or irritants, *second* by inhibiting the astrocyte/microglia activation cascade by both inhibiting the production of stimulant factors such as proinflammatory cytokines and ROS (H_2O_2), and *third* by inhibiting the glia activation associated iNOS induction, thereby preventing NO production. The nitrones appear to have significant potential as therapeutics for neurodegenerative diseases.

REFERENCES

1. Markesbery, W. R. (1997) Oxidative stress hypothesis in Alzheimer's disease. *Free Radical Biol. Med.* **23**, 134–147.
2. Mortimer, J. A., French, L. R., Hutton, J. T., and Schuman, L. M. (1985) Head injury as a risk factor for Alzheimer's disease. *Neurology* **35**, 264–267.

3. Uryu, K., Laurer, H., McIntosh, T., Pratico, D., Martinez, D., Leight, S., et al. (2002) Repetitive mild brain trauma accelerates Aβ deposition, lipid peroxidation, and cognitive impairment in a transgenic mouse model of Alzheimer amyloidosis. *J. Neurosci.* **22,** 446–454.

4. McNeal, M. G., Zareparsi, S., Camicioli, R., Dame, A., Howieson, D., Quinn, J., et al. (2001) Predictors of healthy brain aging. *J. Gerontol.: Biol. Sci.* **56A,** B294–B301.

5. Williams, G. R. (2001) Incidence and characteristics of total stroke in the United States. *BMC Neurol.* **1,** 2.

6. Floyd, R. A. (1990) Role of oxygen free radicals in carcinogenesis and brain ischemia. *FASEB J.* **4,** 2587–2597.

7. Floyd, R. A. and Carney, J. M. (1991) Age influence on oxidative events during brain ischemia/reperfusion. *Arch. Gerontol. Geriatr.* **12,** 155–177.

8. McGeer, P. L., McGeer, E., Rogers, J., and Sibley, J. (1990) Anti-inflammatory drugs and Alzheimer disease. *Lancet* **335,** 8696.

9. Tooyama, I., Kimura, H., Akiyama, H., and McGeer, P. L. (1990) Reactive microglia express class I and class II major histocompatibility complex antigens in Alzheimer's disease. *Brain Res.* **523,** 273–280.

10. McGeer, P. L., Itagaki, S., Tago, H., and McGeer, E. G. (1987) Reactive microglia in patients with senile dementia of the Alzheimer's type are positive for the histocompatability of glycoprotein HLA-DR. *Neurosci. Lett.* **79,** 195–200.

11. McGeer, P. L., Akiyama, H., Itagaki, S., and McGeer, E. G. (1989) Activation of the classical complement pathway in brain tissue of Alzheimer patients. *Neurosci. Lett.* **107,** 341–346.

12. Rogers, J., Cooper, N. R., Webster, S., Schultz, J., McGeer, P. L., Styren, S. D., et al. (1992) Complement activation by β-amyloid in Alzheimer disease. *Proc. Natl. Acad. Sci. USA* **89,** 10,016–10,020.

13. Floyd, R. A. (1999) Neuroinflammatory processes are important in neurodegenerative diseases: an hypothesis to explain the increased formation of reactive oxygen and nitrogen species as major factors involved in neurodegenerative disease development. *Free Radical Biol. Med.* **26,** 1346–1355.

14. Akama, K. T., Albanese, C., Pestell, R. G., and Van Eldik, L. J. (1998) Amyloid β-peptide stimulates nitric oxide production in astrocytes through an NFκB-dependent mechanism. *Proc. Natl. Acad. Sci. USA* **95,** 5795–5800.

15. Cotman, C. W., Tenner, A. J., and Cummings, B. J. (1996) β-Amyloid converts an acute phase injury response to chronic injury responses. *Neurobiol. Aging* **17,** 723–731.

16. Colton, C. A. and Gilbert, D. L. (1987) Production of superoxide anions by a CNS macrophage, the microglia. *Fed. Eur. Biochem. Soc.* **223,** 284–288.

17. Robinson, K. A., Stewart, C. A., Pye, Q. N., Nguyen, X., Kenney, L., Salzman, S., et al. (1999) Redox-sensitive protein phosphatase activity regulates the phosphorylation state of p38 protein kinase in primary astrocyte culture. *J. Neurosci. Res.* **55,** 724–732.

18. Borgerding, R. and Murphy, S. (1995) Cytokine-activated astrocytes induce nitric oxide synthase type II in cerebrovascular endothelial cells. *Soc. Neurosci.* **21,** 1080 (abstract).

19. Ding, M., St. Pierre, B. A., Parkinson, J. F., Medberry, P., Wong, J. L., Rogers, N. E., et al. (1997) Inducible nitric-oxide synthase and nitric oxide production in human fetal astrocytes and microglia. *J. Biol. Chem.* **272,** 11,327–11,335.

20. Bal-Price, A. and Brown, G. C. (2001) Inflammatory neurodegeneration mediated by nitric oxide from activated glia-inhibiting neuronal respiration, causing glutamate release and excitotoxicity. *J. Neurosci.* **21,** 6480–6491.

21. Wallace, M. N., Geddes, J. G., Farquhar, D. A., and Masson, M. R. (1997) Nitric oxide synthase in reactive astrocytes adjacent to β-amyloid plaques. *Exp. Neurol.* **144,** 266–272.

22. Hensley, K., Floyd, R. A., Zheng, N.-Y., Nael, R., Robinson, K. A., Nguyen, X., et al. (1999) p38 Kinase is activated in the Alzheimer's disease brain. *J. Neurochem.* **72,** 2053–2058.

23. Hensley, K., Maidt, M. L., Yu, Z., Markesbery, W. R., and Floyd, R. A. (1998) Electrochemical analysis of protein nitrotyrosine and dityrosine in the Alzheimer brain indicates region-specific accumulation. *J. Neurosci.* **18**, 8126–8132.

24. Langston, J. W., Forno, L. S., Tetrud, J., Reeves, A. G., Kaplan, J. A., and Karluk, D. (1999) Evidence of active nerve cell degeneration in the substantia nigra of humans years after 1-methyl-4-phenyl-1,2,3,6-tetrahydropyridine exposure. *Ann. Neurol.* **46**, 598–605.

25. Schenk, D., Barbour, R., Dunn, W., Gordon, G., Grajeda, H., Guido, T., et al. (1999) Immunization with amyloid-β attenuates Alzheimer-disease-like pathology in the PDAPP mouse. *Nature* **400**, 173–177.

26. Morgan, D., Diamond, D. M., Gottschall, P. E., Ugen, K. E., Dickey, C., Hardy, J., et al. (2000) Aβ peptide vaccination prevents memory loss in an animal model of Alzheimer's disease. *Nature* **408**, 982–985.

27. Cherny, R. A., Atwood, C. S., Xilinas, M. E., Gray, D. N., Jones, W. D., McLean, C. A., et al. (2001) Treatment with a copper–zinc chelator markedly and rapidly inhibits β-amyloid accumulation in Alzheimer's disease transgenic mice. *Neuron* **30**, 665–676.

28. Pasinetti, G. M. (2001) Cyclooxygenase and Alzheimer's disease: implications for preventive initiatives to slow the progression of clinical dementia. *Arch. Gerontol. Geriatr.* **33**, 13–28.

29. Yermakova, A. V. and O'Banion, M. K. (2001) Downregulation of neuronal cyclooxygenase-2 expression in end stage Alzheimer's disease. *Neurobiol. Aging* **22**, 823–836.

30. Iadecola, C., Zhang, F., Casey, R., Nagayama, M., and Ross, M. E. (1997) Delayed reduction of Ischemic brain injury and neurological deficits in mice lacking the inducible nitric oxide synthase gene. *J. Neurosci.* **17**, 9157–9164.

31. Iadecola, C., Zhang, F., Casey, R., Clark, H. B., and Ross, M. E. (1996) Inducible nitric oxide synthase gene expression in vascular cells after transient focal cerebral ischemia. *Stroke* **27**, 1373–1380.

32. McGeer, E. G. and McGeer, P. L. (1999) Brain inflammation in Alzheimer disease and the therapeutic implications. *Curr. Pharm. Design* **5**, 821–836.

33. McGeer, P. L. and McGeer, E. G. (1999) Inflammation of the brain in Alzheimer's disease: implications for therapy. *J. Leukocyte Biol.* **65**, 409–415.

34. Veld, B. A., Ruitenberg, A., Hofman, A., Launer, L. J., van Duijn, C. M., Breteler, M. M., et al. (2001) Nonsteroidal antiinflammatory drugs and the risk of Alzheimer's disease. *N. Engl. J. Med.* **345**, 1515–1521.

35. Hauss-Wegrzyniak, B., Willard, L. B., Soldato, P. D., Pepeu, G., and Wenk, G. L. (1999) Peripheral administration of novel anti-inflammatories can attenuate the effects of chronic inflammation within the CNS. *Brain Res.* **815**, 36–43.

36. Sano, M., Ernesto, M. S., Thomas, R. G., Klauber, M. R., Schafer, K., Grundman, M., et al. (1997) A controlled trial of selegiline, alpha-tocopherol, or both as treatment for Alzheimer's disease. *N. Engl. J. Med.* **336**, 1216–1222.

37. Floyd, R. A. (1999) Antioxidants, oxidative stress, and degenerative neurological disorders. *Proc. Soc. Exp. Biol. Med.* **222**, 236–245.

38. Panigrahi, M., Sadguna, Y., Shivakumar, B. R., Kolluri, S. V. R., Roy, S., Packer, L., et al. (1996) α-Lipoic acid protects against reperfusion injury following cerebral ischemia in rats. *Brain Res.* **717**, 184–188.

39. Kieburtz, K., Schifitto, G., McDermott, M., Bourgeois, K., Palumbo, D., Orme, C., et al. (1998) A randomized, double-blind, placebo-controlled trial of deprenyl and thioctic acid in human immunodeficiency virus-associated cognitive impairment. *Neurology* **50**, 645–651.

40. Melov, S., Ravenscroft, J., Malik, S., Gill, M. S., Walker, D. W., Clayton, P. E., et al. (2000) Extension of life-span with superoxide dismutase/catalase mimetics. *Science* **289**, 1567–1569.

41. Mackensen, G. B., Patel, M., Sheng, H., Calvi, C. L., Batinic-Haberle, I., Day, B. J., et al. (2001) Neuroprotection from delayed postischemic administration of a metalloporphyrin catalytic antioxidant. *J. Neurosci.* **21,** 4582–4592.

42. Floyd, R. A. and Carney, J. M. (1992) Protection against oxidative damage to CNS by α-phenyl-*tert*-butyl nitrone and other spin-trapping agents: a novel series of nonlipid free radical scavengers, in *Emerging Strategies in Neuroprotection* (Marangos, P. J. and Lal, H., eds.), Birkhauser, Boston, pp. 252-272.

43. Floyd, R. A. (1996) The protective action of nitrone-based free radical traps in neurodegenerative diseases, in *Neurodegenerative Diseases '95: Cellular and Molecular and Mechanisms and Therapeutic Advances* (Fiskum, G., ed.), Plenum, New York, pp. 235–245.

44. Floyd, R. A. and Hensley, K. (2000) Nitrone inhibition of age-associated oxidative damage, in *Reactive Oxygen Species from Radiation to Molecular Biology* (Chiueh, C. C., ed.), New York Academy of Sciences, New York, pp. 222–237.

45. Floyd, R. A., Liu, G.-J., and Wong, P. K. (1995) Nitrone radical traps as protectors of oxidative damage in central nervous system, in *Handbook of Synthetic Antioxidants* (Cadenas, E. and Packer, L., eds.), Marcel Dekker, New York.

46. Floyd, R. A. and Carney, J. M. (1996) Nitrone radical traps protect in experimental neurodegenerative diseases, in *Neuroprotective Approaches to the Treatment of Parkinson's Disease and other Neurodegenerative Disorders* (Chapman, C. A., Olanow, C. W., Jenner, P., and Youssim, M., eds.), Academic, London, pp. 69–90.

47. Floyd, R. A., Hensley, K., Forster, M. J., Kelleher-Anderson, J. A., and Wood, P. L. (2002) Nitrones, their value as therapeutics and probes to understand aging. *Mech. Ageing Devel.* **123,** 1021–1031.

48. Floyd, R. A. (1997) Protective action of nitrone-based free radical traps against oxidative damage to the central nervous system. *Adv. Pharmacol.* **38,** 361–378.

49. Janzen, E. G. and Blackburn, B. J. (1969) Detection and identification of short-lived free radicals by electron spin resonance trapping techniques (spin trapping). Photolysis of organolead, -tin, and -mercury compounds. *J. Am. Chem. Soc.* **91,** 4481–4490.

50. Janzen, E. G. (1971) Spin trapping. *Acc. Chem. Res.* **4,** 31–40.

51. Chen, G., Janzen, E. G., Bray, T. M., and McCay, P. B. (1995) PBN and its applications in biology, in *The Oxygen Paradox* (Davies, K. J. A. and Ursini, F., eds.), Cleup University Press, Padova, Italy, pp. 790–800.

52. McCay, P. B., Poyer, J. L., Floyd, R. A., Fong, K.-L., and Lai, E. K. (1976) Spin- trapping of radicals produced during enzymic NADPH-dependent and CCl4-dependent microsomal lipid peroxidation. *Fed. Proc.* **35,** 421.

53. Poyer, J. L., Floyd, R. A., McCay, P. B., Janzen, E. G., and Davis, E. R. (1978) Spin trapping of the trichloromethyl radical produced during enzymic NADPH oxidation in the presence of carbon tetrachloride or carbon bromotrichloromethane. *Biochim. Biophys. Acta* **539,** 402–409.

54. Janzen, E. G. (1980) A critical review of spin trapping in biological systems, in *Free Radicals in Biology* (Pryor, W. A., ed.), Academic, New York, pp. 115–154.

55. Bolli, R., Patel, B. S., Jeroudi, M. O., Lai, E. K., and McCay, P. B. (1988) Demonstration of free radical generation on "stunned" myocardium of intact dogs with the use of the spin trap α-phenyl *N*-tert-butyl nitrone. *J. Clin. Invest.* **82,** 476–485.

56. Janzen, E. G., Stronks, H. J., DuBose, C. M., Poyer, J. L., and McCay, P. B. (1985) Chemistry and biology of spin-trapping radicals associated with halocarbon metabolism in vitro and in vivo. *Environ. Health Perspect.* **64,** 151–170.

57. Lai, E. K., McCay, P. B., Noguchi, T., and Fong, K.-L. (1979) *In vivo* spin-trapping of trichloromethyl radicals formed from CCl4. *Biochem. Pharmacol.* **28,** 2231–2235.

58. Lai, E. K., Crossley, C., Sridhar, R., Misra, H. P., Janzen, E. G., and McCay, P. B. (1986) In vivo spin trapping of free radicals generated in brain, spleen, and liver during gamma radiation of mice. *Arch. Biochem. Biophys.* **244,** 156–160.

59. Bolli, R., Jeroudi, M. O., Patel, B. S., DuBose, C. M., Lai, E. K., Roberts, R., et al. (1989) Direct evidence that oxygen-derived free radicals contribute to postischemic myocardial dysfunction in the intact dog. *Proc. Natl. Acad. Sci. USA* **86,** 4695–4699.

60. Novelli, G. P., Angiolini, P., Tani, R., Consales, G., and Bordi, L. (1985) Phenyl-*t*-butyl-nitrone is active against traumatic shock in rats. *Free Radical Res. Commun.* **1,** 321–327.

61. McKechnie, K., Furman, B. L., and Parratt, J. R. (1986) Modification by oxygen free radical scavengers of the metabolic and cardiovascular effects of endotoxin infusion in conscious rats. *Circ. Shock* **19,** 429–439.

62. Hamburger, S. A. and McCay, P. B. (1989) Endotoxin-induced mortality in rats is reduced by nitrones. *Circ. Shock* **29,** 329–334.

63. Pogrebniak, H. W., Merino, M. J., Hahn, S. M., Mitchell, J. B., and Pass, H. I. (1992) Spin trap salvage from endotoxemia: the role of cytokine down-regulation. *Surgery* **112,** 130–139.

64. Phillis, J. W. and Clough-Helfman, C. (1990) Protection from cerebral ischemic injury in gerbils with the spin trap agent *N-tert*-butyl-α-phenylnitrone (PBN). *Neurosci. Lett.* **116,** 315–319.

65. Clough-Helfman, C. and Phillis, J. W. (1991) The free radical trapping agent *N-tert*-butyl-α-phenylnitrone (PBN) attenuates cerebral ischaemic injury in gerbils. *Free Radical Res. Commun.* **15,** 177–186.

66. Phillis, J. W. and Clough-Helfman, C. (1990) Free radicals and ischaemic brain injury: protection by the spin trap agent PBN. *Med. Sci. Res.* **18,** 403–404.

67. Zhao, Q., Pahlmark, K., Smith, M.-I., and Siesjo, B. K. (1994) Delayed treatment with the spin trap α-phenyl-*N-tert*-butyl nitrone (PBN) reduces infarct size following transient middle cerebral artery occlusion in rats. *Acta Physiol. Scand.* **152,** 349–350.

68. Tsuji, M., Inanami, O., and Kuwabara, K. (2000) Neuroprotective effect of α-phenyl-*N-tert*-butylnitrone in gerbil hippocampus is mediated by the mitogen-activated protein kinase pathway and heat shock proteins. *Neurosci. Lett.* **282,** 41–44.

69. Kuroda, S., Tsuchidate, R., Smith, M.-L., Maples, K. R., and Siesjo, B. K. (1999) Neuroprotective effects of a novel nitrone, NXY-059, after transient focal cerebral ischemia in the rat. *J. Cereb. Blood Flow Metab.* **19,** 778–787.

70. Marshall, J. W. B., Duffin, K. J., Green, A. R., and Ridley, R. M. (2001) NXY-059, a free radical-trapping agent, substantially lessens the functional disability resultying from cerebral ischemia in a primate species. *Stroke* **32,** 190–198.

71. Rao, D. and Fechter, L. D. (2000) Protective effects of phenyl-*N-tert*-butylnitrone on the potentiation of noice-induced hearing loss by carbon monoxide. *Toxicol. Appl. Pharmacol.* **167,** 125–131.

72. Ranchon, I., Chen, S., Alvarez, K., and Anderson, R. E. (2001) Systemic administration of phenyl-*N-tert*-butylnitrone protects the retina from light damage. *Invest. Ophthalmol. Vis. Sci.* **42,** 1375–1379.

73. Floyd, R. A., Hensley, K., and Bing, G. (2000) Evidence for enhanced neuro-inflammatory processes in neurodegenerative diseases and the action of nitrones as potential therapeutics. *J. Neural. Transm.* **60,** 337–364.

74. Floyd, R. A., Hensley, K., Forster, M. J., Kelleher-Anderson, J. A., and Wood, P. L. (2002) Nitrones as neuroprotectants and anti-aging drugs. *Ann. NY Acad. Sci.* **959,** 321–329.

75. Nakae, D., Hideki, K., Osamu, K., Ayumi, D., Kotake, Y., Hensley, K., et al. (2000) Inhibition by phenyl *N-tert*-butyl nitrone of the development of hepatocellular carcinoma by a choline-deficient, ʟ-amino acid-defined diet in male Fischer 344 rats, American Association for Cancer Research Book of Abstracts (Late Breaking Absts/Program Supplement) (abstract).

76. Floyd, R. A., Kotake, Y., Hensley, K., Nakae, D., and Konishi, Y. (2002) Reactive oxygen species in choline deficiency induced carcinogenesis and nitrone inhibition. *Mol. Cell. Biochem.* **234/235,** 195–203.

77. Carney, J. M., Starke-Reed, P. E., Oliver, C. N., Landrum, R. W., Chen, M. S., Wu, J. F., et al. (1991) Reversal of age-related increase in brain protein oxidation, decrease in enzyme activity, and loss in temporal and spacial memory by chronic administration of the spin-trapping compound *N-tert*-butyl-α-phenylnitrone. *Proc. Natl. Acad. Sci. USA* **88,** 3633–3636.

78. Floyd, R. A. and Carney, J. M. (1995) Nitrone radicals traps (NRTs) protect in experimental neurodegenerative disesases, in *Neuroprotective Approaches to the Treatment of Parkinson's Disease and Other Neurodegenerative Disorders* (Chapman, C. A., Olanow, C. W., Jenner, P., and Youssim, M., eds.), Academic, London, pp. 69–90.

79. Chen, G., Bray, T. M., Janzen, E. G., and McCay, P. B. (1990) Excretion, metabolism and tissue distribution of a spin trapping agent, α-phenyl-*N-tert*-butyl-nitrone (PBN) in rats. *Free Radical Res. Commun.* **9,** 317–323.

80. Gordon, M. N., Schreier, W. A., Ou, X., Holcomb, L. A., and Morgan, D. G. (1997) Exaggerated astrocyte reactivity after nigrostriatal deafferentation in the aged rat. *J. Comp. Neurol.* **388,** 106–119.

81. Morgan, T. E., Xie, Z., Goldsmith, S., Yoshida, T., Lanzrein, A.-S., Stone, D., et al. (1999) The mosaic of brain glial hyperactivity during normal ageing and its attenuation by food restriction. *Neuroscience* **89,** 687–699.

82. Janzen, E. G., West, M. S., and Poyer, J. L. (1994) Comparison of antioxidant activity of PBN with hindered phenols in initiated rat liver microsomal lipid peroxidation, in *Frontiers of Reactive Oxygen Species in Biology and Medicine* (Asada, K. and Toshikawa, T., eds.), Elsevier Science, New York, pp. 431–446.

83. Kotake, Y., Sang, H., Miyajima, T., and Wallis, G. L. (1998) Inhibition of NF-κB, iNOS mRNA, COX2 mRNA, and COX catalytic activity by phenyl-*N-tert*-butylnitrone (PBN). *Biochim. Biophys. Acta* **1448,** 77–84.

84. Tabatabaie, T., Graham, K. L., Vasquez, A. M., Floyd, R. A., and Kotake, Y. (2000) Inhibition of the cytokine-mediated inducible nitric oxide synthase expression in rat insulinoma cells by phenyl *N-tert*-butylnitrone. *Nitric Oxide: Biol. Chem.* **4,** 157–167.

85. Hensley, K., Maidt, M. L., Pye, Q. N., Stewart, C. A., Wack, M., Tabatabaie, T., et al. (1997) Quantitation of protein-bound 3-nitrotyrosine and 3,4-dihydroxyphenylalanine by high performance liquid chromatography with electrochemical array detection. *Anal. Biochem.* **251,** 187–195.

86. Endoh, H., Kato, N., Fujii, S., Suzuki, Y., Sato, S., Kayama, T., et al. (2001) Spin trapping agent, phenyl *N-tert*-butylnitrone, reduces nitric oxide production in the rat brain during experimental meningitis. *Free Radical Res. Commun.* **35,** 583–591.

87. Hawthorne, M. F. and Strahm, R. D. (1957) Kinetics of the thermal isomeriazation of 2-*tert*-butyl-3-phenyloxazirane. *J. Org. Chem.* **22,** 1263–1264.

88. Harbour, J. R. and Bolton, J. R. (1975) Superoxide formation in spinach chloroplasts: electron spin resonance detection by spin trapping. *Biochem. Biophys. Res. Commun.* **64,** 803–807.

89. Saito, K., Yoshioka, H., and Cutler, R. G. (1998) A spin trap, *N-tert*-butyl-α-phenylnitrone extends the life span of mice. *Biosci. Biotechnol. Biochem.* **62,** 792–794.

II
Stroke and TBI

9
Inflammation and Potential Anti-Inflammatory Approaches in Stroke

Jari Koistinaho and Juha Yrjänheikki

1. BACKGROUND

Ischemic stroke develops when a major brain artery, most commonly the middle cerebral artery, is obstructed by an embolus or thrombosis and, consequently, blood flow in a restricted area of the brain is severely reduced. Even though animal research conducted on acute ischemic injury has revealed several pathophysiological cascades, which contribute to the enlargement of brain infarction, no major breakthroughs in developing clinically relevant stroke therapy have been achieved. There are several reasons for failed translation of successful basic research approaches into effective treatment modalities for stroke in humans (1). One important factor is that the animal models do not reflect well enough the ischemic damage in both permanent (no reperfusion) and transient (with reperfusion) ischemia in aged female and male victims, which may suffer from other age-related diseases, such as hypertension, diabetes, and Alzheimer's disease (2). Another, even a more valid, fact is that the therapeutic time window for treating human stroke is delayed beyond the hours after the insult and, therefore, the pathophysiological events targeted pharmacologically need to occur several hours or days later (2). For these reasons, targeting mechanisms such as various glutamate receptors and calcium channels may not be considered for treatment of most of the stroke patients.

Increasing evidence suggests that cerebral ischemia elicits an unrestrained inflammation, which maturates in a delayed manner and significantly contributes to the evolution of tissue injury. Inflammation is therefore thought to offer an opening for development of novel therapies against ischemic brain injury (for reviews, see refs. 1 and 3–6). Ischemic inflammation is a complex phenomenon characterized by the production and interplay of cytokines, chemokines, adhesion molecules, free radicals, and destructive enzymes such as cyclo-oxygenase-2 (COX-2), inducible nitric oxide synthase (iNOS), and proteinases. In addition to circulating neutrophils and monocytes/macrophages, also resident microglia, astrocytes, endothelial cells and neurons are involved in *in situ* inflammatory reactions (7). Although a wide variety of inflammatory factors contributing to ischemic injury are well documented, only the inflammatory processes that are still active several hours or days after the insult may have realistic potential as the pharmacologic target. In

From: *Neuroinflammation, 2nd Edition: Mechanisms and Management*
Edited by: P. L. Wood © Humana Press Inc., Totowa, NJ

this chapter, the inflammatory mechanisms involved in stroke and some therapeutic anti-inflammatory strategies that have therapeutic potential are described.

2. PERIPHERAL LEUKOCYTES IN STROKE

The contribution of peripheral leukocytes to ischemic brain injury has been extensively reviewed *(3–6)*.

2.1. Mechanisms of Damage

Following ischemia, leukocytes can induce damage by several mechanisms. *First*, they may physically obstruct vascular capillaries, causing reduced local tissue perfusion *(8)*. This obstructive phenomenon takes place in a majority of capillaries within 1 h of reperfusion following middle cerebral artery occlusion (MCAO) in baboons *(9)*. The role of this microvascular plugging in ischemia is controversial, as this process was not found to have a contributory role in cortical hypoperfusion during global cerebral ischemia *(10)*. The prothrombotic role of leukocytes is the *second* potentially damaging mechanism. Monocytes and polymorphonuclear leukocytes (PMNLs) exhibit clot-promoting activity in vitro, aggregates of platelets and leukocytes are observed in cerebral thrombosis, and coagulation machinery in stroke patients is activated *(11–13)*. Also, the prothrombosis hypothesis is unclear, because monocytes and leukocytes are capable of producing antithrombotic agents, including plasminogen-activating factor and proteases such as elastase *(11,12)*. The *third* injury promoting function of leukocytes is the production of free radicals and other toxic compounds such as hypochlorous acid and various harmful enzymes *(14)*. Neutropenia completely abolishes the production of free radicals in cortex following 60 min of MCAO *(15)*. Neutrophils produce free radicals via activation of NADPH oxidase, and infarction volume is reduced by 50% in transgenic mice lacking NADPH oxidase, indicating that NADPH oxidase contributes to the damaging effect of neutrophils *(16)*. However, selective elimination of either leukocyte or parenchymal NADPH oxidase is not sufficient alone to provide protection, suggesting that elimination of free-radical production both in neutrophils and in glial cells is needed to provide neuroprotection *(16)*. Another free-radical-producing agent in neutrophils is iNOS, an enzyme producing high concentrations of reactive nitric oxide, which causes damage to cellular macromolecules. Expression of iNOS peaks at 12–48 h following permanent ischemia in the rat and both knockout mouse experiments and studies with iNOS inhibitory compounds demonstrates a contributory role of this enzyme to infarction *(17,18)*. Finally, proteases, including matrix metalloproteases (MMPs) are released from activated leukocytes and degrade components of extracellular matrix *(19)*.

2.2. Leukocyte Infiltration

Ischemia-induced leukocyte accumulation and infiltration require leukocyte–endothelial interaction through intracellular adhesion molecules (ICAMs), which bind to β_2-integrin receptors, and selectins, which bind to mucinlike glycoproteins present on leukocytes or endothelial cells. Selectins mediate the initial binding and rolling behavior of leukocytes, whereas ICAM-1 and β_2-integrins are implicated in the aggregation and transmigration of leukocytes across the endothelium *(20)*. E-selectin (ELAM) is expressed by stimulated endothelial cells and mediates neutrophil as well as monocyte adhesion. P-Selectin is recruited to the cell surface upon pro-inflammatory stimulation and mediates leukocyte rolling and local inflammation. L-Selectin is constitutively expressed on the surface of neutrophils and monocytes and is implicated in leukocyte adherence. In baboons, P-selectin is induced to its maximum level 2 h after the onset of 3 h MCAO and the induction persists at least for 24 h *(21)*. E-Selectin mRNA levels peak at 12 h and stay upregulated for 2 d following permanent MCAO in rats *(22)*.

ICAM-1 expression is induced in endothelial cells 2–4 h after transient MCAO and stays upregulated until 24 h. In permanent MCAO, ICAM-1 induction occurs as early as 3 h after ischemia, but is less pronounced *(21–24)*. ICAM-1 expression can be induced by pro-inflammatory cytokines interleukin (IL)-1β and tumor necrosis factor (TNF)-α. In addition to ICAMs, the vascular cell adhesion molecules (VCAMs) are upregulated by ischemia-induced cytokines in endothelial cells. VCAM-1 is induced within hours following ischemia and the expression persists for several days. VCAM-1 binds to its integrin ligand VLA4 ($\alpha_4\beta_1$-integrin) that is present in lymphocytes and monocytes, suggesting a role in recruiting lymphocytes and monocytes into inflammatory zones *(25)*.

2.3. Neutrophils

Neutrophils are initially the predominant leukocytes at the site of inflammation, followed by infiltration of mononuclear phagocytes *(26)*. Neutrophils accumulate in infarcted tissue 12–72 h after permanent or transient ischemia. The temporal profile of infiltration depends on the experimental model and the duration of ischemia used. Usually, the PMNL's accumulation is greater in transient ischemia than in permanent models *(27)*.

Several experimental studies show that PMNLs accumulate in blood vessels within 4 h of the onset of permanent or transient ischemia, but the transmigration does not occur until at least 12 h in rats *(13,28,29)*. In dogs, neutrophils progressively accumulate in blood vessels during the first 4-h period *(30)*. Neutrophil accumulation is maximal in the ischemic lesion at 24–48 h after transient MCAO and at 72 h after the onset of permanent MCAO *(31,32)*. In human stroke and in primate MCAO, neutrophils accumulate at the periphery of the infarcted core *(26)*, but in humans, accumulation occurs early in severe cases and is almost exclusively confined to the vessels even after 24 h following stroke *(33,34)*. Most importantly, neutrophil accumulation correlates with reduced blood flow and severity of insult in the rat *(35)* and with both the infarct size and severity of neurological outcome in humans *(33)*.

Overall, the contribution of neutrophil infiltration to the ischemia still remains controversial. The infiltration occurs only after 12 h, by which time there is already major cell necrosis in the parenchyma. In one study, distal MCAO for 2 h caused infarction at 8 h, whereas significant neutrophil accumulation occurred at 21 h *(36)*. Two different studies report significant accumulation of neutrophils at 6 h following 1–2 h of MCAO, but the location of the early accumulation does not involve the cortex, where antileukocyte treatment is beneficial *(27,37)*. It should be noted that neutrophils may contribute to the tissue injury even when they are sitting in vessels in close apposition to the endothelial cell surface *(6)*. However, recent observations that the infarcted lesion enlarges significantly after 24 h of permanent MCAO and that this late increase is mediated by inflammatory processes *(17,18)* support the significant role of neutrophil infiltration in stroke.

Table 1
Antiadhesion Therapy on the Rat MCAO Model

Target	MCAO model	Reduction in ischemic volume	Ref.
CD18 or	Transient	Yes	Matsuo et al., 1994
CD11a or ICAM-1		Yes	
		Yes	
CD18 or CD11b	Transient	Yes	Zhang et al., 1995
CD11b	Transient	Yes	Chen et al., 1994
CD18/CD11b	Transient	Yes	Chopp et al., 1994
CD11b	Transient	Yes	Jiang et al., 1994
	Permanent	No	
ICAM-1	Transient	Yes	Zhang et al., 1994
ICAM-1	Transient	Yes	Zhang et al., 1995
	Permanent	No	
P-Selectin	Permanent	Yes	Suzuki et al., 1999
Neutrophil	Transient	Yes	Jiang et al., 1998
Permanent		No	

2.4. Monocytes/Macrophages

The exact role of monocytes contributing to ischemic damage is not well documented. In human stroke, infiltration of macrophages occurs approx 48–72 h after the insult, but clinical data for earlier time-points are not very extensive *(38)*. In permanent MCAO in rats, monocytes are seen in vessels at 4 h postischemia, but do not infiltrate into parenchyma until 12 h after the insult. Massive infiltration takes place at 72 h following ischemia and the infiltration peaks after 7 d *(28,29)*. According to Lehrmann et al. *(39)*, macrophages are observed mainly in infarcted caudate putamen at 3–7 d following transient MCAO in spontaneously hypertensive rats. In addition to neutrophils and monocytes, infiltration of T-cells following MCAO and infiltration of cytotoxic T-cells and natural killer cells following photothrombotic lesion have been reported *(40,41)*.

2.5. Antileukocyte Therapy

An antileukocyte strategy has been shown to provide neuroprotection. It is very effective in experimental models of transient focal cerebral ischemia, but not in permanent focal cerebral ischemia (Table 1). Furthermore, despite the promising experimental data, a recent clinical study using murine anti-ICAM-1 monoclonal antibody (enlimomab) in human stroke failed *(42)*, possibly because of the ability of enlimomab to activate human neutrophils *(43)*. Although systemic depletion of neutrophils and other leukocytes is unlikely to be of clinical importance due to the severe immunosuppression, targeted antiadhesion strategies might still be valuable tools *(44–46)*. One interesting approach is synthetic fibronectin and laminin peptides, which decrease infarction volume in MCAO through inhibition of leukocyte adhesion *(47,48)*. Chopp and colleagues *(49)* showed that blocking adhesion molecules in the rat reduces apoptosis induced by focal ischemia and, most importantly, that a prolonged time window and improved outcome can be achieved using a combination of tissue-type plasminogen activator (t-PA) and anti-CD18 antibody *(50)*.

3. RESIDENT INFLAMMATORY CELLS

3.1. Microglia

Microglia represent the resident tissue macrophages of the central nervous system (CNS) and play an active role in brain inflammatory processes of brain injury. The magnitude of the microglial reaction in stroke may depend on the concurrence of other local factors, such as medication, nutrition, sex, and the presence of other brain diseases. The repertoire of secretory microglial products involved in inflammatory and/or repair processes is rapidly expanding and includes cytokines, coagulation factors, complement factors, lipid mediators, free radicals, neurotoxins, enzymes, extracellular matrix (ECM) components and growth factors *(51,52)*. Most measurements concerning microglial secretory products are made in vitro and therefore need to be interpreted carefully.

In ischemia and other trauma, microglia start to proliferate and migrate into the damaged area. They retract processes, become more amoeboid shaped, and finally transform into reactive phagocytic brain macrophages *(39,53)*. The activation of microglia is displayed in a graded fashion starting with subtle morphological changes, increased expression of the constitutive complement type-3 receptor (CR3, Mac-1, CD11b/CD18), and the induction of major histocompatibility complex (MHC) class I antigen and of transforming growth factor-β_1 (TGF-β_1), and, finally, more potent and localized alterations in morphology, surface antigens, and cytokine mRNA expression, which accompany actual neuronal degeneration *(39,53–56)*.

3.2. Microglial Activation in Ischemia

In focal cerebral ischemia, microglial activation occurs very fast and persists several weeks. In transient 60-min MCAO, phosphotyrosine induction is detected already at 3–6 h postischemia in microglial cells with long branches and enlarged cell bodies. Thereafter, microglial cells bearing short and thick processes begin to appear and amoeboid and round microglia can be detected after 12–24 h of reperfusion *(57)*. In transient MCAO of rats, progressive microglial response starts at 6 h, increases at 3–7 d, and subsides at 3 wk to 3 mo *(39)*. Recruitment and activation of microglia/macrophages is first seen in striatum penumbra, and is separated from the delayed and protracted microglial reaction along cortical penumbra, with induction of a wide repertoire of microglial activation markers. It is possible that differences in microglial activation/leukocyte infiltration between the striatal and cortical perifocal areas could partially explain why many therapeutic interventions save degenerating neurons mainly in neocortical but not in striatal penumbra *(39)*.

In permanent MCAO in mice, Rupalla and co-workers *(58)* showed marked induction of microglial CR3 and morphological changes in perifocal area already at 30 min after the onset of ischemia in the perifocal area. With increasing time, a progressive rise in the intensity and number of activated microglia was detected. The CR3 immunostaining peaked at 48 h in the inner boundary zone of infarction and decreased 4 d after MCAO. In summary, regardless of ischemic model and markers used, microglial activation occurs prior to or in parallel with neuronal degeneration and correlates with the ischemic damage.

3.3. Mechanisms Activating Microglia

Several possibilities exist concerning the stimuli leading to microglial activation in ischemia. Originally, it was believed that manifest neuronal damage activates microglial cells to proliferate into tissue macrophages that clean the neuronal debris *(59,60)*. Even though neuronal destruction still might be one trigger for the response, the microglial activation preceding neuronal damage suggests that other mechanisms also exist. Microglia are a very sensitive cell population that responds rather quickly to minor changes in the cellular microenvironment *(3,60)*. Therefore, it is speculated that early microglial activation is triggered by initial ischemic events.

Early ultrastructural alterations in ischemia, including disintegration of cytoskeletal proteins *(61)*, changes in synaptic membrane *(62)*, and accumulations of endoplasmic cisternae *(63,64)*, precede the neuronal loss and may trigger changes in cell–cell contacts between stressed neurons and perineuronal microglia. Moreover, the early metabolic changes, including reduced protein synthesis and altered polyamine metabolism *(54)* as well as increased cytokine and reactive oxygen species (ROS) formation or immediate early genes occur within minutes to hours *(4,65)* and could be sufficient to induce microglial activation. Interleukin (IL)-1β, TNF-α, interferon (INF)-α, and colony-stimulating factor-1, as well as IL-3 can induce gliosis *(66–70)*. Enhanced electrical activity and depolarizations of CA1 neurons in global ischemia as well as spreading depression-like depolarizations and anoxic depolarizations in focal ischemia cause changes in extracellular K^+ concentration, which can activate resting microglial cells that lack voltage-gated channels or express only inwardly rectifying potassium channels *(71–73)*. A possible rapid activation via cotransmitters such as ATP has also been suggested *(72)*. In a study by Rogove et al. *(74)*, it was demonstrated that t-PA-deficient microglia exhibit attenuated activation and that exogenous application of t-PA can restore the activation. Furthermore, it was shown that even proteolytically inactive t-PA could restore the microglial activation in vitro and in vivo *(74)*. Altogether, there are multiple pathways that may trigger ischemia-induced activation of microglia.

3.4. Functional Role of Microglia in Ischemia

The functional role of activated microglia in evolution of ischemic damage is diversified. When the critical point of activation is crossed, microglia launches the cytotoxic machinery that mainly includes, in addition to the phagocytic property, inflammatory cytokines, destructive enzymes with ROS and nitric oxide (NO) production, proteolytic enzymes, and neurotoxins *(60)*.

First, microglia can act as phagocytes, which involves direct cell–cell contact. Phagocytic microglia can be found next to dying CA1 neurons, their dendrites, and synapses. This reaction appears to be strictly controlled and site directed only to degenerating synapses, because adjacent synapses of healthy neurons are left intact *(59)*. On the contrary, in more acute MCAO, massive accumulation of fully maturated microglia takes place in necrotic tissue, unspecifically phagocytosing the tissue debris. However, a more selective phagocytosis may take place in the penumbra *(38,59)*.

Second, both human autopsy material and experimental animal models show that activated microglia produce proinflammatory cytokines such as IL-1β and TNF-α *(75–79)*. The very early and sustained induction of IL-1β and TNF-α (between 2 and 24 h) in microglia following ischemia suggests a primary role for these pro-inflammatories in microglial cytotoxicity. Other cytokines with diverse effects are produced by activated microglia. IL-6 is induced in these cells in penumbra, starting at 1 d and persisting up to 2 wk following transient MCAO *(80)*. IL-6 may have a dual role in ischemia. It can act as a cytokine brake by increasing the levels of IL-1 receptor antagonist (IL-1ra) and soluble TNF-α receptor p55 *(81)*. On the other hand, overexpression of IL-6 is reported to stimulate gliosis and blood brain barrier leakage *(82–83)*.

Third, a variety of destructive enzymes are induced in microglial cells following ischemia. Some of these enzymes, such as iNOS, NADPH oxidase, and COX-2, produce reactive nitrogen and oxygen species that cause damage to cellular macromolecules. In experimental ischemia, as in patients with stroke, mainly neutrophils produce iNOS, an enzyme that produces toxic amounts of nitric oxide *(84)*. In one experimental ischemia and trauma study, iNOS reaction product was located to activated microglia, as detected by electron microscopy. However, the positive cells were located rather distally to the injury *(85)*. Nakashima and coworkers *(86)* showed, with NADPH diaphorase staining, NOS activity containing microglia/macrophages in ischemic striatum after MCAO. Although NOS isoform specificity was not shown, the expression correlated with neuronal death, and upregulation indicates that toxic NO was produced. The pathogenic role of NO produced by iNOS is emphasized by the fact that pharmacological inhibition of iNOS reduces ischemic damage and that iNOS knockout mice show decreased infarct size *(17,18)*.

Although in vivo evidence is lacking, cultured microglia produce toxic amounts of oxygen radicals via NADPH oxidase upon stimulation with phorbol esters *(87)*. Furthermore, autopsy material from stroke patients shows that COX-2 protein is induced, not only in neurons as expected but also perinuclearly in microglia of acute cases *(88)*. COX-2 is an ischemia-inducible enzyme, mainly expressed by neurons in experimental MCAO, producing superoxide and pro-inflammatory prostanoids from arachidonic acid *(89–91)*. In addition, in ischemic brain, cPLA$_2$ is produced in microglial cells within the precise region of neuronal damage *(92)*. cPLA$_2$ hydrolyzes membrane phospholipids, and its activity liberates bioactive metabolites such as platelet activating factor and free fatty acids, including arachidonic acid, leading directly to the production of pro-inflammatory eicosanoids *(92–94)*.

Abnormal activation of the extracellular proteolytic cascade in ischemia leads to the degradation of ECM components such as laminin, loss of structural microenvironment, and ultimately to cell death. Moreover, the key enzymes in this destructive cascade, t-PA, plasmin, and MMPs, are produced by activated microglia suggesting that this is an important mechanism of microglial cytotoxicity *(95)*.

Finally, there is mounting evidence that microglia are capable of producing various neurotoxins. Giulian and colleagues described, in a cell culture system and in MCAO, an unidentified neurotoxin that is secreted by activated microglia and that causes neuronal death, which can be blocked by NMDA antagonists *(96,97)*. Also, quinolinic acid, an endogenous neurotoxin that originates from kynurenine pathway of tryptohan metabolism, is dramatically induced in activated microglia and this induction coincides temporally and spatially with actual neuronal death *(98)*, suggesting an executionary role for it. In addition, Piani et al. *(99)* showed that microglial cells in vitro can produce high amounts of glutamate and thus cause excitotoxicity. However, firm in vivo evidence for this is lacking. Table 2 summarizes the wide range of microglial products contributing to ischemic injury and recovery.

Interestingly, microglial activation does not necessarily mean neuronal death. Banati and colleagues *(59,60)*, showed, in an ultrastructural study of global ischemia, a rapid activation of microglia next to neurons that do not die. They refer to soft pathology; a process also called synaptic stripping, where microglia displace synapses, leading to deafferentiation of the stressed neuron. Furthermore, microglial program is activated even in spreading depression *(100)*, where there is no cell death and antigen representation. According to Kreutzberg *(101)*, an unspecific response takes place, where the sensitive microglia become quickly activated without knowing what is happening in tissue. Microglia have the complete program ready, including the cytotoxic tools, but if no destruction or necrosis occurs, they retain the resting state again *(59)*. This could be the case also in ischemia. Microglia is reactive very early in ischemia starting within 30 min, and gradually preparing the repertoire up to 24–48 h when the cytotoxic machinery is finally launched in areas of degeneration, but may be not in areas of survival.

3.5. Antimicroglia Therapy

Whereas several antileukocyte strategies exist and show neuroprotection in experimental stroke, only a few studies have concentrated on targeting activated microglia. This is mainly because of the difficulties in directing the treatment specifically to microglia. Oostveen et al. *(102)* showed, in a gerbil bicarotic artery occlusion model, that pyrrolopyrimidine lipid peroxidation inhibitors are neuroprotective and reduce microglial activation in the CA1 area, but the inhibition achieved is probably not the result of the direct effect on microglia. Giulian and colleagues *(96,97)* detected reduced microglial activation, diminished production of inflammatory mediators, reduced loss of neurons, and preserved motor function in rabbit spinal cord ischemia using a combination of chloroquine and colchicine. Although Giulian and colleagues failed to show a reduced infarct size in MCAO with the same immunosupressant therapy, they showed a 75% reduction in the number of active mononuclear phagocytes in the infarcted cortex at 48 h. In addition, they showed a marked decrease of neurotoxin production. These studies suggest that acute injury, reflected by MCAO lesion size, is not affected by this immunotherapy, but secondary inflammation and functional recovery may be reduced by such treatment *(96)*.

Several growth factors, including TGF-β and certain interleukins, have been demonstrated to inhibit microglia. For example, TGF-β administration following hypoxic ischemic injury reduces microglial activation and infarct size, as well as improving behavioral outcome *(103)*. Pro-inflammatory cytokines are mainly produced by activated microglia in ischemia. Various anticytokine approaches, considered also as an antimicroglia therapy, have been successfully introduced. IL-1ra *(104)*, soluble TNF-α receptor I *(105)*, and as TNF-α mRNA synthesis inhibitors pentoxifylline and rolipram *(106,107)* are potent neuroprotectants in experimental brain injury.

Tetracycline derivatives doxycycline and minocycline have anti-inflammatory functions independent of their antimicrobial action *(108)*. These compounds provide significant neuroprotection in several brain ischemia models *(109–112)* and reduce microglial activation and induction of both interleukin-converting enzyme (ICE) and iNOS in stroke *(111,112)* and Huntington's disease models *(113)*, suggesting that inhibition of microglial activation is beneficial in brain injuries, including stroke. The hypothesis is further supported by the findings that nanomolar concentrations of minocycline inhibit activation of microglial cells in culture, thereby providing protection against excitotoxic neuronal death *(114)*. Even though minocycline seems to inhibit microglial activation and may thereby reduce ischemic neuronal injury, the compound also reduces induction of MMPs in stroke and may reduce leukocyte adhesion and free-radical production by leukocytes (*108*; Yrjänheikki et al., unpublished observation).

3.6. Astrocytes in Ischemia

Astrocytes are ubiquitous in the brain and play an important role in monitoring the microenvironment in normal conditions and in response to CNS injury *(115)*. These cells may have several beneficial roles in stroke, such as removal of extracellular glutamate and other toxic compounds, release of trophic factors, buffering the extracellular K^+ concentration, and regulation of capillary microcirculation *(116–122)*. However, astrocytes may also contribute to neuronal damage by participating in inflammatory processes, because in global brain ischemia, reactive astrocytes express iNOS protein and produce NO and the spatial and temporal profile correlated with delayed neuronal death *(123)*. Moreover, astrocytes produce $cPLA_2$ in global and focal cerebral ischemia *(92)* and thus possess pro-inflammatory potential.

4. CYTOKINES AND CHEMOKINES

Tumor necrosis factor-α and IL-1β are two pro-inflammatory cytokines, which play a very significant role in stroke *(6)*. In addition, cytokine IL-6 and chemokines such as IL-8 for neutrophils and monocyte chemoattractant protein-1 (MCP-1) for monocytes, as well as RANTES (regulated on activation, normal T expressed and secreted) and γ-interferon-inducible protein-10 (IP-10) are increased in ischemia and involved in inflammation, gliosis and leukocyte activation and adhesion *(4)*.

Upregulation of IL-1β and TNF-α in ischemia occurs either through release of preexisting stores from cells such as macrophages or through synthesis. The synthesis is coupled so that each activates synthesis of the other or, alternatively, synthesis can be activated independently. The upstream events suggested to induce synthesis and also release of IL-1β and TNF-α include free radicals, possibly via nuclear factor (NF)-κB and mitogen-activated protein kinases *(124)*. Even more distally, NMDA receptor activation and increased intracellular Ca^{2+} are possibly involved in IL-1β and TNF-α induction *(125)*.

4.1. TNF-α

Tumor necrosis factor-α induction is demonstrated in several experimental models of brain injury and in clinical neurodegenerative disorders and brain trauma *(4,126)*. The pleotrophic cytokine exerts many biological functions, including acute-phase protein secretion and vascular permeability *(127,128)*. TNF-α appears in the circulation within 20 min after the onset of focal ischemia and persists for at least 2 h of reperfusion *(129)*. This very early response might reflect a pre-existing pool of TNF-α. TNF-α is upregulated in microglia and macrophages within 30 min after initiation of focal ischemia, and the levels are peaking at 8 h *(78)*. In parallel, the TNF receptor (TNFR1), which mediates toxic effects and includes the "death domain," is induced approximately within 6 h following MCAO *(130)*.

Table 2
Products of Activated Microglia

Product	Species	Induced in ischemia	Ref.
Growth factors, etc.			
Nerve growth factor	Rodent		Mallat et al., 1989
Neurotrophin 3	Rodent		Elkabes et al., 1996
Basic fibroblast growth factor	Rodent, human		Shimojo et al., 1991, Presta et al., 1995
Transforming growth factors α and β	Human		Walker et al., 1995
	Rodent	MCAO	Wiessner et al., 1993
Bcl-2	Rodent	MCAO	Urabe et al., 1998
Thymosin β(4)	Rodent	MCAO	Vartiainen et al., 1996
Osteopontin	Rodent	MCAO	Ellison et al., 1998
Cytokines and other inflammatories			
IL-1α	Human		Walker et al., 1995
IL-1β	Rodent	MCAO, BCAO	Zhang et al., 1998,
	Human		Sairanen et al., 1997, Walker et al., 1995, Hetier et al., 1991
IL-3	Human		Walker et al., 1995
IL-5	Rodent		Sawada et al., 1993
IL-6	Rodent	MCAO	Block et al., 2000, Gottschall et al., 1995
IL-10	Rodent, human		Mizuno et al., 1994, Sheng et al., 1995
IL-12	Rodent		Aloisi et al., 1997, Lodge and Sriram, 1996
TNF-α	Rodent	MCAO	Buttini et al., 1996, Gregersen et al., 2000
Macrophage inflammatory protein-1	Rodent	MCAO	Gourmala et al., 1999
Transcription factors			
NF-κB (p65)	Rodent	MCAO	Gabriel et al., 1999
Coagulation factors			
Urokinase-PA	Rodent		Nakajima et al., 1992
Plasminogen	Rodent		Nakajima et al., 1992
Tissue-PA	Rodent		Rogowe and Tsirka, 1998
Complement factors			
C1q	Rodent	BCAO	Schafer et al., 2000
C1, C3, C4	Human		Walker et al., 1995
Receptors			
GluR4, NR1, AMPA, NMDA	Rodent	BCAO	Gottlieb and Matute, 1997
Macrophage scavenger receptor	Human	Ischemic lesion	Honda et al., 1998

Table 2 (Continued)

Product	Species	Induced in ischemia	Ref.
Receptors (cont.)			
Peripheral benzodiazepine-R	Rodent	BCAO	Stephenson et al., 1995
Epidermal growth factor-R	Rodent	MCAO	Planas et al., 1998
Lipid mediators			
Arachidonic acid	Rodent		Minghetti and Levi, 1998
PAF	Human		Jaranowska et al., 1995
Prostaglandins:			Minghetti and Levi, 1998
PGE_2, PGD_2, $PGF_{2\alpha}$	Rodent, human		
Thromboxane B_2	Rodent		Giulian et al., 1996
Leukotriene B_4	Rodent		Matsuo et al., 1995
Free radicals			
Superoxide anions	Rodent, human		Colton et al., 1996, Chao et al., 1995
NO	Rodent		Minghetti and Levi, 1998
Neurotoxins			
Ntox	Rodent	MCAO	Giulian et al., 1997
APP	Rodent	MCAO, BCAO	Banati et al., 1996
Quinolinic acid	Rodent	BCAO	Baratte et al., 1998
	Human		Heyes et al., 1996
Enzymes			
Gelatinase	Rodent		Gottschall et al., 1995
Elastase	Rodent		Nakajima et al., 1992
ICE	Rodent	BCAO	Bhat et al., 1996
iNOS, NOS	Rodent	Ischemia, trauma	Walski and Gajkowska, 1999
		MCAO	Nakashima et al., 1995
$cPLA_2$	Rodent	BCAO, MCAO	Stephenson et al., 1999
COX-2	Human	Human stroke	Sairanen et al., 1998
PYK2 (nonreceptor tyrosine kinase)	Rodent	MCAO	Tian et al., 2000
Protein kinase C δ	Rodent	BCAO	Koponen et al., 2000
p38-Mitogen-activated protein kinase	Rodent	MCAO, BCAO	Tian et al., 2000, Walton et al., 1998
Ectonucleotidase	Rodent	BCAO	Braun et al., 1998
Heat shock proteins			
HSP-27	Rodent	MCAO	Kato et al., 1995
HSP-32	Rodent	MCAO	Koistinaho et al., 1996
HSP-72	Rodent	MCAO	Soriano et al., 1994

Note: BCAO indicates global ischemia produced either with four-vessel occlusion in rat, BCAO + hypotension in mice/rat, or BCAO in gerbils. MCAO indicates transient or permanent focal ischemia produced in mice or rat.

Both TNF-α mRNA and protein are elevated within 1–3 h after MCAO in rats *(131, 132)*. In these studies, TNF-α mRNA levels were reported to be induced in ischemic cortex as early as 1 h, peaking at 12 h, and persisting upregulated for 5 d. Liu et al. *(133)* reported that initial TNF-α induction is in ischemic neurons and later also in inflammatory cells. The early phase of TNF-α induction precedes leukocyte infiltration, and microinjection of TNF-α into the rat cortex causes leukocyte adhesion to the capillary endothelium *(4)*. On the other hand, intracerebroventricular injection of TNF-α 24 h before MCAO exacerbates the subsequent ischemic damage *(31)*. The exacerbating effect was reversed by contralateral cerebroventricular administration of an anti-TNF-α monoclonal antibody. This evidence suggests that TNF-α could predispose the brain for the subsequent damage by causing a pro-adhesive state of the capillary endothelium, possibly via upregulation of adhesion molecules *(4)*.

In transient and permanent focal cerebral ischemia, injection of anti-TNF-α antibodies into the circulation, cortex, or ventricle *(129,134,135)* achieves remarkable protection. In other studies, a soluble TNF-α receptor, acting as a TNF-α-binding protein, was found to be protective in MCAO *(31,136,137)*. An extremely large, 80%, reduction in infarct volume was reached by a drug (CNI-1493) that prevents TNF-α translation in permanent MCAO *(134)*.

There is actually some data supporting a neuroprotective role for TNF-α. Cheng et al. *(138)* showed that TNF-α protected neurons against metabolic–excitotoxic insult. In addition, TNF-α receptor null mice are more susceptible to ischemia than littermates *(139)*, suggesting that low levels of TNF-α might actually be beneficial.

4.2. IL-1β

Interleukin-1β is a pro-inflammatory cytokine, which is produced as a precursor protein (pro-IL-1β) that requires cleavage by ICE (caspase-1) to become biologically active. Focal cerebral ischemia induces IL-1β immunoreactivity within 15 min of the onset of insult, and the levels maximize at 2 h. This induction is seen mainly in endothelial cells and microglia *(77)*. In another study, the rapid increase of IL-1β mRNA in MCAO peaked at 12 h and returned to basal levels at 5 d *(140)*. IL-1β peptide is localized to cerebral vessels, microglia, and macrophages after stroke *(81,141)*. Furthermore, exogenous administration of IL-1β exacerbates ischemic brain injury *(142,143)*.

Interleukin-1ra is a naturally occurring peptide inhibitor of IL-1β activity. It competes with IL-1 for occupancy of the IL-1 receptor without inducing a signal of its own. IL-1ra is induced following focal cerebral ischemia and the temporal profile parallels with IL-1β expression, except that IL-1ra expression is prolonged *(144)*. Both overexpression of IL-1ra and administration of IL-1ra decrease neuronal death in the perifocal area and reduce infarct volume by 50% in permanent MCAO *(145–147)*. In addition, an antibody against IL-1ra exacerbates ischemic damage *(148)*. Furthermore, mice deficient in ICE show attenuated neuronal damage in permanent MCAO *(149)*, suggesting that ICE manipulation leads to neuroprotection through reduced levels of active IL-1β.

4.3. Mechanisms of Damage by Cytokines

Interleukin-1β and TNF-α interact with each other and influence a series of signal transduction pathways, including those affecting gene activation *(6,150)*. Several enzymes activated by IL-1β and TNF-α and contributing to ischemic damage include COX-2, iNOS, and MMP-9. There exists a close relationship between cytokine production and gliosis. IL-1β and TNF-α activate glial cells, which then produce IL-1β and TNF-α *(4)*. IL-1β can stimulate astrocyte proliferation *(151)* and cerebral edema *(152)*, induce leukocyte infiltration *(13,153)*, and increase *de novo* synthesis of endothelial adhesion molecules ICAM-1 and ELAM *(154,155)*. TNF-α increases blood-brain-barrier (BBB) permeability, causes pial artery constriction, and has a direct toxic effect on the capillaries *(156,157)*. TNF-α is toxic to oligodendrocytes, stimulates astrocyte proliferation, and is involved in demyelination and gliosis in brain injury *(78,158,159)*. Furthermore, TNF-α activates the endothelium for leukocyte adherence and promotes procoagulation activity by increasing the levels of tissue factor, von Willebrand factor, and platelet-activating factor (PAF) *(160)*. TNF-α also stimulates expression of leukocyte–endothelial adhesion molecules *(66)*, activates neutrophil free-radical release *(4)*, and causes mitochondrial free-radical production and apoptosis *(161,162)*.

5. MATRIX METALLOPROTEASES

Matrix metalloproteases (MMPs) are zinc- and calcium-dependent neutral proteases capable of degrading the constituents of basal lamina and extracellular matrix, thus playing a central role in extracellular matrix turnover during biological processes such as morphogenesis, development, wound healing, and tissue resorption *(162)*. Recently, involvement of MMP-2 and MMP-9 has been implicated in CNS disorders, including focal brain *(163,164)*. In cell culture systems, astrocytes, microglia, endothelial cells, and fetal neurons produce gelatinases either constitutively or upon stimulation *(165–168)*. As reviewed in Chapter 11, MMP-2 and especially MMP-9 have a contributory role in ischemic stroke, possibly because they break down the BBB and promote the infiltration of leukocytes.

6. PHARMACOLOGIC TARGETS IN SIGNALING PATHWAYS

Expression of many pro-inflammatory genes is regulated by the same signaling pathways, which therefore may involve mediators that represent potential pharmacologic targets. These include extracellular signal-regulated (ERK-1 and ERK-2) and stress-activated (p38) MAPKs (mitogen-activated protein kinases), transcription factor NF-κB (nuclear factor κB), and PPARγ (peroxisome proliferator-activated receptor γ). ERKs are known to mediate neuronal responses to neurotrophic factors and may be less important for induction of inflammatory responses than p38 *(169–173)*. However, strong activation of ERKs takes place in vulnerable neurons after focal ischemia, and inhibition of ERKs appears to be neuroprotective when given intraventricularly *(174)*. p38 can be induced in all cells present in the brain *(175–178)* and is thought to mediate both apoptotic and inflammatory signals. It is also a crucial kinase for expression of inflammatory genes such as IL-1β, TNF-α, iNOS, and COX-2 *(169–171,179–183)*. NF-κB regulates directly the expression of these inflammatory genes, usually in concert with other transcription factors such as activating protein 1 (AP-1) *(184)*. DNA binding of PPARγ counteracts the effects of NF-κB and AP-1 and thereby reduces expression of the key inflammatory genes *(185,186)*.

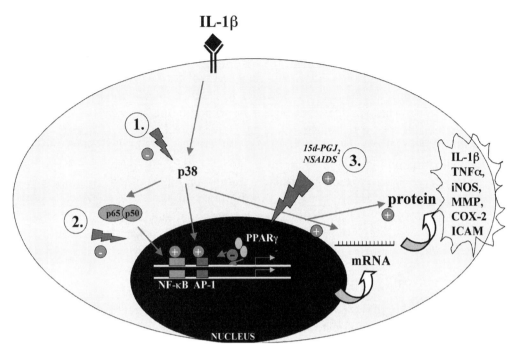

Fig. 1. Potential targets of pharmacologic intervention in ischemic inflammation. The model is simplified and does not illustrate other kinase pathways mediated, for example, by Jun N-terminal kinase (JNK) and ERKs. Activation of the p38 MAPK pathway induces transcription through phosphorylation of transcription factors, stabilizes mRNA, and may also upregulate translation. The p65/p50 NF-κB heterodimer is a transcription factor that upregulates several pro-inflammatory genes and represents one potential site of pharmacologic intervention. PPARγ activators reduce induction of these genes by inhibiting NF-κB and AP-1 DNA binding. (**1**) p38 MAPK or upstream kinase inhibitors; (**2**) NF-κB inhibitors; (**3**) PPARγ agnosits.

Preliminary results indicate that p38 MAPK inhibitors reduce infarction volume when given orally or intravenously prior to or immediately after permanent ischemia in the rat *(187)*, but it is not yet known whether the protection is based on inhibition of immediate p38 MAPK activation in neurons or on inhibition of delayed p38 MAPK activation in glia. Similarly, ischemic injury is reduced in mice lacking the p50 subunit of NF-κB *(188)* and unspecific NF-κB inhibitors are still protective when given immediately after transient ischemia *(189)*. Even though p38 MAPK and NF-κB are activated in glia several days after ischemic insult *(178,190)*, it is obvious that additional experiments are needed to find out whether these compounds are beneficial when administered within a clinically relevant time window. Finally, several classical nonsteroidal anti-inflammatory drugs also activate PPAR-γ and reduce ischemic damage, but studies on the specific activation of PPAR-γ in stroke models are not yet available. Nevertheless, several in vitro studies indicate that specific PPAR-γ agonists act as strong anti-inflammatories and may have potential in treatment of human stroke (*see* Chapter 7). Figure 1 summarizes the potential pharmacologic targets in intracellular signaling of inflammation.

7. CONCLUSIONS

Experimental evidence strongly suggests that inflammation contributes not only to early phases but also the delayed progression of ischemic damage. Even though the whole genomic response triggered by ischemia is not yet fully characterized, some of the induced genes, such as IL-1β, TNFα, iNOS, and COX-2, are among potential targets for pharmacologic intervention. Inflammation is a complex process that involves the activation of characteristic intracellular signaling pathways, and inhibition of these molecules mediating the induction of proinflammatory genes may be especially beneficial by diminishing simultaneously several inflammatory responses. In addition, combination therapy against, for example, cytokines, iNOS, and COX-2 may eventually be an efficient alternative, which deserves to be explored further in animal models.

REFERENCES

1. Dirnagl, U., Iadecola, C., and Moskowitz, M. A. (1999) Pathobiology of ischaemic stroke: an integrated view. *Trends Neurosci.* **22,** 391–397.
2. Stroke Therapy Academic Industry Roundtable (STAIR) (1999) Recommendations for standards regarding preclinical neuroprotective and restorative drug development. *Stroke* **30,** 2752–2758.
3. Sharkey, J., Kelly, J. S., and Butcher, S.P. (1997) Inflammatory responses to cerebral ischemia, in *Clinical Pharmacology of Cerebral Ischemia* (Ter Horst, G. J. and Korf, J., eds.), Humana, Totowa, NJ, pp. 235–265.
4. Barone, F. C. and Feuerstein, G. Z. (1999) Inflammatory mediators and stroke: new opportunities for novel therapeutics. *J. Cereb. Blood Flow Metab.* **19,** 819–834.
5. Iadecola, C. and Alexander, M. (2001) Cerebral ischemia and inflammation. *Curr. Opin. Neurol.* **14,** 89–94.
6. Lipton, P. (1999) Ischemic cell death in brain neurons. *Physiol. Rev.* **79,** 1431–1568.
7. Perry, V. H. and Gordon, S. (1988) Macrophages and microglia in the nervous system. *Trends Neurosci.* **11,** 273–277.
8. Ames, A. III, Wright, R. L., Kowada, M., Thurston, J. M., and Majno, G. (1968) Cerebral ischemia. II. The no-reflow phenomenon. *Am. J. Pathol.* **52,** 437–453.
9. del Zoppo, G. J., Schmid-Schonbein, G. W., Mori, E., Copeland, B. R., and Chang, C. M. (1991) Polymorphonuclear leukocytes occlude capillaries following middle cerebral artery occlusion and reperfusion in baboons. *Stroke* **22,** 1276–1283.
10. Dirnagl, U., Niwa, K., Sixt, G., and Villringer, A. (1994) Cortical hypoperfusion after global forebrain ischemia in rats is not caused by microvascular leukocyte plugging. *Stroke* **25,** 1028–1238.
11. Grau, A. J., Graf, T., and Hacke, W. (1994) Altered influence of polymorphonuclear leukocytes on coagulation in acute ischemic stroke. *Thromb. Res.* **76,** 541–549.
12. Grau, A. J., Sigmund, R., and Hacke, W. (1994) Modification of platelet aggregation by leukocytes in acute ischemic stroke. *Stroke* **25,** 2149–2152.
13. Kochanek, P. M. and Hallenbeck, J. M. (1992) Polymorphonuclear leukocytes and monocytes/macrophages in the pathogenesis of cerebral ischemia and stroke. *Stroke* **23,** 1367–1329.
14. Prasad, K., Kapoor, R., and Kalra, J. (1992) Methionine in protection of hemorrhagic shock: role of oxygen free radicals and hypochlorous acid. *Circ. Shock* **36,** 265–276.
15. Matsuo, Y., Kihara, T., Ikeda, M., Ninomiya, M., Onodera, H., and Kogure, K. (1995) Role of neutrophils in radical production during ischemia and reperfusion of the rat brain: effect of neutrophil depletion on extracellular ascorbyl radical formation. *J. Cereb. Blood Flow. Metab.* **15,** 941–947.

16. Walder, C. E., Green, S. P., Darbonne, W. C., Mathias, J., Rae, J., Dinauer, M. C., et al. (1997) Ischemic stroke injury is reduced in mice lacking a functional NADPH oxidase. *Stroke* **28,** 2252–2258.

17. Iadecola, C., Zhang, F., Casey, R., Nagayama, M., and Ross, M. E. (1997) Delayed reduction of ischemic brain injury and neurological deficits in mice lacking the inducible nitric oxide synthase gene. *J. Neurosci.* **17,** 9157–9164.

18. Nagayama, M., Zhang, F., and Iadecola, C. (1998) Delayed treatment with aminoguanidine decreases focal cerebral ischemic damage and enhances neurologic recovery in rats. *J. Cereb. Blood Flow Metab.* **18,** 1107–1113.

19. Romanic, A. M., Whitem R. F., Arleth, A. J., Ohlstein, E. H., and Barone, F. C. (1998) Matrix metalloproteinase expression increases after cerebral focal ischemia in rats: inhibition of matrix metalloproteinase-9 reduces infarct size. *Stroke* **29,** 1020–1030.

20. Kishimoto, T. K. and Rothlein, R. (1994) Integrins, ICAMs, and selectins: role and regulation of adhesion molecules in neutrophil recruitment to inflammatory sites. *Adv. Pharmacol.* **25,** 117–169.

21. Okada, Y., Copeland, B. R., Mori, E., Tung, M. M., Thomas, W. S., and del Zoppo, G. J. (1994) P-selectin and intercellular adhesion molecule-1 expression after focal brain ischemia and reperfusion. *Stroke* **25,** 202–211.

22. Wang, X. and Feuerstein, G. Z. (1995) Induced expression of adhesion molecules following focal brain ischemia. *J. Neurotrauma* **12,** 825–832.

23. Matsuo, Y., Onodera, H., Shiga, Y., Shozuhara, H., Ninomiya, M., Kihara, T., et al. (1994) Role of cell adhesion molecules in brain injury after transient middle cerebral artery occlusion in the rat. *Brain Res.* **656,** 344–352.

24. Wang, X., Siren, A. L., Liu, Y., Yue, T. L., Barone, F. C., and Feuerstein, G. Z. (1994) Upregulation of intercellular adhesion molecule 1 (ICAM-1) on brain microvascular endothelial cells in rat ischemic cortex. *Mol. Brain Res.* **26,** 61–68.

25. Feuerstein, G. (1997) Inflammatory mediators in brain microvessels, in *Primer on Cerebrovascular Diseases* (Welch, K. M. A., Kaplan, L. R., Reis, D. J., Siesjö, B. K., and Weir, B., eds.), Academic, New York, pp. 220–222.

26. Garcia, J. H. and Kamijyo, Y. (1974) Cerebral infarction. Evolution of histopathological changes after occlusion of a middle cerebral artery in primates. *J. Neuropathol. Exp. Neurol.* **33,** 408–421.

27. Zhang, R. L., Chopp, M., Chen, H., and Garcia, J. H. (1994) Temporal profile of ischemic tissue damage, neutrophil response, and vascular plugging following permanent and transient (2H) middle cerebral artery occlusion in the rat. *J. Neurol. Sci.* **125,** 3–10.

28. Garcia, J. H., Liu, K. F., Yoshida, Y., Chen, S., and Lian, J. (1994) Brain microvessels: factors altering their patency after the occlusion of a middle cerebral artery (Wistar rat). *Am. J. Pathol.* **145,** 728–740.

29. Garcia, J. H., Liu, K. F., Yoshida, Y., Lian, J., Chen, S., and del Zoppo, G. J. (1994) Influx of leukocytes and platelets in an evolving brain infarct (Wistar rat). *Am. J. Pathol.* **144,** 188–199.

30. Hallenbeck, J. M., Dutka, A. J., Tanishima, T., Kochanek, P. M., Kumaroo, K. K., Thompson, C. B., et al. (1986) Polymorphonuclear leukocyte accumulation in brain regions with low blood flow during the early postischemic period. *Stroke* **17,** 246–253.

31. Barone, F. C., Schmidt, D. B., Hillegassm, L. M., Price, W. J., White, R. F., Feuerstein, G. Z., et al. (1992) Reperfusion increases neutrophils and leukotriene B4 receptor binding in rat focal ischemia. *Stroke* **123,** 1337–1347.

32. Zhang, R. L., Chopp, M., Li, Y., Zaloga, C., Jiang, N., Jones, M. L., et al. (1994) Anti-ICAM-1 antibody reduces ischemic cell damage after transient middle cerebral artery occlusion in the rat. *Neurology* **44,** 1747–1751.

33. Akopov, S. E., Simonian, N. A., and Grigorian, G. S. (1996) Dynamics of polymorpho-nuclear leukocyte accumulation in acute cerebral infarction and their correlation with brain tissue damage. *Stroke* **27,** 1739–1743.

34. Lindsberg, P. J., Carpen, O., Paetau, A., Karjalainen-Lindsberg, M. L., and Kaste, M. (1996) Endothelial ICAM-1 expression associated with inflammatory cell response in human ischemic stroke. *Circulation* **94,** 939–945.

35. Kochanek, P. M., Dutka, A. J., Kumaroo, K. K., and Hallenbeck, J. M. (1987) Platelet activating factor receptor blockade enhances recovery after multifocal brain ischemia. *Life Sci.* **41,** 2639–2644.

36. Hayward, N. J., Elliott, P. J., Sawyer, S. D., Bronson, R. T., and Bartus, R. T. (1996) Lack of evidence for neutrophil participation during infarct formation following focal cerebral ischemia in the rat. *Exp. Neurol.* **139,** 188–202.

37. Clark, R. K., Lee, E. V., White, R. F., Jonak, Z. L., Feuerstein, G. Z., and Barone, F. C. (1994) Reperfusion following focal stroke hastens inflammation and resolution of ische-mic injured tissue. *Brain Res. Bull.* **35,** 387–392.

38. Giulian, D., Corpuz, M., Chapman, S., Mansouri, M., and Robertson, C. (1993) Reactive mononuclear phagocytes release neurotoxins after ischemic and traumatic injury to the central nervous system. *J. Neurosci. Res.* **36,** 681–693.

39. Lehrmann, E., Christensen, T., Zimmer, J., Diemer, N. H., and Finsen, B. (1997) Micro-glial and macrophage reactions mark progressive changes and define the penumbra in the rat neocortex and striatum after transient middle cerebral artery occlusion. *J. Comp. Neu-rol.* **386,** 461–476.

40. Morioka, T., Kalehua, T., and Streit, W. J. (1993) Characterization of microglial reaction after middle cerebral artery occlusion in rat brain. *J. Comp. Neurol.* **327,** 123–132.

41. Jander, S., Kraemer, M., Schroeter, M., Witte, O. W., and Stoll, G. (1995) Lymphocytic infiltration and expression of intercellular adhesion molecule-1 in photochemically induced ischemia of the rat cortex. *J. Cereb. Blood Flow Metab.* **15,** 42–51.

42. DeGraba, T. J. (1998) The role of inflammation after acute stroke: utility of pursuing anti-adhesion molecule therapy. *Neurology* **51,** S62–S68.

43. Vuorte, J., Lindsberg, P. J., Kaste, M., Meri, S., Jansson, S. E., Rothlein, R., et al. (1999) Anti-ICAM-1 monoclonal antibody R6.5 (Enlimomab) promotes activation of neutrophils in whole blood. *J. Immunol.* **162,** 2353–2357.

44. Chopp, M., Zhang, R. L., Chen, H., Li, Y., Jiang, N., and Rusche, J. R. (1994) Postische-mic administration of an anti-Mac-1 antibody reduces ischemic cell damage after tran-sient middle cerebral artery occlusion in rats. *Stroke* **25,** 869–876.

45. Chen, H., Chopp, M., and Bodzin, G. (1992) Neutropenia reduces the volume of cerebral infarct after transient middle cerebral artery occlusion in the rat. *Neurosci. Res. Commun.* **11,** 93–99.

46. Chen, Z. L. and Strickland, S. (1997) Neuronal death in the hippocampus is promoted by plasmin-catalyzed degradation of laminin. *Cell* **91,** 917–925.

47. Yanaka, K., Camarata, P. J., Spellman, S. R., McCarthy, J. B., Furcht, L. T., and Low, W. C. (1997) Antagonism of leukocyte adherence by synthetic fibronectin peptide V in a rat model of transient focal cerebral ischemia. *Neurosurgery* **40,** 557–564.

48. Yanaka, K., Camarata, P. J., Spellman, S. R., Skubitz, A. P., Furcht, L. T., and Low, W. C. (1997) Laminin peptide ameliorates brain injury by inhibiting leukocyte accumu-lation in a rat model of transient focal cerebral ischemia. *J. Cereb. Blood Flow Metab.* **17,** 605–611.

49. Chopp, M., Li, Y., Jiang, N., Zhang, R. L., and Prostak, J. (1996) Antibodies against adhesion molecules reduce apoptosis after transient middle cerebral artery occlusion in rat brain. *J. Cereb. Blood Flow Metab.* **16,** 578–584.

50. Zhang, R. L., Zhang, Z. G., and Chopp, M. (1999) Increased therapeutic efficacy with rt-PA and anti-CD18 antibody treatment of stroke in the rat. *Neurology* **52,** 273–279.

51. Minghetti, L. and Levi, G. (1998) Microglia as effector cells in brain damage and repair: focus on prostanoids and nitric oxide. *Prog. Neurobiol.* **54,** 99–125.

52. McGeer, P. L. and McGeer, E. G. (1995) The inflammatory response system of brain: implications for therapy of Alzheimer and other neurodegenerative diseases. *Brain Res. Rev.* **21,** 195–218.

53. Giulian, D. (1997) Reactive microglia and ischemic injury, in *Primer on Cerebrovascular Diseases* (Welch, K. M. A., Kaplan, L. R., Reis, D. J., Siesjö, B. K., and Weir, B., eds.), Academic, New York, pp. 117–124.

54. Gehrmann, J., Bonnekoh, P., Miyazawa, T., Hossmann, K. A., and Kreutzberg, G. W. (1992) Immunocytochemical study of an early microglial activation in ischemia. *J. Cereb. Blood Flow. Metab.* **12,** 257–269.

55. Morioka, T., Kalehua, A. N., and Streit, W. J. (1992) Progressive expression of immunomolecules on microglial cells in rat dorsal hippocampus following transient forebrain ischemia. *Acta Neuropathol. (Berl.)* **83,** 149–157.

56. Finsen, B. R., Tonder, N., Xavier, G. F., Sorensen, J. C., and Zimmer, J. (1993) Induction of microglial immunomolecules by anterogradely degenerating mossy fibres in the rat hippocampal formation. *J. Chem. Neuroanat.* **6,** 267–725.

57. Korematsu, K., Goto, S., Nagahiro, S., and Ushio, Y. (1994) Microglial response to transient focal cerebral ischemia: an immunocytochemical study on the rat cerebral cortex using anti-phosphotyrosine antibody. *J. Cereb. Blood Flow Metab.* **14,** 825–830.

58. Rupalla, K., Allegrini, P. R., Sauer, D., and Wiessner, C. (1998) Time course of microglia activation and apoptosis in various brain regions after permanent focal cerebral ischemia in mice. *Acta Neuropathol. (Berl.)* **96,** 172–178.

59. Banati, R. B., Gehrmann, J., and Kreutzberg, G. W. (1996) Early glial reactions in ischemic lesions. *Adv. Neurol.* **71,** 329–337.

60. Banati, R. B., Gehrmann, J., Schubert, P., and Kreutzberg, G. W. (1993) Cytotoxicity of microglia. *Glia* **7,** 111–118.

61. Yamamoto, K., Morimoto, K., and Yanagihara, T. (1986) Cerebral ischemia in the gerbil: transmission electron microscopic and immunoelectron microscopic investigation. *Brain Res.* **384,** 1–10.

62. Von Lubitz, D. K. and Diemer, N. H. (1983) Cerebral ischemia in the rat: ultrastructural and morphometric analysis of synapses in stratum radiatum of the hippocampal CA-1 region. *Acta Neuropathol. (Berl.)* **61,** 52–60.

63. Kirino, T. and Sano, K. (1984) Fine structural nature of delayed neuronal death following ischemia in the gerbil hippocampus. *Acta Neuropathol. (Berl.)* **62,** 209–218.

64. Kirino, T. and Sano, K. (1984) Selective vulnerability in the gerbil hippocampus following transient ischemia. *Acta Neuropathol. (Berl.)* **62,** 201–208.

65. Sharp, F. R., Lu, A., Tang, Y., and Millhorn, D. E. (2000) Multiple molecular penumbras after focal cerebral ischemia. *J. Cereb. Blood Flow Metab.* **20,** 1011–1032.

66. Balasingam, V., Tejada-Berges, T., Wright, E., Bouckova, R., and Yong, V. W. (1994) Reactive astrogliosis in the neonatal mouse brain and its modulation by cytokines. *J. Neurosci.* **14,** 846–856.

67. Selmaj, K. W., Farooq, M., Norton, W. T., Raine, C. S., and Brosnan, C. F. (1990) Proliferation of astrocytes *in vitro* in response to cytokines. A primary role for tumor necrosis factor. *J. Immunol.* **144,** 129–135.

68. Sawada, M., Kondo, N., Suzumura, A., and Marunouchi, T. (1989) Production of tumor necrosis factor-alpha by microglia and astrocytes in culture. *Brain Res.* **49,** 394–397.

69. Sawada, M., Suzumura, A., Yamamoto, H., and Marunouchi, T. (1990) Activation and proliferation of the isolated microglia by colony stimulating factor-1 and possible involvement of protein kinase C. *Brain Res.* **509,** 119–124.

70. Frei, K., Bodmer, S., Schwerdel, C., and Fontana, A. (1986) Astrocyte-derived interleukin 3 as a growth factor for microglia cells and peritoneal macrophages. *J. Immunol.* **137,** 3521–3527.
71. Suzuki, R., Yamaguchi, T., Li, C. L., and Klatzo, I. (1983) The effects of 5-minute ischemia in Mongolian gerbils: II. Changes of spontaneous neuronal activity in cerebral cortex and CA1 sector of hippocampus. *Acta Neuropathol. (Berl.)* **60,** 217–222.
72. Kettenmann, H., Banati, R., and Walz, W. (1993) Electrophysiological behavior of microglia. *Glia* **7,** 93–101.
73. Nedergaard, M. (1987) Neuronal injury in the infarct border: a neuropathological study in the rat. *Acta Neuropathol. (Berl.)* **73,** 267–274.
74. Rogove, A. D., Siao, C., Keyt, B., Strickland, S., and Tsirka, S. E. (1999) Activation of microglia reveals a non-proteolytic cytokine function for tissue plasminogen activator in the central nervous system. *J. Cell Sci.* **112,** 4007–4016.
75. Tomimoto, H., Akiguchi, I., Wakita, H., Kinoshita, A., Ikemoto, A., Nakamura, S., et al. (1996) Glial expression of cytokines in the brains of cerebrovascular disease patients. *Acta Neuropathol. (Berl.)* **92,** 281–287.
76. Sairanen, T. R., Lindsberg, P. J., Brenner, M., and Siren, A. L. (1997) Global forebrain ischemia results in differential cellular expression of interleukin-1beta (IL-1beta) and its receptor at mRNA and protein level. *J. Cereb. Blood Flow Metab.* **17,** 1107–1120.
77. Zhang, Z., Chopp, M., Goussev, A., and Powers, C. (1998) Cerebral vessels express interleukin 1beta after focal cerebral ischemia. *Brain Res.* **784,** 210–217.
78. Buttini, M., Appel, K., Sauter, A., Gebicke-Haerter, P. J., and Boddeke, H. W. (1996) Expression of tumor necrosis factor alpha after focal cerebral ischaemia in the rat. *Neuroscience* **71,** 1–16.
79. Gregersen, R., Lambertsen, K., and Finsen, B. (2000) Microglia and macrophages are the major source of tumor necrosis factor in permanent middle cerebral artery occlusion in mice. *J. Cereb. Blood Flow Metab.* **20,** 53–65.
80. Block, F., Peters, M., and Nolden-Koch, M. (2000) Expression of IL-6 in the ischemic penumbra. *Neuroreport* **11,** 963–967.
81. Tilg, H., Trehu, E., Atkins, M. B., Dinarello, C. A., and Mier, J. W. (1994) Interleukin-6 (IL-6) as an anti-inflammatory cytokine: induction of circulating IL-1 receptor antagonist and soluble tumor necrosis factor receptor p55. *Blood* **83,** 113–118.
82. Chiang, C. S., Stalder, A., Samimi, A., and Campbell, I. L. (1994) Reactive gliosis as a consequence of interleukin-6 expression in the brain: studies in transgenic mice. *Dev. Neurosci.* **16,** 212–221.
83. Brett, F. M., Mizisin, A. P., Powell, H. C., and Campbell, I. L. (1995) Evolution of neuropathologic abnormalities associated with blood-brain barrier breakdown in transgenic mice expressing interleukin-6 in astrocytes. *J. Neuropathol. Exp. Neurol.* **54,** 766–775.
84. Forster, C., Clark, H. B., Ross, M. E., and Iadecola, C. (1999) Inducible nitric oxide synthase expression in human cerebral infarcts. *Acta Neuropathol. (Berl.)* **97,** 215–220.
85. Walski, M. and Gajkowska, B. (1999) Electron microscopic studies on NO-synthase activity in brain phagocytes of rat cerebral cortex after ischemic and traumatic brain injury. *J. Hirnforsch.* **39,** 455–463.
86. Nakashima, M. N., Yamashita, K., Kataoka, Y., Yamashita, Y. S., and Niwa, M. (1995) Time course of nitric oxide synthase activity in neuronal, glial, and endothelial cells of rat striatum following focal cerebral ischemia. *Cell. Mol. Neurobiol.* **15,** 341–349.
87. Banati, R. B., Rothe, G., Valet, G., and Kreutzberg, G. W. (1991) Respiratory burst activity in brain macrophages: a flow cytometric study on cultured rat microglia. *Neuropathol. Appl. Neurobiol.* **17,** 223–230.
88. Sairanen, T., Ristimaki, A., Karjalainen-Lindsberg, M. L., Paetau, A., Kaste, M., and Lindsberg, P. J. (1998) Cyclooxygenase-2 is induced globally in infarcted human brain. *Ann. Neurol.* **43,** 738–747.

89. Miettinen, S., Fusco, F. R., Yrjänheikki, J., Keinanen, R., Hirvonen, T., Roivainen, R., et al. (1997) Spreading depression and focal brain ischemia induce cyclooxygenase-2 in cortical neurons through *N*-methyl-D-aspartic acid-receptors and phospholipase A2. *Proc. Natl. Acad. Sci. USA* **94,** 6500–6505.

90. Yamagata, K., Andreasson, K. I., Kaufmann, W. E., Barnes, C. A., and Worley, P. F. (1993) Expression of a mitogen-inducible cyclooxygenase in brain neurons: regulation by synaptic activity and glucocorticoids. *Neuron* **11,** 371–386.

91. Nogawa, S., Zhang, F., Ross, M. E., and Iadecola, C. (1997) Cyclo-oxygenase-2 gene expression in neurons contributes to ischemic brain damage. *J. Neurosci.* **17,** 2746–2755.

92. Stephenson, D., Rash, K., Smalstig, B., Roberts, E., Johnstone, E., Sharp, J., et al. (1999) Cytosolic phospholipase A2 is induced in reactive glia following different forms of neurodegeneration. *Glia* **27,** 110–128.

93. Bazan, N. G., Rodriguez de Turco, E. B., and Allan, G. (1995) Mediators of injury in neurotrauma: intracellular signal transduction and gene expression. *J. Neurotrauma* **12,** 791–814.

94. Shimizu, T. and Wolfe, L. S. (1990) Arachidonic acid cascade and signal transduction. *J. Neurochem.* **55,** 1–15.

95. Anthony, D. C. and Perry, V. H. (1998) Stroke: a double-edged sword for cleaving clots? *Curr. Biol.* **8,** R274–R277.

96. Giulian, D. and Vaca, K. (1993) Inflammatory glia mediate delayed neuronal damage after ischemia in the central nervous system. *Stroke* **24,** 184–190.

97. Giulian, D. and Lachman, L. B. (1985) Interleukin-1 stimulation of astroglial proliferation after brain injury. *Science* **228,** 497–499.

98. Baratte, S., Molinari, A., Veneroni, O., Speciale, C., Benatti, L., and Salvati, P. (1998) Temporal and spatial changes of quinolinic acid immunoreactivity in the gerbil hippocampus following transient cerebral ischemia. *Mol. Brain Res.* **59,** 50–57.

99. Piani, D., Frei, K., Do, K. Q., Cuenod, M., and Fontana, A. (1991) Murine brain macrophages induced NMDA receptor mediated neurotoxicity *in vitro* by secreting glutamate. *Neurosci. Lett.* **133,** 159–162.

100. Caggiano, A. O. and Kraig, R. P. (1996) Eicosanoids and nitric oxide influence induction of reactive gliosis from spreading depression in microglia but not astrocytes. *J. Comp. Neurol.* **369,** 93–108.

101. Kreutzberg, G. W. (1996) Microglia: a sensor for pathological events in the CNS. *Trends Neurosci.* **19,** 312–318.

102. Oostveen, J. A., Dunn, E., Carter, D. B., and Hall, E. D. (1998) Neuroprotective efficacy and mechanisms of novel pyrrolopyrimidine lipid peroxidation inhibitors in the gerbil forebrain ischemia model. *J. Cereb. Blood Flow Metab.* **18,** 539–547.

103. McNeill, H., Williams, C., Guan, J., Dragunow, M., Lawlor, P., Sirimanne, E., et al. (1994) Neuronal rescue with transforming growth factor-beta 1 after hypoxic-ischaemic brain injury. *Neuroreport* **5,** 901–904.

104. Relton, J. K., Martin, D., Thompson, R. C., and Russell, D. A. (1996) Peripheral administration of interleukin-1 receptor antagonist inhibits brain damage after focal cerebral ischemia in the rat. *Exp. Neurol.* **138,** 206–213.

105. Barone, F. C., Arvin, B., White, R. F., Miller, A., Webb, C. L., Willette, R. N., et al. (1997) Tumor necrosis factor-alpha. A mediator of focal ischemic brain injury. *Stroke* **28,** 1233–1244.

106. Shohami, E., Bass, R., Wallach, D., Yamin, A., and Gallily, R. (1996) Inhibition of tumor necrosis factor alpha (TNFalpha) activity in rat brain is associated with cerebroprotection after closed head injury. *J. Cereb. Blood Flow Metab.* **16,** 378–384.

107. Semmler, J., Wachtel, H., and Endres, S. (1993) The specific type IV phosphodiesterase inhibitor rolipram suppresses tumor necrosis factor-alpha production by human mononuclear cells. *Int. J. Immunopharmacol.* **15,** 409–413.

108. Golub, L. M., Lee, H. M., Ryan, M. E., Giannobile, W. V., Payne, J., and Sorsa, T. (1998) Tetracyclines inhibit connective tissue breakdown by multiple non-antimicrobial mechanisms. *Adv. Dent. Res.* **12,** 12–26.

109. Clark, W. M., Calcagno, F. A., Gabler, W. L., Smith, J. R., and Coull, B. M. (1994) Reduction of central nervous system reperfusion injury in rabbits using doxycycline treatment. *Stroke* **25,** 1411–1415.

110. Clark, W. M., Lessov, N., Lauten, J. D., and Hazel, K. (1997) Doxycycline treatment reduces ischemic brain damage in transient middle cerebral artery occlusion in the rat. *J. Mol. Neurosci.* **9,** 103–108.

111. Yrjänheikki, J., Keinänen, R., Pellikka, M., Hökfelt, T., and Koistinaho, J. Tetracyclines inhibit microglial activation and are neuroprotective in global brain ischemia. *Proc. Natl. Acad. Sci. USA* **95,** 15,769–15,774.

112. Yrjänheikki, J., Tikka, T., Keinänen, R., Goldsteins, G., Chan, P. H., and Koistinaho, J. (1999) A tetracycline derivative, minocycline, reduces inflammation and protects against focal cerebral ischemia with a wide therapeutic window. *Proc. Natl. Acad. Sci. USA* **96,** 13,496–13,500.

113. Chen, M., Ona, V. O., Li, M., Ferrante, R. J., Fink, K. B., Zhu, S., et al. (2000) Minocycline inhibits caspase-1 and caspase-3 expression and delays mortality in a transgenic mouse model of Huntington disease. *Nat. Med.* **6,** 797–801.

114. Tikka, T., Fiebich, B. L., Goldsteins, G., Keinänen, R., and Koistinaho, J. (2001) Minocycline, a tetracycline derivative, is neuroprotective against excitotoxicity by inhibiting activation and proliferation of microglia. *J. Neurosci.* **21,** 2580–2588.

115. Tacconi, M. T. (1998) Neuronal death: is there a role for astrocytes? *Neurochem. Res.* **23,** 759–765.

116. Rothstein, J. D., Dykes-Hoberg, M., Pardo, C. A., Bristol, L. A., Jin, L., Kuncl, R.W., et al. (1996) Knockout of glutamate transporters reveals a major role for astroglial transport in excitotoxicity and clearance of glutamate. *Neuron* **16,** 675–686.

117. Tanaka, K., Watase, K., Manabe, T., Yamada, K., Watanabe, M., Takahashi, K., et al. (1997) Epilepsy and exacerbation of brain injury in mice lacking the glutamate transporter GLT-1. *Science* **276,** 1699–1702.

118. Eddleston, M. and Mucke, L. (1993) Molecular profile of reactive astrocytes—implications for their role in neurologic disease. *Neuroscience* **54,** 15–36.

119. Brenner, M. (1994) Structure and transcriptional regulation of the GFAP gene. *Brain Pathol.* **4,** 245–257.

120. Brenner, M., Kisseberth, W. C., Su, Y., Besnard, F., and Messing, A. (1994) GFAP promoter directs astrocyte-specific expression in transgenic mice. *J. Neurosci.* **14,** 1030–1037.

121. Yamasaki, Y. and Kogure, K. (1997) Cytokines, growth factors, adhesion molecules, and inflammation after ischemia, in *Primer on Cerebrovascular Diseases* (Welch, K. M. A., Caplan, L. R., Reis, D. J., Siesjö, B. K., and Weir, B. eds.), Academic, New York, pp. 265–269.

122. Sporbert, A., Mertsch, K., Smolenski, A., Haseloff, R. F., Schonfelder, G., Paul, M., et al. (1999) Phosphorylation of vasodilator-stimulated phosphoprotein: a consequence of nitric oxide- and cGMP-mediated signal transduction in brain capillary endothelial cells and astrocytes. *Mol. Brain Res.* **67,** 258–266.

123. Endoh, M., Maiese, K., and Wagner, J. (1994) Expression of the inducible form of nitric oxide synthase by reactive astrocytes after transient global ischemia. *Brain Res.* **651,** 92–100.

124. Meldrum, D. R., Dinarello, C. A., Clevelandm J. C. Jr., Cain, B. S., Shames, B. D., Meng, X., et al. (1998) Hydrogen peroxide induces tumor necrosis factor alpha-mediated cardiac injury by a P38 mitogen-activated protein kinase-dependent mechanism. *Surgery* **124,** 291–297.

125. Bertorelli, R., Adami, M., Di, S. E., and Ghezzi, P. (1998) MK-801 and dexamethasone reduce both tumor necrosis factor levels and infarct volume after focal cerebral ischemia in the rat brain. *Neurosci. Lett.* **246,** 41–44.
126. Feuerstein, G., Wang, X., and Barone, F. C. (1998) Cytokines in brain ischemia—the role of TNF alpha. *Cell. Mol. Neurobiol.* **18,** 695–701.
127. Tracey, K. J. and Cerami, A. (1993) Tumor necrosis factor: an updated review of its biology. *Crit. Care Med.* **21,** S415–S422.
128. Tracey, K. J. and Cerami, A. (1993) Tumor necrosis factor, other cytokines and disease. *Annu. Rev. Cell. Biol.* **9,** 317–343.
129. Lavine, S. D., Hofman, F. M., and Zlokovic, B. V. (1998) Circulating antibody against tumor necrosis factor-alpha protects rat brain from reperfusion injury. *J. Cereb. Blood Flow Metab.* **18,** 52–58.
130. Botchkina, G. I., Meistrell, M. E. III, Botchkina, I. L., and Tracey, K. J. (1997) Expression of TNF and TNF receptors (p55 and p75) in the rat brain after focal cerebral ischemia. *Mol. Med.* **3,** 765–781.
131. Saito, K., Suyama, K., Nishida, K., Sei, Y., and Basile, A. S. (1996) Early increases in TNF-alpha, IL-6 and IL-1 beta levels following transient cerebral ischemia in gerbil brain. *Neurosci. Lett.* **206,** 149–152.
132. Wang, X., Yue, T. L., Barone, F. C., White, R. F., Gagnon, R. C., and Feuerstein, G. Z. (1994) Concomitant cortical expression of TNF-alpha and IL-1 beta mRNAs follows early response gene expression in transient focal ischemia. *Mol. Chem. Neuropathol.* **23,** 103–114.
133. Liu, T., Clark, R. K., McDonnell, P. C., Young, P. R., White, R. F., Barone, F. C., et al. (1994) Tumor necrosis factor-alpha expression in ischemic neurons. *Stroke* **25,** 1481–1488.
134. Meistrell, M. E. III, Botchkina, G. I., Wang, H., Di, S. E., Cockroft, K. M., Bloom, O., et al. (1997) Tumor necrosis factor is a brain damaging cytokine in cerebral ischemia. *Shock* **8,** 341–348.
135. Yang, G. Y., Gong, C., Qin, Z., Ye, W., Mao, Y., and Betz, A. L. (1998) Inhibition of TNF-alpha attenuates infarct volume and ICAM-1 expression in ischemic mouse brain. *Neuroreport* **9,** 2131–2134.
136. Nawashiro, H., Martin, D., and Hallenbeck, J. M. (1997) Neuroprotective effects of TNF binding protein in focal cerebral ischemia. *Brain Res.* **778,** 265–271.
137. Nawashiro, H., Martin, D., and Hallenbeck, J. M. (1997) Inhibition of tumor necrosis factor and amelioration of brain infarction in mice. *J. Cereb. Blood Flow Metab.* **17,** 229–232.
138. Cheng, B., Christakos, S., and Mattson, M. P. (1994) Tumor necrosis factors protect neurons against metabolic-excitotoxic insults and promote maintenance of calcium homeostasis. *Neuron* **12,** 139–153.
139. Bruce, A. J., Boling, W., Kindy, M. S., Peschon, J., Kraemer, P. J., Carpenter, M. K., et al. (1996) Altered neuronal and microglial responses to excitotoxic and ischemic brain injury in mice lacking TNF receptors. *Nat. Med.* **2,** 788–794.
140. Liu, T., McDonnell, P. C., Young, P. R., White, R. F., Siren, A. L., Hallenbeck, J. M., et al. (1993) Interleukin-1 beta mRNA expression in ischemic rat cortex. *Stroke* **24,** 1746–1750.
141. Davies, C. A., Loddick, S. A., Toulmond, S., Stroemer, R. P., Hunt, J., and Rothwell, N. J. (1999) The progression and topographic distribution of interleukin-1beta expression after permanent middle cerebral artery occlusion in the rat. *J. Cereb. Blood Flow Metab.* **19,** 87–98.
142. Yamasaki, Y., Matsuura, N., Shozuhara, H., Onodera, H., Itoyama, Y., and Kogure, K. (1995) Interleukin-1 as a pathogenetic mediator of ischemic brain damage in rats. *Stroke* **26,** 676–681.
143. Loddick, S. A. and Rothwell, N. J. (1996) Neuroprotective effects of human recombinant interleukin-1 receptor antagonist in focal cerebral ischaemia in the rat. *J. Cereb. Blood Flow Metab.* **16,** 932–940.

144. Wang, X., Barone, F. C., Aiyar, N. V., and Feuerstein, G. Z. (1997) Interleukin-1 receptor and receptor antagonist gene expression after focal stroke in rats. *Stroke* **28,** 155–162.

145. Garcia, J. H., Liu, K. F., and Relton, J. K. (1995) Interleukin-1 receptor antagonist decreases the number of necrotic neurons in rats with middle cerebral artery occlusion. *Am. J. Pathol.* **147,** 1477–1486.

146. Relton, J. K. and Rothwell, N. J. (1992) Interleukin-1 receptor antagonist inhibits ischaemic and excitotoxic neuronal damage in the rat. *Brain Res. Bull.* **29,** 243–246.

147. Yang, G. Y., Zhao, Y. J., Davidson, B. L., and Betz, A. L. (1997) Overexpression of interleukin-1 receptor antagonist in the mouse brain reduces ischemic brain injury. *Brain Res.* **751,** 181–188.

148. Loddick, S. A., Wong, M. L., Bongiorno, P. B., Gold, P. W., Licinio, J., and Rothwell, N. J. (1997) Endogenous interleukin-1 receptor antagonist is neuroprotective. *Biochem. Biophys. Res. Commun.* **234,** 211–215.

149. Friedlander, R. M., Gagliardini, V., Hara, H., Fink, K. B., Li, W., MacDonald, G., et al. (1997) Expression of a dominant negative mutant of interleukin-1 beta converting enzyme in transgenic mice prevents neuronal cell death induced by trophic factor withdrawal and ischemic brain injury. *J. Exp. Med.* **185,** 933–940.

150. Benveniste, E. N. and Benos, D. J. (1995) TNF-alpha- and IFN-gamma-mediated signal transduction pathways: effects on glial cell gene expression and function. *FASEB J.* **9,** 1577–1584.

151. Giulian, D. and Lachman, L. B. (1985) Interleukin-1 stimulation of astroglial proliferation after brain injury. *Science* **228,** 497–499.

152. Gordon, C. R., Merchant, R. S., Marmarou, A., Rice, C. D., Marsh, J. T., and Young, H. F. (1990) Effect of murine recombinant interleukin-1 on brain oedema in the rat. *Acta Neurochir. (Wien.)* **51(Suppl.),** 268–270.

153. Rothwell, N. J. and Luheshi, G. (1994) Pharmacology of interleukin-1 actions in the brain. *Adv. Pharmacol.* **25,** 1–20.

154. Kishimoto, T. K. and Rothlein, R. (1994) Integrins, ICAMs, and selectins: role and regulation of adhesion molecules in neutrophil recruitment to inflammatory sites. *Adv. Pharmacol.* **25,** 117–169.

155. Montefort, S. and Holgate, S. T. (1991) Adhesion molecules and their role in inflammation. *Respir. Med.* **85,** 91–99.

156. Beutler, B. and Cerami, A. (1987) Cachectin: more than a tumor necrosis factor. *N. Engl. J. Med.* **316,** 379–385.

157. Goldblum, S. E. and Sun, W. L. (1990) Tumor necrosis factor-alpha augments pulmonary arterial transendothelial albumin flux *in vitro. Am. J. Physiol.* **258,** L57–L67.

158. Selmaj, K. and Raine, C. S. (1988) Tumor necrosis factor mediates myelin damage in organotypic cultures of nervous tissue. *Ann. NY Acad. Sci.* **540,** 568–570.

159. Selmaj, K. W. and Raine, C. S. (1988) Tumor necrosis factor mediates myelin and oligodendrocyte damage *in vitro. Ann. Neurol.* **23,** 339–346.

160. Pober, J. S. and Cotran, R. S. (1990) Cytokines and endothelial cell biology. *Physiol. Rev.* **70,** 427–451.

161. Schulze-Osthoff, K., Bakker, A. C., Vanhaesebroeck, B., Beyaert, R., Jacob, W. A., and Fiers, W. (1992) Cytotoxic activity of tumor necrosis factor is mediated by early damage of mitochondrial functions. Evidence for the involvement of mitochondrial radical generation. *J. Biol. Chem.* **267,** 5317–5323.

162. Van Antwerp, D. J., Martin, S. J., Verma, I. M., and Green, D. R. (1998) Inhibition of TNF-induced apoptosis by NF-kappa B. *Trends Cell. Biol.* **8,** 107–111.

163. Yong, V. W., Krekoski, C. A., Forsyth, P. A., Bell, R., and Edwards, D. R. (1998) Matrix metalloproteinases and diseases of the CNS. *Trends Neurosci.* **21,** 75–80.

164. Lukes, A., Mun-Bryce, S., Lukes, M., and Rosenberg, G. A. (1999) Extracellular matrix degradation by metalloproteinases and central nervous system diseases. *Mol. Neurobiol.* **19,** 267–284.

165. Gottschall, P. E. and Yu, X. (1995) Cytokines regulate gelatinase A and B (matrix metallo-proteinase 2 and 9) activity in cultured rat astrocytes. *J. Neurochem.* **64,** 1513–1520.

166. Gottschall, P. E., Yu, X., and Bing, B. (1995) Increased production of gelatinase B (matrix metalloproteinase-9) and interleukin-6 by activated rat microglia in culture. *J. Neurosci. Res.* **42,** 335–342.

167. Sang, Q. X., Jia, M. C., Schwartz, M. A., Jaye, M. C., Kleinman, H. K., Ghaffari, M. A., et al. (2000) New thiol and sulfodiimine metalloproteinase inhibitors and their effect on human microvascular endothelial cell growth. *Biochem. Biophys. Res. Commun.* **274,** 780–786.

168. Vecil, G. G., Larsen, P. H., Corley, S. M., Herx, L. M., Besson, A., Goodyer, C. G., et al. (2000) Interleukin-1 is a key regulator of matrix metalloproteinase-9 expression in human neurons in culture and following mouse brain trauma in vivo. *J. Neurosci. Res.* **61,** 212–224.

169. Robinson, M. J. and Cobb, M. H. (1997) Mitogen-activated protein kinase pathways. *Curr. Opin. Cell. Biol.* **9,** 180–186.

170. Kummer, J. L., Rao, P. K., and Heidenreich, K. A. (1997) Apoptosis induced by with-drawal of trophic factors is mediated by p38 mitogen-activated protein kinase. *J. Biol. Chem.* **272,** 20,490–20,490.

171. Lee, J. C., Laydon, J. T., McDonnell, P. C., Gallagher, T. F., Kumar, S., Green, D., et al. (1994) A protein kinase involved in the regulation of inflammatory cytokine biosynthe-sis. *Nature* **372,** 739–746.

172. Lee, J. C. and Young, P. R. (1996) Role of CSB/p38/RK stress response kinase in LPS and cytokine signaling mechanisms. *J. Leukocyte Biol.* **59,**152–157.

173. Xia, Z., Dickens, M., Raingeaud, J., Davis, R. J., and Greenberg, M. E. (1995) Opposing effects of ERK and JNK-p38 MAP kinases on apoptosis. *Science* **270,** 1326–1331.

174. Alessandrini, A., Namura, S., Moskowitz, M. A., and Bonventre, J. V. (1999) MEK1 pro-tein kinase inhibition protects against damage resulting from focal cerebral ischemia. *Proc. Natl. Acad. Sci. USA* **96,** 12,866–12,869.

175. Walton, K. M., DiRocco, R., Bartlett, B. A., Koury, E., Marcy, V. R., Jarvis, B., et al. (1998) Activation of p38MAPK in microglia after ischemia. *J. Neurochem.* **70,** 1764–1767.

176. Irving, E. A., Barone, F. C, Reith, A. D., Hadingham, S. J., and Parson, A. A. (2000) Differential activation of MAPK/ERK and p38/SAPK in neurones and glia following focal cerebral ischaemia in the rat. *Mol. Brain Res.* **77,** 65–75.

177. Yrjänheikki, J. and Koistinaho, J. (1999) Induction of p38-MAPK in microglia following focal cerebral ischemia in the rat, in *The 1st Biennial Kuopio Symposium of Ischaemic Stroke* (Jolkkonen J, ed.), University of Kuopio, Finland, p. 28.

178. Tian, D., Litvak, V., and Lev, S. (2000) Cerebral ischemia and seizures induce tyrosine phosphorylation of PYK2 in neurons and microglial cells. *J. Neurosci.* **20,** 6478–6487.

179. Badger, A. M., Cook, M. N., Lark, M. W., Newman-Tarr, T. M., Swift, B. A., Nelson, A. H., et al. (1998) SB 203580 inhibits p38 mitogen-activated protein kinase, nitric oxide production, and inducible nitric oxide synthase in bovine cartilage-derived chondrocytes. *J. Immunol.* **161,** 467–473.

180. Bhat, N. R., Zhang, P., Lee, J. C., and Hogan, E. L. (1998) Extracellular signal-regulated kinase and p38 subgroups of mitogen-activated protein kinases regulate inducible nitric oxide synthase and tumor necrosis factor-alpha gene expression in endotoxin-stimulated primary glial cultures. *J. Neurosci.* **18,** 1633–1644.

181. Kristof, A. S., Marks-Konczalik, J., and Moss, J. (2001) Mitogen-activated protein kinases mediate AP-1-dependent human inducible nitric oxide synthase promoter activation. *J. Biol. Chem.* **276,** 8445–8452.

182. Chen, C. C., Sun, Y. T., Chen, J. J., and Chang, Y. J. (2001) Tumor necrosis factor-alpha-induced cyclooxygenase-2 expression via sequential activation of ceramide-dependent mitogen-activated protein kinases, and IkappaB kinase 1/2 in human alveolar epithelial cells. *Mol. Pharmacol.* **59,** 493–500.

183. van den Blink, B., Juffermans, N. P., ten Hove, T., Schultz, M. J., van Deventerm, S. J., van der Poll, T., et al. (2001) p38 mitogen-activated protein kinase inhibition increases cytokine release by macrophages in vitro and during infection in vivo. *J. Immunol.* **166,** 582–387.
184. Mattson, M. P. and Camandola, S. (2001) NF-kappaB in neuronal plasticity and neurodegenerative disorders. *J. Clin. Invest.* **107,** 247–254.
185. Lemberger, T., Desvergne, B., and Wahli, W. (1996) Peroxisome proliferator-activated receptors: a nuclear receptor signaling pathway in lipid physiology. *Annu. Rev. Cell. Dev. Biol.* **12,** 335–363.
186. Ricote, M., Li, A. C., Willson, T. M., Kelly, C. J., and Glass, C. K. (1998) The peroxisome proliferator-activated receptor-gamma is a negative regulator of macrophage activation. *Nature* **391,** 79–82.
187. Barone, F. C., Irving, E. A., Ray, A. M., Lee, J. C., Kassis, S., Kumar, S., et al. (2001) SB 239063, a second-generation p38 mitogen-activated protein kinase inhibitor, reduces brain injury and neurological deficits in cerebral focal ischemia. *J. Pharmacol. Exp. Ther.* **296,** 312–321.
188. Schneider, A., Martin-Villalba, A., Weih, F., Vogel, J., Wirth, T., and Schwaninger, M. (1999) NF-kappaB is activated and promotes cell death in focal cerebral ischemia. *Nat. Med.* **5,** 554–559.
189. Buchan, A. M., Li, H., and Blackburn, B. (2000) Neuroprotection achieved with a novel proteasome inhibitor which blocks NF-kappaB activation. *Neuroreport* **11,** 427–430.
190. Gabriel, C., Justicia, C., Camins, A., and Planas, A. M. (1999) Activation of nuclear factor-kappaB in the rat brain after transient focal ischemia. *Mol. Brain Res.* **65,** 61–69.

10

Neuroinflammation as an Important Pathogenic Mechanism in Spinal Cord Injury

Yuji Taoka and Kenji Okajima

1. INTRODUCTION

Acute traumatic spinal cord injury (SCI) is an unexpected, catastrophic event, the consequences of which often persist for the life of the patient and influence in diverse ways not only the patient but also family members and society at large. Only limited therapeutic measures are currently available for its treatment (1). SCI induced by trauma is a consequence of an initial physical insult followed by a progressive injury process that involves various pathochemical events that lead to tissue destruction (1,2). Although prevention programs have been initiated, there is no evidence that the incidence is declining. Therefore, during the acute phase, therapeutic intervention in SCI should be directed at reducing or alleviating this secondary process. Although the mechanisms involved in the secondary injury process are not fully understood, the activated leukocyte-induced vascular damage leading to neuroinflammation has been postulated to be one of the important pathomechanisms of acute spinal cord injury (3–6).

This chapter will describe the possible mechanisms by which neuroinflammation can be induced in the pathologic process of spinal cord injury and some potential therapeutic measures for the spinal cord injury by alleviating the neuroinflammation.

2. POSTULATED PATHOMECHANISMS OF SPINAL CORD INJURY

2.1. Concept of the Two-Step Mechanism in Acute SCI: the Primary and Secondary Injury

It has been hypothesized that there are two steps in the pathologic process leading to acute traumatic spinal cord injury: the primary mechanical injury and a secondary injury resulting from one or more additional damaging processes initiated by the primary injury (7–9). The concept of secondary injury was first postulated by Allen (7), when he found that myelotomy and removal of the posttraumatic hematomyelia resulted in improvement of neurologic function in dogs subjected to experimental acute spinal cord injury. Allen hypothesized that there was a noxious agent present in the hemorrhagic necrotic material that might cause further damage to the spinal cord and that the injurous agent was a biochemical factor (8). This was the first experimental evidence of posttraumatic autodestruction. Since then, numerous other pathophysiological mechanisms have been postulated to explain the progressive posttraumatic destruction of spinal cord tissue.

From: *Neuroinflammation, 2nd Edition: Mechanisms and Management*
Edited by: P. L. Wood © Humana Press Inc., Totowa, NJ

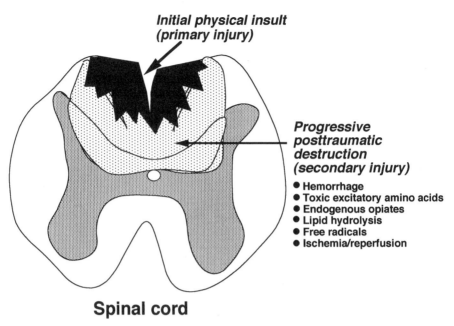

Spinal cord

Fig. 1. Possible mechanisms of spinal cord injury. (Modified from ref. *12*.)

With respect to this point, important reviews are given by Tator and Fehling *(9)*, Simpson et al. *(10)* and Young *(11)*. The possible mechanisms for secondary injury summerized by Zeidman et al. are shown in Fig. 1 *(12)*. Similar theories have been used to explain the progressive loss of neural tissue in other conditions such as head injury, ischemia, and subarachinoidal hemorrhage. Neuroinflammation might play a role in the secondary progressive tissue damage.

2.2. Histopathologic Changes in Spinal Cord Injury

After acute injury, the spinal cord undergoes sequential pathologic changes including hemorrhage, edema, axonal and neuronal necrosis, and demyelination, followed by cyst formation and infarction *(13,14)*. Within 15 min of acute injury, petechial hemorrhages occurred in the gray matter and edema of the white matter. During the first 2 h, the hemorrhages in the gray matter increased, and at 4 h, there were "numerous swollen axis cylinders" *(7)*. Several studies have shown that edema develops at the injury site and spreads into adjacent segments of the cord. The injury site is necrotic, especially the central zone previously occupied by hemorrhage 24–48 h after major trauma *(15)*. Several days later, the hemorrhagic zone shows cavitation and the adjacent areas exhibit patchy necrosis, often with sharply defined margins. These progressive changes, consisting of cavitation and coagulative necrosis at the injury site and in adjacent areas *(14)*, have pathological features of infarcts, this process is termed "posttraumatic infarction" *(16)*.

The amount of spared spinal cord tissue has been shown to be closely related to behavioral recovery after spinal cord injury. Young et al. *(11,17)* have demonstrated in animal models that motor function can recover to normal levels after a spinal cord injury if as few as 4–6% of the cortical motor neurons regain physiologic connection through the injured spinal cord segment to the caudal spinal cord. These experiments support a threshold effect in the return of motor function after the recovery of the surviving neurons in these tracts. An increase in axonal survival at the injury site of from less than 3% to more than 6% allows neurologic function to return through the site and converts paralyzed muscles to muscles caudal to the injury with normal function. Basso et al. *(18)* have also shown that sparing as few as 5–10% of the fibers at the lesion center is sufficient to help drive the segmental circuits involved in the production of basic locomotion observed following spinal cord contusion injury in rats. Thus, these observations suggest that any treatment that can salvage or prevent tissue damage to even a small percentage of spinal axons may dramatically improve functional recovery. However, the role of neuroinflammation in the development of these histopathologic changes are not fully understood.

3. ROLE OF NEUROINFLAMMATION IN THE SECONDARY INJURY PROCESS OF SPINAL CORD INJURY

3.1. Pathologic Events Leading to Tissue Injury in the Inflammatory Process

In general, inflammation can be induced by some insults to tissues. Pro-inflammatory stimuli such as oxygen radicals generated during the ischemia/reperfusion activate monocytes to induce the production of pro-inflammatory cytokines such as tumor necrosis factor-α (TNF-α) and interleukin-1β (IL-1β) *(19)*. These pro-inflammatory cytokines activate neutrophils and endothelial cells to promote the neutrophil adhesion to the endothelial cell *(20)*. Activated neutrophils, in turn, adhere to the endothelial cell by interacting with the endothelial cell leukocyte adhesion molecules such as P-selectin, E-selectin, and intercellular adhesion molecule-1 (ICAM-1) *(19)*. Activated neutrophils adherent to the endothelial cell damage endothelial cells by releasing various inflammatory mediators such as oxygen free radicals and neutrophil elastase *(21)*. Because the intercellular clefts between tightly adherent activated neutrophils and the endothelial cell form a microenvironment protected from circulating antiproteases and antioxidants, the neutrophil–endothelial cell interaction represents a critical aspect of activated neutrophil-induced endothelial cell damage *(22)*. The interaction of activated neutrophils with endothelial cells mediated by ICAM-1 could be a prerequisite for neutrophil migration *(23)*. Extravasated neutrophils might damage tissues by releasing their inflammatory mediators *(21)*. During these process, the activated neutrophil-induced endothelial cell damage results in an increase in the vascular permeability leading to local hemoconcentration and the subsequent microthrombus formation *(24,25)*. Thus, the pathologic changes induced by activated neutrophils might contribute to the tissue injury during the local inflammatory responses in the damaged area of the organ *(19)* (Fig. 2). Local inflammatory responses in the injured segment of the spinal cord (i.e., neuroinflammation) might play an important role in the development of SCI. Detailed mechanisms in the development of neuroinflammation in the secondary injury process of SCI are described next.

3.2. Role of Tumor Necrosis Factor-α

Recently, we demonstrated that the production of TNF-α at the site of SCI is implicated in the secondary damage to tissue in SCI *(26)*. In that study, we found that the level of this protein in the traumatized spinal cord tissue was significantly increased after com-

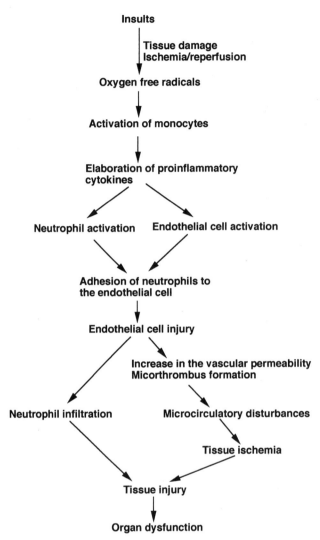

Fig. 2. Pathologic events leading to organ dysfunction in the inflammatory process.

pression trauma, with a peak in levels seen after 4 h. These results are consistent with other studies. Wang et al. *(27)* showed the presence of TNF-α at the sites of traumatic spinal cord lesions, but did not detect this factor in cerebrospinal fluid or in serum. In addition, Yakovlev and Faden *(28)* demonstrated that spinal cord impact in rats caused an elevation of TNF-α mRNA levels at the site of trauma 30 min after the injury, with the level of the TNF-α message proportional to the severity of the injury. Bartholdi and Schwab *(29)* have shown that using *in situ* hybridization, the expression of TNF-α mRNA can be detected shortly after spinal cord damage, but is downregulated after 6 h. However, these observations could not elucidate the causal relationship between TNF-α production and SCI. Using our rat model of SCI, we measured the TNF-α mRNA expression in traumatized segments over time, as determined by reverse transcription–polymerase chain reaction (RT-PCR) *(30)*. An increase in the expression of TNF-α mRNA was found

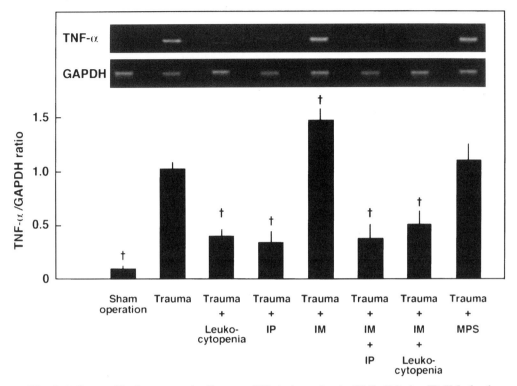

Fig. 3. Effects of leukocytopenia, iloprost (IP), indomethacin (IM), IM plus IP, IM plus leukocytopenia and methyl prednisolone (MPS) on the expression of TNF-α mRNA in traumatized spinal cord. The expressions of TNF-α mRNA were measured 1 h after compression trauma or sham operation. Data are expressed as mean ± SD of five experiments. $p < .01$ vs trauma.

in traumatized spinal cord, which peaked 1 h after trauma *(31)*. Tissue levels of TNF-α mRNA 1, 2, 3, and 4 h posttrauma were significantly increased in the animals subjected to compression trauma compared with sham-operated animals *(31)*. Although the precise mechanisms by which the transcriptional process of TNF-α synthesis is enhanced are not fully known, ischemia–reperfusion mechanisms may be involved *(32,33)*. Reactive oxygen intermediates produced after ischemia–reperfusion have been shown to activate nuclear regulatory factor kappa B (NF-κB), which leads to an increase in TNF-α production by enhancing the transcription of TNF-α mRNA in monocytes *(34)*. In the pathologic process of ischemia–reperfusion, hypoxia followed by reoxygenation induces monocytes to produce interleukin-1β (IL-1β) *(34)*. IL-1β induces monocytes to produce TNF-α by activating NF-κB *(34)*.

Tumor necrosis factor-α contributes to the leukocyte-induced endothelial cell damage not only by activating neutrophils *(20)* but also by increasing the expression of endothelial leukocyte adhesion molecules, such as E-selectin, causing activated neutrophils to adhere to the endothelial cell surface *(35)*. TNF-α could directly cause endothelial cell damage even in the absence of neutrophils *(36)*. We have also found that both the increase in TNF-α mRNA expression and the subsequent increases in tissue levels of TNF-α in the injured segments of spinal cord were significantly reduced in leukocytopenic animals (Fig. 3); the level of motor disturbances were also markedly reduced

(31). These findings indicate that an increase in the level of TNF-α at the site of injury may contribute to the secondary injury process by activation of neutrophils, leading to endothelial damage.

3.3. Role of Neutrophils

The presence of neutrophils in ischemic brain and spinal tissue has been considered to represent a pathophysiological response to existing injury. Because neutrophils were observed mainly in areas of hemorrhagic necrosis after acute SCI, these neutrophilic responses have been recognized as phagocytes of red blood cells and necrotic tissue. Recent evidence, however, suggests that neutrophils may also be directly involved in the pathogenesis and extension of SCI in rats *(5,37,38)*. Neutrophils have been shown to release neutrophil proteases and reactive oxygen species *(21)*. Reactive oxygen species increase the expression of endothelial leukocyte adhesion molecules by which activated neutrophils interact to damage endothelial cells *(19)*.

3.3.1. Neutrophil Elastase

We recently reported that L-658 758, a specific neutrophil elastase inhibitor *(39)*, prevents compression trauma-induced SCI in our animal model *(40)*. In that study, L-658 758 attenuated motor disturbances in rats after compression trauma to the spinal cord. This neutrophil elastase inhibitor also inhibited both neutrophil accumulation and intramedullary hemorrhages in the injured spinal cord segment. These findings implicate neutrophil elastase in compression trauma-induced SCI.

Neutrophil elastase is an enzyme capable of damaging endothelial cells *(21)*. Neutrophil elastase increases vascular permeability and damages endothelial integrity *(19,39)*. Because intramedullary hemorrhages were markedly reduced by the neutrophil elastase inhibitor, hemorrhage may be a consequence of neutrophil elastase-induced endothelial damage. The endothelial damage also may induce microcirculatory disturbances, leading to tissue ischemia, an important cause of motor disturbances. Neutrophil elastase influences neutrophil infiltration by increasing the expression of the CD18 molecule on the cell surface, by which neutrophils can adhere to endothelial cells *(44)*. This sequence explains how neutrophil elastase inhibitors prevent neutrophil accumulation in the damaged spinal cord segment. Inhibition of the adhesion of neutrophils to endothelial cells by neutrophil elastase inhibitor also might contribute to prevent endothelial cell damage that would have been induced by inflammatory mediators released from the activated neutrophils.

Neutrophil elastase inhibits production of prostacyclin by cultured porcine aortic endothelial cells in response to extracelluar ATP *(42)*. Neutrophil elastase also suppresses the release of thrombin-induced prostacyclin by cultured human umblical vein endothelial cells *(43)*. We recently demonstrated that neutrophil elastase inhibits the endothelial production of prostacyclin in rats subjected to stress *(44)*. Because iloprost, a stable derivative of prostacyclin, reduces the severity of the motor disturbances after compression trauma-induced SCI *(45)*, a decrease in the level of prostacyclin in the injured segment of the spinal cord induced by neutrophil elastase might be a cause of motor disturbances observed following SCI. Indeed, our preliminary study indicated that L-658 758 significantly attenuated the decrease in prostacyclin levels at the injured site of the spinal cord (unpublished data).

3.3.2. Reactive Oxygen Species

Activated neutrophils have also been shown to release reactive oxygen species that are capable of damaging endothelial cells *(21,22)*. Reactive oxygen species (ROS) have been reported to be involved in the progressive injury process in traumatic SCI *(46)*. Nakauchi et al. *(47)* demonstrated that lecithinized superoxide dismutase (SOD), a long-acting SOD, markedly attenuated the motor function following compression trauma to the spinal cord in rats and significantly suppressed MPO activity in the tissue and spinal cord edema. We have previously shown that SM-SOD, another long-acting SOD, improved mortality and significantly attenuated orthostatic hypotension in rats subjected to compression trauma-induced SCI *(48)*. Thus, it is probable that ROS may also contribute to the progressive injury process of SCI. Because ROS and neutrophil elastase have been shown to synergistically damage endothelial cell integrity *(49,50)*, both may be significantly involved in the pathological process leading to SCI.

Lipid peroxidation has been shown to contribute to SCI in rats *(31)*. Reactive oxygen species can be generated either by the conversion of xanthine dehydrogenase (XDH) to xanthine oxidase (XO) or by activated leukocytes *(21,51)*. Endothelial cells are a major site of XDH conversion in free-radical-mediated injury *(51)*. Thus, it is possible that in traumatic SCI, conversion of XDH to XO in endothelial cells at the site of injury may be an important mechanism leading to increased ROS generation, which can cause lipid peroxidation. This possibility, however, appears to be less in SCI. Xu et al. *(52)* demonstrated that, in rats, allopurinol, an inhibitor of XO, does not affect posttraumatic edema or MPO activity in traumatized spinal cord. Consistent with this findings are our preliminary observations that allopurinol did not inhibit MPO activity and lipid peroxidation in our animal model of SCI. Recently, we demonstrated that leukocytopenia and anti-P-selectin monoclonal antibody almost completely inhibited lipid peroxidation *(53)*. These findings suggested that ROS produced by activated leukocytes may play a more important role in lipid peroxidation at the site of injury than those derived from XO. They also suggest that in the injured section of the spinal cord, lipid peroxidation could be induced in the endothelial cell membrane by activated leukocytes.

3.4. Role of Endothelial Leukocyte Adhesion Molecules

As described earlier, leukocytes release cytokines and other inflammatory mediators, including neutrophil proteases and ROS *(21)*. These inflammatory mediators increase the expression of endothelial leukocyte adhesion molecules, which play an important role in the activated neutrophil-induced endothelial cell damage *(19)*. Because the intercellular clefts between tightly adherent activated neutrophils and the endothelium form a microenvironment protected from circulating antiproteases and antioxidants, the neutrophil–endothelial cell interaction is an important aspect of activated neutrophil–induced endothelial cell damage *(22)*. P-Selectin, a member of the endothelial leukocyte adhesion molecule family, is rapidly expressed on the endothelial cell surface in response to stimuli such as thrombin, histamine, and oxygen free radicals *(22)*. P-Selectin mediates early neutrophil–endothelial cell interactions that lead to the rolling of neutrophils on the endothelium and facilitates subsequent cellular adhesion, migration, and tissue injury. We recently demonstrated that a P-selectin-mediated interaction between activated neutrophils and endothelial cells may be a critical step in endothelial cell damage, leading to spinal cord injury *(38,40,54)*. In these studies, administration of an anti-P-selectin monoclonal antibody significantly reduced the motor disturbances observed following spinal cord compression in rats. Histological examination revealed that intramedullary hemorrhages observed 24 h after compression at the 12th thoracic vertebra of the spinal cord were significantly attenuated in animals administered the anti-P-selectin monoclonal antibody, suggesting that activated neutrophil-induced endothelial cell damage may play a role in these hemorrhages. The accumulation of neutrophils at the site of compression, as evaluated by measuring tissue MPO activity, significantly increased with time following the compression, peaking 3 h postcompression. Spinal cord MPO activity did not increase in sham-operated animals. Administration of the anti-P-selectin monoclonal antibody reduced the accumulation of neutrophils in the damaged spinal cord segment 3 h posttrauma. These observations strongly suggest that the interaction between activated neutrophils and endothelial cells may be a critical step in endothelial cell injury leading to SCI. This hypothesis is consistent with an observation by Hamada et al. *(37)* that intercellular adhesion molecule-1 (ICAM-1) mRNA expression correlated with the severity of the injury, and that posttraumatic administration of anti-ICAM-1 monoclonal antibody significantly reduced motor disturbances and MPO activity following compression trauma to the spinal cord in rats.

In contrast to the observations reported here, Holtz et al. *(55)* demonstrated that neutrophil depletion by antineutrophil serum (ANS) does not improve the outcome of compression trauma-induced neurologic deficits in rats. In that study, although the number of neutrophils was reduced to only 2% of the pre-ANS level, the number of mononuclear cells was reduced to 60% that of control animals. In our studies, the number of mononuclear cells was reduced to 18.6% of that of control rats by treatment with nitrogen mustard *(54)*. Monocytes have been shown to release cytokines upon their activation by platelet-activating factor when they are tethered to P-selectin *(56)*. Because these cytokines increase the vascular permeability *(57,58)*, which may, in turn, result in a microcirculatory disturbance leading to tissue ischemia, the failure by ANS to reduce the number of monocytes might explain why treatment did not attenuate spinal cord injury. In addition, it is possible that the anti-P-selectin monoclonal antibody might inhibit the endothelial adhesion of monocytes as well as neutrophils, thereby preventing endothelial damage induced by cytokines. Thus, together, monocytes and cytokines may play a more important role than neutrophils alone in SCI induced by compression trauma.

4. THERAPEUTIC INTERVENTION DIRECTED AT ATTENUATION OF NEUROINFLAMMATION BY INHIBITING LEUKOCYTE ACTIVATION

We have recently reported that some therapeutic agents such as gabexate mesilate, activated protein C, and recombinant human soluble thrombomodulin (r-TM), which inhibit leukocyte activation, attenuate the motor disturbances induced by compression trauma of the spinal cord in rats *(26,59,60)*. These agents could possibly be used as new pharmacological treatments for patients with acute traumatic SCI, because methylprednisolone (MPS) and monosialtetrahexosyl ganglioside (GM1), which are the only two agents currently being used in SCI clinically, do not inhibit leukocyte activation *(61,62)*.

4.1. Gabexate Mesilate

Gabexate mesilate, a synthetic protease inhibitor, inhibits various serine proteases generated during the coagulation cascade and the inflammatory process *(63)*. Based on these inhibitory properties, gabexate mesilate has been used for treating patients with disseminated intravascular coagulation resulting from sepsis and acute pancreatitis *(64,65)*. A previous study from our laboratory demonstrated that gabexate mesilate markedly suppressed TNF-α production in lipopolysaccharide (LPS)-stimulated human monocytes in a dose-dependent fashion in vitro and also significantly inhibited the LPS-induced increase in TNF-α concentration in vivo *(66)*. Thus, it is possible that gabexate mesilate may be effective in posttraumatic SCI by inhibiting leukocyte activation.

Recently, we investigated the effect of gabexate mesilate on traumatic compression trauma-induced SCI in rats and we found that posttraumatic SCI was significantly attenuated by gabexate mesilate *(59)*. In this study, the motor disturbances observed following traumatic spinal cord compression and the accumulation of leukocytes in the injured tissue, evaluated by measuring tissue MPO activity, were markedly reduced by leukocyte depletion induced by nitrogen mustard and by pretreatment (10 mg/kg) or posttreatment (20 mg/kg) of animals with gabexate mesilate. Histological examination revealed that intramedullary hemorrhages observed 24 h after trauma at the injury site were significantly reduced by nitrogen-mustard-induced leukocytopenia and the administration of gabexate mesilate. Because neither the inactive derivative of activated factor X (DEGR-Xa), a selective inhibitor of thrombin generation *(67)*, nor heparin prevented these pathologic changes, gabexate mesilate appeared to prevent posttraumatic SCI, not by its anticoagulant properties but by its inhibition of leukocyte accumulation at the site of injury *(59)*. We previously demonstrated that gabexate mesilate prevents both endotoxin-induced pulmonary vascular injury *(66)* and ischemia–reperfusion-induced hepatic injury *(68)* by inhibiting TNF-α production by monocytes. In that study, gabexate mesilate in concentrations of $(0.5–1.0) \times 10^{-3}$ *M* inhibited the release of neutrophil elastase and superoxide radicals in vitro. The plasma concentrations of gabexate mesilate in our animal model were estimated to be below 3.0×10^{-4} *M* and 6.0×10^{-4} *M* in animals given 10 mg/kg and 20 mg/kg of gabexate mesilate, respectively, assuming that intraperitoneally administered gabexate mesilate is absorbed immediately into the circulation. In contrast, gabexate mesilate significantly inhibited TNF-α production by endotoxin-stimulated monocytes in vitro at a concentration of 2.0×10^{-8} *M (66)*. Thus, gabexate mesilate most likely prevented posttraumatic SCI by inhibiting production of TNF-α, which plays an important role in activated leukocyte-induced tissue injury *(36,57)* (Fig. 3).

Gabexate mesilate is a safe drug that is already used in the clinical setting and has few known side effects, even at relatively high dosages *(69)*. Posttreatment of rats with this agent as well as the pretreatment prevented posttraumatic SCI in our animal model, suggesting that gabexate mesilate may be useful in treating patients with traumatic SCI in the clinical setting.

4.2. Activated Protein C and Thrombomodulin

Activated protein C (APC), a serine protease, is an important natural anticoagulant that is generated from protein C by the action of the thrombin–thrombomodulin complex on the endothelial cells *(70)*. APC inactivates factors Va and VIIIa, thereby regulating the coagulation system *(70,71)*. APC is also involved in the regulation of the inflammatory process by its inhibition of TNF-α production by monocytes *(72)*. We previously demonstrated that APC prevents activated leukocyte-induced endothelial cell injury primarily by inhibiting the ability of monocytes to produce TNF-α *(73,74)*. Because TNF-α is a potent activator of neutrophils *(20)*, it is possible that APC may also prevent the secondary injury of trauma-induced SCI by inhibiting neutrophil activation. We, therefore, have evaluated the effects of APC in a rat model of compression trauma-induced SCI.

We found that APC significantly reduced the effects of SCI in rats *(26)*. When administered before or after the induction of SCI, APC reduced the number of intramedullary hemorrhages as well as the severity of motor disturbances. Although the motor disturbances in control animals evaluated using the inclined plane test were not completely recovered 56 d posttrauma, those in animals administered APC before or after the trauma were recovered 35 d posttrauma, suggesting that APC might promote the functional recovery of the motor disturbances in this animal model of SCI. Because DEGR-Xa, a selective inhibitor of thrombin generation, did not have any effect, those findings with APC suggest that the efficacy of this protein was not mediated by inhibition of thrombin generation. The accumulation of neutrophils in traumatized segments of the spinal cord as reflected by tissue MPO activity was also significantly inhibited in animals treated with APC. Because no increase in the level of TNF-α induced by SCI was found in the animals that had received APC in that study, APC may inhibit the accumulation of neutrophils at the site of traumatic SCI primarily by inhibiting TNF-α production (Fig. 3). This notion is supported by our earlier finding that APC also inhibits the in vivo and in vitro production of TNF-α by monocytes *(73,74)*. More recently, we demonstrated that recombinant human soluble thrombomodulin (r-TM) prevents compression trauma-induced SCI by inhibiting leukocyte accumulation by reducing the expression of TNF-α m-RNA at the site of injury and such therapeutic effects of r-TM could be dependent on its protein C activation capacity *(60)*. Because administration of APC and r-TM after injury was as effective as its administration prior to injury in preventing the secondary injury of SCI, these agents may have a potential for clinical use in alleviating the effects of traumatic compression injury to the spinal cord *(26,60)* (Fig. 3). APC also reduced the ischemia–reperfusion-induced spinal cord injury, a pathologic event sometimes seen in patients undergoing surgery to repair aortic aneurysms *(75)*. The therapeutic mechanisms also involved the inhibition of leukocyte activation induced by ischemia–reperfusion *(75)*.

As described earlier, although gabexate mesilate, APC, and r-TM are well known as anticoagulant agents, they may attenuate motor dysfunction following compression trauma to the spinal cord in rats by inhibiting leukocyte activation. However, these anticoagulant activities may provide additional benefits for patients with SCI. Thromboembolism is a major cause of morbidity and mortality in patients with SCI *(76,77)*. Thromboembolism has been reported to occur in 70–100% of patients who have complete motor paralysis after SCI *(78,79)*. Thus, gabexate mesilate, APC, or r-TM therapy may be beneficial to patients with SCI, not only by improving the motor recovery but also by preventing thromboembolism through its anticoagulant potency.

4.3. Prostacyclin

Prostacyclin is a powerful vasodilator and a potent inhibitor of platelet aggregation *(80)*. Recently, prostacyclin and its analogs have been shown to inhibit neutrophil activation and the production of cytokines by monocytes *(81,82)*. Thus, it is possible that prostacyclin may be effective in the treatment of posttraumatic SCI in which activated leukocytes appear to play an important role. Therefore, we investigated the effect of iloprost, a stable analog of prostacyclin *(81)* on compression trauma-induced SCI in rats *(45)*. In that study, the motor disturbances observed following compression trauma of the spinal cord in rats were significantly attenuated in animals that received iloprost. Histologic examination of traumatized spinal cord segments further demonstrated that the number of intramedullary hemorrhages was significantly reduced in animals that received iloprost compared with control animals 24 h posttrauma. The accumulation of leukocytes in traumatized segments of the spinal cord 3 h posttrauma was significantly inhibited in animals that received iloprost. In that study, mean arterial blood pressure fell from 106.8 ± 10.5 to 91.2 ± 2.0 mm Hg after iloprost infusion in animals prior to trauma. However, no significant changes in physiological parameters such as blood pressure or body temperature were found in the postinjury period between control animals and animals administered iloprost. These physiological parameters were not significantly different between animals with and without leukocytopenia. These observations suggest that iloprost may attenuate the compression trauma-induced SCI by inhibition of the accumulation of leukocytes at the site of spinal cord damage. Consistent with this notion is the report by Maria-Riva et al. *(83)*, who demonstrated that iloprost attenuates neutrophil-induced lung injury not by its vasodilatory effect but by inhibiting leukocyte activation. Simpson et al. *(84)* have shown that iloprost (0.1–100 μ*M*) inhibited the in vitro production of the superoxide anion by canine neutrophils in a concentration-dependent manner. Okumura et al. *(85)* also demonstrated that beraprost, another analog of prostacyclin, inhibited in a dose-dependent manner formyl-methionyl-leucyl- phenyl-alanine-induced superoxide generation in human neutrophils. Thus, iloprost may attenuate the SCI in this animal model not by its vasodilatory effect but by its inhibitory effect on leukocyte activation *(45)*.

Because prostacyclin inhibits the production of TNF-α *(86)*, it is possible that maintaining a proper level of prostacyclin at the injury site may attenuate the severity of SCI in rats by reducing TNF-α production. This hypothesis is supported by the following observations: (1) Indomethacin (IM), a potent inhibitor of prostacyclin synthesis, itself enhanced both the increases of the TNF-α production and mRNA TNF-α expression in the injured spinal cord tissue induced by compressive trauma (Fig. 3) and (2) iloprost inhibited these increases, and both iloprost and leukocytopenia also inhibited the IM-induced enhancements (Fig. 3). These observations strongly suggest that maintaining the prostacyclin level at the injury site plays an important role in preventing trauma-induced activation of leukocytes in SCI through the inhibition of TNF-α production. Indeed, our preliminary experiments indicated that gabexate mesilate, APC, and r-TM, which reduced the motor disturbances induced by SCI, prevented the decreases of prostacyclin in injured spinal cord tissue. This hypothesis is consistent with our previous findings demonstrating that iloprost prevented the stress-induced gastric mucosal lesion formation in rats by inhibiting the decrease in the prostacyclin level at the injury site *(44)*.

As discussed earlier, prostacyclin may have a major role in the protection of motor disturbances observed following spinal cord trauma by inhibiting the activation of leukocytes. Because the chemical instability of prostacyclin in solution or in biologic fluids makes it a less than ideal agent for SCI *(80)*, iloprost, which is a stable analog of prostacyclin, may be a more useful agent for such application. However, the clinical dosage toler-ance of iloprost does not exceed 2 ng/kg/min because of side effects such as facial flushing, headache, and gastrointestinal distress *(86)*. Thus, pharmacological agents such as gabexate mesilate, APC and r-TM that inhibits the decrease in endogenous prostacy-clin production at the site of injury through inhibiting leukocyte activation may be more effective and safer (Fig. 4).

5. EFFECTS OF STEROID AND GM1 GANGLIOSIDE ON NEUROINFLAMMATION OF THE SPINAL CORD

5.1. Methylprednisolone

Until recently, no treatment was available to improve neurologic recovery in patients with traumatic SCI. Recently, a randomized controlled study in patients with SCI showed that high-dose intravenous administration of MPS (total dose, 154.2 mg/kg/24 h) to parti-ally improve neurologic function when therapy began within 8 h of the injury and con-tinued for 24 h, representing considerable promise for pharmacologic treatment of acute SCI *(1)*. Thus, at present, MPS is the only therapeutic agent for traumatic SCI in humans during the acute phase. However, little is known about the precise mechanisms of action of MPS in SCI. MPS, like other glucocorticoids, has an anti-inflammatory effect. Acute inflammation is a complex hormonal and cellular response that includes neutrophil infiltration, deposition of platelets, and alteration of endothelial cell function, culmi-nating in increased vascular permeability and edema formation. Eisosanoids, free radi-cals, kinins, proteolytic enzymes, and other inflammatory mediators are implicated in the activation of inflammatory processes *(87)*. Although the exact mechanisms of action of MPS in SCI remain to be defined, an antioxidant effect of high-dose glucocorticoids has been proposed *(1,88)*. Glucocorticoids, including MPS, are potent anti-inflammatory agents. The anti-inflammatory action of glucocorticoids has been studied extensively. Their findings indicated that glucocorticoids suppress inflammation via the inhibition of leukocyte functions, including chemotaxis *(89)*, phagocytosis *(90)*, synthesis of inflam-matory mediators, and release of lysosomal enzymes *(91)*. The anti-inflammatory effects may be, at least partly, the result of a stimulating effect on the synthesis of such anti-inflammatory polypeptides as lipocortin, vasocortin, and angiotensin-converting enzymes *(92)*, with resultant inhibition of phospholipase A2, which catalyzes the release of ara-chidonic acid from membrane phospholipids and the subsequent formation of eicosa-noids and free radicals *(93)*.

Numerous dosing schedules of MPS and dexamethasone have been investigated in SCI models, with the majority reporting benefits *(94)*. MPS is one of a class of synthe-tic steroids with a glucocorticoid potency greater than cortisone but considerably less than dexamethasone. 21-Aminosteroids, which inhibit iron-dependent lipid peroxida-tion but do not possess glucocorticoid action *(88)*, were reported to provide benefit in SCI models *(95)*. Improvement in the patient with SCI occurs after a bolus dose of 30 mg/ kg (iv), and this is 1000-fold the amount necessary to activate glucocorticosteroid recep-tors in the body *(96)*. Hence, MPS may have a direct chemical action unrelated to its hormonal action. Demopoulos et al. *(46)* theorized that steroids play an important role in stabilizing membranes by inhibiting free-radical reactions induced by trauma. Braughler and Hall *(97)* found that a large dose of MPS (30 mg/kg, iv) reduced lipid peroxidation,

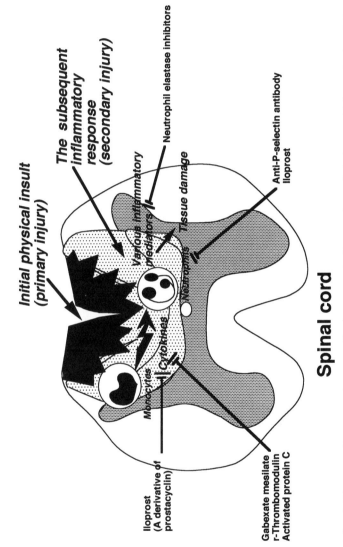

Fig. 4. Therapeutic agents that can alleviate neuroinflammation might be useful for reducing SCI.

227

protected membrane-bound enzymes, such as ATPase, and intracellular molecular assemblies, such as neurofilaments, and reversed the biologically hazardous increase in lactic acid after SCI. Recently, we also reported that a large dose of MPS significantly reduced the motor disturbances and increased lipid peroxidation in rat SCI *(53)*. Thus, presently, the most likely explanation for the protective effect of MPS in spinal cord injury is that MPS suppresses membrane breakdown by inhibiting lipid peroxidation and hydrolysis at the site of injury. The doses required for therapeutic results in clinical use (154.2 mg/kg/24 h) are similar to those shown to be most effective in inhibiting lipid peroxidation and the breakdown of neurofilaments in animal models *(98)*. This breakdown of membrane peaks within 8 h of injury *(97)*. These findings are consistent with the clinical observations of Bracken et al. *(1)* that patients treated with MPS more than eight 8 h after the injury did not differ in their neurologic outcomes from those administered placebo. A secondary effect of the inhibition of lipid peroxidation is that vasoreactive byproducts of arachidonic acid metabolism are reduced, which improves the flow of blood at the injury site *(99)*.

A higher dose of MPS (total dose, 889 mg/kg/24 h) did not produce a significant decrease in neutrophil accumulation, TNF-α, and TNF-α mRNA in our rat model of SCI *(53)*. This observation is consistent with the findings of Xu et al. *(62)*. They showed that high-dose MPS (total dose, 889 mg/kg/24 h) significantly reduced vascular permeability and tissue edema but did not decrease the increase in MPO activity in injured spinal cord tissue observed following traumatic spinal cord injury in rats. Hall et al. *(100)* showed that posttraumatic increase in leukotrien B4, which is a major eicosanoid synthesized by neutrophils *(101)*, at the injury site, was significantly enhanced by MPS in SCI models in cats. Indeed, Katori et al. *(102,103)* demonstrated that dexamethasone with a glucocorticoid potency greater than MPS did not inhibit the adhesion of neutrophils on endothelial cells and only inhibited the penetration of the pericyte basement membrane. Thus, the protective effects of MPS on SCI may involve factors other than reduction of neutrophil-mediated endothelial cell injury *(53)*.

5.2. GM1 Ganglioside

Geisler et al. *(61)* initially examined the recovery of neurologic function after SCI in humans with a prospective, randomized, placebo-controlled, double-blinded trial of GM1 (the Maryland GM1 clinical trial). This was a pilot study testing whether the addition of GM1 ganglioside to the initial medical and surgical care of patients with acute SCI would, in any way, alter neurologic recovery. Thirty-four patients completed the test-drug protocol (100 mg/d of GM1 sodium salt or placebo intravenously for 18–32 doses, with the first dose taken within 72 h of injury) and a 1-yr follow-up period. Neurologic recovery was assessed using the Frankel scale *(104)* and the American Spinal Injury Association *(105)* motor score. The GM1-treated patients had significantly better improvement on both Frankel grade and ASIA motor score from baseline to the 1-yr follow-up than did the placebo-treated patients. Analysis of individual muscle recoveries revealed that the increased recovery in the GM1 group was attributable to initially paralyzed muscles regaining motor strength rather than to the strengthening of weakened muscles. The study provides evidence that GM1 enhances the recovery of neurologic function after SCI. Further analysis of the recovery of motor function for each of the 10 neurologic levels assessed in the study was noted in a subsequent study *(106,107)*. This provided additional evidence that the largest enhanced recovery of motor function in the GM1 treatment group occurred in the muscles of the lower extremities.

Gangliosides are complex acidic glycolipids present in high concentrations in cells of the mammalian central nervous system (CNS). They are a major component of the cell membrane and are located predominantly in the outer leaflet of the cell membrane's phospholipid bilayer *(108)*. Although the natural functions of the gangliosides remain largely unknown, possible biochemical mechanisms by which GM1 ganglioside may enhance neuronal survival in the white-matter tracts or increase the response of the distal spinal cord to the attenuated signal in the damaged white-matter tracts after a SCI and other CNS insults have been covered in recent review articles *(109)*. Beneficial effects of GM1 reported in CNS injury are more likely to be related to the effects of long-term neural recovery than to acute injury processes, that is, the prevention of anterograde and retrograde neuronal degeneration and the stimulation of neuronal repair produced by GM1 ganglioside *(110)* mechanistically implies that it must act much later in the process that follows acute traumatic SCI *(111)*. It remains to be shown whether GM1 ganglioside inhibits leukocyte activation.

6. CONCLUSIONS

In addition to the immediate damage to the spinal cord following injury, a delayed progressive destructive process that follows causes even more spinal cord damage. During the last two decades, considerable research has been done to determine the pathology and pathophysiology of SCI. Based on this knowledge, new modalities for treatment of acute SCI have been developed. In this review, we have briefly outlined a new concept: that leukocytes are critically involved in the progressive destructive process of the spinal cord after primary mechanical injury in the rat model of SCI. Thus, neuroinflammation possibly plays a role in the secondary injury process. Some potential therapeutic agents such as gabexate mesilate, APC, and r-TM that inhibit leukocyte activation significantly attenuate the motor disturbances induced by compression trauma of the spinal cord in rats. The efficacy of these agents should further be examined to clarify the therapeutic window (i.e., the time period after injury during which the therapy can be given), as well as the dosage and duration of treatment. Furthermore, combination of these agents with MPS or GM1 may provide greater enhancement of motor recovery for patients with acute spinal cord trauma, because the pharmacological mechanisms of these agents are apparently different from those of MPS and GM1. New therapeutic strategies pointing to alleviation of neuroinflammation by some new therapeutic agents will benefit victims of SCI by improving their quality of life through reducing their physical disabilities.

REFERENCES

1. Bracken, M. B., Shepard, M. J., Collins, W. F., Holord, T. R., Young, W., Baskin, D. S., et al. (1990) A randomized, controlled trial of methylprednisolone or naloxone in the treatment of acute spinal-cord injury. *N. Engl. J. Med.* **322,** 1405–1411.
2. Young, W. (1988) Secondary CNS injury. *J. Neurotrauma* **5,** 219–221.
3. Demopoulos, H. B., Yoder, M., Gutman, E. G., Seligman, M. L., Flamm, E. S., and Ransohoff, J. (1978) The fine structure of endothelial surfaces in the microcirculation of experimentally injured feline spinal cords. *Scan. Electron Microsc.* **2,** 677–680.

4. Means, E. D. and Anderson, D. (1983) Neuronophagia by leukocytes in experimental spinal cord injury. *J. Neuropathol. Exp. Neurol.* **42,** 707–719.

5. Xu, J., Hsu, C. Y., Liu, T. H., Hogan, E. L., Perot, E., and Tai, H. (1990) Leukotriene B4 release and polymorphonuclear cell infiltration in spinal cord injury. *J. Neurochem.* **55,** 907–912.

6. Blight, A. R. (1985) Delayed demyelination and macrophage invasion: a candidate for secondary cell damage in spinal cord injury. *Central Nervous Syst. Trauma* **2,** 299–315.

7. Allen, A. R. (1911) Surgery of experimental lesions of spinal cord equivalent to crush injury of fracture dislocation of spinal column. A preliminary report. *JAMA* **57,** 878–880.

8. Allen, A. R. (1914) Remarks on the histopathological changes in the spinal cord due to impact. An experimental study. *J. Nerv. Ment. Dis.* **31,** 141–147.

9. Tator, C. H. and Fehlings, M. G. (1991) Review of the secondary injury theory of acute spinal cord trauma with emphasis on vascular mechanisms. *J. Neurosurg.* **75,** 15–26.

10. Simpson, R. K., Hsu, C. Y., and Dimitrijevic, M. R. (1991) The experimental basis for early pharmacological intervention in spinal cord injury. *Paraplegia* **29,** 364–372.

11. Young, W. (1993) Secondary injury mechanisms in acute spinal cord injury. *J. Emerg. Med.* **11,** 13–22.

12. Zeidman, S. M., Ling, G. S., Ducker, T. B., and Ellenbogen, R. G. (1996) Clinical applications of pharmacologic therapies for spinal cord injury. *J. Spinal Disord.* **9,** 367–380.

13. Blight, A. R. (1983) Axonal physiology of chronic spinal cord injury in the cat: intracellular recording in vitro. *Neuroscience* **10,** 1471–1486.

14. Wallace, M. C., Tator, C. H., and Lewis, A. J. (1987) Chronic regenerative changes in the spinal cord after cord compression injury in rats. *Surg. Neurol.* **27,** 209–219.

15. Tator, C. H. and Rowed, D. W. (1979) Current concepts in the immediate management of acute spinal cord injuries. *Can. Med. Assoc. J.* **121,** 1453–1464.

16. Sandler, A. N. and Tator, C. H. (1976) Review of the effect of spinal cord trauma on the vessels and blood flow in the spinal cord. *J. Neurosurg.* **45,** 638–646.

17. Blight, A. R. and Young, W. (1989) Central axons in injured cat spinal cord recover electrophysiological function following remylination by Schwann cells. *J. Neurol. Sci.* **91,** 15–34.

18. Basso, D. M., Beattie, M. S., and Bresnahan, J. C. (1996) Graded histological and locomotor outcomes after spinal cord contusion using the NYU weight-drop device versus transection. *Exp. Neurol.* **139,** 244–256.

19. Carlos, T. M. and Harlan, J. M. (1994) Leukocyte–endothelial adhesion molecules. *Blood* **84,** 2068–2101.

20. Klebanoff, S. J., Vadas, M. A., and Harlan, J. M. (1986) Stimulation of neutrophils by tumor necrosis factor. *J. Immunol.* **136,** 4220–4225.

21. Harlan, J. M. (1987) Consequences of leukocytes-vessel wall interactions in inflammatory and immune reactions. *Semin. Thromb. Hemost.* **13,** 425–433.

22. Campbell, E. J., Senior, R. M., McDonald, J. A., and Cox, D. L. (1982) Proteolysis by neutrophils. *J. Clin. Invest.* **70,** 845–852.

23. Kishimoto, T. K. and Rothlein, R. (1994) Integrins, ICAMs and selectins: role and regulation of adhesion molecules in neutrophil recruitment to inflammatory sites. *Adv. Pharmacol.* **25,** 117–169.

24. Suttop, N., Nolte, A., Wilke, A., and Drenckhalm, D. (1993) Human neutrophil elastase increases permeability of cultured pulmonary endothelial cell monolayer. *Int. Microcirculation* **13,** 187–203.

25. Shasby, D. M., Yorec, M., and Shasby, S. S. (1988) Exogenous oxidants initiates hydrolysis of endothelial cell inositol phospholipids. *Blood* **72,** 491–499.

26. Taoka, Y., Okajima, K., Uchiba, M., Murakami, K., Harada, N., Johno, S., et al. (1998) Activated protein C reduces the severity of compression-induced spinal cord injury in rats by inhibiting activation of leukocytes. *J. Neurosci.* **18,** 1393–1398.

27. Wang, C. X., Nuttin, B., Heremans, H., and Gybels, R. (1996) Production of tumor necrosis factor in spinal cord following traumatic injury in rats. *J. Neuroimmunol.* **89,** 151–156.

28. Yakovlev, A. G. and Faden, A. I. (1994) Sequential expression of c-fos protooncogene, TNF-α, and dynorphin genes in spinal cord following experimental traumatic injury. *Mol. Chem. Neuropathol.* **23,** 179–190.

29. Bartholdi, D. and Schwab, M. E. (1997) Expression of pro-inflammatory cytokine and chemokine mRNA upon experimental spinal cord injury in mouse: an in situ hybridization study. *Eur. J. Neurosci.* **9,** 1422–1438.

30. Liu, S., Adcock, I. M., Old, R. V., Barnes, P. J., and Evans, T. W. (1993) Lipopolysaccharide treatment in vivo induces wide-spread tissue expression of inducible nitric oxide synthase m-RNA. *Biochem. Biophys. Res. Commun.* **196,** 1208–1213.

31. Taoka, Y. and Okajima, K. (2000) Role of leukcoytes in spinal cord injury in rats. *J. Neurotrauma* **17,** 219–229.

32. Iwasa, K., Ikata, T., and Fukuzawa, K. (1989) Protective effect of vitamin E on spinal cord injury by compression and concurrent lipid peroxidation. *Free Radical Biol. Med.* **6,** 599–606.

33. Nystrom, B. and Berglund, J.-E. (1988) Spinal cord restitution following compression injuries in rats. *Acta Neurol. Scand.* **78,** 467–472.

34. Schreck, R., Albermann, K., and Baeuerle, P. A. (1992) Nuclear factor kappa B: an oxidative stress-responsive transcription of eukaryotic cells. [review]. *Free Radical Res. Commun.* **12,** 221–237.

35. Mulligan, M. S., Varani, J., and Dame, M. K. (1991) Role of endothelial-leukocyte adhesion molecule 1 (ELAM-1) in neutrophil-mediated lung injury in rats. *J. Clin. Invest.* **88,** 1396–1406.

36. Zheng, H., Crowley, J. J., and Chan, J. C. (1990) Attenuation of tumor necrosis factor-induced endothelial cell cytotoxicity and neutrophil chemiluminescence. *Am. Rev. Respir. Dis.* **142,** 1073–1078.

37. Hamada, Y., Ikata, T., Katoh, S., Nakauchi, K., Niwa, M., Kawai, Y., et al. (1996) Involvement of an intracellular adhesion molecule 1-dependent pathway in the pathogenesis of secondary changes after spinal cord injury in rats. *J. Neurochem.* **66,** 1525–1531.

38. Taoka, Y. and Okajima, K. (1998) Spinal cord injury in the rat. *Prog. Neurobiol.* **56,** 341–358.

39. Zimmerman, B. J. and Granger, D. N. (1990) Reperfusion-induced leukocyte infiltration: role of elastase. *Am. J. Physiol.* **259,** H390–H394.

40. Taoka, Y., Okajima, K., Jono, M., and Naruo, M. (1998) Role of neutrophil elastase in compression-induced spinal cord injury in rats. *Brain Res.* **799,** 264–269.

41. Woodman, R. C., Reinhardt, P. H., Kanwa, S., Johnston, F. L., and Kubes, P. (1993) Effects of human neutrophil elastase (HNE) on neutrophil function in vitro and in inflamed microvessels. *Blood* **82,** 2188–2195.

42. Leroy, E. C., Ager, A., and Gordon, J. L. (1984) Effects of neutrophil elastase and other proteases on porcine aortic endothelial prostaglandin I2 production, adenine nucleotide release, and responses to vasoactive agents. *J. Clin. Invest.* **74,** 1003–1010.

43. Weksler, B. B., Jaffe, E. A., Brower, M. S., and Cole, O. F. (1989) Human leukocyte cathepsin G and elastase specifically suppress thrombin-induced prostacyclin production in human endothelial cells. *Blood* **74,** 1627–1634.

44. Harada, N., Okajima, K., and Murakami, K. (1997) Leukocyte depletion and ONO-5046, a specific inhibitor of granulocoyte elastase, prevent a stress-induced decrease in gastric prostaglandin I2 in rats. *Biochem. Biophys. Res. Commun.* **231,** 52–55.

45. Taoka, Y., Okajima, K., Uchiba, M., Murakami, K., Kushimoto, S., Johno, M., et al. (1997) Reduction of spinal cord injury by administration of iloprost, a stable prostacyclin analog. *J. Neurosurg.* **86,** 1007–1011.

46. Demopoulos, H. B., Flamm, E. S., Pietronigro, D. D., and Seligman, M. L. (1980) The free radical pathology and the microcirculation in the major central nervous system disorders. *Acta Physiol. Scand.* **492(Suppl.),** 91–119.

47. Nakauchi, K., Ikata, T., Katoh, S., Hamada, Y., Tsuchiya, K., and Fukuzawa, K. (1996) Effects of lecithinized superoxide dismutase on rat spinal cord injury. *J. Neurotrauma* **13,** 573–582.

48. Taoka, Y., Naruo, M., Koyanabi, E., Urakado, M., and Inoue, M. (1995) Superoxide radicals play important roles in the pathogenesis of spinal cord injury. *Paraplegia* **33,** 450–453.

49. Kusner, D. V. and King, C. H. (1989) Protease-modulation of neutrophil superoxide response. *J. Immunol.* **143,** 1696–1702.

50. Abe, H., Okajima, K., and Okabe, H. (1994) Granulocyte proteases and hydrogen peroxide synergistically inactivate thrombomodulin of endothelial cells in vitro. *J. Lab. Clin. Med.* **23,** 874–881.

51. McCord, J. M. (1985) Oxygen-derived free radicals in post-ischemic tissue injury. *N. Engl. J. Med.* **312,** 159–163.

52. Xu, J., Beckman, J. S., Hogan, E. L., and Hsu, C. Y. (1991) Xanthine oxidase in experimental spinal cord injury. *J. Neurotrauma* **8,** 11–18.

53. Taoka, Y., Okajima, K., Uchiba, M., and Johno, M. (2001) Methylprednisolone reduces spinal cord injury in rats without affecting tumor necrosis factor-a production. *J. Neurotrauma* **18,** 533–543.

54. Taoka, Y., Okajima, K., Uchiba, M., Murakami, K., Kushimoto, S., Johno, M., et al. (1997) Role of neutrophils in spinal cord injury in the rat. *Neuroscience* **79,** 1177–1182.

55. Holtz, A., Nystrom, B., and Gerdin, B. (1990) Relation between spinal cord blood flow and functional recovery after blocking weight-induced spinal cord injury in rats. *Neurosurgery* **26,** 952–957.

56. Weyrich, A. S., McIntyre, T. M., Mcever, R. P., Prescott, S. N., and Zimmerman, G. A. (1995) Monocyte tethering by P-selectin regulates monocyte chemotactic protein-1 and tumor necrosis factor-a secretion. *J. Clin. Invest.* **95,** 2297–2302.

57. Stephens, K. E., Ishizaka, A., and Larrick, J. W. (1988) Tumor necrosis factor causes increased pulmonary permeability and edema. *Am. Rev. Respir. Dis.* **137,** 1364–1370.

58. Watson, M. L., Lewis, G. P., and Westwick, J. (1989) Increased vascular permeability and polymorphonuclear leukocyte accumulation in vivo in response to recombinant cytokine and supernatant from cultures of human synovial cells treated with interleukin 1. *Br. J. Exp. Pathol.* **70,** 93–101.

59. Taoka, Y., Okajima, K., Uchiba, M., Murakami, K., Kushimoto, S., Johno, M., et al. (1997) Gabexate mesilate, a synthetic protease inhibitor, prevents compression-induced spinal cord injury by inhibiting activated leukocytes. *Crit. Care Med.* **25,** 874–879.

60. Taoka, Y., Okajima, K., Uchiba, M., and Johno, M. (2000) Neuroprotection by recombinant human soluble thrombomodulin. *Thromb. Haemost.* **83,** 462–468.

61. Geisler, F. H., Dorsey, F. C., and Coleman, W. P. (1991) Recovery of motor function after spinal cord injury—a randomized, placebo-controlled trial with GM-1 ganglioside. *N. Engl. J. Med.* **324,** 1829–1838.

62. Xu, J., Qu, Z. X., Hogan, E. L., and Perot, P. L. Jr. (1992) Protective effect of methylprednisolone on vascular injury in rat spinal cord injury. *J. Neurotrauma* **9,** 245–253.

63. Tamura, K., Hirado, M., and Okamura, K. (1977) Synthetic inhibitors of trypsin, plasmin, kallikrein, thrombin, C1r, and C1 esterase. *Biochim. Biophys. Acta* **484,** 417–422.

64. Taenaka, N., Shimada, Y., and Hirata, T. (1983) Gabexate mesilate (FOY) therapy of disseminated intravascular coagulation due to sepsis. *Crit. Care. Med.* **9,** 735–738.

65. Messori, A., Rampazzo, R., and Scroccaro, G. (1995) Effectiveness of gabexate mesilate in acute pancreatitis. *Digest. Dis. Sci.* **40,** 734–738.

66. Murakami, K., Okajima, K., and Uchiba, M. (1996) Gabexate mesilate, a synthetic protease inhibitor, attenuates endotoxin-induced pulmonary vascular injury by inhibiting activated leukocytes. *Crit. Care Med.* **24,** 1047–1053.

67. Nesheim, M. E., Kettner, C., Shaw, E., and Mann, K. G. (1981) Cofactor dependence of factor Xa incorporation into the prothrombinase complex. *J. Biol. Chem.* **256,** 6537–6540.

68. Harada, N., Okajima, K., and Kushimoto, S. (1999) Gabexate mesilate, a synthetic protease inhibitor, reduces ischemia/reperfusion injury of rat liver by inhibiting leukocyte activation. *Crit. Care Med.* **27,** 1958–1964.

69. Okamura, T., Niho, Y., Itoga, T., Chiba, S., Miyake, M., Kotsuru, M., et al. (1993) Treatment of disseminated intravascular coagulation and its prodromal stage with gabexate mesilate (FOY): a multicenter trial. *Acta Haematol. (Basel)* **90,** 120–124.

70. Walker, F. J., Sexton, P. W., and Esmon, C. T. (1979) The inhibition of blood coagulation by activated protein C through the selective inactivation of activated factor V. *Biochim. Biophys. Acta* **571,** 333–342.

71. Esmon, C. T. (1992) The protein C anticoagulant pathway. *Arterioscler. Thromb.* **12,** 135–145.

72. Grey, S., Hau, H., Salem, H. H., and Hancock, W. W. (1993) Selective effects of protein C on activation of human monocytes by lipopolysaccharide, interferon-g, or PMA: modulation of effects on CD11b and CD14 but not CD25 or CD54 induction. *Transplant. Proc.* **25,** 2913–2914.

73. Murakami, K., Okajima, K., Uchiba, M., Johno, M., Nakagaki, T., Okabe, H., et al. (1996) Activated protein C attenuates endotoxin-induced pulmonary vascular injury by inhibiting activated leukocytes in rats. *Blood* **87,** 642–647.

74. Murakami, K., Okajima, K., Uchiba, M., Johno, M., Nakagaki, T., Okabe, H., et al. (1997) Activated protein C prevents LPS-induced pulmonary vascular injury by inhibiting cytokine production. *Am. J. Physiol.* **272,** L197–L202.

75. Hirose, K., Okajima, K., Taoka, Y., Uchiba, M., Takami, H., Nakano, K., et al. (2000) Activated protein C reduces the ischemia-induced spinal cord inury in rats by inhibiting neutrophil activation. *Ann. Surg.* **232,** 272–280.

76. Silver, J. R. (1974) Prophylactic use of anticoagulant therapy in prevention of pulmonary emboli in one hundred consecutive spinal injury patients. *Paraplegia* **12,** 188–196.

77. Frisbie, J. H. and Sharma, G. V. R. K. (1992) Circadian rhythm of pulmonary embolism in patients with acute spinal cord injury. *Am. J. Cardiol.* **70,** 847–848.

78. Todd, J. W., Frisbie, J. H., and Rossier, A. B. (1976) Deep vein thrombosis in acute spinal injury: a comparison of ^{125}I-fibrinogen scanning, impedance plethysmorgraphy, and venography. *Paraplegia* **14,** 50–57.

79. Myllynen, P., Kammonen, M., and Rokkanen, P. (1985) Deep venous thrombosis and pulmonary embolism in patients with acute spinal cord injury: a comparison with nonparalyzed patients immobilized due to spinal fracture. *J. Trauma* **25,** 541–546.

80. Moncada, S. and Vane, J. R. (1979) Arachidonic acid metabolites and their interactions between platelets and blood vessel walls. *N. Engl. J. Med.* **300,** 1142–1147.

81. Muller, B., Sturzebechr, S., and Krais, T. (1989) The experimental and clinical pharmacology of iloprost, in *The Pathophysiology of Critical Limb Ischaemia and Pharmacological Intervention with a Stable Prostacyclin Analogue* (Dormandy, J. A., ed.), Iloprost International Congress and Symposium Series No. 159. Royal Society of Medicine Services, London, pp. 33–50.

82. Eisenhut, T., Sinha, B., and Grottrup-Wolfers, E. (1993) Prostacyclin analog suppress the synthesis of tumor necrosis factor-a in LPS-stimulated human peripheral blood mononuclear cells. *Immunopharmacology* **26,** 259–264.

83. Maria-Riva, C., Morganroth, M. L., and Ljungman, A. G. (1990) Iloprost inhibits neutrophil-induced lung injury and neutrophil adherence to endothelial monocytes. *Am. J. Respir. Cell Mol. Biol.* **3,** 301–309.

84. Simpson, P. J., Mickelson, J., and Fantone, J. (1987) Iloprost inhibits neutrophil function in vitro and in vivo and limits experimental infarct size in canine heart. *Circ. Res.* **60,** 666–673.

85. Okuyama, M., Kambayashi, J., Sakon, M., Kawasaki, T., and Monden, M. (1995) PGI2 analogue, sodium deraprost, suppresses superoxide generation in human neutrophils by inhibiting p47phox phosphorylation. *Life Sci.* **57,** 1051–1059.

86. Grant, S. M. and Goa, K. L. (1992) Iloprost, a review of its pharmacodynamic and pharmacokinetic properties, and therapeutic potential in peripheral vascular diseases, myocardial ischemia and extracorporeal circulation procedures. *Drugs* **43,** 889–924.

87. Hsu, C. Y. and Dimitrijevic, M. R. (1990) Methylprednisolone in spinal cord injury: the possible mechanism of action. *J. Neurotruma* **7,** 115–119.

88. Hall, E. D., Yonkers, P. A., McCall, J. M., and Braughler, J. M. (1988) Effects of the 21-aminosteroid U7400F, on experimental head injury in mice. *J. Neurosurg.* **68,** 456–461.

89. Espersen, G. T., Ernst, E., and Vestergaard, M. (1989) Changes in PMN leukocyte migration activity and complement C3d levels in RA patients with high disease activity during steroid treatment. *Scand. J. Rheumatol.* **18,** 51–56.

90. Becker, J. and Grasso, R. J. (1985) Suppression of phagocytosis by dexamethasone in macrophage culture: inability of arachidonic acid, indomethacin and nordihydroguaiaretic acid to reverse the inhibitory response mediated by a steroid-inducible factor. *Int. J. Immunopharmacol.* **7,** 839–847.

91. Schleimer, R. P., Freeland, H. S., and Peters, S. P. (1989) An assessment of the effects of glucocorticoids on degranulation, chemotaxis, binding to vascular endothelium and formation of leukotriene B4 by purified human neutrophils. *J. Pharmacol. Exp. Ther.* **250,** 598–605.

92. Hargreaves, K. M. and Costella, A. (1990) Glucocorticoids suppress levels of immunoreactive bradykinin in inflamed tissue as evaluated by micro dialysis probes. *Clin. Pharmacol. Ther.* **48,** 168–178.

93. Flowers, R. J. (1989) Glucocorticoids and inhibition of phospholipase A2, in *Anti-inflammatory Steroid Action: Basic and Clinical Aspects* (Schleimer, R. P., Ford-Clamam, H. N., and Oronsky, A. L., eds.), Academic, New York, pp. 48–64.

94. Anderson, D. K., Hsu, C. Y., Michel, M. E., and Stokes, B. T. (1992) NIH workshop on experimental spinal cord injury: modeling and criteria. *J. Neurotrauma* **9,** 113–186.

95. Behrmann, D. L., Bresnahan, J. C., and Beattie, M. S. (1994) Modeling of acute spinal cord injury in the rat: neuroprotection and enhanced recovery with methylprednisolone, U74006F and YM-14673. *Exp. Neurol.* **126,** 1–16.

96. Young, W., DeCrescito, V., Flamm, E. S., Blight, A. R., and Gruner, J. A. (1990) Pharmacological therapy of acute spinal cord injury: studies of high dose methyprednisolone and naloxone. *Clin. Neurosurg.* **38,** 657–697.

97. Braughler, J. M. and Hall, E. D. (1984) Effects of multidose methylprednisolone sodium succinate administration to an injured cat spinal cord neurofilament degradation and energy metabolism. *J. Neurosurg.* **61,** 290–295.

98. Braughler, J. M., Hall, E. D., Means, E. D., Waters, T. R., and Anderson, D. K. (1987) Evaluation of an intensive methylprednisolone sodium succinate dosing regimen in experimental spinal cord injury. *J. Neurosurg.* **67,** 102–105.

99. Young, W. (1985) Blood blow, metabolic and neurophysiological mechanisms in spinal cord injury, in *Central Nervous System Trauma Status Report* (Becker, D. and Povlishock, J. T., eds.), National Institutes of Health, Rockville, MD, pp. 463–473.

100. Hall, E. D., Yonkers, P. A., Taylor, B. M., and Sun, F. F. (1995) Lack of effect of post-injury treatment with methylprednisolone or tirilazad mesylate on the increase in eicosanoid levels in the acutely injured cat spinal cord. *J. Neurotrauma* **3,** 245–258.

101. Suttorp, N., Seeger, W., Zucker-Reimann, J., Roka, L., and Bhakdi, S. (1987) Mechanism of leukotriene generation in polymorphonuclear leukocytes by staphylococcal alpha-toxin. *Infect. Immun.* **55,** 104–110.

102. Katori, M., Oda, T., and Nagai, K. (1990) Asite action of dexamethasone on leukocyte extravasation in microcirculation. *Agents Action* **29(1/2),** 24–26.
103. Oda, T. and Katori, M. (1992) Inhibition site of dexamethasone on extravasation of poly-morphonuclear leukocytes in the hamster cheek pouch microcirculation. *J. Leukocyte Biol.* **52,** 337–342.
104. Frankel, H. L., Hancock, D. O., and Hyslop, G. (1969) The value of postural reduction in the initial management of closed injuries to the spine with paraplegia and tetraplegia. *Paraplegia* **7,** 179–192.
105. American Spinal Injury Association (1984) *Standards for Neurological Classification of Spinal Injury Patients*, American Spinal Injury Association, Chicago.
106. Geisler, F. H. (1993) GM-1 ganglioside and motor recovery following human spinal cord injury. *J. Emerg. Med.* **1,** 49–55.
107. Geisler, F. H., Dorsey, F. C., and Coleman, W. P. (1993) Past and current clinical studies with GM-1 ganglioside in acute spinal cord injury. *Ann. Emerg. Med.* **22,** 1041–1047.
108. Ledeen, R. W. (1978) Ganglioside structures and distribution: are they localized at the nerve ending? *J. Supramol. Struct.* **8,** 1–17.
109. Gorio, A., Di Giulio, A. M., Young, W., Gruner, J., Blight, A., De Crescito, V., et al. (1986) GM1 effects on chemical, traumatic and peripheral nerve induce lesions to the spinal cord, in *Development and Platicity of the Mammalian Spinal Cord, Volume 3* (Goldberger, M. E., Gorio, A., and Murray, M., eds.), Liviana Press, Padova, pp. 227–242.
110. Sabel, B. A., CelMastro, R., Dunbar, G. L., and Stein, D. G. (1987) Reduction of anterograde degeneration in brain damaged rats by GM1ganglioside. *Neurosci. Lett.* **77,** 360–366.
111. Greene, K. A., Marciano, F. F., and Sonntag, V. K. H. (1996) Pharmacological manage-ment of spinal cord injury: current status of drugs designed to augment functional recov-ery of the injured human spinal cord. *J. Spinal Disord.* **9,** 355–366.

11

Type IV Collagenases and Blood-Brain Barrier Breakdown in Brain Ischemia

Yvan Gasche, Jean-Christophe Copin, and Pak H. Chan

1. INTRODUCTION

Stability of the neuronal microenvironment is indispensable for maintaining the normal function of the brain. The blood-brain barrier (BBB) is dedicated to the preservation of cerebral homeostasis. The functional features of BBB are a low diffusional permeability to water-soluble molecules, a low hydraulic conductivity, a high reflection coefficient, and a high electrical resistance. These characteristics of BBB permit a highly selective exchange between blood and brain and an optimally controlled environment in the physiological state. As a functional entity, BBB includes several cell types and the extracellular matrix (ECM). Microvascular endothelial cells (EC) sealed by tight junctions and featuring only a very few endocytotic vesicles are primarily responsible for the permeability properties of BBB *(1)*. Astrocytes, in discontinuous contact with endothelial cells through their end-feet, actively participate in BBB phenotype *(2)*, although some controversy exists regarding the importance of astrocytes in the in vivo maintenance of BBB function *(3,4)*. Pericytes also have been shown to change endothelial behavior *(5–7)*. These perivascular cells seem to control angiogenesis by inhibiting EC proliferation *(8)* and to regulate microvessel permeability by contributing to basal lamina synthesis *(9,10)*. Finally, the endothelial basal lamina represents the noncellular component of BBB. Produced by EC and pericytes, the basal lamina is a specialized ECM composed of type IV collagen, fibronectin, laminin, and various proteoglycans *(11)*. The ECM components are linked to endothelial cells via integrins and regulate specific biological processes such as cellular morphology, differentiation, survival, adhesion, and gene expression *(12–15)*.

When BBB integrity is lost, inflammatory cells and fluid penetrate the brain, causing vasogenic edema and secondary brain damage *(16,17)*. BBB permeability is altered in various brain pathologies such as bacterial meningitis *(18)*, multiple sclerosis *(19)*, and ischemic stroke *(20,21)*.

2. MECHANISMS OF BBB DYSFUNCTION DURING BRAIN ISCHEMIA: PROTEOLYTIC DISRUPTION AS A KEY EVENT

Various mechanisms of BBB disruption have been implicated in stroke-related vasogenic edema. The role of chemical mediators in the opening of the BBB has been evaluated on

From: *Neuroinflammation, 2nd Edition: Mechanisms and Management*
Edited by: P. L. Wood © Humana Press Inc., Totowa, NJ

the basis of their ability to increase cerebral vascular permeability and to induce vaso-motor responses and edema formation. Among various mediator candidates, bradykinin has been widely studied *(22)* and consistently shown to influence BBB function in trau-matic *(23)* and ischemic *(24,25)* brain injury. Although intracerebral concentration of histamine has been shown to increase during brain ischemia *(26,27)*, conflicting results have been reported regarding the benefit of its inhibition *(28–31)*. Arachidonic acid and its metabolites *(32–35)* as well as oxygen free radicals *(36,37)* are also involved in stroke-related BBB disruption. Recently, proinflammatory cytokines such as tumor necrosis factor-α (TNF-α) *(38–40)*, interleukin-1β (IL-1β) *(41–44)*, and cytokine-associated cas-cades appear to participate in the pathophysiological processes, leading to secondary brain injury.

Although our knowledge of the different pathways and mediators potentially involved in the opening of the BBB is growing, the exact molecular and cellular mechanisms lead-ing to the disruption of the BBB is still not fully elucidated. The potential regulatory role of ECM on capillary morphogenesis *(15)* has emerged as a key factor of BBB imperme-ability maintenance. During cerebral ischemia–reperfusion, the ECM is disrupted. Major components of the endothelial basal lamina such as laminin, type IV collagen, and fibro-nectin start to disappear as soon as 2 h after the onset of ischemia *(45)*. Consistently, the first signs of BBB leakage are observed between 2 and 8 h after the onset of ischemia *(46–49)*. By 24 h of ischemia–reperfusion, dissolution of microvascular structures leads to clear interruption of microvessels *(45,50)* and local hemorrhage *(50)*. Thus, proteo-lysis seems to be a critical process of stroke-related BBB disruption.

Various proteases have been shown to increase BBB permeability *(51)*. Intracerebral injection of type IV collagenase *(52)*, elastase *(53)*, and plasmin induces BBB disrup-tion and multifocal hemorrhage *(52,53)*. Although, the nature of the proteolytic media-tors disrupting the BBB during ischemic stroke is still a subject of debate, experimental evidences suggest that matrix metalloproteinases play a key role in this process.

3. TYPE IV COLLAGENASES
AND BBB DISRUPTION DURING BRAIN ISCHEMIA

All endothelial basal lamina components can be digested by matrix metalloproteinases (MMPs), which are proteolytic enzymes (Zn^{2+}-endopeptidases) secreted as zymogen and cleaved to their fully active form in the interstitial space. MMPs are involved in the remodeling of the ECM in a variety of physiological and pathophysiological con-ditions *(54–56)*. Among MMPs, gelatinases also known as type IV collagenases, are expressed by neurons *(57,58)*, astrocytes *(59)*, microglial cells *(60)*, endothelial cells *(61)*, and oligodendrocytes *(62)*. Gelatinase A (MMP-2) and gelatinase B (MMP-9) spe-cifically digest type IV collagen in the basal lamina. Several in vivo studies, investigat-ing the role of MMPs in stroke, have shown that gelatinase expression was induced during experimental focal ischemia. Early studies, conducted in spontaneously hyper-tensive rats, raised doubt regarding the role of MMP in early BBB disruption, because they could not show any upregulation or activation of the enzymes during the first hours after permanent middle cerebral artery occlusion (pMCAO) *(61,63)*. Later studies, using permanent or transient models of MCAO in mice, showed that gelatinase B was already upregulated 1–2 h after ischemia *(47,64)*. This upregulation was rapidly followed by the appearance of the active form of gelatinase B *(47,65)*. One study, conducted in baboons, showed an early upregulation of gelatinase A *(66)*. These discrepancies can be explained by methodological particularities of the experimental models with variations in MMP measurement techniques and species-related specificities. Two studies, carried out in humans, confirmed that MMP-9 expression after stroke is mainly increased in acute ische-mic lesions (less than 1 wk after stroke onset), whereas MMP-2 and matrilysin (MMP-7) are increased in chronic lesions (more than 1 wk after stroke onset) *(67,68)*. Although MMPs form a growing family, only gelatinases have been consistently studied in stroke. One study has examined the expression of MMP-1 and MMP-3 (by Western blotting) after pMCAO, but no expression of these proteins was detected *(61)*.

Matrix metalloproteinase activation is a tightly regulated and complex process. Gelati-nase B can be activated in vitro by other metalloproteinases such as Gelatinase A *(69)*, MMP-3 *(70)*, and MMP-7 or other enzymes such as trypsin. The natural inhibitor of MMP-9 (TIMP-1) prevents its activation. No modification of TIMP-1 at the protein level has been observed during the first 24 h following experimental stroke *(47,49,61,63,66)*. MMP-2 activation is specifically mediated by a membrane associated MMP (MT-MMP). TIMP-2, the natural inhibitor of MMP-2, plays a dual role in its interaction with the gel-atinase. Indeed, when TIMP-2 complexes with proMMP-2, cell-surface-mediated acti-vation of the enzyme is facilitated, whereas interaction of TIMP-2 with the active enzyme results in inhibition *(71–75)*.

Recently, Rosenberg et al. confirmed the role of MMPs in BBB disruption by demon-strating the ability of the nonselective MMP inhibitor BB-1101 (British Biotechnology, UK) to reduce the early BBB leakage following transient focal ischemia *(49)*. The pos-sible mechanisms of MMP-related BBB alteration is explained by the disruption of the endothelial basal lamina, which prevents the anchorage of the endothelial cells onto the ECM. This loss of connection between endothelial cells and the ECM may lead to endo-thelial apoptosis (anoikis) and vascular involution *(14,15,76)*. Recent in vitro studies suggest that MMPs directly affect endothelial tight junctions *(77,78)*. Whereas Wachtel et al. showed that under specific conditions, occludin cleavage and subsequent endothe-lial leakage are under MMP control *(78)*, Harkness et al. observed by immunofluores-cence that MMP-9 was able to alter another junctional protein, ZO-1, in brain microvascular EC cultures *(77)*.

Matrix metalloproteinases may also indirectly affect BBB permeability by interfering with inflammatory pathways triggered by ischemia–reperfusion. IL-1β is an essential medi-ator of inflammation and plays a significant role in ischemic brain injury *(41,43,44,79)* and vasogenic edema *(42)*. From in vitro data, we know that MMPs can process IL-1β into its biologically active form. Whereas MMP-2 activates IL-1β in 24 h, MMP-3 takes 1 h and MMP-9 only a few minutes to process the cytokine. Thus, MMPs may promote inflammatory processes, which will, in a positive feedback loop increase MMP produc-tion by resident or migrating cells. Eventually, MMPs will also contribute to the increase of endothelial permeability induced by TNF *(80)*. The fundamental role played by MMPs in the development of vasogenic edema during stroke is further substantiated by the fact that these enzymes mediate the capillary leakage triggered by oxidative stress. It is well established that the oxidative unbalance during focal cerebral ischemia is a major con-tributor to BBB disruption, secondary brain injury, and hemorrhagic transformation *(48, 81,82)*. Our laboratory has shown that after traumatic brain injury, gelatinase expression and activation were reduced in mice overexpressing CuZn-superoxide dismutase as compared to wild-type animals *(83)*. These results confirm previous in vitro data suggesting that oxygen free radicals participate in MMP activity regulation *(84)*. During focal ischemia–reperfusion, we have observed that oxidative-stress-associated BBB disruption was prevented by the nonselective inhibition of MMPs (Gasche et al., personal communication). In this study, gelatinase-mediated *in situ* proteolysis, observed at the capillary level, correlated with the local production of oxygen free radicals. On the basis of these overall results and considering that growing evidence implicates MMPs in endothelial cell apoptosis *(85)*, one can speculate that oxidative stress-related endothelial cell anoikis *(86)* might be mediated by MMPs.

Matrix metalloproteinases seem to be essential effectors of proteolytic BBB disruption during stroke, but the respective role of gelatinases or other MMPs in BBB disruption has not yet been elucidated. Nevertheless, several studies have shown that MMP-9 is an important mediator of secondary brain injury following permanent focal ischemia. In these studies, the infarct volume was reduced in mice treated with an anti-MMP-9 antibody *(61)* or in animals lacking MMP-9 *(64)* as compared to controls. BBB disruption was studied in a model of transient focal ischemia in MMP-9 knockout mice (Gidday et al., personal communication). The authors observed a reduced BBB leakage in MMP-9-deficient mice during the first hours following the ischemic insult.

4. CONSEQUENCE OF PROTEOLYTIC BBB DISRUPTION DURING BRAIN ISCHEMIA

Proteolysis is a critical process of stroke-related BBB disruption. As already mentioned, major components of the endothelial basal lamina such as laminin, type IV collagen, and fibronectin rapidly disappear after the onset of experimental ischemia *(45)*. Dissolution of capillary walls leads to local hemorrhage *(50)*. Hemorrhagic transformation of the ischemic tissue, at the time of reperfusion, is a major complication of ischemic stroke in humans. This potentially lethal complication has dramatically limited the indication of active reperfusion by plasminogen activators (tPA) *(87)*. As a consequence of BBB disruption, the hemorrhagic transformation of ischemic brain tissues follows the same mechanistic pathways. Free-radical spin-trapping has been shown to reduce tPA-induced hemorrhage in a rat model of embolic stroke *(82)* and the nonselective inhibition of MMPs seems to have the same effect in a rabbit model of embolic stroke *(88)*.

Our understanding of the exact involvement of MMPs in the pathophysiological cascades governing cerebral ischemia–reperfusion is growing but still fragmentary. Future studies, specifically targeting the different enzymes of this family, are needed to unravel the proteolytic mechanisms leading to BBB disruption. The preservation of BBB integrity by the inhibition of MMPs, with a consequent reduction in the risk of hemorrhagic transformation of the infarcted tissue, might provide a new strategy destined to increase the therapeutic window between the onset of ischemia and thrombolytic reperfusion.

REFERENCES

1. Reese, T. S. and Karnowsky, M. J. (1967) Fine ultrastructural localization of a blood-brain barrier to exogenous peroxidase. *J. Cell Biol.* **34,** 207–217.
2. Janzer, R. C. and Raff, M. C. (1987) Astrocytes induce blood-brain barrier properties in endothelial cells. *Nature* **325,** 253–257.

3. Krum, J. M. and Rosenstein, J. M. (1993) Effect of astroglial degeneration on the blood-brain barrier to protein in neonatal rats. *Brain Res. Dev. Brain Res.* **74(1)**, 41–50.

4. Krum, J. M. and Rosenstein, J. M. (1989) The fine structure of vascular–astroglial relations in transplanted fetal neocortex. *Exp. Neurol.* **103(3)**, 203–212.

5. Larson, D. M., Carson, M. P., and Haudenschild, C. C. (1987) Junctional transfer of small molecules in cultured bovine brain microvascular endothelial cells and pericytes. *Microvasc. Res.* **38**, 184–199.

6. Fujimoto, K. (1995) Pericyte–endothelial gap junctions in developing rat cerebral capillaries. *Anat. Rec.* **242**, 562–565.

7. Antonelli-Orlidge, A., Saunders, K., Smith, S. R., and D'Amor, P. A. (1989) An activated form of transforming growth factor β is produced by cocultures of endothelial cells and pericytes. *Proc. Natl. Acad. Sci. USA* **86**, 4544–4548.

8. Wakui, S., Furusato, M., Muto, T., Ohshige, H., Takahashi, H., and Ushigome, S. (1997) Transforming growth factor-beta and urokinase plasminogen activator presents at endothelial cell–pericyte interdigitation in human granulation tissue. *Microvasc. Res.* **54(3)**, 262–269.

9. Cohen, M. P., Frank, R. N., and Khalifa, A. A. (1980) Collagen production by cultured retinal capillary pericytes. *Invest. Ophthalmol. Vis. Sci.* **19(1)**, 90–94.

10. Stramm, L. E., Li, W., Aguirre, G. D., and Rockey, J. H. (1987) Glycosaminoglycan synthesis and secretion by bovine retinal capillary pericytes in culture. *Exp. Eye Res.* **44(1)**, 17–28.

11. Yurchenco, P. D. and Schittny, J. C. (1990) Molecular architecture of basement membranes. *FASEB J.* **4**, 1577–1590.

12. Chen, C. S., Mrksich, M., Huang, S., Whitesides, G. M., and Ingber, D. E. (1997) Geometric control of cell life and death. *Science* **276**, 1425–1428.

13. Maniotis, A. J., Chen, C. S., and Ingber, D. E. (1997) Demonstration of mechanical connections between integrins cytoskeletal filaments and nucleoplasm that stabilize nuclear structure. *Proc. Natl. Acad. Sci. USA* **94**, 849–854.

14. Brooks, P. C., Montgomery, A. M., Rosenfeld, M., Reisfeld, R. A., Hu, T., Klier, G., et al. (1994) Integrin alpha v beta 3 antagonists promote tumor regression by inducing apoptosis of angiogenic blood vessels. *Cell* **79(7)**, 1157–1164.

15. Ingber, D. E. and Folkman, J. (1989) How does extracellular matrix control capillary morphogenesis? *Cell* **58**, 803–805.

16. Chen, H., Chopp, M., and Bodzin, G. (1992) Neutropenia reduces the volume of cerebral infarct after transient middle cerebral artery occlusion in the rat. *Neurosci. Res. Commun.* **11**, 93–99.

17. Fishman, R. A. (1975) Brain edema. *N. Engl. J. Med.* **293**, 706–711.

18. Pfister, H. W., Borasio, G. D., Dirnagl, U., Bauer, M., and Einhaupl, K. M. (1992) Cerebrovascular complications of bacterial meningitis in adults. *Neurology* **42(8)**, 1497–1504.

19. Sears, E. S., Tindall, R. S., and Zarnow, H. (1978) Active multiple sclerosis. Enhanced computerized tomographic imaging of lesions and the effect of corticosteroids. *Arch. Neurol.* **35(7)**, 426–434.

20. Hatashita, S. and Hoff, J. T. (1990) Brain edema and cerebrovascular permeability during cerebral ischemia in rats. *Stroke* **21(4)**, 582–588.

21. Hatashita, S. and Hoff, J. T. (1990) Role of blood-brain barrier permeability in focal ischemic brain edema. *Adv. Neurol.* **52**, 327–333.

22. Wahl, M., Unterberg, A., Baethmann, A., and Schilling, L. (1988) Mediators of blood-brain barrier dysfunction and formation of vasogenic brain edema. *J. Cereb. Blood Flow Metab.* **8(5)**, 621–634.

23. Unterberg, A., Dautermann, C., Baethmann, A., and Muller-Esterl, W. (1986) The kallikrein–kinin system as mediator in vasogenic brain edema. Part 3: inhibition of the kallikrein–kinin system in traumatic brain swelling. *J. Neurosurg.* **64(2)**, 269–276.

24. Relton, J. K., Beckey, V. E., Hanson, W. L., and Whalley, E. T. (1997) CP-0597, a selective bradykinin B2 receptor antagonist, inhibits brain injury in a rat model of reversible middle cerebral artery occlusion. *Stroke* **28(7),** 1430–1436.

25. Kamiya, T., Katayama, Y., Shimizu, J., Soeda, T., Nagazumi, A., and Terashi, A. (1990) Studies on relation of bradykinin to ischemic brain edema in stroke-resistant spontaneously hypertensive rat. *Adv. Neurol.* **52,** 543.

26. Adachi, N., Itoh, Y., Oishi, R., and Saeki, K. (1992) Direct evidence for increased continuous histamine release in the striatum of conscious freely moving rats produced by middle cerebral artery occlusion. *J. Cereb. Blood Flow Metab.* **12(3),** 477–483.

27. Subramanian, N., Theodore, D., and Abraham, J. (1981) Experimental cerebral infarction in primates: regional changes in brain histamine content. *J. Neural. Transm.* **50(2–4),** 225–232.

28. Adachi, N., Oishi, R., Itano, Y., Yamada, T., Hirakawa, M., and Saeki, K. (1993) Aggravation of ischemic neuronal damage in the rat hippocampus by impairment of histaminergic neurotransmission. *Brain Res.* **602(1),** 165–168.

29. Adachi, N., Seyfried, F. J., and Arai, T. (2001) Blockade of central histaminergic H2 receptors aggravates ischemic neuronal damage in gerbil hippocampus. *Crit. Care Med.* **29(6),** 1189–1194.

30. Nemeth, L., Deli, M. A., Falus, A., Szabo, C. A., and Abraham, C.S. (1998) Cerebral ischemia reperfusion-induced vasogenic brain edema formation in rats: effect of an intracellular histamine receptor antagonist. *Eur. J. Pediatr. Surg.* **8(4),** 216–219.

31. Tosaki, A., Szerdahelyi, P., and Joo, F. (1994) Treatment with ranitidine of ischemic brain edema. *Eur. J. Pharmacol.* **264(3),** 455–458.

32. Chan, P. H. and Fishman, R. A. (1984) The role of arachidonic acid in vasogenic brain edema. *Fed. Proc.* **43(2),** 210–213.

33. Chan, P. H., Fishman, R. A., Caronna, J., Schmidley, J. W., Prioleau, G., and Lee, J. (1983) Induction of brain edema following intracerebral injection of arachidonic acid. *Ann. Neurol.* **13(6),** 625–632.

34. Chan, P. H., Fishman, R. A., Longar, S., Chen, S., and Yu, A. (1985) Cellular and molecular effects of polyunsaturated fatty acids in brain ischemia and injury. *Prog. Brain Res.* **63,** 227–235.

35. Hillered, L. and Chan, P. H. (1988) Role of arachidonic acid and other free fatty acids in mitochondrial dysfunction in brain ischemia. *J. Neurosci. Res.* **20(4),** 451–456.

36. Chan, P. H., Fishman, R. A., Schmidley, J. W., and Chen, S. F. (1984) Release of polyunsaturated fatty acids from phospholipids and alteration of brain membrane integrity by oxygen-derived free radicals. *J. Neurosci. Res.* **12(4),** 595–605.

37. Chan, P. H., Schmidley, J. W., Fishman, R. A., and Longar, S. M. (1984) Brain injury, edema, and vascular permeability changes induced by oxygen-derived free radicals. *Neurology* **34(3),** 315–320.

38. Sairanen, T., Carpen, O., Karjalainen-Lindsberg, M. L., Paetau, A., Turpeinen, U., Kaste, M., et al. (2001) Evolution of cerebral tumor necrosis factor-alpha production during human ischemic stroke. *Stroke* **32(8),** 1750–1758.

39. Yang, G. Y., Schielke, G. P., Gong, C., Mao, Y., Ge, H. L., Liu, X. H., et al. (1999) Expression of tumor necrosis factor-alpha and intercellular adhesion molecule-1 after focal cerebral ischemia in interleukin-1beta converting enzyme deficient mice. *J. Cereb. Blood Flow Metab.* **19(10),** 1109–1117.

40. Yang, G. Y., Gong, C., Qin, Z., Ye, W., Mao, Y., and Bertz, A. L. (1998) Inhibition of TNF-alpha attenuates infarct volume and ICAM-1 expression in ischemic mouse brain. *Neuroreport* **9(9),** 2131–2134.

41. Clark, E. T., Desai, T. R., Hynes, K. L., and Gewertz, B. L. (1995) Endothelial cell response to hypoxia–reoxygenation is mediated by IL-1. *Surg. Res.* **58(6),** 675–681.

42. Holmin, S. and Mathiesen, T. (2000) Intracerebral administration of interleukin-1beta and induction of inflammation, apoptosis, and vasogenic edema. *J. Neurosurg.* **92(1),** 108–120.

43. Liu, T., McDonnell, P. C., Young, P. R., White, R. F., Siren, A. L., Hallenbeck, J. M., et al. (1993) Interleukin-1 beta mRNA expression in ischemic rat cortex. *Stroke* **24(11),** 1746–1750.

44. Liu, X. H., Kwon, D., Schielke, G. P., Yang, G. Y., Silverstein, F. S., and Barks, J. D. (1999) Mice deficient in interleukin-1 converting enzyme are resistant to neonatal hypoxic-ischemic brain damage. *J. Cereb. Blood Flow Metab.* **19(10),** 1099–1108.

45. Hamann, G. F., Okada, Y., Fitridge, R., and Del Zoppo, G. J. (1995) Microvascular basal lamina antigens disappear during cerebral ischemia and reperfusion. *Stroke* **26,** 2120–2126.

46. Belayev, L., Busto, R., Zhao, W., and Ginsberg, M. D. (1996) Quantitative evaluation of blood-brain barrier permeability following middle cerebral artery occlusion in rats. *Brain Res.* **739,** 88–96.

47. Gasche, Y., Fujimura, M., Morita-Fujimura, Y., Copin, J. C., Kawase, M., Massengale, J., et al. (1999) Early appearance of activated matrix metalloproteinase-9 after focal cerebral ischemia in mice: a possible role in blood-brain barrier dysfunction. *J. Cereb. Blood Flow Metab.* **19(9),** 1020–1028.

48. Kondo, T., Reaume, A. G., Huang, T. T., Carlson, E., Murakami, K., Chen, S. F., et al. (1997) Reduction of CuZn-superoxide dismutase activity exacerbates neuronal cell injury and edema formation after transient focal cerebral ischemia. *J. Neurosci.* **17,** 4180–4189.

49. Rosenberg, G. A., Estrada, E. Y., and Dencoff, J. E. (1998) Matrix metalloproteinases and TIMPs are associated with blood-brain barrier opening after reperfusion in rat brain. *Stroke* **29(10),** 2189–2195.

50. Hamann, G. F., Okada, Y., and Del Zoppo, G. J. (1996) Hemorrhagic transformation and microvascular integrity during focal cerebral ischemi/reperfusion. *J. Cereb. Blood Flow Metab.* **16,** 1373–1378.

51. Robert, A. M. and Godeau, G. (1974) Action of proteolytic and glycolytic enzymes on the permeability of the blood-brain barrier. *Biomedicine* **21(1),** 36–39.

52. Rosenberg, G. A., Kornfeld, M., Estrada, E., Kelley, R. O., Liotta, L. A., and Stetler-Stevenson, W. G. (1992) TIMP-2 reduces proteolytic opening of blood brain barrier by type IV collagenase. *Brain Res.* **576,** 203–207.

53. Armao, D., Kornfeld, M., Estrada, E. Y., Grossetete, M., and Rosenberg, G. A. (1997) Neutral proteases and disruption of the blood-brain barrier in rat. *Brain Res.* **767(2),** 259–264.

54. Migita, K., Eguchi, K., Kawabe, Y., Ichinose, Y., Tsukada, T., Aoyagi, T., et al. (1996) TNF-alpha-mediated expression of membrane-type matrix metalloproteinase in rheumatoid synovial fibroblasts. *Immunology* **89,** 553–557.

55. Nikkari, S. T., Hoyhtya, M., Isola, J., and Nikkari, T. (1996) Macrophages contain 92-kd gelatinase (MMP-9) at the site of degenerated internal elastic lamina in temporal arteritis. *Am. J. Pathol.* **149,** 1427–1433.

56. Clements, J. M., Cossins, J. A., Wells, G. M., Corkill, D. J., Helfrich, K., Wood, L. M., et al. (1997) Matrix metalloproteinase expression during experimental autoimmune encephalomyelitis and effects of a combined matrix metalloproteinase and tumour necrosis factor-alpha inhibitor. *J. Neuroimmunol.* **74,** 85–94.

57. Pagenstecher, A., Stalder, A. K., Kincaid, C. L., Shapiro, S. D., and Campbell, I. L. (1998) Differential expression of matrix metalloproteinase and tissue inhibitor of matrix metalloproteinase genes in the mouse central nervous system in normal and inflammatory states. *Am. J. Pathol.* **152,** 729–741.

58. Backstrom, J. R., Lim, G. P., Cullen, M. J., and Tokes, Z. A. (1996) Matrix metalloproteinase-9 (MMP-9) is synthesized in neurons of the human hippocampus and is capable of degrading the amyloid-beta peptide (1–40). *J. Neurosci.* **16,** 7910–7919.

59. Gottschall, P. E. and Yu, X. (1995) Cytokines regulate gelatinase A and B (matrix metalloproteinase 2 and 9) activity in cultured rat astrocytes. *J. Neurochem.* **64,** 1513–1520.

60. Gottschall, P. E., Yu, X., and Bing, B. (1995) Increased production of gelatinase B (matrix metalloproteinase-9) and interleukin-6 by activated rat microglia in culture. *J. Neurosci. Res.* **42,** 335–342.

61. Romanic, A. M., White, R. F., Arleth, A. J., Ohlstein, E. H., and Barone, F. C. (1998) Matrix metalloproteinase expression increases after cerebral focal ischemia in rats: inhibition of matrix metalloproteinase-9 reduces infarct size. *Stroke* **29,** 1020–1030.

62. Oh, L. Y. S., Larsen, P. H., Krekoski, C. A., Edwards, D. R., Donovan, F., Werb, Z., et al. (1999) Matrix metalloproteinase-9/Gelatinase B is required for process outgrowth by oligodendrocytes. *J. Neurosci.* **19(19),** 8464–8475.

63. Rosenberg, G. A., Navratil, M., Barone, F., and Feuerstein, G. (1996) Proteolytic cascade enzymes increase in focal cerebral ischemia in rat. *J. Cereb. Blood Flow Metab.* **16,** 360–366.

64. Asahi, M., Asahi, K., Jung, J. C., del Zoppo, G. J., Fini, M. E., and Lo, E. H. (2000) Role for matrix metalloproteinase 9 after focal cerebral ischemia: effects of gene knockout and enzyme inhibition with BB-94. *J. Cereb. Blood Flow Metab.* **20(12),** 1681–1689.

65. Fujimura, M., Gasche, Y., Morita-Fujimura, Y., Massengale, J., Kawase, M., and Chan, P. H. (1999) Early appearance of activated matrix metalloproteinase-9 and blood-brain barrier disruption in mice after focal cerebral ischemia and reperfusion. *Brain Res.* **842,** 92–100.

66. Heo, J. H., Lucero, J., Abumiya, T., Koziol, J. A., Copeland, B. R., and del Zoppo, G. J. (1999) Matrix metalloproteinases increase very early during experimental focal cerebral ischemia. *J. Cereb. Blood Flow Metab.* **19(6),** 624–633.

67. Anthony, D. C., Ferguson, B., Matyzak, M. K., Miller, K. M., Esiri, M. M., and Perry, V. H. (1997) Differential matrix metalloproteinase expression in cases of multiple sclerosis and stroke. *Neuropathol. Appl. Neurobiol.* **23,** 406–415.

68. Clark, A.W., Krekoski, C. A., Bou, S. S., Chapman, K. R., and Edwards, D. R. (1997) Increased gelatinase A (MMP-2) and gelatinase B (MMP-9) activities in human brain after focal ischemia. *Neurosci. Lett.* **238(1–2),** 53–56.

69. Fridman, R., Toth, M., Pena, D., and Mobashery, S. (1995) Activation of progelatinase B (MMP-9) by gelatinase A (MMP-2). *Cancer Res.* **55,** 2548–2555.

70. Okada, Y., Gonoji, Y., Naka, K., Tomita, K., Nakanishi, I., Iwata, K., et al. (1992) Matrix metalloproteinase 9 (92 kDa gelatinase/type IV collagenase) from HT 1080 human firborsarcoma cells. Purification and activation of the precursor and enzymatic properties. *J. Biol. Chem.* **267,** 21,712–21,719.

71. Bergmann, U., Tuuttila, A., Stetler-Stevenson, W. G., and Tryggvason, K. (1995) Autolytic activation of recombinant human 72 kilodalton type IV collagenase. *Biochemistry* **34,** 2819–2825.

72. Goldberg, G. I., Marmer, B. L., Grant, J. A., Eisen, A. Z., Wilhelm, S., and He, C. (1989) Human 72k type IV collagenase forms a complex with a tissue inhibitor of metalloproteinase designed TIMP-2. *Proc. Natl. Acad. Sci. USA* **86,** 8207–8211.

73. Strongin, A. Y., Collier, Y., Bannikov, G., Marmer, B. L., Grant, G. A., and Goldberg, G. I. (1995) Mechanism of cell surface activation of 72-kDa type IV collagenase. *J. Biol. Chem.* **270,** 5331–5338.

74. Strongin, A. Y., Marmer, B. L., Grant, G. A., and Goldberg, G. I. (1993) Plasma membrane-dependent activation of the 72-kDa type IV collagenase is prevented by complex formation with TIMP-2. *J. Biol. Chem.* **268(19),** 14,033–14,039.

75. Willenbrock, F. and Murphy, G. (1994) Structure–function relationships in the tissue inhibitors of metalloproteinases. *Am. J. Respir. Crit. Care Med.* **150,** S165–S170.

76. Ingber, D. E., Madri, J. A., and Folkman, J. (1986) A possible mechanism for inhibition of angiogenesis by angiostatic steroids: induction of capillary basement membrane dissolution. *Endocrinology* **119(4),** 1768–1775.

77. Harkness, K. A., Adamson, P., Sussman, J. D., Davies-Jones, G. A., Greenwood, J., and Woodroofe, M. N. (2000) Dexamethasone regulation of matrix metalloproteinase expression in CNS vascular endothelium. *Brain* **123(Pt. 4),** 698–709.

78. Wachtel, M., Frei, K., Ehler, E., Fontana, A., Winterhalter, K., and Gloor, S. M. (1999) Occludin proteolysis and increased permeability in endothelial cells through tyrosine phosphatase inhibition. *J. Cell. Sci.* **112(Pt. 23),** 4347–4356.

79. Yang, G. Y., Liu, X. H., Kadoya, C., Zhao, Y. J., Mao, Y., Davidson, B. L., et al. (1998) Attenuation of ischemic inflammatory response in mouse brain using an adenoviral vector to induce overexpression of interleukin-1 receptor antagonist. *J. Cereb. Blood Flow Metab.* **18(8),** 840–847.

80. Partridge, C. A., Jeffrey, J. J., and Malik, A. B. (1993) A 96-kDa gelatinase induced by TNFα contributes to increased microvascular endothelial permeability. *Am. J. Physiol.* **265,** L438–L447.

81. Lewen, A., Matz, P., and Chan, P. H. (2000) Free radical pathways in CNS injury. [in process citation]. *J. Neurotrauma* **17(10),** 871–890.

82. Asahi, M., Asahi, K., Wang, X., and Lo, E. H. (2000) Reduction of tissue plasminogen activator-induced hemorrhage and brain injury by free radical spin trapping after embolic focal cerebral ischemia in rats. *J. Cereb. Blood Flow Metab.* **20,** 452–457.

83. Morita-Fujimura, Y., Fujimura, M., Gasche, Y., Copin, J. C., and Chan, P. H. (2000) Overexpression of copper and zinc superoxide dismutase in transgenic mice prevents the induction and activation of matrix metalloproteinases after cold injury-induced brain trauma. *J. Cereb. Blood Flow Metab.* **20(1),** 130–138.

84. Weiss, S. J., Peppin, G., Ortiz, X., Ragsdale, C., and Test, S. T. (1985) Oxidative autoactivation of latent collagenase by human neutrophils. *Science* **227,** 747–749.

85. Kuzuya, M., Satake, S., Ramos, M. A., Kanda, S., Koike, T., Yoshino, K., et al. (1999) Induction of apoptotic cell death in vascular endothelial cells cultured in three-dimensional collagen lattice. *Exp. Cell Res.* **248(2),** 498–508.

86. Li, A. E., Ito, H., Rovira, I. I., Kim, K. S., Takeda, K., Yu, Z. Y., et al. (1999) A role for reactive oxygen species in endothelial cell anoikis. *Circ. Res.* **85(4),** 304–310.

87. The National Institute of Neurological Disorders and Stroke rt-PA Stroke Study Group (1995) Tissue plasminogen activator for acute ischemic stroke. *N. Engl. J. Med.* **333(24),** 1581–1587.

88. Lapchak, P. A., Chapman, D. F., and Zivin, J. A. (2000) Metalloproteinase inhibition reduces thrombolytic (tissue plasminogen activator)-induced hemorrhage after thromboembolic stroke. *Stroke* **31(12),** 3034–3040.

III
Alzheimer's Disease

12

Neuroinflammatory Environments Promote Amyloid-β Deposition and Posttranslational Modification

Craig S. Atwood, Mark A. Smith, Ralph N. Martins, Rudolph E. Tanzi, Alex E. Roher, Ashley I. Bush, and George Perry

1. INTRODUCTION

Indisputable evidence indicates that an inflammatory response is associated with neuron and neurite damage and the deposition of amyloid β (Aβ) and neurofibrillary tangles (NFT) in Alzheimer disease (AD) (*see* ref. *1* for a comprehensive review). Just as in the periphery, where degenerating tissue and insoluble materials (resulting from trauma, embolism, and rupture) promote inflammation, these classical stimulants also promote inflammation in the AD brain. From a spatio-temporal perspective, the stimuli promoting neuroinflammation are microlocalized and are present from early preclinical to the terminal stages of AD. Likewise, the upregulation of acute-phase proteins, complement, cytokines, and other inflammatory mediators also is microlocalized and chronic.

An integral component of the inflammatory response is the localization of microglia to the inflammatory stimuli (protein deposits) and their activation, resulting in the release of the oxidizing system of O_2^-, myeloperoxidase (MPO), and nitric oxide synthase (NOS). Although the respiratory burst of macrophages during inflammation in peripheral tissues is a response designed to kill invading pathogens or tumor cells, such a (chronic) response within the brain containing a resident population of nonrenewing cells could have a devastating impact on the function and the survival of the organism. Nowhere is this more apparent than in AD, where the loss of large hippocampal neurons leads to a progressive loss in memory (*see* ref. *2*) and is associated with numerous inflammatory and oxidative changes (*3,4*). Similar inflammatory responses to protein accumulation are observed in other neurodegenerative diseases such as Parkinson's disease, amyotrophic lateral sclerosis (*5*), following head trauma, and Down's syndrome (*6*), and are also likely to exacerbate neuronal cell loss in these conditions. In this chapter, we will review the acute-phase response to inflammation and the resultant oxidative environments that lead to the posttranslational modification of amyloid deposits and neurotoxicity in the AD brain.

2. ACUTE-PHASE RESPONSE TO NEURONAL INFLAMMATION

Like other inflammatory diseases, AD is associated with the upregulation of a diverse set of acute-phase proteins that arise early in inflammation. Acute-phase mechanisms

From: *Neuroinflammation, 2nd Edition: Mechanisms and Management*
Edited by: P. L. Wood © Humana Press Inc., Totowa, NJ

involved in the initiation, clearance, and subsequent tissue rebuilding process after injury are coordinated by the pleiotrophic actions of numerous molecules. Acute-phase molecules that signal the pro-inflammatory mechanisms of wound healing include interleukin-1 (IL-1), IL-6, tumor necrosis factor (TNF)-α, cell adhesion molecules (ICAM-1), colony-stimulating factors (M-CSFs) and acute-phase proteins such as C-reactive protein, serum amyloid A, and transthyretin (TTR) (1). Simultaneous signaling from other molecules control and assuage inflammation toward the end of the wound healing process. For example, transforming growth factor-β1 (TGF-β1) has been implicated in the alleviation of inflammation and the tissue rebuilding process, whereas α$_2$-macroglobulin (α$_2$M), together with its protease inhibitory and removal activity, when bound to LRP acts as a clearance system for inflammatory proteins (7) such as apolipoprotein E (ApoE), amyloid β protein precursor (AβPP), Aβ (8), lactoferrin, tissue plasminogen activator, urokinase-type plasminogen activator, plasminogen activator inihibitor-1, lipoprotein lipase, receptor-associated protein (9,10), IL-1β, TGF-β, platelet-derived growth factor, and fibroblast growth factor (7,11,12). The oxidative environment induced by inflammatory processes also may act as a signaling mechanism to neurons, at this time, as has been observed with the activation of c-Jun N-terminal kinase/stress-activated protein kinase and p38 (13,14). Indeed, human neurons oxidatively challenged with H$_2$O$_2$ upregulate the expression of proteins involved in neurite outgrowth and synapse formation (proliferating cell nuclear antigen, GAP-43, nitric oxide synthase 3, and neuronal thread protein) (15) and implicates oxidative mechanisms in the healing process.

2.1. Pleiotrophic Properties of AβPP and Aβ

The amyloid β protein precursor is another acute-phase reactant upregulated in neurons, astrocytes, and microglial cells in response to inflammation, energetic stress, and a multitude of associated cellular stresses (reviewed in ref. 16). Depending on its concentration, AβPP may be either toxic or trophic. At subnanomolar concentrations, soluble AβPP (sAβPP) stimulates NF-κB activity, IL-1, inducible nitric oxide synthase (iNOS) expression, and neurotoxicity (17). Conversely, sAβPP has neurotrophic properties, including protection from transient ischemia (18) and acute and chronic excitotoxic injury (19, 20). In addition, sAβPP has been shown to increase neuronal survival and growth by mediating nerve growth factor (NGF)-induced neurite extension (21–23), increasing synaptic density (24), having synaptotrophic properties (25), regulating cell growth (26), and showing general trophic responses (27,28). Both NGF and neuronal differentiation regulate AβPP expression (29–31). Because of these pleiotrophic properties, it is likely that AβPP is involved in the initial clearance and subsequent rebuilding of tissue after injury (27). Indeed, the prominent growth response induced by sAβPP may reflect attempted regeneration of viable, healthy neurons following synaptic disconnection resulting from death of other neurons, a situation that might be expected during the course of AD and following head injury.

Amyloid β also exhibits neurotrophic properties when present in physiological (low nanomolar) concentrations. The increased generation of Aβ under conditions of energetic stress (*see* ref. 16) may be a response to the oxidative challenge observed in the brain in AD and following injury. We have recently found that Aβ has significant antioxidant (superoxide dismutase) activity (32) and that nanomolar concentrations of Aβ can block neuronal apoptosis following trophic factor withdrawal (33). These findings are consistent with the trophic and neuroprotective action of Aβ at physiological concentrations in deprived conditions and neonatal cells that have been reported during the last decade *(34–44)*. In support of this role of Aβ as an antioxidant, the Aβ burden of the AD-affected brain has been shown to be significantly negatively correlated with oxidative stress markers *(45–47)* and *in situ* soluble Aβ levels are inversely correlated with synaptic loss *(48)*. Aβ also has been shown to protect lipoproteins from oxidation in cerebrospinal fluid and plasma (the mechanism of which is thought to involve metal ion sequestration) *(49,50)*. Moreover, Andorn and Kalaria *(51)* have recently shown that low concentrations of Aβ possess significant antioxidant activity in an ascorbate-stimulated lipid-peroxidation assay of postmortem human brain membrane preparations. Together, these data provide a plausible physiological explanation for the increased generation of Aβ in AD and following head trauma, one that is aimed at reducing oxidative damage (thereby preventing reactive oxygen species [ROS]-mediated neuronal apoptosis), promoting neurite outgrowth, and maintaining structural integrity (*see* below).

2.2. Aβ Functions Integral to the Inflammatory Response to Injury

A number of the physiochemical properties of Aβ support the above-mentioned functions of Aβ. We have previously shown that Aβ, unlike most proteins, binds Cu under acidotic conditions, in keeping with a role of Aβ as an antioxidant or molecule involved in maintaining structural integrity under stress conditions *(50)*. Because Aβ can bind to heparan sulfate of the extracellular matrix (via residues 12–17, VHHQKL) *(52–54)*, a structure that regulates adhesive events such as neurite outgrowth and synaptogenesis, Aβ may be rapidly assembled in the extracellular space by Cu and Zn that are known to be mobilized to sites of inflammation (reviewed in refs. *50* and *55*). In this way, Aβ may dampen oxidative insults by binding excess or loosely bound redox active metal ions *(56, 57)*, which, at the same time, act to switch on its neurotrophic properties (antioxidant activity/sealant properties). The small size of Aβ together with its hydrophobic and hydrophilic bridging structure that can span both the plasma membrane and bind to extracellular matrix molecules such as heparan sulfate, together with its ability to self-aggregate make Aβ an excellent candidate as a molecule that could form an intracranial "scab." The aggregated Cu,Zn-Aβ would serve as an O_2^- scavenging solid-phase matrix (which disassembles when Zn and Cu levels lower, as the tissue damage resolves). This may explain the rapid cortical deposition of Aβ in stroke and following head injury *(58)*. It should be noted that the brain maintains high concentrations of both Cu (approx 70 µ*M*) and Zn (approx 350 µ*M*) *(59)*.

Because cerebrovascular hemorrhage has serious consequences for neuronal survival and given the vast vasculature that nourishes the brain, it would seem likely that mechanisms to prevent and/or limit vascular rupture must have evolved, particularly in higher vertebrates of greater life-span where there is an increased time-related potential for vascular rupture. Acute-phase molecules responding to inflammation associated with injury are prime candidates for such a function. Aβ may have evolved as such a molecule and Aβ's vascular deposition supports a role in maintaining vascular structure *(60)*. However, abnormal deposition of Aβ, such as in individuals carrying the Glu-Gln substitution at position 22 of Aβ, results in cerebrovascular hemorrhage (e.g., ref. *61*). The release of molecules from the VHHQKL region (i.e., as inflammation resides and pH increases) would be expected to promote Aβ-induced microglial activation and the engulfing and removal of the "scab." The interplay among metal ions, extracellular matrix proteins, and microglia, which all bind to the VHHQKL region of Aβ, may have important consequences for the recruitment and activation of microglia (*see* Section 3) and AD pathology. The above activities of Aβ would ensure that minimal numbers of neurons are lost after head trauma, an important physiological response that would limit the loss of terminally differentiated neurons. Thus, the acute-phase generation of Aβ may have evolved as a secondary antioxidant defense and/or sealant system required during times of excessive ROS generation and/or trauma *(50,55)*, whereas the upregulation of sAβPP/Aβ may act to promote neuronal growth and neurite extension.

3. STIMULI PROMOTING MICROGLIAL ACTIVATION

Unlike nondemented elderly individuals in whom resting microglia are typically found in the white matter, the AD brain contains clusters of activated microglia in white and gray matter *(62,63)*. Microglia have been shown to cluster at sites of Aβ deposition in AD brain *(62,64,65)*, at sites of Aβ deposition in AβPP transgenic mice *(66,67)* and with extracellular NFTs *(68)*. Quantitative histopathological analyses of AD brains indicate that >80% of core plaques are associated with clusters of reactive microglia, whereas <50% of diffuse plaques show such an association *(69,70)*. Interestingly, diffuse Aβ deposits in elderly individuals contain only quiescent microglia, suggesting a role for activated microglia *(63,71–74)*. Thus, like peripheral macrophages in systemic amyloidosis *(75)*, microglia appear to be intimately involved in plaque homeostasis, although further work is needed to confirm this view.

The upregulated expression of cell adhesion molecules *(76)*, increased cytokine production, and activation of the complement system and microglial cells in the brain parenchyma of AD patients are closely associated with Aβ deposits *(77,78)*. It has been suggested that one mechanism by which Aβ may drive AD pathogenesis is via its ability to stimulate inflammation *(77)*. This concept has gained strong support from the results of in vitro studies that show that Aβ is capable of priming and/or triggering the respiratory burst of cultured rat microglia and human phagocytes *(79–81)*. These cells, which exist in a quiescent state in normal brain tissue, can become activated in response to neuronal damage or aggregated Aβ *(78,79,82–87)*. Studies by Giulian, Roher, and colleagues have shown that the HHQK domain (residues 13–16) of Aβ binds microglia and promotes microglial activation and neuron cell death via toxins released into the culture media *(82,84)*. This sequence itself was not neurotoxic and reduced inflammation elicited by injection of Aβ peptides in the rat brain *(84)*. This same cluster of basic amino acids also is known to bind with high affinity to heparan sulfate *(53,54,88)*. Interestingly, fibrillized Aβ can induce cortical glial cells in vitro and in vivo to locally deposit chondroitin sulfate containing proteoglycan *(89)*. The activation of astroglial cells by this mechanism has been postulated to inhibit cortical neuron adhesion or growth *(89)*. Thus, the overproduction of Aβ by the stresses described above or the lack of sufficient heparan sulfate molecules would be predicted to lead to Aβ (plaque) accessibility to glia and their activation. Indeed, it has been demonstrated that infusions of Aβ peptides or implantation of native peptide fragments into the rat neocortex induces reactive microgliosis beyond that of simple needle trauma *(84,90,91)*. Aβ has been shown to directly activate the NADPH–oxidase complex of inflammatory cells with the increased production of $O_2^{-\bullet}$ and increased generation of H_2O_2 *(92–97)*. Thus, the release of ROS from activated

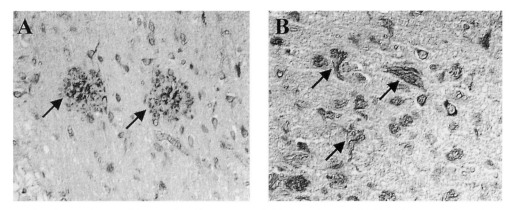

Fig. 1. Myeloperoxidase colocalizes to amyloid deposits and NFTs. Tissue sections from AD brain were fixed in methacarn, sectioned (8 μm), and stained with an antibody to MPO (Chemicon International, Inc., CA). Numerous amyloid plaques (**A**) and neurofibrillary tangles (**B**) (determined with Congo red) are immunopositive for MPO (arrows). Nonimmune staining and control sections were negative (not shown).

microglia likely contributes in a significant way to the oxidative challenge observed in the AD brain.

The respiratory burst of activated phagocytes also is accompanied by the release of granule-containing proteins such as the myeloid-specific enzyme myeloperoxidase (MPO), a heme protein of 150 kDa *(98,99)*. Together with H_2O_2, MPO acts as an anti-microbial agent and is present in very high concentrations in monocytes, neutrophils, and some reactive macrophages, where it comprises up to 2–5% of the cell mass. The release of MPO and ROS during the respiratory burst of human phagocytes, designed to generate potent cytotoxins that kill invading pathogens and tumor cells, promotes an environment that oxidizes, crosslinks, nitrates, and chlorinates amino acids, cholesterol, lipids, lipo(proteins), nucleotides, and DNA. MPO–H_2O_2 systems promote the synthesis of tyrosine crosslinked species such as dityrosine *(100)* and catalyze the formation of nitrotyrosine-modified proteins *(101)* as well as advanced end-product modifications *(102)*. In the presence of a halide ion (usually Cl), MPO catalyzes the reaction between H_2O_2 and Cl to generate hypochlorous acid (HOCl), a potent oxidant that chlorinates and oxidizes lipoproteins *(103)*, ApoE *(104,105)*, and cholesterol *(106)*, contributes to the formation of reactive nitrogen species *(107)*, and crosslinks proteins *(100)*.

Myeloperoxidase has been colocalized to both amyloid plaques (Fig. 1A) *(108)* and NFT (Fig. 1B) in AD-affected brains, colocalizing with Aβ1-42, but it is not found in the large neurons of the hippocampus or in age-matched control sections *(108)*. The highest levels of MPO are detected in the brains of individuals carrying the ApoE4 gene *(108)*. Interestingly, Aβ1-42 has been shown to enhance MPO mRNA expression in rat microglial cultures *(108)*. Furthermore, inheritance of a polymorphism on MPO has been associated with increased incidence of AD in females, but of decreased incidence in males *(108)*. Interestingly, enhanced peroxidase immunoreactivity has been observed following the activation of astrocytes and microglia in rat brain *(109)*.

Similar inflammatory environments exist in the atherosclerotic lesions of the heart, where MPO has been colocalized to HOCL-modified epitopes of human atheroma (type

IV), fibroatheroma (type [V] and complicated [type VI]) lesions *(110)* and selective enrichment of protein dityrosine (DT) crosslinks and 3-chlorotyrosine *(111)*. In vitro, high-density lipoprotein (HDL) exposed to peroxidase-generated tyrosyl radical undergoes tyrosylation and crosslinking of its apolipoproteins *(112)*. Therefore, the MPO–H_2O_2–Cl system may play a critical role in converting low-density lipoprotein (LDL) into an atherogenic form *(113)*, important in the formation of vascular atherosclerotic lesions.

4. OXIDATIVE POSTTRANSLATIONAL MODIFICATIONS OF Aβ

Although Aβ is a normally soluble and constitutive protein found in tissue and biological fluids of healthy individuals (*see* ref. *50* and references therein), Aβ aggregates together with other proteins such as ApoE, MPO, and amyloid P-component to form diffuse amorphous deposits and dense, focal, extracellular deposits in AD *(60,108,114)*. Aβ extracted from biological systems normally migrates as an apparent approx 4-kDa monomer on sodium dodecyl sulfate–polyacrylamide gel electrophoresis (SDS-PAGE) *(115)*; however, Aβ extracted from the AD-affected postmortem brain specimens migrates on SDS-PAGE as SDS-, urea-, and formic acid-resistant oligomers *(116–119)*.

4.1. Aβ Crosslinking

Oligomerized Aβ is a characteristic posttranslational modification of Aβ extracted from amyloid plaques of AD brains that may account for up to 10–20% of total Aβ species *(116–119)*. Oligomerized Aβ is not normally detected or is found in very low concentrations in soluble Aβ extracted from cell lines *(120,121)*. Matrix-assisted laser desorption ionization–mass spectrometry of these SDS-resistant oligomers extracted from neuritic plaque and vascular amyloid indicate the presence of covalently crosslinked dimeric and trimeric Aβ species *(118)*. We recently reported that these SDS-resistant oligomers of Aβ extracted from amyloid core plaques are tyrosine crosslinked at position 10, forming dimeric and trimeric Aβ oligomers *(122*; Atwood et al., unpublished data). Our finding may contribute to the elevation of total DT (fivefold to eightfold) in the AD brain compared to normal control tissue *(123)*. In support of this, Hensley and colleagues only found DT in those regions affected by senile plaque pathology (hippocampus, inferior parietal lobule, superior/middle temporal gyri). Both peroxidase- and metal-catalyzed oxidation systems promote the crosslinking of human Aβ *(122,124, 125)*, but not rat Aβ in vitro. The formation of tyrosine crosslinked human Aβ in vivo indicates that tyrosine residues of Aβ are proximate within the amyloid deposit. Benzinger et al. *(126)* has recently provided evidence that the tyrosines of fibrillized Aβ may be in close proximity.

Tyrosine crosslinking is present in mammalian systems, most notably in connective tissues and structural proteins *(127)*. The DT crosslink is resistant to cleavage (resistant to 6 *N* HCl at 110°C for 24 h and to protease digestion, trypsin, chymotrypsin, and pronase) *(128)*. Therefore, it is not surprising that tyrosine crosslinkage of proteins is a mechanism normally utilized by various organisms to increase the structural strength of cell membranes and walls for protection against proteolysis and physical trauma. For example, a DT content of only 5 to 8 residues per 10,000 amino acid residues is sufficient to make the fertilization membrane of the sea urchin embryo resistant to proteolysis and physical trauma. Likewise, DT in the cell wall of ascospores of *Saccharomyces cerevisiae* makes these organisms 30- to 100-fold more resistant to heating to 55.5°C, 100-fold more resistant to ether, and extremely resistant to physical trauma and proteolytic degradation by numerous proteases *(128–132)*. Although this protective mechanism is advantageous in single-cell organisms and peripheral tissues, increased tyrosine crosslinkage may not be a usual feature of proteins within the cellular structures of the brain except perhaps during head trauma, when it may serve a structural function. The DT load of Aβ suggests amyloid plaques should be very resistant to resolubilization (e.g., ref. *116*) and may greatly stabilize Aβ deposits.

4.2. Other Posttranslational Modifications

Other modifications identified in Aβ extracted from amyloid core plaques include the identification of pyroglutamated and racemized amino acids *(133–135)* and carbonyl groups and alterations in the amino acid composition of human Aβ *(136)*. Alterations in the composition of certain amino acids (decreases in histidine and tyrosine residues) are consistent with metal and MPO catalyzed oxidation of Aβ. We have previously reported that Cu is coordinated to the histidine residues of Aβ *(50)* and this complex is redox active *(137,138)*. Therefore, it is likely that OH• generation in this coordination site promotes histidine oxidation.

5. MICROGLIAL ACTIVATION AND Aβ POSTTRANSLATIONAL MODIFICATIONS

All of the components (peroxidases, H_2O_2, ROCl, and Cu) necessary for the oxidative modification of Aβ and associated proteins in amyloid deposits are present in the neurochemical environment promoted by the chronic inflammatory response to amyloidosis and trauma. Our results indicate a central role for this neurochemical environment in the posttranslational modification of amyloid plaques. The activation of microglia in response to Aβ accumulation may promote tyrosine crosslinkage of depositing Aβ, thereby inhibiting its clearance and leading to a vicious cycle of enhanced microglial activation. HOCl-oxidized LDL has been shown to induce and amplify inflammatory reactions by the induction of chemokine synthesis and to induce chemoattraction of neutrophils *(139)*. HOCl–LDL also stimulates enhanced production of ROS, adhesion to endothelial cells *(140)* and alters the aggregation and release reactions of activated platelets *(141)*.

It is not completely clear why microglia, the resident macrophages of the brain, do not clear amyloid deposits in vivo. It is possible that glia in vivo receive multiple (or incomplete) signals from their interaction with amyloid plaques and astrocytes, one that activates and one that prevents phagocytosis (and the removal of normal structures). It is known that this process is influenced by resident astrocyes *(142)*. Amyloid plaque clearance is markedly suppressed by the presence of astrocytes and is, in part, the result of a diffusible factor(s) because astrocyte-conditioned but not fibroblast-conditioned media reduces microglial phagocytosis. Thus, astrocytes that lie in close proximity to microglia and senile plaques may modulate and may prevent the efficient clearance of senile plaque material and allow them to persist in Alzheimer's disease *(142)*. Alterations in the recognition sequence of Aβ as a result of its oxidative modification also may prevent the normal uptake and clearance of Aβ by microglia. Both amyloid-derived monomeric and oligomeric Aβ extracted from AD-affected brains promote neuron cell death at concentrations of around 100 nM, but only in the presence of microglia, indicating the cytotoxic actions of Aβ may be mediated by activating microglia. Therefore, these inflammatory mechanisms may be central not only to the formation of amyloid plaques but also to the progressive neuronal cell loss of the AD brain that result from the oxidative attack. Whether the oxidative crosslinking of Aβ is a means of concentrating and insulating toxic-soluble Aβ from neuronal damage is unclear.

The chronic activation of oxidant-producing inflammatory enzyme systems could act as an important link between the development of amyloid plaques and atherosclerotic plaques in artery walls and explain the deposition of proteins in other degenerative conditions. For example, the catabolic resistance of DT modifications of proteins could explain the contribution of tyrosine crosslink polymers to the crosslinking of α-crystallin in fluorescent cataract formation *(143)*, the oligomerization of Mn-superoxide dismutase (SOD) detected in renal graft rejection *(144)*, the crosslinking of proteins in sputum from cystic fibrosis patients *(145)*, and lipofuscin formation *(146)*. MPO also has been implicated as playing a role in multiple sclerosis, where increased MPO expression in brain macrophages accelerates damage to the myelin sheath *(147)*. SOD also is known to deposit in the nervous system in amyotrophic lateral sclerosis *(148)*, although it is unknown if this is a result of DT crosslinkage. Recently, DT crosslinking of recombinant human α-synuclein induced by MPO oxidative conditions was proposed as a mechanism for the formation of Lewy body intraneuronal inclusions seen in Parkinson's disease *(149)*.

6. Aβ-INDUCED NEUROTOXICITY MEDIATED BY MICROGLIA

Amyloid β at high concentrations (generally >10 μM) has been found to act as a potent neurotoxic agent to both neuronal (e.g., refs. *36, 47, 150,* and *151*) and non-neuronal cells *(37,152)*, particularly when present in the fibrillar form. Although there has been much debate as to whether such concentrations are physiologically relevant, support for an indirect action of Aβ on neuronal survival is indicated by the facts that neuritic/core plaques are not directly neurotoxic, as neurons can be successfully grown atop Aβ peptides *(39,89,153)*, neuritic/core plaques added directly to neurons do not cause neuron damage *(70,142)*, and Aβ peptides infused into the brain do not cause tissue injury *(154–157)*. Interestingly, injection of soluble Aβ into the brain has been shown to protect against redox metal-ion-induced neuronal lesions in rats *(158)*. In support of this idea, transgenic mice with massive accumulations of Aβ deposits (approx 150–200 μM) (Atwood et al., unpublished data) generally show no neuronal death, in marked contrast to in vitro findings of Aβ mediated neurotoxicity and apoptosis *(159,160)*. In contrast, total Aβ levels in the AD brain are approximately 10 μM *(119)*.

Support for Aβ-induced neurotoxicity mediated via microglial activation has arisen from in vitro studies, indicating concentrations of Aβ as low as 1–2 μM can induce neuron cell death but only in the presence of microglia *(82,118)*. Aβ has been shown to directly induce an inflammatory response when injected into the rat brain *(84)* and, as described earlier, microglial activation by Aβ is dependent on the open HHQK domain (*see* Section 3.1).

7. SUMMARY

Aβ, a soluble constituent of cells has numerous beneficial properties, however its aggregation into soluble deposits may drive neuroinflammation and oxidative stress. The induction of inflammation by protein deposits may explain the crosslinking and deposition of proteins in AD and other inflammatory diseases. It is becoming clear that the activation of oxidant-producing inflammatory enzyme systems in the AD brain plays a central role in the posttranslational modification of amyloid plaques. Blocking inflammatory-induced Aβ modifications, Aβ generation, and Aβ interaction with microglia may prevent activation of microglia, plaque deposition, and neuronal cell death and could, therefore, provide an important therapeutic target.

REFERENCES

1. Neuroinflammation Working Group. Akiyama, H., Barger, S., Barnum, S., Bradt, B., Bauer, J., Cole, G. M., et al. (2000) Inflammation and Alzheimer's disease. *Neurobiol. Aging* **21,** 383–421.
2. Atwood, C. S. (2000) Evidence that oxidative challenges promote neuronal sprouting and cell cycle re-entry. *J. Alzheimer's Dis.* **2,** 283–287.
3. Smith, M. A., Perry, G., Richey, P. L., Sayre, L. M., Anderson, V. E., Beal, M. F., et al. (1996) Oxidative damage in Alzheimer's. *Nature* **382,** 120–121.
4. Sayre, L. M., Zelasko, D. A., Harris, P. L. R., Perry, G., Salomon, R. G., and Smith, M. A. (1997) 4-Hydroxynonenal-derived advanced lipid peroxidation end products are increased in Alzheimer's disease. *J. Neurochem.* **68,** 2092–2097.
5. McGeer, P. L. and McGeer, E. G. (1998) Glial cell reactions in neurodegenerative diseases: pathophysiology and therapeutic interventions. *Alzheimer Dis. Assoc. Disord.* **12,** S1–S6.
6. Eikelenboom, P., Rozemuller, J. M., and van Muiswinkel, F. L. (1998) Inflammation and Alzheimer's disease: relationships between pathogenic mechanisms and clinical expression. *Exp. Neurol.* **154,** 89–98.
7. Borth W. (1992) Alpha 2-macroglobulin, a multifunctional binding protein with targeting characteristics. *FASEB J.* **6,** 3345–3353.
8. Narita, M., Holtzman, D. M., Schwartz, A. L., and Bu, G. (1997) Alpha2-macroglobulin complexes with and mediates the endocytosis of beta-amyloid peptide via cell surface low-density lipoprotein receptor-related protein. *J. Neurochem.* **69,** 1904–1911.
9. Williams, S. E., Kounnas, M. Z., Argraves, K. M., Argraves, W. S., and Strickland, D. K. (1994) The alpha 2-macroglobulin receptor/low density lipoprotein receptor-related protein and the receptor-associated protein. An overview. *Ann. NY Acad. Sci.* **737,** 1–13.
10. Kounnas, M. Z., Moir, R. D., Rebeck, G. W., Bush, A. I., Argraves, W. S., Tanzi, R. E., et al. (1995) LDL receptor-related protein, a multifunctional ApoE receptor, binds secreted beta-amyloid precursor protein and mediates its degradation. *Cell* **82,** 331–340.
11. Du, Y., Ni, B., Glinn, M., Dodel, R. C., Bales, K. R., Zhang, Z., et al. (1997) α2-Macroglobulin as a beta-amyloid peptide-binding plasma protein. *J. Neurochem.* **69,** 299–305.
12. Hughes, S. R., Khorkova, O., Goyal, S., Knaeblein, J., Heroux, J., Riedel, N. G., et al. (1998) Alpha2-macroglobulin associates with beta-amyloid peptide and prevents fibril formation. *Proc. Natl. Acad. Sci. USA* **95,** 3275–3280.
13. Zhu, X., Raina, A. K., Rottkamp, C. A., Aliev, G., Perry, G., Boux, H., et al. (2001) Activation and redistribution of c-Jun N-terminal kinase/stress activated protein kinase in degenerating neurons in Alzheimer's disease. *J. Neurochem.* **76,** 435–441.
14. Zhu, X., Raina, A. K., Boux, H., Takeda, A., Perry, G., and Smith, M. A. (2000) Activation of p38 pathway links tau phosphorylation, oxidative stress and cell cycle related events in Alzheimer disease. *J. Neuropathol. Exp. Neurol.* **59,** 880–888.
15. de la Monte, S. M., Ganju, N., Feroz, N., Luong, T., Banerjee, K., Cannon, J., et al. (2000) Oxygen free radical injury is sufficient to cause some Alzheimer-type molecular abnormalities in human CNS neuronal cells. *J. Alzheimer's Dis.* **2,** 261–281.

16. Atwood, C. S., Huang, X., Moir, R. D., Smith, M. A., Tanzi, R. E., Roher, A. E., et al. (2001) Neuroinflammatory responses in the Alzheimer's disease brain promote the oxidative post-translational modification of amyloid deposits, in *Alzheimer's Disease: Advances in Etiology, Pathogenesis and Therapeutics* (Iqbal, K., Sisodia, S. S., and Winblad, B., eds.), Wiley, Chichester, pp. 341–361.

17. Banati, R. B., Gehrmann, J., Czech, C., Monning, U., Jones, L. L., Konig, G., et al. (1993) Early and rapid de novo synthesis of Alzheimer beta A4-amyloid precursor protein (APP) in activated microglia. *Glia* **9**, 199–210.

18. Smith-Swintosky, V. L., Pettigrew, L. C., Craddock, S. D., Culwell, A. R., Rydel, R. E., and Mattson, M. P. (1994) Secreted forms of beta-amyloid precursor protein protect against ischemic brain injury. *J. Neurochem.* **63**, 781–784.

19. Mattson, M. P., Cheng, B., Culwell, A. R., Esch, F. S., Lieberburg, I., and Rydel, R. E. (1993) Evidence for excitoprotective and intraneuronal calcium-regulating roles for secreted forms of the beta-amyloid precursor protein. *Neuron* **10**, 243–254.

20. Masliah, E., Westland, C. E., Rockenstein, E. M., Abraham, C. R., Mallory, M., Veinberg, I., et al. (1997) Amyloid precursor proteins protect neurons of transgenic mice against acute and chronic excitotoxic injuries in vivo. *Neuroscience* **78**, 135–146.

21. Milward, E. A., Papadopoulos, R., Fuller, S. J., Moir, R. D., Small, D., Beyreuther, K., et al. (1992) The amyloid protein precursor of Alzheimer's disease is a mediator of the effects of nerve growth factor on neurite outgrowth. *Neuron* **9**, 129–137.

22. Schubert, D. and Behl, C. (1993) The expression of amyloid beta protein precursor protects nerve cells from beta-amyloid and glutamate toxicity and alters their interaction with the extracellular matrix. *Brain Res.* **629**, 275–282.

23. Akar, C. A. and Wallace, W. C. (1998) Amyloid precursor protein modulates the interaction of nerve growth factor with p75 receptor and potentiates its activation of trkA phosphorylation. *Mol. Brain Res.* **56**, 125–132.

24. Roch, J. M., Masliah, E., Roch-Levecq, A. C., Sundsmo, M. P., Otero, D. A., Veinbergs, I., et al. (1994) Increase of synaptic density and memory retention by a peptide representing the trophic domain of the amyloid beta/A4 protein precursor. *Proc. Natl. Acad. Sci. USA* **91**, 7450–7454.

25. Mucke, L., Masliah, E., Johnson, W. B., Ruppe, M. D., Alford, M., Rockenstein, E. M., et al. (1994) Synaptotrophic effects of human amyloid beta protein precursors in the cortex of transgenic. *Brain Res.* **666**, 151–167.

26. Saitoh, T., Sundsmo, M., Roch, J. M., Kimura, N., Cole, G., Schubert, D., et al. (1989) Secreted form of amyloid beta protein precursor is involved in the growth regulation of fibroblasts. *Cell* **58**, 615–622.

27. Araki, W., Kitaguchi, N., Tokushima, Y., Ishii, K., Aratake, H., Shimohama, S., et al. (1991) Trophic effect of beta-amyloid precursor protein on cerebral cortical neurons in culture. *Biochem. Biophys. Res. Commun.* **181**, 265–271.

28. Yamamoto, K., Miyoshi, T., Yae, T., Kawashima, K., Araki, H., Hanada, K., et al. (1994) The survival of rat cerebral cortical neurons in the presence of trophic APP peptides. *J. Neurobiol.* **25**, 585–594.

29. Yoshikawa, K., Aizawa, T., and Maruyama, K. (1990) Neural differentiation increases expression of Alzheimer amyloid protein precursor gene in murine embryonal carcinoma cells. *Biochem. Biophys. Res. Commun.* **171**, 204–209.

30. Fukuchi, K., Deeb, S. S., Kamino, K., Ogburn, C. E., Snow, A. D., Sekiguchi, R. T., et al. (1992) Increased expression of beta-amyloid protein precursor and microtubule-associated protein tau during the differentiation of murine embryonal carcinoma cells. *J. Neurochem.* **58**, 1863–1873.

31. Cosgaya, J. M., Latasa, M. J., and Pascual, A. (1996) Nerve growth factor and ras regulate beta-amyloid precursor protein gene expression in PC12 cells. *J. Neurochem.* **67**, 98–104.

32. Bush, A. I., Lynch, T., Cherny, R. C., Atwood, C. S., Goldstein, L. E., Moir, R. D., et al. (1999) Alzheimer Aβ functions as a superoxide antioxidant in vitro and in vivo. *Soc. Neurosci. Abstr.* **25,** 14.

33. Chan, C.-W., Dharmarajan, A., Atwood, C. S., Huang, X., Tanzi, R. E., Bush, A. I., et al. (1999) Anti-apoptotic action of Alzheimer Aβ. *Alzheimer's Rep.* **2,** 1–6.

34. Whitson, J. S., Selkoe, D. J., and Cotman, C. W. (1989) Amyloid beta protein enhances the survival of hippocampal neurons in vitro. *Science* **243,** 1488–1490.

35. Whitson, J. S., Glabe, C. G., Shintani, E., Abcar, A., and Cotman, C. W. (1990) Beta-amyloid protein promotes neuritic branching in hippocampal cultures. *Neurosci. Lett.* **110,** 319–324.

36. Yankner, B. A., Duffy, L. K., and Kirschner, D. A. (1990) Neurotrophic and neurotoxic effects of amyloid beta protein: reversal by tachykinin neuropeptides. *Science* **250,** 279–282.

37. Behl, C., Davis, J. B., Cole, G. M., and Schubert, D. (1994) Vitamin E protects nerve cells from amyloid beta protein toxicity. *Biochem. Biophys. Res. Commun.* **186,** 944–950.

38. Stephenson, D. T., Rash, K., and Clemens, J. A. (1992) Amyloid precursor protein accumulates in regions of neurodegeneration following focal cerebral ischemia in the rat. *Brain Res.* **593,** 128–135.

39. Koo, E. H., Park, L., and Selkoe, D. J. (1993) Amyloid beta-protein as a substrate interacts with extracellular matrix to promote neurite outgrowth. *Proc. Natl. Acad. Sci. USA* **90,** 4748–4752.

40. Singh, V. K., Cheng, J. F., and Leu, S. J. (1994) Effect of substance P and protein kinase inhibitors on beta-amyloid peptide-induced proliferation of cultured brain cells. *Brain Res.* **660,** 353–356.

41. Takenouchi, T. and Munekata, E. (1995) Trophic effects of substance P and beta-amyloid peptide on dibutyryl cyclic AMP-differentiated human leukemic (HL-60) cells. *Life Sci.* **56,** PL479–PL484.

42. Luo, Y., Sunderland, T., Roth, G. S., and Wolozin, B. (1996) Physiological levels of beta-amyloid peptide promote PC12 cell proliferation. *Neurosci. Lett.* **217,** 125–128.

43. Kaltschmidt, B., Uherek, M., Wellmann, H., Volk, B., and Kaltschmidt, C. (1999) Inhibition of NF-kappaB potentiates amyloid beta-mediated neuronal apoptosis. *Proc. Natl. Acad. Sci. USA* **96,** 9409–9414.

44. Postuma, R. B., He, W., Nunan, J., Beyreuther, K., Masters, C. L., Barrow, C. J., et al. (2000) Substrate-bound beta-amyloid peptides inhibit cell adhesion and neurite outgrowth in primary neuronal cultures. *J. Neurochem.* **74,** 1122–1130.

45. Nunomura, A., Perry, G., Hirai, K., Aliev, G., Takeda, A., Chiba, S., et al. (1999) Neuronal RNA oxidation in Alzheimer's disease and Down's syndrome. *Ann. NY Acad. Sci.* **893,** 362–364.

46. Nunomura, A., Perry, G., Pappolla, M. A., Friedland, R. P., Hirai, K., Chiba, S., et al. (2000) Neuronal oxidative stress precedes amyloid-β deposition in Down syndrome. *J. Neuropathol. Exp. Neurol.* **59,** 1011–1017.

47. Cuajungco, M. P., Goldstein, L. E., Nunomura, A., Smith, M. A., Lim, J. T., Atwood, C. S., et al. (2000) Evidence that the β-amyloid plaques of Alzheimer's disease represent the redox-silencing and entombment of Aβ by zinc. *J. Biol. Chem.* **275,** 19,439–19,442.

48. Lue, L. F., Kuo, Y. M., Roher, A. E., Brachova, L., Shen, Y., Sue, L., et al. (1999) Soluble amyloid beta peptide concentration as a predictor of synaptic change in Alzheimer's disease. *Am. J. Pathol.* **155,** 853–862.

49. Kontush, A., Berndt, C., Arlt, S., Schippling, S., and Beisiegel, U. (1998) Amyloid-β is a physiological antioxidant for lipoproteins in cerebrospinal fluid and plasma. *Neurobiol. Aging* **19,** S43.

50. Atwood, C. S., Huang, X., Moir, R. D., Bacarra, N. M., Romano, D., Tanzi, R. E., et al. (1998) Dramatic aggregation of Alzheimer Aβ by Cu(II) is induced by conditions representing physiological acidosis. *J. Biol. Chem.* **273,** 12,817–12,826.

51. Andorn, A. C. and Kalaria, R. N. (2000) Factors affecting pro- and anti-oxidant properties of fragments of the β-protein precursor (βPP): implication for Alzheimer's disease. *J. Alzheimer's Dis.* **2**, 69–78.

52. Narindrasorasak, S., Lowery, D., Gonzalez-DeWhitt, P., Poorman, R. A., Greenberg, B., and Kisilevsky, R. (1991) High affinity interactions between the Alzheimer's beta-amyloid precursor proteins and the basement membrane form of heparan sulfate proteoglycan. *J. Biol. Chem.* **266**, 12,878–12,883.

53. Brunden, K. R., Richter-Cook, N. J., Chaturvedi, N., and Frederickson, R. C. (1993) pH-dependent binding of synthetic beta-amyloid peptides to glycosaminoglycans. *J. Neurochem.* **61**, 2147–2154.

54. Snow, A. D., Sekiguchi, R., Nochlin, D., Fraser, P., Kimata, K., Mizutani, A., et al. (1994) An important role of heparan sulfate proteoglycan (Perlecan) in a model system for the deposition and persistence of fibrillar A beta-amyloid in rat brain. *Neuron* **12**, 219–234.

55. Atwood, C. S., Huang, X., Moir, R. D., Tanzi, R. E., and Bush, A. I. (1999) The role of free radicals and metal ions in the pathogenesis of Alzheimer's disease, in *Interrelations Between Free Radicals and Metal Ions in Life Processes* (Sigel, A. and Sigel, H., eds.), Metal Ions in Biological Systems, vol. 36. Marcel Dekker, New York, pp. 309–364.

56. Smith, M. A., Harris, P. L. R., Sayre, L. M., and Perry, G. (1997) Iron accumulation in Alzheimer disease is a source of redox-generated free radicals. *Proc. Natl. Acad. Sci. USA* **94**, 9866–9868.

57. Sayre, L. M., Perry, G., Harris, P. L. R., Liu, Y., Schubert, K. A., and Smith, M. A. (2000) In situ oxidative catalysis by neurofibrillary tangles and senile plaques in Alzheimer's disease: a central role for bound transition metals. *J. Neurochem.* **74**, 270–279.

58. Roberts, G. W., Gentleman, S. M., Lynch, A., Murray, L., Landon, M., and Graham, D. I. (1994) Beta amyloid protein deposition in the brain after severe head injury: implications for the pathogenesis of Alzheimer's disease. *J. Neurol. Neurosurg. Psychiatry* **57**, 419–425.

59. Lovell, M. A., Robertson, J. D., Teesdale, W. J., Campbell, J. L., and Markesbery, W. R. (1998) Copper, iron and zinc in Alzheimer's disease senile plaques. *J. Neurol. Sci.* **158**, 47–52.

60. Glenner, G. G. and Wong, C. W. (1984) Alzheimer's disease: initial report of the purification and characterization of a novel cerebrovascular amyloid protein. *Biochem. Biophys. Res. Commun.* **120**, 885–890.

61. Bornebroek, M., Haan, J., and Roos, R. A. (1999) Hereditary cerebral hemorrhage with amyloidosis—Dutch type (HCHWA-D): a review of the variety in phenotypic expression. *Amyloid* **6**, 215–224.

62. Styren, S. D., Civin, W. H., and Rogers, J. (1990) Molecular, cellular, and pathologic characterization of HLA-DR immunoreactivity in normal elderly and Alzheimer's disease brain. *Exp. Neurol.* **10**, 93–104.

63. Eikelenboom, P., Hack, C. E., Kamphorst, W., van der Valk, P., van Nostrand, W. E., et al. (1993) The sequence of neuroimmunological events in cerebral amyloid plaque formation in Alzheimer's disease, in *Alzheimer's Disease: Advances in Clinical and Basic Research* (Corain, B., Iqbal, K., Nicoline, M., Winblad, B., Wisniewski, H., and Zatta, P., eds.), Wiley, Chichester, pp. 165–170.

64. Rogers, J., Luber-Narod, J., Styren, S. D., and Civin, W. H. (1988) Expression of immune system-associated antigens by cells of the human central nervous system: relationship to the pathology of Alzheimer's disease. *Neurobiol. Aging* **9**, 339–349.

65. McGeer, P. L., Akiyama, H., Itagaki, S., and McGeer, E. G. (1989) Activation of the classical complement pathway in brain tissue of Alzheimer patients. *Neurosci. Lett.* **107**, 341–346.

66. Frautschy, S. A., Yang, F., Irrizarry, M., Hyman, B., Saido, T. C., Hsiao, K., et al. (1998) Microglial response to amyloid plaques in APPsw transgenic mice. *Am. J. Pathol.* **152**, 307–317.

67. Stalder, M., Phinney, A., Probst, A., Sommer, B., Staufenbiel, M., and Jucker, M. (1999) Association of microglia with amyloid plaques in brains of APP23 transgenic mice. *Am. J. Pathol.* **154**, 1673–1684.

68. Cras, P., Kawai, M., Siedlak, S., and Perry, G. (1991) Microglia are associated with the extracellular neurofibrillary tangles of Alzheimer disease. *Brain Res.* **558,** 312–314.

69. Cras, P., Kawai, M., Siedlak, S., Mulvihill, P., Gambetti, P., Lowery, D., et al. (1990) Neuronal and microglial involvement in beta-amyloid protein deposition in Alzheimer's disease. *Am. J. Pathol.* **137,** 241–246.

70. Giulian, D., Haverkamp, L. J., Li, J., Karshin, W. L., Yu, J., Tom, D., et al. (1995) Senile plaques stimulate microglia to release a neurotoxin found in Alzheimer brain. *Neurochem. Int.* **26,** 119–137.

71. Mann, D. M. A. (1993) The role of microglial cells in plaque formation in Alzheimer's disease, in *Alzheimer's Disease: Advances in Clinical and Basic Research* (Corain, B., Iqbal, K., Nicoline, M., Winblad, B., Wisniewski, H., and Zatta, P., eds.), Wiley, Chichester, pp. 159–164.

72. Sheng, J. G., Mrak, R. E., and Griffin, W. S. (1995) Microglial interleukin-1 alpha expression in brain regions in Alzheimer's disease: correlation with neuritic plaque distribution. *Neuropathol. Appl. Neurobiol.* **21,** 290–301.

73. Sheng, J. G., Mrak, R. E., and Griffin, W. S. (1996) Apolipoprotein E distribution among different plaque types in Alzheimer's disease: implications for its role in plaque progression. *Neuropathol. Appl. Neurobiol.* **22,** 334–341.

74. Wisniewski, H. M., Wegiel, J., and Kotula, L. (1996) David Oppenheimer Memorial Lecture 1995: some neuropathological aspects of Alzheimer's disease and its relevance to other disciplines. *Neuropathol. Appl. Neurobiol.* **22,** 3–11.

75. Shirahama, T., Miura, K., Ju, S. T., Kisilevsky, R., Gruys, E., and Cohen, A. S. (1990) Amyloid enhancing factor-loaded macrophages in amyloid fibril formation. *Lab. Invest.* **62,** 61–68.

76. Gerst, J. L., Raina, A. K., Pirim, I., McShea, A., Harris, P. L. R., Siedlak, S. L., et al. (2000) Altered cell-matrix associated ADAM proteins in Alzheimer disease. *J. Neurosci. Res.* **59,** 680–684.

77. McGeer, P. L. and McGeer, E. G. (1995) The inflammatory response of brain: implications for therapy of Alzheimer and other neurodegenerative diseases. *Brain Res.* **21,** 195–218.

78. Eikelenboom, P. and Veerhuis, R. (1996) The role of complement and activated microglia in the pathogenesis of Alzheimer's disease. *Neurobiol. Aging* **17,** 673–680.

79. Meda, L., Cassatella, M. A., Szendrei, G. I., Otvos, L. Jr., Baron, P., Villalba, M., et al. (1995) Activation of microglial cells by beta-amyloid protein and interferon-gamma. *Nature* **374,** 647–650.

80. Van Muiswinkel, F. L., Veerhuis, R., and Eikelenboom, P. (1996) Amyloid beta protein primes cultured rat microglial cells for an enhanced phorbol12-myristate 13-acetate-induced respiratory burst activity. *J. Neurochem.* **66,** 2468–2476.

81. Akama, K. T., Albanese, C., Pestell, R. G., and Van Eldik, L. J. (1998) Amyloid beta-peptide stimulates nitric oxide production in astrocytes through an NFkappaB-dependent mechanism. *Proc. Natl. Acad. Sci. USA* **95,** 5795–5800.

82. Giulian, D., Haverkamp, L. J., Yu, J. H., Karshin, W. L., Tom, D., Li, J., et al. (1996) Specific domains of β-amyloid from Alzheimer plaque elicit neuron killing in human microglia. *J. Neurosci.* **16,** 6021–6037.

83. Paresce, D. M., Chung, H., and Maxfield, F. R. (1997) Slow degradation of aggregates of the Alzheimer's disease amyloid beta-protein by microglial cells. *J. Biol. Chem.* **272,** 29,390–29,397.

84. Giulian, D., Haverkamp, L. J., Yu, J. H., Karshin, W. L., Tom, D., Li, J., et al. (1998) The HHQK domain of β-amyloid provides a structural basis for the immunopathology of Alzheimer's disease. *J. Biol. Sci.* **273,** 29,719–29,726.

85. Colton, C. A. and Gilbert, D. L. (1987) Production of superoxide anions by a CNS macrophage, the microglia. *FEBS Lett.* **223,** 284–288.

86. Colton, C. A. and Gilbert, D. L. (1993) Microglia, an in vivo source of reactive oxygen species in the brain. *Adv. Neurol.* **59,** 321–326.

87. Colton, C. A., Chernyshev, O. N., Gilbert, D. L., and Vitek, M. P. (2000) Microglial contribution to oxidative stress in Alzheimer's disease. *Ann. NY Acad. Sci.* **899,** 292–307.

88. Narindrasorasak, S., Lowery, D., Gonzalez-DeWhitt, P., Poorman, R. A., Greenberg, B., and Kisilevsky, R. (1991) High affinity interactions between the Alzheimer's beta-amyloid precursor proteins and the basement membrane form of heparan sulfate proteoglycan. *J. Biol. Chem.* **266,** 12,878–12,883.

89. Canning, D. R., McKeon, R. J., DeWitt, D. A., Perry, G., Wujek, J. R., Frederickson, R. C., et al. (1993) Beta-amyloid of Alzheimer's disease induces reactive gliosis that inhibits axonal outgrowth. *Exp. Neurol.* **124,** 289–298.

90. Giulian, D., Chen, J., Ingeman, J. E., George, J., and Noponen, M. (1989) The role of mononuclear phagocytes in wound healing after traumatic injury to the adult mammalian brain. *J. Neurosci.* **9,** 4416–4429.

91. Frautschy, S. A., Cole, G. M., and Baird, A. (1992) Phagocytosis and deposition of vascular beta-amyloid in rat brains injected with Alzheimer beta-amyloid. *Am. J. Pathol.* **140,** 1389–1399.

92. Klegeris, A., Walker, D. G., and McGeer, P. L. (1994) Activation of macrophages by Alzheimer beta amyloid peptide. *Biochem. Biophys. Res. Commun.* **199,** 984–991.

93. McDonald, D. R., Brunden, K. R., and Landreth, G. E. (1997) Amyloid fibrils activate tyrosine kinase-dependent signaling and superoxide production in microglia. *J. Neurosci.* **17,** 2284–2294.

94. Klegeris, A. and McGeer, P. L. (1997) Beta-amyloid protein enhances macrophage production of oxygen free radicals and glutamate. *J. Neurosci. Res.* **49,** 229–235.

95. Della-Bianca, V., Dusi, S., Bianchini, E., Dal Pra, I., and Rossi, F. (1999) Beta-amyloid activates the O-2 forming NADPH oxidase in microglia, monocytes, and neutrophils. A possible inflammatory mechanism of neuronal damage in Alzheimer's disease. *J. Biol. Chem.* **274,** 15,493–15,499.

96. Van Muiswinkel, F. L., Raupp, S. F., de Vos, N. M., Smits, H. A., Verhoef, J., Eikelenboom, P., et al. (1999) The amino-terminus of the amyloid-beta protein is critical for the cellular binding and consequent activation of the respiratory burst of human macrophages. *J. Neuroimmunol.* **96,** 121–130.

97. Van Muiswinkel, F. L., DeGroot, C., Rozemuller-Kwakkel, J., and Eikelenboom, P. (1999) Enhanced expression of microglial NADPH-oxidase (p22-phox), in *Alzheimer's Disease and Related Disorders* (Iqbal, K., Swaab, D. F., Winblad, B., and Wisniewski, H. M., eds.), Wiley, West Sussex, England, pp. 451–456.

98. Pembe, S. O. and Kinkade, J. M. (1983) Differences in myeloperoxidase activity from neutrophilic polymorphonuclear leukocytes of differing density: relationship to selective exocytosis of distinct forms of the enzyme. *Blood* **61,** 1116–1124.

99. Albrecht, D. and Jungi, T. W. (1993) Luminol-enhanced chemiluminescence induced in peripheral blood-derived human phagocytes: obligatory requirement of myeloperoxidase exocytosis by monocytes. *J. Leukocyte Biol.* **54,** 300–306.

100. Jacob, J. S., Cistola, D. P., Hsu, F. F., Muzaffar, S., Mueller, D. M., Hazen, S. L., et al. (1996) Human phagocytes employ the myeloperoxidase-hydrogen peroxide system to synthesize dityrosine, trityrosine, pulcherosine, and isodityrosine by a tyrosyl radical-dependent pathway. *J. Biol. Chem.* **271,** 19,950–19,956.

101. Podrez, E. A., Schmitt, D., Hoff, H. F., and Hazen, S. L. (1999) Myeloperoxidase-generated reactive nitrogen species convert LDL into an atherogenic form *in vitro. J. Clin. Invest.* **103,** 1547–1560.

102. Anderson, M. M., Requena, J. R., Crowley, J. R., Thorpe, S. R., and Heinecke, J. W. (1999) The myeloperoxidase system of human phagocytes generates Nepsilon-(carboxymethyl)lysine on proteins: a mechanism for producing advanced glycation end products at sites of inflammation. *J. Clin. Invest.* **104,** 103–113.

103. Hazen, S. L., Hsu, F. F., Duffin, K., and Heinecke, J. W. (1996) Molecular chlorine generated by the myeloperoxidase–hydrogen peroxide–chloride system of phagocytes converts

low density lipoprotein cholesterol into a family of chlorinated sterols. *J. Biol. Chem.* **271**, 23,080–23,088.

104. Jolivalt, C., Leininger-Muller, B., Drozdz, R., Naskalski, J. W., and Siest, G. (1996) Apolipoprotein E is highly susceptible to oxidation by myeloperoxidase, an enzyme present in the brain. *Neurosci. Lett.* **210**, 61–64.

105. Jolivalt, C., Leininger-Muller, B., Bertrand, P., Herber, R., Christen, Y., and Siest, G. (2000) Differential oxidation of apolipoprotein E isoforms and interaction with phospholipids. *Free Radical Biol. Med.* **28**, 129–140.

106. Byun, J., Mueller, D. M., Fabjan, J. S., and Heinecke, J. W. (1999) Nitrogen dioxide radical generated by the myeloperoxidase-hydrogen peroxide-nitrite system promotes lipid peroxidation of low density lipoprotein. *FEBS Lett.* **455**, 243–246.

107. Eiserich, J. P., Hristova, M., Cross, C. E., Jones, A. D., Freeman, B. A., Halliwell, B., et al. (1998) Formation of nitric oxide-derived inflammatory oxidants by myeloperoxidase in neutrophils. *Nature* **391**, 393–397.

108. Reynolds, W. F., Rhees, J., Maciejewski, D., Paladino, T., Sieburg, H., Maki, R. A., et al. (1999) Myeloperoxidase polymorphism is associated with gender specific risk for Alzheimer's disease. *Exp. Neurol.* **155**, 31–41.

109. Lindenau, J., Noack, H., Asayama, K., and Wolf, G. (1998) Enhanced cellular glutathione peroxidase immunoreactivity in activated astrocytes and in microglia during excitotoxin induced neurodegeneration. *Glia* **24**, 252–256.

110. Malle, E., Waeg, G., Schreiber, R., Grone, E. F., Sattler, W., and Grone, H.-J. (2000) Immunohistochemical evidence for the myeloperoxidase/H_2O_2/halide system in human atherosclerotic lesions. *Eur. J. Biochem.* **267**, 1–10.

111. Leeuwenburgh, C., Rasmussen, J. E., Hsu, F. F., Mueller, D. M., Pennathur, S., et al. (1997) Mass spectrometric quantification of markers for protein oxidation by tyrosyl radical, copper, and hydroxyl radical in low density lipoprotein isolated from human atherosclerotic plaques. *J. Biol. Chem.* **272**, 3520–3526.

112. Francis, G. A., Mendez, A. J., Bierman, E. L., and Heinecke, J. W. (1993) Oxidative tyrosylation of high density lipoprotein by peroxidase enhances cholesterol removal from cultured fibroblasts and macrophage foam cells. *Proc. Natl. Acad. Sci. USA* **90**, 6631–6635.

113. Hazen, S. L. and Heinecke, J. W. (1997) 3-Chlorotyrosine, a specific marker of myeloperoxidase-catalyzed oxidation, is markedly elevated in low density lipoprotein isolated from human atherosclerotic intima. *J. Clin. Invest.* **99**, 2075–2081.

114. Maury, C. P. (1995) Molecular pathogenesis of beta-amyloidosis in Alzheimer's disease and other cerebral amyloidosis. *Lab. Invest.* **72**, 4–16.

115. Shoji, M., Golde, T. E., Ghiso, J., Cheung, T. T., Estus, S., Shaffer, L. M., et al. (1992) Production of the Alzheimer amyloid beta protein by normal proteolytic processing. *Science* **258**, 126–129.

116. Masters, C. L., Simms, G., Weinman, N. A., Multhaup, G., McDonald, B. L., and Beyreuther, K. (1985) Amyloid plaque core protein in Alzheimer disease and Down syndrome. *Proc. Natl. Acad. Sci. USA* **82**, 4245–4249.

117. Kuo, Y. M., Emmerling, M. R., Vigo-Pelfrey, C., Kasunic, T. C., Kirkpatrick, J. B., Murdoch, G. H., et al. (1996) Water-soluble Abeta (N-40, N-42) oligomers in normal and Alzheimer disease brains. *J. Biol. Chem.* **271**, 4077–4081.

118. Roher, A. E., Chaney, M. O., Kuo, Y. M., Webster, S. D., Stine, W. B., Haverkamp, L. J., et al. (1996) Morphology and toxicity of Abeta-(1-42) dimer derived from neuritic and vascular amyloid deposits of Alzheimer's disease. *J. Biol. Chem.* **271**, 20,631–20,635.

119. Cherny, R. A., Legg, J. T., McLean, C. A., Fairlie, D. P., Huang, X., Atwood, C. S., et al. (1999) Aqueous dissolution of Alzheimer's disease Aβ amyloid deposits by biometal depletion. *J. Biol. Chem.* **274**, 23,223–23,228.

120. Podlisny, M. B., Ostaszewski, B. L., Squazzo, S. L., Koo, E. H., Rydell, R. E., Teplow, D. B. et al. (1995) Aggregation of secreted amyloid beta-protein into sodium dodecyl sulfate-stable oligomers in cell culture. *J. Biol. Chem.* **270**, 9564–9570.

121. Podlisny, M. B., Walsh, D. M., Amarante, P., Ostaszewski, B. L., Stimson, E. R., Maggio, J. E., et al. (1998) Oligomerization of endogenous and synthetic amyloid beta-protein at nanomolar levels in cell culture and stabilization of monomer by Congo red. *Biochemistry* **37,** 3602–3611.

122. Atwood, C. S., Scarpa, R. C., Huang, X., Moir, R. D., Jones, W. D., Fairlie, D. P., et al. (2000) Characterization of copper interactions with Alzheimer amyloid beta peptides: identification of an attomolar-affinity copper binding site on amyloid beta1-42. *J. Neurochem.* **75,** 1219–1233.

123. Hensley, K., Maidt, M. L., Yu, Z., Sang, H., Markesbery, W. R., and Floyd, R. A. (1998) Electrochemical analysis of protein nitrotyrosine and dityrosine in the Alzheimer brain indicates region-specific accumulation. *J. Neurosci.* **18,** 8126–8132.

124. Galeazzi, L., Ronchi, P., Franceschi, C., and Giunta, S. (1999) *In vitro* peroxidase oxidation induces stable dimers of beta-amyloid (1-42) through dityrosine bridge formation. *Amyloid* **6,** 7–13.

125. Atwood, C. S., Moir, R. D., Jones, W. D., Huang, X., Perry, G., Tanzi, R. E., et al. (2000) Human amyloid-derived Aβ contains tyrosine cross-linked oligomers. *Neurobiol. Aging* **21,** S199.

126. Benzinger, T. L., Gregory, D. M., Burkoth, T. S., Miller-Auer, H., Lynn, D. G., Botto, R. E., et al. (2000) Two-dimensional structure of beta-amyloid (10–35) fibrils. *Biochemistry* **39,** 3491–3499.

127. Amado, R., Aeschbach, R., and Neukom, H. (1984) Dityrosine: *in vitro* production and characterization. *Methods Enzymol.* **107,** 377–388.

128. Smail, E. H., Briza, P., Panagos, A., and Berenfeld, L. (1995) Candida albicans cell walls contain the fluorescent crosslinking amino acid dityrosine. *Infect. Immun.* **63,** 4078–4083.

129. Deits, T., Farrance, M., Kay, E. S., Medill, L., Turner, E. E., Weidman, P. J., et al. (1984) Purification and properties of ovoperoxidase, the enzyme responsible for hardening the fertilization membrane of the sea urchin egg. *J. Biol. Chem.* **259,** 13,525–13,533.

130. Nomura, K., Suzuki, N., and Matsumoto, S. (1990) Pulcherosine, a novel tyrosine-derived, trivalent cross-linking amino acid from the fertilization envelope of sea urchin embryo. *Biochemistry* **29,** 4525–4534.

131. Briza, P., Breitenbach, P. M., Ellinger, A., and Segall, J. (1990) Isolation of two developmentally regulated genes involved in spore wall maturation in *Saccharomyces cerevisiae*. *Genes Dev.* **4,** 1775–1789.

132. Briza, P., Ellinger, A., Winkler, G., and Segall, J. (1990) Characterization of a DL-dityrosine-containing macromolecule from yeast ascospore walls. *J. Biol. Chem.* **265,** 15,118–15,123.

133. Shapira, R., Austin, G. E., and Mirra, S. S. (1988) Neuritic plaque amyloid in Alzheimer's disease is highly racemized. *J. Neurochem.* **50,** 69–74.

134. Mori, H., Takio, K., Ogawara, M., and Selkoe, D. J. (1992) Mass spectrometry of purified amyloid beta protein in Alzheimer's disease. *J. Biol. Chem.* **267,** 17,082–17,086.

135. Roher, A. E., Lowenson, J. D., Clarke, S., Wolkow, C., Wang, R., Cotter, R. J., et al. (1993) Structural alterations in the peptide backbone of beta-amyloid core protein may account for its deposition and stability in Alzheimer's disease. *J. Biol. Chem.* **268,** 3072–3083.

136. Atwood, C. S., Huang, X., Khatri, A., Scarpa, R. C., Kim, Y.-S., Moir, R. D., et al. (2000) Copper catalyzed oxidation of Alzheimer Aβ. *Cell. Mol. Biol.* **46,** 777–783.

137. Huang, X., Atwood, C. S., Hartshorn, M. A., Multhaup, G., Goldstein, L. E., Scarpa, R. C., et al. (1999) The Aβ peptide of Alzheimer's disease directly produces hydrogen peroxide through metal ion reduction. *Biochemistry* **38,** 7609–7616.

138. Huang, X., Cuajungco, M. P., Atwood, C. S., Hartshorn, M. A., Tyndall, J., Hanson, G. R., et al. (1999) Alzheimer Aβ interaction with Cu(II) induces neurotoxicity, radicalization, metal reduction and hydrogen peroxide formation. *J. Biol. Chem.* **274,** 37,111–37,116.

139. Woenckhaus, C., Kaufmann, A., Bussfeld, D., Gemsa, D., Sprenger, H., and Grone, H. J. (1998) Hypochlorite-modified LDL: chemotactic potential and chemokine induction in human monocytes. *Clin. Immunol. Immunopathol.* **86,** 27–33.

140. Kopprasch, S., Leonhardt, W., Pietzsch, J., and Kuhne, H. (1998) Hypochlorite-modified low-density lipoprotein stimulates human polymorphonuclear leukocytes for enhanced production of reactive oxygen metabolites, enzyme secretion, and adhesion to endothelial cells. *Atherosclerosis* **136,** 315–324.

141. Zabe, M., Feltzer, R. E., Malle, E., Sattler, W., and Dean, W. L. (1999) Effects of hypochlorite-modified low-density and high-density lipoproteins on intracellular Ca2+ and plasma membrane Ca(2+)-ATPase activity of human platelets. *Cell Calcium* **26,** 281–287.

142. DeWitt, D. A., Perry, G., Cohen, M., Doller, C., and Silver, J. (1998) Astrocytes regulate microglial phagocytosis of senile plaque cores of Alzheimer's disease. *Exp. Neurol.* **149,** 329–340.

143. Kikugawa, K., Kato, T., Beppu, M., and Hayasaka, A. (1991) Development of fluorescence and cross-links in eye lens crystallin by interaction with lipid peroxy radicals. *Biochim. Biophys. Acta* **1096,** 108–114.

144. MacMillan-Crow, L. A., Crow, J. P., and Thompson, J. A. (1998) Peroxynitrite-mediated inactivation of manganese superoxide dismutase involves nitration and oxidation of critical tyrosine residues. *Biochemistry* **37,** 1613–1622.

145. Van Der Vliet, A., Nguyen, M. N., Shigenaga, M. K., Eiserich, J. P., Marelich, G. P., and Cross, C. E. (2000) Myeloperoxidase and protein oxidation in cystic fibrosis. *Am. J. Physiol. Lung Cell Mol. Physiol.* **279,** L537–L546.

146. Kato, Y., Maruyama, W., Naoi, M., Hashizume, Y., and Osawa, T. (1998) Immunohistochemical detection of dityrosine in lipofuscin pigments in the aged human brain. *FEBS Lett.* **439,** 231–234.

147. Nagra, R. M., Becher, B., Tourtellotte, W. W., Antel, J. P., Gold, D., Paladino, T., et al. (1997) Immunohistochemical and genetic evidence of myeloperoxidase involvement in multiple sclerosis. *J. Neuroimmunol.* **78,** 97–107.

148. Bruijn, L. I., Houseweart, M. K., Kato, S., Anderson, K. L., Anderson, S. L., Ohama, E., et al. (1998) Aggregation and motor neuron toxicity of an ALS-linked SOD1 mutant independent from wild-type SOD1. Science 281, 1851–1854.

149. Souza, J. M., Giasson, B. I., Chen, Q., Lee, V. M., and Ischiropoulos, H. (2000) Dityrosine cross-linking promotes formation of stable alpha-synuclein polymers. Implication of nitrative and oxidative stress in the pathogenesis of neurodegenerative synucleinopathies. *J. Biol. Chem.* **275,** 18,344–18,349.

150. Pike, C. J., Walencewicz, A. J., Glabe, C. G., and Cotman, C. W. (1991) Aggregation-related toxicity of synthetic beta-amyloid protein in hippocampal cultures. *Eur. J. Pharmacol.* **207,** 367–368.

151. Cotman, C. W., Pike, C. J., and Copani, A. (1992) Beta-amyloid neurotoxicity: a discussion of *in vitro* findings. *Neurobiol. Aging* **13,** 587–590.

152. Pollack, S. J., Sadler, I. I., Hawtin, S. R., Tailor, V. J., and Shearman, M. S. (1995) Sulfonated dyes attenuate the toxic effects of beta-amyloid in a structure-specific fashion. *Neurosci. Lett.* **197,** 211–214.

153. Wujek, J. R., Dority, M. D., Frederickson, R. C., and Brunden, K. R. (1996) Deposits of A beta fibrils are not toxic to cortical and hippocampal neurons *in vitro. Neurobiol. Aging* **17,** 107–113.

154. Games, D., Khan, K. M., Soriano, F. G., Keim, P. S., Davis, D. L., Bryant, K., et al. (1992) Lack of Alzheimer pathology after beta-amyloid protein injections in rat brain. *Neurobiol. Aging* **13,** 569–576.

155. Podlisny, M. B., Stephenson, D. T., Frosch, M. P., Lieberburg, I., Clemens, J. A., and Selkoe, D. J. (1992) Synthetic amyloid beta-protein fails to produce specific neurotoxicity in monkey cerebral cortex. *Neurobiol. Aging* **13,** 561–567.

156. Podlisny, M. B., Stephenson, D. T., Frosch, M. P., Tolan, D. R., Lieberburg, I., Clemens, J. A., et al. (1993) Microinjection of synthetic amyloid beta-protein in monkey cerebral cortex fails to produce acute neurotoxicity. *Am. J. Pathol.* **142,** 17–24.

157. Clemens, J. A. and Stephenson, D. T. (1992) Implants containing beta-amyloid protein are not neurotoxic to young and old rat brain. *Neurobiol. Aging* **13,** 581–586.
158. Bishop, G. M. and Robinson, S. R. (2000) β-Amyloid helps to protect neurons from oxidative stress. *Neurobiol. Aging* **21(Suppl. 1S),** S226.
159. Irizarry, M. C., McNamara, M., Fedorchak, K., Hsiao, K., and Hyman, B. T. (1997) APPSw transgenic mice develop age-related A beta deposits and neuropil abnormalities, but no neuronal loss in CA1. *J. Neuropathol. Exp. Neurol.* **56,** 965–973.
160. Irizarry, M. C., Soriano, F., McNamara, M., Page, K. J., Schenk, D., Games, D., et al. (1997) Aβ deposition is associated with neuropil changes, but not with overt neuronal loss in the human amyloid precursor protein V717F (PDAPP) transgenic mouse. *J. Neurosci.* **17,** 7053–7059.

13

Microglial Responses in Alzheimer's Disease

Recent Studies in Transgenic Mice and Alzheimer's Disease Brains

Douglas G. Walker and Lih-Fen Lue

1. INTRODUCTION

This chapter will consider the significance of a range of markers used in studies to describe activated microglia and reactive astrocytes in chronic neurodegenerative disorders, particularly as they are modeled in the recently developed transgenic mice models of Alzheimer's disease (AD). The purpose of this chapter is to discuss the significance of these markers in describing the degenerative mechanisms that are progressing in this disease or models of this disease. Since the completion of the first article dealing with this topic *(1)*, there have been a number of major publications concerning the presence of inflammatory microglia and astrocyte markers in transgenic mice models of AD. These mice develop AD-like amyloid β peptide (Aβ) plaques as a result of the possession of mutated copies of the amyloid precursor protein (APP) genes *(2–6)*. A comparison of the expression of inflammatory markers in these AD models with those expressed in AD brain tissues will be included in this chapter. Significant progress in understanding the appearance of inflammatory markers in these Aβ plaque-developing animals has come from the fact that they can be sacrificed and studied at different time-points during the development of the plaques. This type of study is obviously not possible in human patients. In addition, these animals allow one to study the effects of progressive deposition of Aβ independent of other events proceeding in an aging human brain.

Since the initial descriptions of the presence of activated microglia in AD brain tissue were published about 15 yr ago *(7–10)*, a range of observations and experimental systems showed that activated microglia could be contributing in a significant manner to the neuropathology of AD and other neurodegenerative diseases. Recently, it has been shown that activated microglia could be used to "treat" AD. This has brought a shift in the established concepts of the role of microglia in AD. From the initial observations, the hypothesis that activated microglia may be contributing to the pathology of AD was established because of much data showing that activated microglia/macrophages can cause neurotoxicity. Recently, it has been shown that activating microglia through the Fcγ immunoglobulin receptors may aid in the clearance of the Aβ plaques and in slowing down the progression of the disease. These data have shown that developing

From: *Neuroinflammation, 2nd Edition: Mechanisms and Management*
Edited by: P. L. Wood © Humana Press Inc., Totowa, NJ

circulating antibodies to Aβ in APP transgenic mice by immunization or by injecting these animals with anti-Aβ immunoglobulin, resulted in microglial-mediated clearance of Aβ plaques *(11,12)*. This effect was mediated by microglial Fc receptors *(11)*. The beneficial effects of Aβ immunization on transgenic mice models has been confirmed by others *(13,14)*. Optimism about this strategy for treating human AD cases has come from initial trials showing that human volunteers tolerate immunization with Aβ peptide vaccines well and also start to develop antibody responses. It needs to be stated though that recent data have demonstrated that the Aβ plaques that develop in transgenic mice are more soluble than those found in elderly AD brains *(15,16)*. It has been suggested that this may be the result of the lack of posttranslational modifications of the Aβ *(16)*, which is a prominent feature of Aβ isolated from AD brains *(17,18)*. This new concept on the beneficial outcome of microglial activation in the AD brain will be considered in relation to the significance of markers for identifying microglia that could be causing tissue damage contributing to AD.

2. INFLAMMATION AND ALZHEIMER'S DISEASE

Since the initial observations on activated microglia being associated with pathological structures in AD *(7,9,19)*, extensive data have provided indirect evidence that chronic inflammatory processes could be contributing to the pathology of the disease (reviewed in refs. *20* and *21*). The hypothesis being proposed is that activated microglia (or astrocytes) are producing toxins, which cause further damage to neurons within AD-affected brains and thus are contributing to the rate of mental impairment. For example, it was demonstrated that Aβ-activated microglia can produce a neurotoxic factor that is active on *N*-methyl-D-aspartate (NMDA) receptor positive neurons *(22–24)*. Other studies have shown that Aβ-stimulated microglia could be producing toxic levels of superoxide radicals through the activation of the NADPH oxidase complex, causing the so-called respiratory burst *(25,26)*. There has also been extensive data showing increased complement activation contributing to the inflammatory damage *(27,28)*. This being said, clinical trials aimed at controlling inflammation in AD have generally produced disappointing results *(29,30)*. The transgenic mice models of AD offer great hope in aiding in the study of inflammatory processes in AD because of the ability to study the progression of changes over time. There are some obvious limitations in using mice to model a uniquely human disease; for instance, some of the features of AD, in particular the presence of tangles, have not been recreated in these Aβ-plaque developing models. Recently, it has been shown that tangles can be created in transgenic mice that contain a mutated human tau gene *(31)*. These animals were shown to develop increased numbers of tangles when crossed with Aβ-plaque-developing mice *(31)*. In addition, differences in the immune responses of rodents compared to humans may affect some of the manifestations of progressive plaque development. However, although extensive data have been compiled on inflammatory and biochemical changes proceeding in AD, it is still possible to identify the key feature(s) (inflammatory or others) that link excessive amyloid deposition and dementia. The concepts behind possible neuroinflammatory processes contributing to pathology in AD are illustrated in Fig. 1. As illustrated, the major inflammatory events, namely complement activation and/or microglial activation, are likely to be later events in the pathogenesis of AD, but could be acting as disease accelerators.

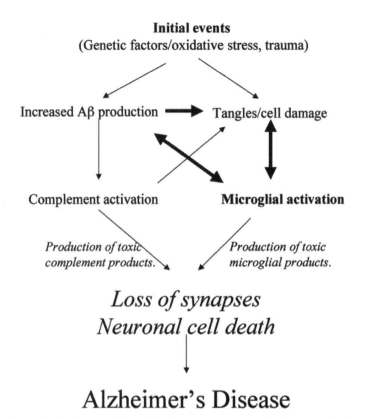

Fig. 1. The hypothetical interactions of inflammatory-associated events in the causation of Alzheimer's disease.

3. MICROGLIAL RESPONSES IN Aβ-PLAQUE-DEVELOPING TRANSGENIC MICE

Research with Aβ-plaque-developing mice has mainly been carried out on three independently developed lines of mice. The first characterized mice were the mice line developed by Elan Pharmaceuticals (formely Athena Neurosciences). This line of animals contained an inserted copy of the human APP gene, with a mutation at amino acid 717, whose expression is driven by the platelet-derived growth factor (PDGF) promoter *(32)*. A number of similar lines employing the PDGF promoter have also been independently developed by the laboratory of Mucke *(33)*. A widely used line (designated Tg2576) developed by Hsiao contains a human APP gene with mutations at amino acids 670 and 671 (Swedish mutation) under transcriptional control of the hamster prion promoter *(6)*. This line has been widely distributed and is the subject of a number of AD-related studies. The third line (designated APP23) contains a copy of the human APP gene with mutations at amino acids 670, 671, and 717 under transcriptional control of the murine Thy-1 promoter *(34)*. This transgenic line also develops prominent cerebrovascular amyloid deposits *(35)*, a common pathological feature in AD. This triple-mutated human APP gene under the control of the Thy-1 promoter demonstrated increased rates of Aβ deposition compared to mice with the Swedish mutations alone or the 717 mutation

Table 1
Markers for Activated Microglia
Characterized in Aβ-Plaque-Developing Transgenic Mice

Antigen	Identity/function	Ref.
F4/80	Cell surface activation marker	*43*
CD11b/Mac-1	Complement receptor III	*40,43*
MHCII	Major histocompatibility complex	*43*
Phosphotyrosine	Product of tyrosine kinases	*37,40,44*
Griffonia simplicifolia lectin	Lectin for GSAII	*37*
MSRA	Class A scavenger receptor	*43*
RAGE	Receptor for advanced glycation end products	*45*
Macrosialin (CD68)	Lysosomal membrane-associated protein	*45*
Fcγ receptors II, III	Immunoglobulin receptors	*43*
M-CSF receptor	Macrophage colony-stimulating factor receptor	*46*
Interleukin-1β	Pro-inflammatory cytokine	*47*
Tumor necrosis factor-α	Pro-inflammatory cytokine	*47*
CD45	Protein phosphatase	*42,47*

(33). An additional mouse line, which develops amyloid plaques at an accelerated rate, is the Tg2576/PS1M146L double transgenic line. This line contains a mutated transgene of presenilin 1 crossed with the APP670/671 transgene containing the line developed by Hsiao *(36)*. Studies have been reported on the microglial responses present in each of these different lines of Aβ-plaque-developing animals *(37–42)*.

Studies on the role of microglia in AD-like pathology have been aided by the use of these transgenic mice models. The results have shown a progressive activation of microglia over time as the deposition of amyloid into plaques progressed. Such direct studies have not been possible using human AD tissues. The features of the progressive activation will be discussed.

4. MICROGLIAL MARKERS IN TRANSGENIC MICE

As listed in Table 1, a range of markers have been employed to characterize microglial responses in Aβ-plaque-developing transgenic mice. Of the three mice lines with APP mutations, the most pronounced amyloid deposition and neuritic changes were found in the APP23 line, which has the APP670, 671, and 717 mutations. These animals showed reactive astrocytes (visualized by glial fibrillary acidic protein [GFAP] immunocytochemistry) and activated microglia (visualized by Mac-1 [complement receptor 3] and phosphotyrosine immunocytochemistry) around congophilic plaques of 12-mo-old animals *(34)*. Use of antibodies that recognize phosphotyrosine-modified proteins has not been extensively used as a marker for describing microglia in human brains, but they appear to recognize activated microglia in AD brains *(48)*, as they do in transgenic mice. Two more detailed follow-up studies on microglial responses were carried out using this line of transgenic mice *(40,43)*. Distinct microglial clusters (more than two cell bodies) were observed around dense-core amyloid deposits in these animals. Approximately 85% of dense-core Aβ plaques had associated activated microglial clusters *(40)*. This was followed up by an examination of other functional markers in APP23 mice. The identification of Mac-1 immunoreactive microglia as activated microglia was confirmed by showing similar patterns of increased immunoreactivity with F4/80, another macrophage/monocyte-specific marker. In addition, more of the microglia around the plaques incorporated bromodeoxyuridine (BrdU, a marker for proliferation) compared to microglia remote from the plaques. The amount of BrdU incorporation in wild-type mice was considerably less than in the plaque-developing mice *(43)*. This study also demonstrated that macrophage scavenger receptor A (MSRA), a known phagocytosis and adhesion-associated receptor, was increased on microglia accumulated around all the compacted plaques *(43)*. It had been suggested that Aβ interactions with MSRA on microglia could mediate many of the pro-inflammatory outcomes associated with microglial activation *(49,50)*. However, a line of Aβ-plaque-developing transgenic mice were prepared that had been crossed with MSRA knockout mice so that microglia did not express this receptor. In these animals, it was shown that there was no difference in number, extent, distribution, or age-dependent accumulation of plaques compared to mice with normal microglial MSRA expression *(51)*. In addition, the MSRA knockout mice showed the same amounts of synaptic degeneration as APP mice with normal levels of MSRA expression. Additional microglial markers were examined in aged APP23 mice *(43)*. A subset of the microglia associated with plaques were immunoreactive for the receptor for advanced glycation end products (RAGE), another receptor that is potentially mediating pro-inflammatory Aβ–microglial interactions *(52)*. These authors also showed increased expression of macrosialin and of the FcγRII and FcγRIII immunoglobulin receptors on microglia around plaques. These proteins are associated with phagocytosis in macrophages/microglia. Increased expression of the major histocompatibility complex class II (MHCII) protein was observed in a subpopulation of activated microglia. This marker has become the most widely used marker to describe activated microglia in AD and other neuropathological studies (discussed in Section 6 and reviewed in ref. *20*). This protein is involved in antigen presentation to lymphocytes; however, similar to what has been observed in AD brains *(53)*, in these transgenic mice there was no evidence for extensive lymphocyte recruitment into the brain *(43)*.

Microglial activation in the Tg2576 mouse line was measured using a quantitative approach *(37)*. Although the progression and amount of Aβ deposition differs between the different lines, it was found that activated microglia were only localized to plaque-developing areas. Similar patterns of reactivity of microglia were demonstrated using the lectin *Griffonia simplicifolia* B4 and with an antibody to phosphotyrosine. The use of antibodies to phosphotyrosine for describing microglial activation in vivo has become an accepted marker, although the proteins being labeled are poorly defined. Activation of multiple tyrosine kinases occurs in macrophage/microglia cells in response to stimuli. This results in the increase in phosphotyrosine reactivity of multiple proteins *(54, 55)*. Tyrosine kinase activation in relation to macrophage/microglia is a relatively transient phenomenon, so, by definition, the labeled microglia are considered activated. In the Tg2578 mice, it was demonstrated that there was a significant increase in the density and area of microglia compared with wild-type animals *(37)*. This difference was greatest in plaque-enriched areas (cortex and hippocampus) compared to a nonplaque area (thalamus). The microglial density was highest in the cortical layers with the highest Aβ plaque density. Quantitative analyses demonstrated a gradient of activation when microglia density and microglia area were measured as a function of the distance from the plaque.

More detailed characterization of the microglial responses of the PD-APP mice was recently published *(46)*. A strong progressive increase in the expression of the macrophage colony-stimulating factor receptor (M-CSFR) was demonstrated on microglia clustered around Aβ plaques. The expression of this marker was shown to colocalize on microglia with immunoreactivity for the markers phosphotyrosine, F4/80, and CD11b. The expressions of these markers progressively increased in transgene-bearing mice from 10–13 mo, when significant amounts of Aβ began to be deposited into plaques *(16,52)*, up to 20 mo, by which time extensive Aβ plaques can be observed. The significance of increased expression of M-CSFR is that this receptor can mediate the signals of the mitogen M-CSF, resulting in localized proliferation of microglia at sites of inflammation. In the AD brain, this can exacerbate the inflammatory response. Increased expression of M-CSFR had been reported on microglia in AD brains *(57)*. Additional studies on inflammatory changes in Tg2576 mice showed proinflammatory cytokines in microglia around plaques *(42,47)*. Interleukin-1β (IL-1β) and tumor necrosis–α (TNF-α) immunoreactive microglia were prominent around plaques *(47)*. Consistent with human studies, the cytokine immunoreactive microglia were only observed around fibrillar Aβ plaques, and their expression increased with the age of the transgene-bearing animal. In this study, the antibodies to CD45 (leukocyte common antigen, a tyrosine protein phosphatase) and CD11b (complement receptor 3), markers for all microglia, preferentially stained activated microglia around fibrillar plaques *(47)*. These authors demonstrated that plaque-associated astrocytes were immunoreactive for IL-6, another pro-inflammatory cytokine *(47)*. This was confirmed in a later study with Tg2576 mice *(42)*, where IL-10 in plaque-associated astrocytes was also observed. In all of these referenced studies, the authors observed an increase in the numbers and degrees of activated microglia and reactive astrocytes as Aβ deposition progressed with age. This progressive age-related increase in the numbers of activated microglia and astrocytes were also observed in PS1/APP–Tg2576 mice *(38)*. Using CD11b as a marker for microglia, prominent activated microglia could be observed by 3 mo of age. These animals show accelerated deposition of Aβ into plaques compared with the single Tg2576 transgenic mice, because of the presence of the mutant gene for presenilin-1.

These mice models have been used to demonstrate the beneficial effects of a nonsteroidal anti-inflammatory agent and an antioxidant in treating Aβ-induced microglial responses *(44,58)*. It was shown that long-term feeding (6 mo) of the nonsteroidal anti-inflammatory drug (NSAID) ibuprofen or the spice curcumin (an antioxidant) to Tg2578 Aβ-plaque-developing mice resulted in significant decreases in indices of inflammation *(44,58)*. These included reductions in the numbers of activated microglia and reactive astrocytes, and lowering of levels of the pro-inflammatory cytokine IL-1β. Curcumin was also effective at lowering the levels of oxidized proteins in the brain. These studies have demonstrated mechanistically that inflammation can be driving much of the pathology in these mice and also, potentially, in AD brains. It was particularly of interest that both agents caused a significant decrease in levels of soluble and insoluble Aβ peptide in the brains of treated animals. These data imply that exacerbation of the inflammation can cause increased Aβ production, potentially as a result of the neuronal damage.

This might indicate that factors that result in increased inflammation could result in accelerated amyloid production. This was also demonstrated using Aβ-plaque-developing mice that overexpressed the anti-inflammatory cytokine transforming growth factor-β1 (TGF-β1) in astrocytes. These mice demonstrated reduced numbers of Aβ plaques in the brain parenchyma, along with reduced levels of brain Aβ, but had increased amounts of Aβ in the blood vessels compared to single transgenic Aβ mice *(59)*. The double transgenic animals appeared to have reduced numbers of activated microglia within plaque-developing regions of their brains and also reduced numbers of dystrophic neurites. It was concluded increased amounts of TGF-β1-induced increased phagocytosis and clearance of Aβ by microglia *(59)*, but reduced the pro-inflammatory features of the activated microglia.

5. MICROGLIAL PHAGOCYTOSIS OF ANTIBODY-OPSONIZED AMYLOID

The concepts proposed in Fig. 1 concerning the neurotoxic consequences of inflammation in AD brains have been reassessed recently with the discoveries that activated microglia can be employed to remove Aβ plaques and thus, potentially, be a strategy for treating AD. Immunization of transgenic plaque-developing PD-APP mice with Aβ such that these animals develop a circulating antibody response to this peptide was effective in reducing the amount of plaque material present in these animals *(12)*. Immunization of young mice, before plaque development had commenced, resulted in the prevention of the development of significant amounts of plaques. If the immunization was given to older animals, it appeared that raising an antibody response to Aβ resulted in clearance of this material. The mechanisms associated with Aβ clearance were investigated using these transgenic mice and ex vivo tissue slices from mice and AD brains. If Aβ antibodies were administered peripherally to mice, the same plaque-clearing effect was observed *(11)* as when the mice were immunized with peptide. These authors demonstrated that the administered antibodies could penetrate the brain and bind to plaques. They also demonstrated that microglia, when cultured with antibody-opsonized plaque material, could clear the Aβ from tissue slices. This effect was not observed when specific antibodies were omitted *(11)*. It is well known that antibody opsonization can promote macrophage/microglia phagocytosis through engagement of the Fc receptors *(60, 61)*; however, stimulation of these receptors can also have pro-inflammatory effects by increasing oxidative bursts and cytotoxicity *(62)*. As the induction of antibody-driven autoimmune responses in elderly AD patients could be pathological, further studies on this novel procedure for treating AD need to be conducted. This concern was addressed in one study whereby mice were immunized with a soluble nonamyloidogenic Aβ homologous peptide for a 7-mo-period, not with the fibrillogenic Aβ(1–42) peptide that can enter the brain *(63)*. This nonamyloidogenic peptide elicited an antibody response in the animals and was effective in reducing the plaque burden. It was observed that the numbers of IL-1β immunoreactive microglia were reduced in the immunized mice. This would indicate that at the end of the treatment period, there was no overall increase in microglial activation *(63)*. The effectiveness of using specific antibodies to remove Aβ plaques in transgenic mice was shown using an in vivo imaging technique *(64,65)*. It had first been demonstrated using this technique that plaques in these animals were stable over a period of 3 mo *(65)*. Direct addition of Aβ antibody to the brains of plaque-developing mice resulted in significant clearance of plaque material within 3 d *(64)*. It was observed that there was a significant microglial response in the region of the brains receiving antibody, indicating that these cells were indeed responsible for the clearance of the plaque material *(64)*. These observations were in agreement with the results using ex vivo tissue slices from AD brains *(11)*.

6. MICROGLIAL ACTIVATION IN ALZHEIMER'S DISEASE BRAINS

6.1. MHC II

As MHCII has been the most widely used marker for describing activated microglia in AD brains, a summary of some of the previous observations is included below. Initial immunohistochemical observations of microglia in AD brains demonstrated increased expression of the MHCII proteins, particularly HLA-DR, on microglia in AD-affected brain regions (examples are provided in refs. *7, 9, 66,* and *67*). Increased expression of this marker on microglia is not restricted to AD. Increased expression on pathological regions of brains from Parkinson's disease (PD) *(68)*, multiple sclerosis (MS), amyotrophic lateral sclerosis (ALS), Pick's disease *(69)*, and frontotemporal dementia *(70)*. This marker has been the most widely used to describe "activated" microglia, although its function in neurodegenerative processes has not been elucidated.

HLA-DR consists of two noncovalently associated polypeptides. The α-chain (34 kDa) and β-chain (29 kDa) have similar structures. Both proteins have single transmembrane domains and are heavily glycosylated. The α-chain and β-chain are coded by separate genes on chromosome 6. Three different genes have been identified for the β-chain. This ensures that the functional HLA-DR protein is highly polymorphic. The normal function of HLA-DR is presentation of antigen to CD8-positive T-lymphocytes. Although the extent of the involvement of lymphocytes in AD pathogenesis is still being studied, it is apparent that most HLA-DR immunoreactive microglia in AD brains do not colocalize with T-lymphocytes *(53,71)*. It would appear that induction of HLA-DR is an indices of inflammatory activation, although it does not necessarily reflect increased antigen presentation. In regions of AD-affected brains, increased HLA-DR expression is quite widespread throughout the tissue, indicative of a diffuse activating agent. The identity of this activating agent or agents is not known. It is known that interferon-γ (IFN-γ) is a potent activator of HLA-DR expression; however, the source of this cytokine in the brain is unclear. It is known that IFN-γ is secreted by lymphocytes and natural killer cells, but they are not prominently accumulated in AD brains. Some recent data have indicated that astrocytes and endothelial cells can secrete IFN-γ or IFN-γ-like molecules *(72,73)*. Determining the stimuli for class II protein expression in AD would aid in understanding the underlying inflammatory processes. It has been observed that Aβ peptide alone is not able to induce class II protein gene expression. This has been shown in in vitro experiments, and it has been frequently observed that there are not activated microglia with increased class II expression associated with diffuse Aβ plaques *(10)*. Biochemical studies have indicated that Aβ activation of microglia can be mediated by activation of the transcription factor NF-κB *(52)*, whereas this transcription factor is not involved in the activation of MHC class II gene expression. It has been recently shown that CIITA and RFX, the transcription factors associated with IFN-γ induction of HLA-DR, are increased in microglia around multiple sclerosis plaques *(74)*. Such

Table 2
Update of Microglial Markers Used
in Immunohistochemical Studies of AD Brain Tissues

Antigen	Function	Expression in AD	Ref.
CD40	Interaction with CD40 L	Increased	76
RAGE	Interaction with AGE, S100 Aβ	Increased	75
CCR3	Chemokine receptor	Increased	78
CCR5	Chemokine receptor	Increased	78

studies with these reagents using AD tissue sections will produce interesting results and possibly point to mechanisms as to how MHCII is increased in AD brains.

As outlined, recent studies by a number of different groups on the role of microglia in AD have focused on studies using Aβ-plaque-developing transgenic mice. Along with studies on tissue-culture-grown microglia, these have been designed to look at activation mechanisms. There have been fewer recent studies identifying novel markers of microglia in AD brains.

In recent studies of interest (Table 2), there has been the characterization of the expression of the receptor for advanced glycation endproducts (RAGE) in microglia in AD brains *(75)*, and the expression of CD40, a cell surface protein with key functions in coordinating pro-inflammatory stimulation *(76)*. In addition, the nature of the cyclooxygenase enzymes being expressed by microglia is being resolved by immunohistochemical staining of tissue *(77)*.

6.2. Micoglial RAGE in AD

Recent studies using both human brain tissues as well as human microglia derived from adult postmortem brains have shown that there was a significant role for RAGE in mediating interactions of microglia with Aβ peptide and also with AD pathology *(75)*. These studies extended the earlier observation of RAGE as a receptor for Aβ on microglia *(52)*. It was demonstrated that there was significantly increased amounts of RAGE protein in pathologically vulnerable regions in AD brains compared to nondemented individuals, and along with this, there was significantly increased amounts of RAGE expressed by microglia *(75)*. The percentage of RAGE immunoreactive microglia in different regions of the hippocampus correlated significantly with the scores for plaques and tangles in these cases *(73)*. In vitro, using human brain-derived microglia, we showed that blocking RAGE with specific immunological reagents significantly reduced the Aβ-stimulated increase in the secretion of the cytokine M-CSF, both when the Aβ was added to the culture medium of the microglia and when the Aβ was presented to the microglia as plaquelike deposits. The significance of increased RAGE expression in AD brains remains to be elucidated. Both neurons and microglia in AD brains were shown to have increased expression of RAGE *(75)*. The effect of this increased expression of RAGE could exacerbate the cellular stress responses induced by Aβ. It has been shown that Aβ binding to RAGE on neurons and microglia induce a range of stress-related genes as a result of the activation of the transcription factor NF-κB *(79)*.

6.3. Microglial CD40 in AD

A series of in vitro and animal studies demonstrated a role for the cell surface protein CD40 in amplifying an Aβ-stimulated inflammatory response in microglia *(80–82)*. Examination of a series of AD and nondemented brain tissue sections demonstrated significantly increased expression of CD40 on microglia in the sections from the AD cases. Increased CD40 staining was also observed on vascular endothelial cell in the AD tissue sections. It was observed, however, that the strongest microglial CD40 immunoreactivity was seen in the AD cases with other complicating pathology, such as bacterial encephalitis *(76)*. This indicated that Aβ was not the only stimuli for inducing CD40 expression in AD brains. There have been no reports on the cellular localization of the CD40 ligand (CD154) in AD brain tissue sections.

6.4. Microglial Cyclo-Oxygenase in AD

As mentioned, the mechanism for inflammation to be exacerbating the pathology in AD remains to be precisely defined. It has been known that patients with a history of taking NSAIDs have lower incidences of AD *(83–85)*. The principal mode of action of these drugs is as inhibitors of cyclo-oxygenase (COX) enzymes, which is responsible for producing prostaglandins from arachidonic acid. The older types of NSAIDs (e.g., indomethacin, ibuprofen) are more effective against COX-1, the constitutive form of the enzyme. It had generally been thought that the newer COX-2-specific inhibitors would be more effective in treating AD-related inflammation. COX-2 is highly induced in macrophages and other cells under conditions of inflammatory stimuli in many peripheral inflammatory diseases. Surprisingly, COX-2 is principally localized to neurons in the human brain and its has not been demonstrated in microglia in AD brains *(77,86)*. COX-2 immunoreactive microglia have been identified in Parkinson's disease brains and in acute inflammatory conditions, but in AD microglia, only COX-1 immunoreactivity has been detected *(77,86–88)*. The degree of induction of COX-1 in microglia around Aβ plaques was not prominent. This leads to the question of whether alteration in expression of microglial COX-1 is important for inflammatory processes in AD, whether changes in expression of neuronal COX-2 results in neurodegeneration *(89)*, or whether the effective biological targets for the NSAIDs are not related to their cyclo-oxygenase inhibiting activity. The NSAID indomethacin is a ligand for the peroxisome proliferator-activated receptor-γ (PPARγ). Activation of this receptor on macrophages/microglia induces anti-inflammatory pathways *(90,91)*. These issues should be resolved once clinical trials on AD patients are carried out using the new COX-2 specific inhibitors. These are underway at several centers in the United States.

7. CONCLUSIONS AND FUTURE DIRECTIONS

With the proposed use of immune therapy for treating Alzheimer's disease, some of the established concepts about inflammation and AD have been revisited. The concept of selectively exacerbating the inflammation in the brains of AD patients would seem to have the potential of accelerating the pathology of the disease. However, the spectacular results of Aβ vaccine therapy in clearing or preventing plaques in Aβ-plaque-developing mice indicate that this approach should be tested in human AD patients. Further studies would be required to understand the biology of microglial Fc receptors in the human brain. Previous concepts have suggested that engagement of these receptors by immune complexes can have serious pro-inflammatory consequences, which would obviously not be beneficial to the AD patients. Overall, with new therapeutic options becoming available, the potential to effectively treat this terrible disease is becoming a reality.

ACKNOWLEDGMENTS

The authors' work is supported by grants from the Alzheimer's Association and National Institutes of Health.

NOTE ADDED IN PROOF

The human trials of the Aβ vaccine were halted due to a number of the patients developing severe inflammatory complications in the brain. Such complications had not been reported in immunized transgenic mice. This highlights the potential limitations of basing this therapy on results from these Aβ-plaque forming transgenic mice. As the plaques in these mice are considerably more soluble than those of AD patients, immunization may have resulted in exacerbated cerebral inflammation in humans when the plaques could not easily be removed.

REFERENCES

1. Walker, D. G. (1998) Inflammatory markers in chronic neurodegenerative disorders, with emphasis on Alzheimer's disease, in *Neuroinflammation: Mechanisms and Management* (Wood, P. L., ed.), Humana, Totowa, NJ, pp. 61–90.
2. Sturchler-Pierrat, C., Abramowski, D., Duke, M., Wiederhold, K. H., Mistl, C., Rothacher, S., et al. (1997) Two amyloid precursor protein transgenic mouse models with Alzheimer disease-like pathology. *Proc. Natl. Acad. Sci. USA* **94**, 13,287–13,292.
3. Masliah, E., Sisk, A., Mallory, M., Mucke, L., Schenk, D., and Games, D. (1996) Comparison of neurodegenerative pathology in transgenic mice overexpressing V717F beta-amyloid precursor protein and Alzheimer's disease. *J. Neurosci.* **16**, 5795–5811.
4. Johnson-Wood, K., Lee, M., Motter, R., Hu, K., Gordon, G., Barbour, R., et al. (1997) Amyloid precursor protein processing and A beta42 deposition in a transgenic mouse model of Alzheimer disease. *Proc. Natl. Acad. Sci. USA* **94**, 1550–1555.
5. Games, D., Adams, D., Alessandrini, R., Barbour, R., Berthelette, P., Blackwell, C., et al. (1995) Alzheimer-type neuropathology in transgenic mice overexpressing V717F beta-amyloid precursor protein. *Nature* **373**, 523–527.
6. Hsiao, K., Chapman, P., Nilsen, S., Eckman, C., Harigaya, Y., Younkin, S., et al. (1996) Correlative memory deficits, Abeta elevation, and amyloid plaques in transgenic mice. *Science* **274**, 99–102.
7. McGeer, P. L., Itagaki, S., Tago, H., and McGeer, E. G. (1987) Reactive microglia in patients with senile dementia of the Alzheimer type are positive for the histocompatibility glycoprotein HLA-DR. *Neurosci. Lett.* **79**, 195–200.
8. McGeer, P. L., Itagaki, S., Tago, H., and McGeer, E. G. (1988) Occurrence of HLA-DR reactive microglia in Alzheimer's disease. *Ann. NY Acad. Sci.* **540**, 319–323.
9. Luber-Narod, J. and Rogers, J. (1988) Immune system associated antigens expressed by cells of the human central nervous system. *Neurosci. Lett.* **94**, 17–22.
10. Mattiace, L. A., Davies, P., and Dickson, D. W. (1990) Detection of HLA-DR on microglia in the human brain is a function of both clinical and technical factors. *Am. J. Pathol.* **136**, 1101–1114.

11. Bard, F., Cannon, C., Barbour, R., Burke, R. L., Games, D., Grajeda, H., et al. (2000) Peripherally administered antibodies against amyloid beta-peptide enter the central nervous system and reduce pathology in a mouse model of Alzheimer disease. *Nat. Med.* **6,** 916–919.

12. Schenk, D., Barbour, R., Dunn, W., Gordon, G., Grajeda, H., Guido, T., et al. (1999) Immunization with amyloid-beta attenuates Alzheimer-disease-like pathology in the PDAPP mouse. *Nature* **400,** 173–177.

13. Morgan, D., Diamond, D. M., Gottschall, P. E., Ugen, K. E., Dickey, C., Hardy, J., et al. (2000) A beta peptide vaccination prevents memory loss in an animal model of Alzheimer's disease. *Nature* **408,** 982–985.

14. Janus, C., Pearson, J., McLaurin, J., Mathews, P. M., Jiang, Y., Schmidt, S. D., et al. (2000) A beta peptide immunization reduces behavioural impairment and plaques in a model of Alzheimer's disease. *Nature* **408,** 979–982.

15. Kuo, Y. M., Kokjohn, T. A., Beach, T. G., Sue, L. I., Brune, D., Lopez, J. C., et al. (2001) Comparative analysis of A{beta} chemical structure and amyloid plaque morphology of transgenic mice and Alzheimer disease brains. *J. Biol. Chem.* **276,** 12,991–12,998.

16. Kawarabayashi, T., Younkin, L. H., Saido, T. C., Shoji, M., Ashe, K. H., and Younkin, S. G. (2001) Age-dependent changes in brain, CSF, and plasma amyloid (beta) protein in the Tg2576 transgenic mouse model of Alzheimer's disease. *J. Neurosci.* **21,** 372–381.

17. Kuo, Y. M., Webster, S., Emmerling, M. R., De Lima, N., and Roher, A. E. (1998) Irreversible dimerization/tetramerization and post-translational modifications inhibit proteolytic degradation of Abeta peptides of Alzheimer's disease. *Biochim. Biophys. Acta* **1406,** 291–298.

18. Kuo, Y. M., Emmerling, M. R., Woods, A. S., Cotter, R. J., and Roher, A. E. (1997) Isolation, chemical characterization, and quantitation of A beta 3-pyroglutamyl peptide from neuritic plaques and vascular amyloid deposits. *Biochem. Biophys. Res. Commun.* **237,** 188–191.

19. Itagaki, S., McGeer, P. L., Akiyama, H., Zhu, S., and Selkoe, D. (1989) Relationship of microglia and astrocytes to amyloid deposits of Alzheimer disease. *J. Neuroimmunol.* **24,** 173–182.

20. Neuroinflammation Working Group. (2000) Inflammation and Alzheimer's disease. *Neurobiol. Aging* **21,** 383–421.

21. Eikelenboom, P., Rozemuller, A. J., Hoozemans, J. J., Veerhuis, R., and van Gool, W. A. (2000) Neuroinflammation and Alzheimer disease: clinical and therapeutic implications. *Alzheimer Dis. Assoc. Disord.* **14(Suppl. 1),** S54–S61.

22. Giulian, D., Haverkamp, L. J., Yu, J. H., Karshin, W., Tom, D., Li, J., et al. (1996) Specific domains of beta-amyloid from Alzheimer plaque elicit neuron killing in human microglia. *J. Neurosci.* **16,** 6021–6037.

23. Giulian, D., Haverkamp, L. J., Li, J., Karshin, W. L., Yu, J., Tom, D., et al. (1995) Senile plaques stimulate microglia to release a neurotoxin found in Alzheimer brain. *Neurochem. Int.* **27,** 119–137.

24. Giulian, D., Haverkamp, L. J., Yu, J., Karshin, W., Tom, D., Li, J., et al. (1998) The HHQK domain of beta-amyloid provides a structural basis for the immunopathology of Alzheimer's disease. *J. Biol. Chem.* **273,** 29,719–29,726.

25. Shimohama, S., Tanino, H., Kawakami, N., Okamura, N., Kodama, H., Yamaguchi, T., et al. (2000) Activation of NADPH oxidase in Alzheimer's disease brains. *Biochem. Biophys. Res. Commun.* **273,** 5–9.

26. Van Muiswinkel, F. L., Raupp, S. F., de Vos, N. M., Smits, H. A., Verhoef, J., Eikelenboom, P., et al. (1999) The amino-terminus of the amyloid-beta protein is critical for the cellular binding and consequent activation of the respiratory burst of human macrophages. *J. Neuroimmunol.* **96,** 121–130.

27. Yasojima, K., Schwab, C., McGeer, E. G., and McGeer, P. L. (1999) Up-regulated production and activation of the complement system in Alzheimer's disease brain. *Am. J. Pathol.* **154,** 927–936.

28. Webster, S., Lue, L. F., Brachova, L., Tenner, A. J., McGeer, P. L., Terai, K., et al. (1997) Molecular and cellular characterization of the membrane attack complex, C5b-9, in Alzheimer's disease. *Neurobiol. Aging* **18,** 415–421.

29. Aisen, P. S. (2000) Anti-inflammatory therapy for Alzheimer's disease: implications of the prednisone trial. *Acta Neurol. Scand.* **176(Suppl.),** 85–89.

30. Hull, M. H., Fiebich, B. L., and Lieb, K. (2000) Strategies to delay the onset of Alzheimer's disease. *EXS* **89,** 211–225.

31. Lewis, J., Dickson, D. W., Lin, W. L., Chisholm, L., Corral, A., Jones, G., et al. (2001) Enhanced neurofibrillary degeneration in transgenic mice expressing mutant tau and APP. *Science* **293,** 1487–1491.

32. Games, D., Adams, D., Alessandrini, R., Barbour, R., Berthelette, P., Blackwell, C., et al. (1995) Alzheimer-type neuropathology in transgenic mice overexpressing V717F beta-amyloid precursor protein. *Nature* **373,** 523–527.

33. Mucke, L., Masliah, E., Yu, G. Q., Mallory, M., Rockenstein, E. M., Tatsuno, G., et al. (2000) High-level neuronal expression of abeta 1-42 in wild-type human amyloid protein precursor transgenic mice: synaptotoxicity without plaque formation. *J. Neurosci.* **20,** 4050–4058.

34. Sturchler-Pierrat, C., Abramowski, D., Duke, M., Wiederhold, K. H., Mistl, C., Rothacher, S., et al. (1997) Two amyloid precursor protein transgenic mouse models with Alzheimer disease-like pathology. *Proc. Natl. Acad. Sci. USA* **94,** 13,287–13,292.

35. Calhoun, M. E., Burgermeister, P., Phinney, A. L., Stalder, M., Tolnay, M., Wiederhold, K. H., et al. (1999) Neuronal overexpression of mutant amyloid precursor protein results in prominent deposition of cerebrovascular amyloid. *Proc. Natl. Acad. Sci. USA* **96,** 14,088–14,093.

36. Holcomb, L., Gordon, M. N., McGowan, E., Yu, X., Benkovic, S., Jantzen, P., et al. (1998) Accelerated Alzheimer-type phenotype in transgenic mice carrying both mutant amyloid precursor protein and presenilin 1 transgenes. *Nat. Med.* **4,** 97–100.

37. Frautschy, S. A., Yang, F., Irrizarry, M., Hyman, B., Saido, T. C., Hsiao, K., et al. (1998) Microglial response to amyloid plaques in APPsw transgenic mice. *Am. J. Pathol.* **152,** 307–317.

38. Matsuoka, Y., Picciano, M., Malester, B., LaFrancois, J., Zehr, C., Daeschner, J. M., et al. (2001) Inflammatory responses to amyloidosis in a transgenic mouse model of Alzheimer's disease. *Am. J. Pathol.* **158,** 1345–1354.

39. Irizarry, M. C., Soriano, F., McNamara, M., Page, K. J., Schenk, D., Games, D., et al. (1997) Abeta deposition is associated with neuropil changes, but not with overt neuronal loss in the human amyloid precursor protein V717F (PDAPP) transgenic mouse. *J. Neurosci.* **17,** 7053–7059.

40. Stalder, M., Phinney, A., Probst, A., Sommer, B., Staufenbiel, M., and Jucker, M. (1999) Association of microglia with amyloid plaques in brains of APP23 transgenic mice. *Am. J. Pathol.* **154,** 1673–1684.

41. Benzing, W. C., Wujek, J. R., Ward, E. K., Shaffer, D., Ashe, K. H., Younkin, S. G., et al. (1999) Evidence for glial-mediated inflammation in aged APP(SW) transgenic mice. *Neurobiol. Aging* **20,** 581–589.

42. Apelt, J. and Schliebs, R. (2001) Beta-amyloid-induced glial expression of both pro- and anti-inflammatory cytokines in cerebral cortex of aged transgenic Tg2576 mice with Alzheimer plaque pathology. *Brain Res.* **894,** 21–30.

43. Bornemann, K. D., Wiederhold, K. H., Pauli, C., Ermini, F., Stalder, M., Schnell, L., et al. (2001) Aβ-induced inflammatory processes in microglia cells of APP23 transgenic mice. *Am. J. Pathol.* **158,** 63–73.

44. Lim, G. P., Yang, F., Chu, T., Chen, P., Beech, W., Teter, B., et al. (2000) Ibuprofen suppresses plaque pathology and inflammation in a mouse model for Alzheimer's disease. *J. Neurosci.* **20,** 5709–5714.

45. Bornemann, K. D., Wiederhold, K. H., Pauli, C., Ermini, F., Stalder, M., Schnell, L., et al. (2001) Aβ-induced inflammatory processes in microglia cells of APP23 transgenic mice. *Am. J. Pathol.* **158,** 63–73.

46. Murphy, G. M., Zhao, F., Yang, L., and Cordell, B. (2000) Expression of macrophage colony-stimulating factor receptor is increased in the AbetaPP(V717F) transgenic mouse model of Alzheimer's disease. *Am. J. Pathol.* **158,** 63–73.

47. Benzing, W. C., Wujek, J. R., Ward, E. K., Shaffer, D., Ashe, K. H., Younkin, S. G., et al. (1999) Evidence for glial-mediated inflammation in aged APP(SW) transgenic mice. *Neurobiol. Aging* **20,** 581–589.

48. Wood, J. G. and Zinsmeister, P. (1991) Tyrosine phosphorylation systems in Alzheimer's disease pathology. *Neurosci. Lett.* **121,** 12–16.

49. El Khoury, J., Hickman, S. E., Thomas, C. A., Cao, L., Silverstein, S. C., and Loike, J. D. (1996) Scavenger receptor-mediated adhesion of microglia to beta-amyloid fibrils. *Nature* **382,** 716–719.

50. El Khoury, J., Hickman, S. E., Thomas, C. A., Loike, J. D., and Silverstein, S. C. (1998) Microglia, scavenger receptors, and the pathogenesis of Alzheimer's disease. *Neurobiol. Aging* **19,** S81–S84.

51. Huang, F., Buttini, M., Wyss-Coray, T., McConlogue, L., Kodama, T., Pitas, R. E., et al. (1999) Elimination of the class A scavenger receptor does not affect amyloid plaque formation or neurodegeneration in transgenic mice expressing human amyloid protein precursors. *Am. J. Pathol.* **155,** 1741–1747.

52. Yan, S. D., Chen, X., Fu, J., Chen, M., Zhu, H., Roher, A., et al. (1996) RAGE and amyloid-beta peptide neurotoxicity in Alzheimer's disease. *Nature* **382,** 685–691.

53. Itagaki, S., McGeer, P. L., and Akiyama, H. (1988) Presence of T-cytotoxic suppressor and leucocyte common antigen positive cells in Alzheimer's disease brain tissue. *Neurosci. Lett.* **91,** 259–264.

54. Combs, C. K., Johnson, D. E., Cannady, S. B., Lehman, T. M., and Landreth, G. E. (1999) Identification of microglial signal transduction pathways mediating a neurotoxic response to amyloidogenic fragments of beta-amyloid and prion proteins. *J. Neurosci.* **19,** 928–939.

55. Combs, C. K., Karlo, J. C., Kao, S. C., and Landreth, G. E. (2001) Beta-amyloid stimulation of microglia and monocytes results in TNFalpha-dependent expression of inducible nitric oxide synthase and neuronal apoptosis. *J. Neurosci.* **21,** 1179–1188.

56. Johnson-Wood, K., Lee, M., Motter, R., Hu, K., Gordon, G., Barbour, R., et al. (1997) Amyloid precursor protein processing and A beta42 deposition in a transgenic mouse model of Alzheimer disease. *Proc. Natl. Acad. Sci. USA* **94,** 1550–1555.

57. Akiyama, H., Nishimura, T., Kondo, H., Ikeda, K., Hayashi, Y., and McGeer, P. L. (1994) Expression of the receptor for macrophage colony stimulating factor by brain microglia and its upregulation in brains of patients with Alzheimer's disease and amyotrophic lateral sclerosis. *Brain Res.* **639,** 171–174.

58. Lim, G. P., Chu, T., Yang, F., Beech, W., Frautschy, S. A., and Cole, G. M. (2001) The curry spice curcumin reduces oxidative damage and amyloid pathology in an Alzheimer transgenic mouse. *J. Neurosci.* **21,** 8370–8377.

59. Wyss-Coray, T., Lin, C., Yan, F., Yu, G. Q., Rohde, M., McConlogue, L., et al. (2001) TGF-beta1 promotes microglial amyloid-beta clearance and reduces plaque burden in transgenic mice. *Nat. Med.* **7,** 612–618.

60. Greenberg, S. (1999) Modular components of phagocytosis. *J. Leukocyte Biol.* **66,** 712–717.

61. Aderem, A. and Underhill, D. M. (1999) Mechanisms of phagocytosis in macrophages. *Annu. Rev. Immunol* **17,** 593–623.

62. Ulvestad, E., Williams, K., Matre, R., Nyland, H., Olivier, A., and Antel, J. (1994) Fc receptors for IgG on cultured human microglia mediate cytotoxicity and phagocytosis of antibody-coated targets. *J. Neuropathol. Exp. Neurol.* **53,** 27–36.

63. Sigurdsson, E. M., Scholtzova, H., Mehta, P. D., Frangione, B., and Wisniewski, T. (2001) Immunization with a nontoxic/nonfibrillar amyloid-beta homologous peptide reduces Alzheimer's disease-associated pathology in transgenic mice. *Am. J. Pathol.* **159,** 439–447.

64. Bacskai, B. J., Kajdasz, S. T., Christie, R. H., Carter, C., Games, D., Seubert, P., et al. (2001) Imaging of amyloid-beta deposits in brains of living mice permits direct observation of clearance of plaques with immunotherapy. *Nat. Med.* **7,** 369–372.

65. Christie, R. H., Bacskai, B. J., Zipfel, W. R., Williams, R. M., Kajdasz, S. T., Webb, W. W., et al. (2001) Growth arrest of individual senile plaques in a model of Alzheimer's disease observed by in vivo multiphoton microscopy. *J. Neurosci.* **21,** 858–864.

66. Styren, S. D., Civin, W. H., and Rogers, J. (1990) Molecular, cellular, and pathologic characterization of HLA-DR immunoreactivity in normal elderly and Alzheimer's disease brain. *Exp. Neurol.* **110,** 93–104.

67. Dickson, D. W. and Mattiace, L. A. (1989) Astrocytes and microglia in human brain share an epitope recognized by a B-lymphocyte-specific monoclonal antibody (LN-1). *Am. J. Pathol.* **135,** 135–147.

68. McGeer, P. L., Itagaki, S., Boyes, B. E., and McGeer, E. G. (1988) Reactive microglia are positive for HLA-DR in the substantia nigra of Parkinson's and Alzheimer's disease brains. *Neurology* **38,** 1285–1291.

69. McGeer, P. L., Itagaki, S., and McGeer, E. G. (1988) Expression of the histocompatibility glycoprotein HLA-DR in neurological disease. *Acta Neuropathol. (Berl.)* **76,** 550–557.

70. Arnold, S. E., Han, L. Y., Clark, C. M., Grossman, M., and Trojanowski, J. Q. (2000) Quantitative neurohistological features of frontotemporal degeneration. *Neurobiol. Aging* **21,** 913–919.

71. Rogers, J., Luber-Narod, J., Styren, S. D., and Civin, W. H. (1988) Expression of immune system-associated antigens by cells of the human central nervous system: relationship to the pathology of Alzheimer's disease. *Neurobiol. Aging* **9,** 339–349.

72. Wei, Y. P., Kita, M., Shinmura, K., Yan, X. Q., Fukuyama, R., Fushiki, S., et al. (2000) Expression of IFN-gamma in cerebrovascular endothelial cells from aged mice. *J. Interferon. Cytokine Res.* **20,** 403–409.

73. Lau, L. T. and Yu, A. C. (2001) Astrocytes produce and release interleukin-1, interleukin-6, tumor necrosis factor alpha and interferon-gamma following traumatic and metabolic injury. *J. Neurotrauma* **18,** 351–359.

74. Gobin, S. J., Montagne, L., Van Zutphen, M., Van Der Valk, P., Van Den Elsen, P. J., and De Groot, C. J. (2001) Upregulation of transcription factors controlling MHC expression in multiple sclerosis lesions. *Glia* **36,** 68–77.

75. Lue, L. F., Walker, D. G., Brachova, L., Beach, T. G., Rogers, J., Schmidt, A. M., et al. (2001) Involvement of microglial receptor for advanced glycation endproducts (RAGE) in Alzheimer's disease: identification of a cellular activation mechanism. *Exp. Neurol.* **171,** 29–45.

76. Togo, T., Akiyama, H., Kondo, H., Ikeda, K., Kato, M., Iseki, E., et al. (2000) Expression of CD40 in the brain of Alzheimer's disease and other neurological diseases. *Brain Res.* **885,** 117–121.

77. Hoozemans, J. J., Rozemuller, A. J., Janssen, I., De Groot, C. J., Veerhuis, R., and Eikelenboom, P. (2001) Cyclooxygenase expression in microglia and neurons in Alzheimer's disease and control brain. *Acta Neuropathol. (Berl.)* **101,** 2–8.

78. Xia, M. Q., Qin, S. X., Wu, L. J., Mackay, C. R., and Hyman, B. T. (1998) Immunohistochemical study of the beta-chemokine receptors CCR3 and CCR5 and their ligands in normal and Alzheimer's disease brains. *Am. J. Pathol.* **153,** 31–37.

79. Yan, S. D., Roher, A., Schmidt, A. M., and Stern, D. (1999) Cellular cofactors for amyloid β-peptide-induced cell stress. *Am. J. Pathol.* **155,** 1403–1411.

80. Tan, J., Town, T., Paris, D., Mori, T., Suo, Z., Crawford, F., et al. (1999) Microglial activation resulting from CD40-CD40L interaction after beta-amyloid stimulation. *Science* **286,** 2352–2355.

81. Tan, J., Town, T., and Mullan, M. (2000) CD45 inhibits CD40L-induced microglial activation via negative regulation of the Src/p44/42 MAPK pathway. *J. Biol. Chem.* **275,** 37,224–37,231.

82. Town, T., Tan, J., and Mullan, M. (2001) CD40 signaling and Alzheimer's disease pathogenesis. *Neurochem. Int.* **39,** 371–380.

83. Breitner, J. C., Gau, B. A., Welsh, K. A., Plassman, B. L., McDonald, W. M., Helms, M. J. et al. (1994) Inverse association of anti-inflammatory treatments and Alzheimer's disease: initial results of a co-twin control study. *Neurology* **44,** 227–232.

84. Breitner, J. C., Welsh, K. A., Helms, M. J., Gaskell, P. C., Gau, B. A., Roses, A. D., et al. (1995) Delayed onset of Alzheimer's disease with nonsteroidal anti-inflammatory and histamine H2 blocking drugs. *Neurobiol. Aging* **16,** 523–530.

85. McGeer, P. L., Schulzer, M., and McGeer, E. G. (1996) Arthritis and anti-inflammatory agents as possible protective factors for Alzheimer's disease: a review of 17 epidemiologic studies. *Neurology* **47,** 425–432.

86. Yermakova, A. V., Rollins, J., Callahan, L. M., Rogers, J., and O'Banion, M. K. (1999) Cyclooxygenase-1 in human Alzheimer and control brain: quantitative analysis of expression by microglia and CA3 hippocampal neurons. *J. Neuropathol. Exp. Neurol.* **58,** 1135–1146.

87. Tomimoto, H., Akiguchi, I., Wakita, H., Lin, J. X., and Budka, H. (2000) Cyclooxygenase-2 is induced in microglia during chronic cerebral ischemia in humans. *Acta Neuropathol. (Berl.)* **99,** 26–30.

88. Knott, C., Stern, G., and Wilkin, G. P. (2000) Inflammatory regulators in Parkinson's disease: iNOS, lipocortin-1, and cyclooxygenases-1 and -2. *Mol. Cell. Neurosci.* **16,** 724–739.

89. Ho, L., Purohit, D., Haroutunian, V., Luterman, J. D., Willis, F., Naslund, J., et al. (2001) Neuronal cyclooxygenase 2 expression in the hippocampal formation as a function of the clinical progression of Alzheimer disease. *Arch. Neurol.* **58,** 487–492.

90. Ricote, M., Huang, J. T., Welch, J. S., and Glass, C. K. (1999) The peroxisome proliferator-activated receptor (PPARgamma) as a regulator of monocyte/macrophage function. *J. Leukocyte Biol.* **66,** 733–739.

91. Combs, C. K., Johnson, D. E., Karlo, C., Cannady, S. B., and Landreth, G. E. (2000) Inflammatory mechanisms in Alzheimer's disease: inhibition of β-amyloid stimulated pro-inflammatory responses and neurotoxicity by PPAR-γ agonists. *J. Neurosci.* **20,** 558–567.

14

The Amyloid Hypothesis of Cognitive Dysfunction

Dave Morgan and Marcia N. Gordon

1. INTRODUCTION

The amyloid hypothesis of Alzheimer's disease posits that a critical event in the pathogenesis of the disorder is the deposition of amyloid fibrils containing the amyloid β (Aβ) peptide derived from the processing of the amyloid precursor protein. This hypothesis is the subject of multiple reviews (1,2). The greatest support for this hypothesis derives from the genetics of familial Alzheimer's disease, where all mutations known to cause Alzheimer's disease share the capacity to increase production of the long form of the Aβ peptide (3). Still, acceptance of this hypothesis is far from universal (see ref. 4 and affiliated commentaries). Importantly, Alzheimer's disease is, by definition, a cognitive disorder, requiring a patient to have a graded deterioration in multiple cognitive domains before a diagnosis can be made. In the absence of biopsy, pathological criteria for the disorder can only be met by autopsy. Thus, a critical question in the context of the amyloid hypothesis is the role of amyloid in the cognitive dysfunction of Alzheimer's disease.

2. THE ASSOCIATION OF Aβ DEPOSITS WITH COGNITIVE FUNCTION IN PATIENTS

On a straightforward level, if Aβ deposits are responsible for Alzheimer's dementia, one might expect that the degree of dementia would correlate to some extent with the amounts of amyloid deposited. Early work supported this argument. The performance of patients and controls on the Blessed memory scale prior to death correlated with the numbers of amyloid plaques at autopsy (5). However, one problem with the argument was that the correlations included both demented and nondemented individuals; if the nondemented individuals were removed from the correlation, little association between amyloid deposition and memory function remained. Because the definition of Alzheimer's dementia includes both memory loss and amyloid plaques and the definition of age-matched control cases requires normal memory and no amyloid plaques, it would be almost impossible not to have a correlation when both groups are included.

There have been a number of similar studies since this original description, each using more refined measures of cognitive function and brain pathology. Terry et al. (6) did find correlations between cognitive function and plaques or tangles, but the best correlation was with synaptophysin staining (an index of synaptic density). Cummings et al.

From: *Neuroinflammation, 2nd Edition: Mechanisms and Management*
Edited by: P. L. Wood © Humana Press Inc., Totowa, NJ

(7) performed similar analyses and found that the best pathologic predictor of premorbid cognitive status was Aβ load in the entorhinal cortex. Importantly, these authors mention that significant correlations exist even if control subjects are removed (although all data shown include control subjects). They also mention that if patients with mini-mental status evaluation scores below 15 are included, there is no correlation of cognitive status with any pathologic measurement. The commentaries accompanying the Cummings et al. article clearly indicate lack of consensus regarding the role of amyloid in cognitive declines in Alzheimer's dementia, even after 30 yr of study.

Recently, Naslund et al. *(8)* used a clinical dementia rating (CDR) scale and measured Aβ levels in several brain regions chemically (enzyme-linked immunosorbent assay [ELISA]) rather than histopathologically. They found correlations between Aβ content of several cortical regions and clinical dementia ratings when including all cases with CDRs ranging from 0 to 5 (normal to severely demented) as well as those cases with CDRs of 0 to 2 (normal to moderately impaired). The correlation within the normal to moderately impaired group was argued to indicate that amyloid levels rose early in the onset of dementia. No mention was made whether the correlations remained if the CDR = 0 category was excluded from the analysis.

In general, these correlations between cognitive status and Aβ deposition have not been terribly satisfying for the adherents of the amyloid hypothesis. Indeed the weakness of these relationships has been used as an argument against the amyloid hypothesis *(4,9)*. However, only if a simple linear relationship between amyloid deposits and cognitive dysfunction exists would one expect to detect a correlation between Aβ deposition and cognitive status within the demented groups. This relationship would assume that Aβ deposits gradually accumulate over time, cause neurodegeneration in a purely linear manner (no threshold effects or vicious cycles), and interfere with cognitive status with little modulation by other factors (initial intellectual status, age, etc.). Yet, evidence exists to argue that none of these assumptions hold true in Alzheimer's disease.

First, the notion that amyloid deposits are stable features that continue to accumulate over time is likely not true, at least once the cognitive decline in Alzheimer's disease begins. Two studies *(10,11)* examined the density of plaque pathology in biopsies obtained early in the disease and, ultimately, from autopsies of the same individuals. Even though the dementia in these individuals progressed, the amount of plaque pathology was similar in the biopsy and autopsy results. Hyman et al. *(12)* argue that there is a continuous turnover of Aβ deposits in Alzheimer's disease. This is consistent with the inflammation hypothesis of Alzheimer dementia *(13)*, in which microglia phagocytize Aβ while participating in an inflammatory response that ultimately harms bystander neurons (*see also* Chapter 13). In fact, it may be the removal of the Aβ deposits that is toxic rather than simply their presence. If everyone accumulated Aβ at a similar rate and if amyloid removal caused toxicity, one would predict an *inverse* correlation between memory deficits and Aβ deposition, rather than a direct linear correlation, as predicted earlier.

Second, even if the amyloid is directly neurotoxic, it is likely that the effects on neural function would not be linear. Certainly it is the case that many cognitively intact individuals die with considerable numbers of amyloid deposits (so-called "high-plaque normals") *(14–16)*. Thus, simply having amyloid deposits is not sufficient to cause memory loss. Yet, that cannot be interpreted as meaning that the plaques are unrelated to the cognitive dysfunction of Alzheimer's. Many die with atherosclerotic plaques in coronary arteries without having a heart attack, yet no one would claim that this means that atherosclerosis and heart disease are not related. Obviously, there is a threshold level of virtually any pathology that will be required for symptomatic expression of neural disorders. The brain is blessed with a remarkable degree of functional redundancy, especially for the more integrative functions. Hence, up to a certain level of deposition, it is likely that the amyloid will not correlate with cognitive status.

There have been multiple hypotheses that Aβ deposits participate in a vicious cycle. Although different hypotheses posit different agents at the initiation of the cycle, the major participants are Aβ deposits, acute-phase reactants, and inflammatory cytokines *(13)*. The general argument is that Aβ deposits provoke activation of microglia (through complement, scavenger receptors, AGE receptors or other mechanisms), leading to a release of pro-inflammatory cytokines (interleukin [IL]-1, IL-6, tumor necrosis factor [TNF]-α), and the upregulation of acute-phase proteins (amyloid precursor protein, apolipoprotein E, α-anti-chymotripsin, apolipoprotein J, complement pathway proteins, pentraxins, proteoglycans). The end result is further deposition of Aβ and further input to the vicious cycle *(3,17,18)*. One possibility is that cognitive dysfunction is caused by the inflammation itself, or neurotoxicity is secondary to such inflammation. If the vicious cycle is never initiated (due to the presence of anti-inflammatory medications, for example), Aβ deposits could be present, but associated with minimal cognitive dysfunction. Consistent with this argument, some have found these high-plaque normals do lack signs of neural inflammation *(19)*.

A third problem in correlating cognitive function and amyloid is individual heterogeneity in cognitive function. The presence of genetic and environmental influences on cognitive function have been recognized for centuries. Some individuals are born with or develop intellectual reserves which appear to delay the onset of Alzheimer's disease, at least by delaying the age of onset *(20,21)*. In an earlier review of the impact of age on Alzheimer's pathology *(22)*, it was reported that for virtually every pathological end point, the severity of degeneration in Alzheimer patients coming to autopsy was *inversely* related to patient age. Large amounts of pathology were necessary to cause dementia in young people, whereas relatively little pathology could produce similar dementia ratings in older individuals. This may reflect a declining threshold in the amount of pathology required to elicit dementia (or launch a vicious cycle) with increasing age. Alternatively, Alzheimer cases carrying an apolipoprotein E4 allele have increased amyloid deposition and are overrepresented in the "early-onset" categories of dementia *(23)*. In either event, if age at onset is not a covariate in the analysis, strong correlations might be missed unless the age range chosen is narrow.

3. THE ASSOCIATION OF Aβ DEPOSITS WITH COGNITIVE FUNCTION IN EXPERIMENTAL ANIMALS

A complimentary approach to function–pathology studies in humans is the production of Aβ deposits in experimental animals and the evaluation of learning and memory functions. In general, this approach has been tested in two manners: by injecting amyloid peptides into the brains of mice or rats or by generating transgenic mice that accumulate Aβ deposits over time. The prediction of the amyloid hypothesis of cognitive dysfunction is that the presence of Aβ deposits should lead to impaired memory performance. To a substantial extent, this appears to be the case.

A large number of studies have evaluated the mnemonic influences of Aβ peptides injected into the brains of rodents. For the most part, the results reported favor the conclusion that these agents can disrupt memory (it is uncertain how extensively negative results would be published). The peptide variants found effective include Aβ1–28, Aβ25–35, Aβ1–40, and Aβ1–42 *(24–28)*. Behavioral deficits were found in a variety of mnemonic tasks, including water maze *(29)*, radial arm maze *(30)*, Y maze alternation *(27,31)*, or avoidance *(32–35)*. When investigated, these injections did not cause neurotoxicity *(27,32)*, consistent with other studies evaluating Aβ injections into rodents *(see* ref. *36)* and discussion therein). It also appeared that aggregated peptides were more effective memory-disrupting agents than soluble peptides *(37)*. Thus, several groups have consistently found that intracranial administration of Aβ peptides cause learning and memory deficits. These data support the hypothesis that Aβ can directly interfere with neural function.

A second approach to testing the hypothesis that Aβ peptides cause memory deficits has been to overexpress the amyloid precursor protein (APP) in transgenic mice. In some instances, these transgenic animals overxpress enough of the mutated variants of human APP that Aβ deposits accumulate in the cerebral cortex and hippocampus as the mice age *(38–40)*. In the Tg2576 APP transgenic mouse *(39)*, there was a decline in water maze performance at 11–13 mo of age, a time when the amyloid deposits were appearing. Whether this represented a true learning defect at this age has been questioned *(41)*. However, more recent analyses indicates that with large numbers of older mice, reliable learning and memory deficits are found in the reference memory version of the water maze. This same mouse line also shows deficits in T-maze performance, and this performance correlates with each mouse's capacity for long-term potentiation (LTP) (when nontransgenic mice are included in the correlation) *(42)*. Still, no correlation was reported between the extent of amyloid deposition and mnemonic performance of individual mice.

The PDAPP mouse described by Games et al. *(38)* has been more difficult to test behaviorally. Even at very young ages, before amyloid deposits appear, these mice were deficient on multiple learning and memory tasks *(43,44)*. Part of the problem appears to be developmental, as very young PDAPP mice have 20–40% smaller hippocampi than their nontransgenic littermates *(45)*. Still, these authors are able to detect deficits in an object-recognition task that appear when the mice get older and accumulate Aβ deposits *(43)*. Moreover, these deficits correlate with the number of Aβ deposits, even excluding the nontransgenic mice, when the 3-, 6-, and 10-mo-old animals are combined *(45)*. Most recently, a working memory version of the water maze was applied to the PDAPP mice over the life-span *(46)*. Although the transgenic mice were deficient in this task when compared to young nontransgenic mice, the differential worsened as the mice grew older. Moreover, the learning capacity of these mice correlated with the hippocampal Aβ burden in the 16- to 22-mo-old mice (excluding the nontransgenic and younger animals). These results indicate that in addition to causing an early deficiency in cognitive function related to hippocampal size, sophisticated behavioral analyses can tease out cognitive dysfunctions in these mice that worsen as the mice age and are correlated with the amount Aβ accumulated in the brain (even when excluding the nontransgenic mice). This is powerful support for the notion that Aβ can cause deficits in learning and memory.

Our own work has focused on a doubly transgenic mouse model of amyloid deposition that combines the Tg2576 APP transgenic mouse *(39)* with a presenilin-1 (PS1)1 transgenic mouse *(47)*. The major advantage of this doubly transgenic mouse is a considerably more rapid pace of Aβ deposition than either parental line *(48)*. It also presents a deficit in Y-maze alternation at 3, 6, and 9 mo of age *(49)*, which is no longer present when the mice reach 15 mo of age *(50)*. At 6 and 9 mo of age, these mice demonstrate strong learning and retention for platform location in the reference memory version of the water maze *(51)*, whereas at 15 mo, they exhibit slower acquisition than nontransgenic mice, yet are no different from nontransgenic mice on the probe (retention) trial *(50)*. This difficulty in obtaining clear-cut learning and memory deficits led us to examine a slightly different version of the water maze task, which emphasizes working memory.

The radial arm water maze is a hybrid of the radial arm maze and water maze. It imposes swim alleys on the pool (formed by aluminum wedge barriers) that radiate out from a central swim area. In the working memory version of the radial arm water maze, a submerged platform is placed in one arm and mice are given four consecutive learning trials (each from a different start arm) to find the platform *(52)*. Thirty minutes later, they are given a fifth retention trial. Each incorrect arm entry is counted as an error. The next day, the process is repeated with the platform located in a different arm. This approach retains the advantages of the water maze paradigm (no need for foot shock or food deprivation), with the advantages of the radial arm maze (use of errors rather than swim latency as the dependent measure). Mice initially do poorly on all trials, averaging four to five errors per trial. After several days, the mice begin to acquire the procedural components of the task, with fewer errors on trials 4 and 5 than on trial 1. Finally, between 1 and 2 wk of training, the mice learn the new platform location quickly each day, scoring zero to one errors by trials 4 and 5.

Doubly transgenic APP+PS1 mice are able to learn this task at 6 mo *(50)* and 12 mo *(53)*. However, by 16 mo, there was a clear discrimination between transgenic and nontransgenic mice *(50,54)*, with transgenic mice unable to improve their performance over trials and average greater than three errors on trials 4 and 5 *(50)*. We felt that this difference was sufficient to permit us to use this task to evaluate the functional consequences of treatments designed to reduce the deposition of Aβ (unlike the reference memory version of the water maze). Remarkably, even with a sample size of nine mice, there was a significant correlation within the transgenic group between the amount of fibrillar Aβ (stained by Congo red) and the number of errors on trial 5 of the water maze task ($r = -0.74$) *(54)*. This correlation was also observed when antibody to Aβ1–40 was used to stain fibrillar amyloid deposits.

We have also found that vaccination of doubly transgenic mice with the Aβ peptide is capable of reducing the amount of Congo red-stained amyloid deposits in doubly transgenic mice (by about 20%) *(53)*. This vaccination protocol, first described by Schenk et al. *(55)* is also capable of partially protecting transgenic mice (both APP+PS1 and APP only) from developing deficits in the water maze paradigm (they learn within-day platform location slower than nontransgenic mice, but perform similarly by trials 4 and 5). Interestingly, when the errors on trials 4 and 5 are averaged in the doubly transgenic PS1+APP mice from this study, we also find a correlation between learning and memory and Congo red-stained deposits (Fig. 1). This finding replicates our results from (ref. *54*).

In summarizing the results using transgenic mice, there are learning and memory deficits that develop in transgenic mice as Aβ amyloid accumulates. In addition, there appears to be a threshold level of Aβ deposition required for the functional deficits to

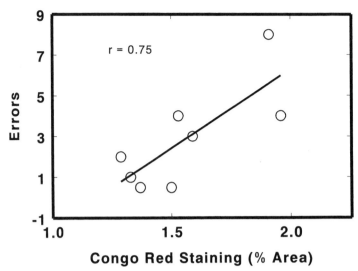

Fig. 1. Correlation between amyloid deposition and working memory performance in the radial arm water maze in 15-mo-old doubly transgenic mice. Mice were all APP+PS1 doubly transgenic, They were trained in the radial arm water maze for 11 d at 15.5 mo of age. The number of errors on trials 4 and 5 from d 10 and 11 of training were averaged for each mouse. Shortly after the training, mice were killed and sections stained with Congo red to identify amyloid deposits. The fractional area of the anterior cortex positively stained by Congo red was quantified by image analysis. Data indicate a positive correlation between number of errors on trials 4 and 5 and the amount of Congo red-stained area for each mouse. This relationship is the same as that found for a different group of mice in (ref. *54*). Three of the mice in the data shown here were vaccinated against the Aβ peptide for 9 mo *(53)*.

be manifest. Moreover, for some task–age combinations, the performance on the learning task is correlated with the amount of deposited amyloid. However, in some APP transgenic mouse lines, even in the absence of fibrillar amyloid deposition there are deficiencies in either Y-maze alternation *(56)* or water maze performance at young ages prior to Aβ deposit formation *(57,58)*. This highlights one complication of interpreting the effects of amyloid in transgenic lines; the transgenes could have developmental influences (as is apparently the case in the hippocampal shrinkage found in PDAPP mice). Additionally, the overexpression of APP itself may have impact, irrespective of its role in elevating Aβ *(59–61)*.

Still, assuming that the transgenic and experimental injection models of amyloid deposition do find that Aβ deposits disrupt memory, there is a disturbing problem in relating this to human Alzheimer's disease. The amounts of neural damage are negligible in the rodent models, but substantial in older humans. Most studies fail to detect differences in the neuron number or synapse density in the transgenic models of amyloid deposition *(62–64)*, and those that do detect differences find modest changes in neuron and synapse number relative to the loss in Alzheimer's disease *(45,65)*.

In the experimental Aβ injection models, neuron loss has been a rare finding *(see* refs. *36* and *66* and references therein). If the argument in Alzheimer's disease is that the effect of Aβ deposits is to cause neurodegeneration, and the loss of neurons and synapses is what leads to the cognitive dysfunction, then the rodents are a poor model, as neurodegeneration is questionable.

What might then be drawn from the aggregate of data regarding Aβ and cognitive function? Ultimately, at least two scenarios may be acting in concert. One is the ability of Aβ deposits to launch a cascade of events leading to neuron and/or synapse loss in the Alzheimer's disease (AD) brain, which is directly responsible for the cognitive dysfunction. In this scenario, intervening in the steps between Aβ deposition and neurotoxic reactions could be effective therapies to slow or arrest the progression of AD. A second scenario, given token support from the animal literature, is that the Aβ peptide itself is capable of disrupting neural function without causing overt neuron or synapse loss. Given the amounts of Aβ found in older transgenic mice relative to AD, it would appear that the amounts of Aβ necessary to directly disrupt neural function are somewhat greater than the quantity capable of launching a neurodegenerative cascade in humans. Still, one must consider the high local Aβ concentrations in key structures of AD brain. Moreover, the loss of plasticity and other damage that accumulates with typical brain aging *(4)* may considerably lower the threshold amounts of amyloid that are required to cause the symptoms of dementia *(22)*.

Addressing these questions is becoming tractable. The transgenic models of amyloid deposition are increasingly being characterized functionally as well as pathologically. Numerous manipulations are being developed to *(1)* modify Aβ deposition and *(2)* evaluate the relationship to cognitive status in the mice. Several of the Aβ-lowering therapies are starting to be tested on AD patients. Combined, the results of these studies will tell us whether either or both of the above-mentioned scenarios are operating in causing the cognitive problems found in AD.

ACKNOWLEDGMENTS

This work was supported by NIH grants AG 18478 to DGM and AG 15490 to MG.

REFERENCES

1. George-Hyslop, P. H. (2000) Piecing together Alzheimer's. *Sci. Am.* **283,** 76–83.
2. Selkoe, D. J. (2000) Toward a comprehensive theory for Alzheimer's disease. Hypothesis: Alzheimer's disease is caused by the cerebral accumulation and cytotoxicity of amyloid beta-protein. *Ann. NY Acad. Sci.* **924,** 17–25.
3. Hardy, J. (1997) Amyloid, the presenilins and Alzheimer's disease. *Trends Neurosci.* **20,** 154–160.
4. Joseph, J. A., Shukhitt-Hale, B. D. N. A., Martin, A., Perry, G., and Smith, M. A. (2001) Copernicus revisited: amyloid beta in Alzheimer's disease. *Neurobiol. Aging* **22,** 131–146.
5. Blessed, G. T. B. E. (1968) The association between quantitative measures of dementia and of senile change in the cerebral grey matter of elderly subjects. *Br. J. Psychiatry* **114,** 797–811.
6. Terry, R. D., Masliah, E., Salmon, D. P., Butter, N., DeTeresa, R., Hill, R., et al. (1991) Physical basis of cogntive alterations in Alzheimer's disease: synapse loss is the major correlate of cognitive impairment. *Ann. Neurol.* **30,** 572–580.
7. Cummings, B. J., Pike, C. J., Shankle, R., and Cotman, C. W. (1996) Beta-amyloid deposition and other measures of neuropathology predict cognitive status in Alzheimer's disease. *Neurobiol. Aging* **17,** 921–933.

8. Naslund, J. H., Haroutunian, V., Mohs, R., Davis, K. L., Davies, P., Greengard, P., et al. (2000) Correlation between elevated levels of amyloid beta-peptide in the brain and cognitive decline. *JAMA* **283(12),** 1571–1577.

9. Roses, A. D. (1994) Apolipoprotein E affects the rate of Alzheimer disease expression: beta-amyloid burden is a secondary consequence dependent on APOE genotype and duration of disease. [see comments]. *J. Neuropathol. Exp. Neurol.* **53,** 429–437.

10. Bennett, D. A., Cochran, E. J., Saper, C. B., Leverenz, J. B., Gilley, D. W., and Wilson, R. S. (1993) Pathological changes in frontal cortex from biopsy to autopsy in Alzheimer's disease. *Neurobiol. Aging* **14,** 589–596.

11. Mann, D. M., Marcyniuk, B., Yates, P. O., Neary, D., and Snowden, J. S. (1988) The progression of the pathological changes of Alzheimer's disease in frontal and temporal neocortex examined both at biopsy and at autopsy. *Neuropathol. Appl. Neurobiol.* **14,** 177–195.

12. Hyman, B. T., Marzloff, K., and Arriagada, P. V. (1993) The lack of accumulation of senile plaques or amyloid burden in Alzheimer's disease suggests a dynamic balance between amyloid deposition and resolution. *J. Neuropathol. Exp. Neurol.* **52,** 594–600.

13. Akiyama, H., Barger, S., Barnum, S., Bradt, B., Bauer, J., Cole, G. M., et al. (2000) Inflammation and Alzheimer's disease. *Neurobiol. Aging* **21,** 383–421.

14. Crystal, H., Dickson, D., Fuld, P., Masur, D., Scott, R., Mehler, M., et al. (1988) Clinicopathological studies in dementia: nondemented subjects with pathologically confirmed Alzheimer's disease. *Neurology* **38,** 1682–1687.

15. Davis, D. G., Schmitt, F. A., Wekstein, D. R., and Markesbery, W. R. (1999) Alzheimer neuropathologic alterations in aged cognitively normal subjects. *J. Neuropathol. Exp. Neurol.* **58,** 376–388.

16. Wolf, D. S., Gearing, M., Snowdon, D. A., Mori, H., Markesbery, W. R., and Mirra, S. S. (1999) Progression of regional neuropathology in Alzheimer disease and normal elderly: findings from the Nun study. *Alzheimer Dis. Assoc. Disord.* **13,** 226–231.

17. Griffin, W. S., Sheng, J. G., Roberts, G. W., and Mrak, R. E. (1995) Interleukin-1 expression in different plaque types in Alzheimer's disease: significance in plaque evolution. *J. Neuropathol. Exp. Neurol.* **54,** 276–281.

18. Rogers, J., Webster, S., Lue, L. F., Brachova, L., Civin, W. H., Emmerling, M., et al. (1996) Inflammation and Alzheimer's disease pathogenesis. *Neurobiol. Aging* **17,** 681–686.

19. Lue, L., Brachova, L., Civin, H., and Rogers, J. (1996) Inflammation, Abeta deposition and neurofibrillary tangle formation as correlates of Alzheimer's disease neurodegeneration. *J. Neuropathol. Exp. Neurol.* **55,** 1083–1088.

20. Katzman, R. (1993) Education and the prevalence of dementia and Alzheimer's disease. *Neurology* **43,** 13–20.

21. Snowdon, D. A., Kemper, S. J., Mortimer, J. A., Greiner, L. H., Wekstein, D. R., and Markesbery, W. R. (1996) Linguistic ability in early life and cognitive function and Alzheimer's disease in late life. Findings from the Nun Study. [see comments]. *JAMA* **275,** 528–532.

22. Morgan, D. G. (1992) Neurochemical changes with aging: predisposition towards age-related mental disorders, in: *Handbook of Mental Health and Aging* (Birren, J. E., Sloane, R. B., and Cohen, G. D., eds.), Academic, San Diego, CA, pp. 175-200.

23. Hyman, B. T., West, H. L., Rebeck, G. W., Buldyrev, S. V., Mantegna, R. N., Ukleja, M., et al. (1995) Quantitative analysis of senile plaques in Alzheimer disease: observation of log-normal size distribution and molecular epidemiology of differences associated with apolipoprotein E genotype and trisomy 21 (Down syndrome). *Proc. Natl. Acad. Sci. USA* **92,** 3586–3590.

24. Dornan, W. A., Kang, D. E., McCampbell, A., and Kang, E. E. (1993) Bilateral injections of beta A(25-35) + IBO into the hippocampus disrupts acquisition of spatial learning in the rat. *NeuroReport* **5,** 165–168.

25. Flood, J. F., Morley, J. E., and Roberts, E. (1991) Amnestic effects in mice of four synthetic peptides homologous to amyloid beta protein from patients with Alzheimer disease. *Proc. Natl. Acad. Sci. USA* **88,** 3363–3366.

26. Maurice, T., Lockhart, B. P., and Privat, A. (1996) Amnesia induced in mice by centrally administered beta-amyloid peptides involves cholinergic dysfunction. *Brain Res.* **706,** 181–193.

27. McDonald, M. P., Dahl, E. E., Overmier, J. B., Mantyh, P., and Cleary, J. (1994) Effects of an exogenous beta-amyloid peptide on retention for spatial learning. *Behav. Neural. Biol.* **62,** 60–67.

28. Nabeshima, T. and Nitta, A. (1994) Memory impairment and neuronal dysfunction induced by beta-amyloid protein in rats. *Tohoku J. Exp. Med.* **174,** 241–249.

29. Yamada, K., Tanaka, T., Mamiya, T., Shiotani, T., Kameyama, T., and Nabeshima, T. (1999) Improvement by nefiracetam of beta-amyloid-(1-42)-induced learning and memory impairments in rats. *Br. J. Pharmacol.* **126,** 235–244.

30. Sweeney, W. A., Luedtke, J., McDonald, M. P., and Overmier, J. B. (1997) Intrahippocampal injections of exogenous beta-amyloid induce postdelay errors in an eight-arm radial maze. *Neurobiol. Learn. Mem.* **68,** 97–101.

31. Maurice, T., Su, T. P., and Privat, A. (1998) Sigma1 (sigma 1) receptor agonists and neurosteroids attenuate B25-35-amyloid peptide-induced amnesia in mice through a common mechanism. *Neuroscience* **83,** 413–428.

32. Alvarez, X. A., Miguel-Hidalgo, J. J., Fernandez-Novoa, L., and Cacabelos, R. (1997) Intrahippocampal injections of the beta-amyloid 1-28 fragment induces behavioral deficits in rats. *Methods Find. Exp. Clin. Pharmacol.* **19,** 471–479.

33. Flood, J. F., Roberts, E., Sherman, M. A., Kaplan, B. E., and Morley, J. E. (1994) Topography of a binding site for small amnestic peptides deduced from structure-activity studies: relation to amnestic effect of amyloid beta protein. *Proc. Natl. Acad. Sci. USA* **91,** 380–384.

34. Maurice, T., Lockhart, B. P., Su, T. P., and Privat, A. (1996) Reversion of beta 25-35-amyloid peptide-induced amnesia by NMDA receptor-associated glycine site agonists. *Brain Res.* **731,** 249–253.

35. Yamada, K., Tanaka, T., Zou, L. B., Senzaki, K., Yano, K., Osada, T., et al. (1999) Long-term deprivation of oestrogens by ovariectomy potentiates beta-amyloid-induced working memory deficits in rats. *Br. J. Pharmacol.* **128,** 419–427.

36. Holcomb, L. A., Gordon, M. N., Benkovic, S. A., and Morgan, D. G. (2000) β and perlecan in rat brain. Glial activation, gradual clearance and limited neurotoxicity. *Mech. Ageing Dev.* **112,** 135–152.

37. Delobette, S., Privat, A., and Maurice, T. (1997) In vitro aggregation facilities beta-amyloid peptide-(25–35)-induced amnesia in the rat. *Eur. J. Pharmacol.* **319,** 1–4.

38. Games, D., Adams, D., Alessandrini, R., Barbour, R., Berthelette, P., Blackwell, C., et al. (1995) Alzheimer-type neuropathology in transgenic mice overexpressing V717F beta-amyloid precursor protein. *Nature* **373,** 523–527.

39. Hsiao, K., Chapman, P., Nilsen, S., Eckman, C., Harigaya, Y., Younkin, S., et al. (1996) Correlative memory deficits, Abeta elevation, and amyloid plaques in transgenic mice. *Science* **274,** 99–102.

40. Sturchler-Pierrat, C., Abramowski, D., Duke, M., Wiederhold, K. H., Mistl, C., Rothacher, S., et al. (1997) Two amyloid precursor protein transgenic mouse models with Alzheimer disease-like pathology. *Proc. Natl. Acad. Sci. USA* **94,** 13,287–13,292.

41. Routtenberg, A. (1997) Measuring memory in a mouse model of Alzheimer's disease. *Science* **277,** 839–840.

42. Chapman, P. F., White, G. L., Jones, M. W., Cooper-Blacketer, D., Marshall, V. J., Irizarry, M., et al. (1999) Impaired synaptic plasticity and learning in aged amyloid precursor protein transgenic mice. *Nat. Neurosci.* **2,** 271–276.

43. Dodart, J. C., Meziane, H., Mathis, C., Bales, K. R., Paul, S. M., and Ungerer, A. (1999) Behavioral disturbances in transgenic mice overexpressing the V717F beta-amyloid precursor protein. *Behav. Neurosci.* **113,** 982–990.

44. Justice, A. and Motter, R. (1997) Behavioral characterization of PDAPP transgenic Alzheimer mice. *Neurosci. Abst.* **23(2),** 1637.

45. Dodart, J. C., Mathis, C., Saura, J., Bales, K. R., Paul, S. M., and Ungerer, A. (2000) Neuroanatomical abnormalities in behaviorally characterized APP(V717F) transgenic mice. *Neurobiol. Dis.* **7,** 71–85.

46. Chen, G., Chen, K. S., Knox, J., Inglis, J., Bernard, A., Martin, S. J., et al. (2000) A learning deficit related to age and beta-amyloid plaques in a mouse model of Alzheimer's disease. *Nature* **408,** 975–979.

47. Duff, K., Eckman, C., Zehr, C., Yu, X., Prada, C. M., Perez-tur, J., et al. (1996) Increased amyloid-beta42(43) in brains of mice expressing mutant presenilin 1. *Nature* **383,** 710–713.

48. Holcomb, L. A., Gordon, M. N., McGowan, E., Yu, X., Benkovic, S., Jantzen, P., et al. (1998) Accelerated Alzheimer-type phenotype in transgenic mice carrying both mutant amyloid precursor protein and presenilin 1 transgenes. *Nat. Med.* **4,** 97–100.

49. Holcomb, L. A., Gordon, M. N., Jantzen, P., Hsiao, K., Duff, K., and Morgan, D. (1999) Behavioral changes in transgenic mice expressing both amyloid precursor protein and presenilin-1 mutations: lack of association with amyloid deposits. *Behav. Gen.* **29,** 177–185.

50. Arendash, G. W., King, D. L., Gordon, M. N., Morgan, D., Hatcher, J. M., Hope, C. E., et al. (2001) Progressive behavioral impairments in transgenic mice carrying both mutant APP and PS1 transgenes. *Brain Res.* **891,** 45–53.

51. Holcomb, L. A., Gordon, M. N., Jantzen, P., Hsiao, K., Duff, K., and Morgan, D. (1999) Behavioral changes in transgenic mice expressing both amyloid precursor protein and presenilin-1 mutations: lack of association with amyloid deposits. *Behav. Gen.* **29,** 177–185.

52. Diamond, D. M., Park, C. R., Heman, K. L., and Rose, G. M. (1999) Exposing rats to a predator impairs spatial working memory in the radial arm water maze. *Hippocampus* **9,** 542–551.

53. Morgan, D., Diamond, D. M., Gottschall, P. E., Ugen, K. E., Dickey, C., Hardy, J., et al. (2000) A beta peptide vaccination prevents memory loss in an animal model of Alzheimer's disease. *Nature* **408,** 982–985.

54. Gordon, M. N., King, D. L., Diamond, D. M., Jantzen, P. T., Boyett, K. L., et al. (2001) Correlation between cognitive deficits and Aβ deposits in transgenic APP+PS1 mice. *Neurobiol. Aging* **22,** 377–385.

55. Schenk, D., Barbour, R., Dunn, W., Gordon, G., Grajeda, H., Guido, T., et al. (1999) Immunization with amyloid-beta attenuates Alzheimer-disease-like pathology in the PDAPP mouse. *Nature* **400,** 173–177.

56. Moran, P. M., Higgins, L. S., Cordell, B., and Moser, P. M. (1995) Age-related learning deficits in transgenic mice expressing human beta-APP751. *Proc. Natl. Acad. Sci. USA* **92,** 5341–5345.

57. D'Hooge, R., Nagels, G., Westland, C. E., Mucke, L., and De Deyn, P. P. (1996) Spatial learning deficit in mice expressing human 751-amino acid beta-amyloid precursor protein. *NeuroReport* **7,** 2807–2811.

58. Hsiao, K., Borchelt, D., Olson, K., Johansdottir, R., Kitt, C., Yunis, W., et al. (1995) Age-related CNS disorder and early death in transgenic FVB/N mice overexpressing Alzheimer amyloid precursor proteins. *Neuron* **15,** 1203–1218.

59. Doyle, E., Bruce, M. T., Breen, K. C., Smith, D. C., Anderton, B., and Regan, C. M. (1990) Intraventricular infusions of antibodies to amyloid-beta-protein precursor impair the acquisition of a passive avoidance response in the rat. *Neurosci. Lett.* **115,** 97–102.

60. Phinney, A. L., Deller, T., Stalder, M., Calhoun, M. E., Frotscher, M., Sommer, B., et al. (1999) Cerebral amyloid induces aberrant axonal sprouting and ectopic terminal formation in amyloid precursor protein transgenic mice. *J. Neurosci.* **19,** 8552–8559.

61. Roch, J. M., Masliah, E., Roch-Levecq, A. C., Sundsmo, M. P., Otero, D. A., Veinbergs, I., et al. (1994) Increase of synaptic density and memory retention by a peptide representing the trophic domain of the amyloid beta/A4 protein precursor. *Proc. Natl. Acad. Sci. USA* **91,** 7450–7454.
62. Irizarry, M. C., McNamara, M., Fedorchak, K., Hsiao, K., and Hyman, B. T. (1997) APPSw transgenic mice develop age-related A beta deposits and neuropil abnormalities, but no neuronal loss in CA1. *J. Neuropathol. Exp. Neurol.* **56,** 965–973.
63. Irizarry, M. C., Soriano, F., McNamara, M., Page, K. J., Schenk, D., Games, D., et al. (1997b) Abeta deposition is associated with neuropil changes, but not with overt neuronal loss in the human amyloid precursor protein V717F (PDAPP) transgenic mouse. *J. Neurosci.* **17,** 7053–7059.
64. Takeuchi, A., Irizarry, M. C., Duff, K., Saido, T. C., Hsiao, A. K., Hasegawa, M., et al. (2000) Age-related amyloid beta deposition in transgenic mice overexpressing both Alzheimer mutant presenilin 1 and amyloid beta precursor protein Swedish mutant is not associated with global neuronal loss. *Am. J. Pathol.* **157,** 331–339.
65. Calhoun, M. E., Wiederhold, K. H., Abramowski, D., Phinney, A. L., Probst, A., Sturchler-Pierrat, C., et al. (1998) Neuron loss in APP transgenic mice. [letter]. *Nature* **395,** 755–756.
66. Weldon, D. T., Rogers, S., Ghilardi, J. R., Finke, M. P., Cleary, J. P., O'Hare, E., et al. (1998) Fibrillar beta-amyloid induces microglial phagocytosis,expression of inducible nitric oxide synthase and loss of a select population of neurons in the rat CNS in vivo. *J. Neurosci.* **18,** 2161–2173.

15

The Cerebellum in AD

A Case for Arrested Neuroinflammation?

Paul L. Wood

1. INTRODUCTION

The concept of a neuroinflammatory etiology in Alzheimer's disease continues to gain increasing credibility since its inception in the McGeer labs over a decade ago. In this brief review, we compare and contrast the published findings for inflammatory mediators in the neocortex, a brain region with extensive neuronal losses, and those in the cerebellum, a brain region with minimal or no neuronal cell death. The basic hypothesis that evolves from this review of the literature is that neuroinflammation is a widespread phenomenon in the brains of Alzheimer's patients, but that in the case of the cerebellum, this is an inflammatory process that is either halted at a critical step or buffered sufficiently by endogenous protective mechanisms to prevent the cascade of events that results in neuronal losses. This hypothesis, therefore, has the corollary that pharmacological treatments that interfere with the neuroinflammatory cascade in the neocortex of Alzheimer's patients should slow the progression of the disease and/or prevent the onset of the disease, if implemented prior to the neuronal degeneration that results from sustained neuroinflammation.

2. CELLULAR CHANGES

The hallmark features of neuronal degeneration in the neocortex of Alzheimer's patients include the following:

- Neuronal degeneration, monitored as senile or neuritic plaques
- Neuofibrillary tangles
- Cerebrovascular amyloidosis
- Microglial activation
- Astrogliosis

In contrast to the neocortex, the cerebellar cortex *(1)* does not demonstrate neuronal degeneration, microgliosis, or astrogliosis (Table 1). Microgliosis, as reflected by interleukin (IL)-1α immunocytochemistry is widely evident in Alzheimer's neocortex but not the cerebellum. However, activated microglia are found in areas of diffuse plaques *(2)*. Astrogliosis as demonstrated by glial fibrillary acidic protein (GFAP) immunocytochemistry, GFAP mRNA and by enzyme-linked immunosorbent assay (ELISA) measurements

From: *Neuroinflammation, 2nd Edition: Mechanisms and Management*
Edited by: P. L. Wood © Humana Press Inc., Totowa, NJ

Table 1
Cellular Markers, in the Neocortex and Cerebellar Cortex,
in Postmortem Alzheimer's Brain; Ratio Relative to Controls

Parameter	Temporal cortex	Cerebellar cortex	Ref.
IL-1α⁺ Microglia	2.9	1.0	*4*
GFAP⁺ Astrocytes	2.9	1.0	*5*
GFAP mRNA	5.0	1.0	*6*
S100β	3.7	1.0	*5*
MAP2 mRNA	0.49	1.0	*6*

Table 2
FE65 mRNA in the Neocortex and Cerebellar Cortex
in Postmortem Alzheimer's Brain; Ratio Relative to Controls

Parameter	Temporal cortex	Cerebellar cortex	Ref.
FE65 mRNA (neuronal)	0.54	1.7	*6*
FE65 mRNA (non-neuronal)	1.9	1.7	*6*

of S100β protein is similarly only widespread in the neocortex *(3)*. Neuronal degeneration, which can be demonstrated in the neocortex via decreases in mitogen-activated protein (MAP) 2 mRNA levels, does not occur in the cerebellum.

3. AMYLOID PLAQUES

Although the cerebellar cortex in Alzheimer's disease (AD) does not possess classical neuritic plaques, diffuse plaques and plaques with an amyloid core are present in the molecular and granular/Purkinje cell layers of Alzheimer's cerebellum, respectively *(7–10)*. This plaque formation is an active ongoing process, as reflected by increases in both the neuronal and non-neuronal mRNAs for FE65 (Table 2), an adapter protein that facilitates β-amyloid genesis *(6)*. The lack of progression to neuritic plaques distinguishes the cerebellum from the neocortex in this disease, but the abundance of diffuse plaques indicates that the inflammatory cascade that ultimately leads to neuronal degeneration in the neocortex is also present in the cerebellum. The question that then arises is whether this lack of cerebellar dense core plaques is the result of one or more of the following:

- Later onset of underlying biochemical changes in the cerebellum that do not progress to neuronal degeneration prior to the death of Alzheimer's patients.
- The cerebellum possesses a greater intrinsic buffering capacity to deal with ongoing neuroinflammation in this brain area.
- A biochemical step present in the cortex is absent in the cerebellum, leading to a roadblock in the inflammatory cascade.
- A biochemical process is present in the cerebellum that is not in the neocortex and acts to inhibit the inflammatory cascade.

Table 3
Heme Oxygenase-1, Protein Oxidation Products,
RNA Oxidation Products, and Advanced Glycation
End Products in the Neocortex and Cerebellar Cortex
in Postmortem Alzheimer's Brain; Ratio Relative to Control

Parameter	Temporal cortex	Cerebellar cortex	Ref.
3,3'-Dityrosine	2.7	1.0	*11*
3-NO$_2$-Tyrosine	5.2	0.17	*12*
8-Hydroxyguanosine	3.0	1.0	*13*
HO-1 mRNA	1.6	1.0	*14*
Advanced glycation end products	++	0	*15*

Note: ++, significantly increased.

Whichever of these alternatives is the reality does not limit the usefulness of studying the cerebellum either as an index of early inflammatory responses in Alzheimer's disease or as a potential model of arrested neuroinflammation.

4. INFLAMMATORY CASCADE: OXIDATION PRODUCTS

A significant component of the neuroinflammatory cascade involves oxidative stress. In this regard, protein and RNA oxidation products have been examined by a number of laboratories and been found to be dramatically increased in the Alzheimer's neocortex but not in the cerebellum (Table 3). Immunocytochemistry has shown that these oxidation products of proteins *(11,12)* and RNA *(13)* are generated in vulnerable neuronal populations in the neocortex of Alzheimer's patients. With regard to chronology, studies in Down's syndrome suggest that this neuronal oxidative stress precedes β-amyloid deposition and is not a secondary phenomenon of amyloid deposition *(16)*. These oxidation products are presumed to result from the superoxide ($O_2^{\bullet-}$), hydrogen peroxide (H_2O_2), and peroxynitrite ($ONOO^-$) released from activated microglia in the neocortex. The lack of activated microglia in the cerebellum may explain the lack of protein and RNA oxidation products in this brain region and further suggests a pivotal role of activated microglia in the neuronal degeneration observed in the neocortex. Heme oxygenase-1 (HO-1, HSP27), which can be upregulated as an endogenous antioxidant system, is also upregulated in AD neocortex (Table 3), presumably to counteract upregulated prooxidant mechanisms. The cerebellar cortex lacks upregulation of either pro-oxidant or antioxidant activities (Table 3).

5. INFLAMMATORY CASCADE: MAPK CELL SIGNALING

The role(s) of MAP kinase (MAPK) in complex regulation of pro-inflammatory gene function is well established (*see* Chapter 3). This signaling cascade is particularly involved in microglial activation and the associated neuroinflammation (*see* Chapter 1). In autopsy studies, upregulation of the MAPKs, pERK, and p38 has been demonstrated in AD temporal cortex but not the cerebellum (Table 4). Upstream regulators of MAPK pathways have also been found to be increased in the AD temporal cortex (Table 4). These data

Table 4
MAPK Markers in the Neocortex
and Cerebellar Cortex in Postmortem Alzheimer's Brain

Parameter	Temporal cortex	Cerebellar cortex	Ref.
Neuronal pERK	++	ND	*17*
Neuronal p38	++	0	*18–19*
MKK6	++	ND	*20*
Neuronal SOS-1	++	0	*21*

Note: ++, significantly increased; ND, not determined; 0, no change.

Table 5
Comparison of Cellular and Biochemical Measures Between
Neocortex and Cerebellar Cortex in Alzheimer's Postmortem Tissues;
Ratio Relative to Controls

Parameter	Temporal cortex	Cerebellar cortex	Ref.
α2-Macroglobulin	1.9	1.0	*22*
α2-Macroglobulin (histochemistry)	++	±	*23*
ICAM (histochemistry)	+	−	*24*
Perlecan[a] (histochemistry)	+	−	*25*
Proteasome activity	0.62	1.0	*26*
Clusterin[b] (histochemistry)	++	±	*27*
Apolipoprotein E	++	±	*28*
α1-antichymotrypsin	++	±	*28*
C1q	3.6	1.6	*29*

[a]Heparin sulfate proteoglycan.
[b]SGP-2; ApoJ.

further support the advanced stage of neuroinflammatory response in AD neocortex relative to the cerebellar cortex.

6. INFLAMMATORY CASCADE: CELL ADHESION MOLECULES AND ACUTE-PHASE PROTEINS

A hallmark feature of AD neocortex is the dramatic increases in the levels of downstream inflammatory end products like acute-phase proteins (Table 5). These proteins have wide extracellular distributions and also concentrate in plaques. In contrast, these end products are not overexpressed in the cerebellum and are only associated with diffuse plaques. Again, these data point to an arrested stage of neuroinflammation in which not all factors are available to allow a full-blown inflammatory response in the cerebellum.

7. SUMMARY

This brief review highlights the hallmark differences between the neocortex and the cerebellar cortex in AD. Despite the presence of many of the components of the neuroinflammatory cascade in AD cerebellum, this tissue never develops a full-blown inflammatory response that results in neuronal destruction. Further studies of this apparent halt in the progression of the neuroinflammatory response in this brain region might prove invaluable in our understanding of the etiology and progression of AD at the molecular level.

REFERENCES

1. Larner, A. J. (1997) The cerebellum in Alzheimer's disease. *Dement. Geriatr. Cogn. Disord.* **8,** 203–209.
2. Mattiace, L. A., Davies, P., Yen, S. H., and Dickson, D. W. (1990) Microglia in cerebellar plaques in Alzheimer's disease. *Acta Neuropathol. (Berl.)* **80,** 493–498.
3. Delacourte, A. (1990) General and dramatic glial reaction in Alzheimer brains. *Neurology* **40,** 33–37.
4. Sheng, J. G., Mrak, R. E., and Griffin, W. S. (1995) Microglial interleukin-1α expression in brain regions in Alzheimer's disease: correlation with neuritic plaque distribution. *Neuropathol. Appl. Neurobiol.* **21,** 290–301.
5. Van Eldik, L. J. and Griffin, W. S. (1994) S100 beta expression in Alzheimer's disease: relation to neuropathology in brain regions. *Biochim. Biophys. Acta* **1223,** 398–403.
6. Hu, Q., Jin, L. W., Starbuck, M. Y., and Martin, G. M. (2000) Broadly altered expression of the mRNA isoforms of FE65, a facilitator of beta amyloidogenesis, in Alzheimer cerebellum and other brain regions. *J. Neurosci. Res.* **60,** 73–86.
7. Cole, G., Neal, J. W., Singhrao, S. K., Jasani, B., and Newman, G. R. (1993) The distribution of amyloid plaques in the cerebellum and brain stem in Down's syndrome and Alzheimer's disease: a light microscopical analysis. *Acta Neuropathol. (Berl.)* **85,** 542–552.
8. Toledano, A., Alvarez, M. I., Rivas, L., Lacruz, C., and Martinez-Rodriguez, R. (1999) Amyloid precursor proteins in the cerebellar cortex of Alzheimer's disease patients devoid of cerebellar beta-amyloid deposits: immunocytochemical study of five cases. *J. Neural Transm.* **106,** 1151–1169.
9. Joachim, C. L., Morris, J. H., and Selkoe, D. J. (1989) Diffuse senile plaques occur commonly in the cerebellum in Alzheimer's disease. *Am. J. Pathol.* **135,** 309–319.
10. Suenaga, T., Hirano, A., Llena, J. F., Ksiezak-Reding, H., Yen, S. H., and Dickson, D. W. (1990) Modified Bielschowsky and immunocytochemical studies on cerebellar plaques in Alzheimer's disease. *J. Neuropathol. Exp. Neurol.* **49,** 31–40.
11. Smith, M. A., Richey-Harris, P., Sayre, L. M., Beckman, J. S., and Perry, G. (1997) Widespread peroxynitrite-mediated damage in Alzheimer's disease. *J. Neurosci.* **17,** 2653–2657.
12. Hensley, K., Maidt, M. L., Yu, Z., Sang, H., Markesbery, W. R., and Floyd, R. A. (1998) Electrochemical analysis of protein nitrotyrosine and dityrosine in the Alzheimer brain indicates region-specific accumulation. *J. Neurosci.* **18,** 8126–8132.
13. Nunomura, A., Perry, G., Pappolla, M. A., Wade, R., Hirai, K., Chiba, S., et al. (1999) RNA oxidation is a prominent feature of vulnerable neurons in Alzheimer's disease. *J. Neurosci.* **19,** 1959–1964.
14. Premkumar, D. R., Smith, M. A., Richey, P. L., Petersen, R. B., Castellani, R., Kutty, R. K., et al. (1995) Induction of heme oxygenase-1 mRNA and protein in neocortex and cerebral vessels in Alzheimer's disease. *J. Neurochem.* **65,** 1399–1402.
15. Takeda, A., Wakai, M., Niwa, H., Dei, R., Yamamoto, M., and Li, M. (2001) Neuronal and glial advanced glycation end product [*N*-epsilon-(carboxymethyl)lysine] in Alzheimer's disease brains. *Acta Neuropathol.* **101,** 27–35.
16. Nunomura, A., Perry, G., Pappolla, M. A., Friedland, R. P., Hirai, K., Chiba, S., et al. (2000) Neuronal oxidative stress precedes amyloid-beta deposition in Down syndrome. *J. Neuropathol. Exp. Neurol.* **59,** 1011–1017.

17. Perry, G., Roder, H., Nunomura, A., Takeda, A., Friedlich, A. L., Zhu, X. W., et al. (1999) Activation of neuronal extracellular receptor kinase (ERK) in Alzheimer disease links oxidative stress to abnormal phosphorylation *Neuroreport* **10,** 2411–2415.

18. Hensley, K., Floyd, R. A., Zheng, N. Y., Nael, R., Robinson, K. A., Nguyen, X., et al. (1999) p38 kinase is activated in the Alzheimer's disease brain. *J. Neurochem.* **72,** 2053–2058.

19. Zhu, X. W., Rottkamp, C. A., Boux, H., Takeda, A., Perry, G., and Smith, M. A. (2000) Activation of p38 kinase links tau phosphorylation, oxidative stress, and cell cycle-related events in Alzheimer disease. *J. Neuropathol. Exp. Neurol.* **59,** 880–888.

20. Zhu, X., Rottkamp, C. A., Hartzler, A., Sun, Z., Takeda, A., Boux, H., et al. (2001) Activation of MKK6, an upstream activator of p38, in Alzheimer's disease. *J. Neurochem.* **79,** 311–318.

21. McShea, A., Zelasko, D. A., Gerst, J. L., and Smith, M. A. (1999) Signal transduction abnormalities in Alzheimer's disease: evidence of a pathogenic stimuli. *Brain Res.* **815,** 237–242.

22. Wood, J. A., Wood, P. L., Ryan, R., Graff-Radford, N. R., Pilapil, C., Robitaille, Y., et al. (1993) Cytokine indices in Alzheimer's temporal cortex: no changes in mature IL-1 beta or IL-1RA but increases in the associated acute phase proteins IL-6, alpha 2-macroglobulin and C-reactive protein. *Brain Res.* **629,** 245–252.

23. Strauss, S., Bauer, J., Ganter, U., Jonas, U., Berger, M., and Volk, B. (1992) Detection of interleukin-6 and α_2-macroglobulin immunoreactivity in cortex and hippocampus of Alzheimer's disease patients. *Lab. Invest.* **66,** 223–230.

24. Verbeek, M. M., Otte-Holler, I., Wesseling, P., Ruiter, D. J., and de Waal, R. M. (1996) Differential expression of intercellular adhesion molecule-1 (ICAM-1) in the Aβ-containing lesions in brains of patients with dementia of the Alzheimer type. *Acta Neuropathol. (Berl.)* **91,** 608–615.

25. Snow, A. D., Sekiguchi, R. T., Nochlin, D., Kalaria, R. N., and Kimata, K. (1994) Heparan sulfate proteoglycan in diffuse plaques of hippocampus but not of cerebellum in Alzheimer's disease brain. *Am. J. Pathol.* **144,** 337–347.

26. Keller, J. N., Hanni, K. B., and Markesbery, W. R. (2000) Impaired proteasome function in Alzheimer's disease. *J. Neurochem.* **75,** 436–439.

27. Lidstrom, A. M., Bogdanovic, N., Hesse, C., Volkman, I., Davidsson, P., and Blennow, K. (1998) Clusterin (apolipoprotein J) protein levels are increased in hippocampus and in frontal cortex in Alzheimer's disease. *Exp. Neurol.* **154,** 511–521.

28. Styren, S. D., Kamboh, M. I., and DeKosky, S. T. (1998) Expression of differential immune factors in temporal cortex and cerebellum: the role of alpha-1-antichymotrypsin, apolipoprotein E, and reactive glia in the progression of Alzheimer's disease. *J. Comp. Neurol.* **296,** 511–520.

29. Brachova, L., Lue, L., Schultz, J., el Rashidy, T., and Rogers, J. (1993) Association cortex, cerebellum, and serum concentrations of C1q and factor B in Alzheimer's disease. *Mol. Brain Res.* **18,** 329–334.

16

The Neuroinflammatory Components of the Trimethyltin (TMT) Model of Hippocampal Neurodegeneration

G. Jean Harry and Christian Lefebvre d'Hellencourt

1. INTRODUCTION

A characteristic feature of brain injury is the rapid reaction of microglia and astrocytes *(1–3)* (Fig. 1). Reactive astrocytes display hypertrophy, elevated glutamine synthetase, oxido-reductive enzyme activity, and accumulation of the astrocyte-specific structural protein, glial fibrillary acidic protein (GFAP) *(2,4)*. Activated microglia are characterized by hypertrophy, proliferation, increased surface expression of immune marker molecules, increased migration, release of oxygen radicals and proteases, and differentiation into a macrophagelike phenotype *(5–7)*. Such responses may be beneficial in the healing phases of central nervous system (CNS) injury by actively monitoring and controlling the extracellular environment, walling off areas of the CNS from non-CNS tissue, and removing dead or damage cells *(3,5,8)*. This gliotic process, however, is also thought to impart detrimental consequences by collateral neuronal damage from released microglial cytotoxins and inhibit neuronal regeneration by physical or biochemical impediments *(6,8–14)*. In various models of nervous system injury, brain ischemia, deafferentation, physical trauma, and excitotoxicity, glial cells become activated upon injury and emit an inflammatory-like response, including the release of pro-inflammatory cytokines.

As is evident in the other chapters in this book, it has been recognized that immune cytokines play important physiological roles in mediating CNS injury. One such mechanism is via mediation of interactions between neurons and glia. Under certain pathological conditions, disturbance of homeostasis by the abnormal expression of cytokines may lead to neuronal damage. For example, elevated levels of pro-inflammatory cytokines in the CNS have been observed under various conditions such as multiple sclerosis *(15–17)*, Alzheimer's disease and Down's syndrome *(11,18,118)* and experimentally induced brain lesions *(19–24)*. Pro-inflammatory cytokines, especially tumor necrosis factor-α (TNF-α) and interleukin-1 (IL-1), appear to be pivotal mediators of brain inflammation *(25,26)*. Although TNF-α can have detrimental effects, it also appears to provide a beneficial action on brain injury response through growth factor production and quenching of damage from generation of free radicals and excitotoxins *(8,27)*. TNFα

From: *Neuroinflammation, 2nd Edition: Mechanisms and Management*
Edited by: P. L. Wood © Humana Press Inc., Totowa, NJ

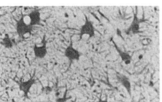

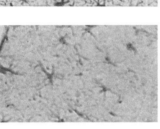

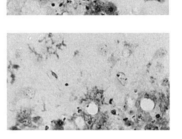

Astrocytes –
hypertrophy
increased expression of GFAP
astrogliosis
glia scarring

Microglia –
hypertrophy
phagocytic activity
MHC II expression
production of: nitric oxide
radical oxygen species
proteases

Fig. 1. Generalized injury response of astrocytes and microglia. Basal level vs injury.

and IL-1 have been demonstrated to be associated with the production of neurotoxic molecules by glia and elicit release of arachidonic acid, nitric oxide, and β-amyloid precursor protein, all which have been implicated in neurodegeneration. Given the early response of glia cells to nervous system injury and their ability to secrete pro-inflammatory cytokines, glia activation and interaction have been proposed as an early signaling mechanism responsible for subsequent neuronal degeneration *(9,12,28–30)*.

Balance between activation and suppression of immune cells is critical in the host response against insult and prolonged injury. Excessive or prolonged activation may result in hypersensitivity resulting in a more severe response to subsequent injury. Some cytokines are secreted upon complete intracellular processing, whereas others, at one stage, are stored intracellularly and additional stimuli are required to trigger secretion. For example, IL-1 and TNF-α mRNA can be induced by monocyte adherence and the corresponding protein produced only upon further stimulation of the cells. In many cases, the level of production induced from primed cells is at a much greater level than that which would have been induced directly *(31–33)*. It is possible that, in certain models, initial brain injury could increase the transcription of pro-inflammatory cytokines, resulting in a "primed" system. The basal level of the system would then be altered and primed cells would overrespond upon subsequent stimulation.

Although astrogliosis and associated factors have been a focus in examining a neurotoxic response to physical injury and chemical exposure, it is only relatively recently that the role of microglia and the inflammatory response has been considered. Initial studies examined such responses in reaction to physical trauma to the brain, models of brain ischemia, and excitotoxicity (for review, *see* refs. *27, 34*, and *35*). Immediately following injury, various antigens trigger distinct phases of the host defense response consisting of cell recruitment and induction of cytokines contributing to activation of pro-inflammatory, immunostimulatory, and catabolic responses. Cytokines initiating these early events include TNF-α and IL-1, which stimulate a set of chemotactive factors whose primary function is to promote recruitment of inflammatory cells. Enhanced expression of these pro-inflammatory cytokines initiates a complex regulatory cytokine network similar to patterns seen in non-neural tissue and can produce multiple CNS effects. Although many models of brain injury continue to provide critical information toward the understanding of the microglia and associated immune-mediated responses in neurodegeneration, they also have some limitations, as does any model system. Although each of the model systems of ischemia, trauma, physical injury, and excitotoxicity is well established and reproducible, the ability to examine a number of factors, including dose and temporal response relationships, reaction of resident microglia cells in the absence of major infiltration of cells from the circulation, or a specific pattern of neuronal degeneration, is not always possible. It is for these reasons that we chose to examine such responses in a model of chemical-induced neurodegeneration using the prototypic hippocampal neurotoxicant, trimethyltin.

Trimethyltin (TMT) is known to be toxic to the limbic system of multiple animal species, with a pattern of hippocampal damage being the most consistent morphological feature *(36–44)*. The acutely intoxicated rodent, either rat or mouse, displays an evolution of pathologic changes in a highly predictable manner over a relatively short period (4–5 d for the rat and 24 h for the mouse). The selective nature of the lesion produced by TMT is not associated with a preferential distribution of the chemical or with other toxicokinetic factors *(45,46)*. In the mouse, the peak level of tin in the brain is reached within 1 h of dosing and remains at this level for a minimum of 16 h *(47)*. In the rat, Cook and coworkers *(45)* demonstrated that the half-life of tin in the brain is approximately 10 d. Thus, exposure of brain tissue to TMT continues at a relatively constant level for over a week following one acute systemic injection.

Spontaneous tremors and occasional seizures accompany the histopathological effects of TMT on the limbic system in a dose-dependent manner. In a temporal relationship to the onset of neuronal degeneration, animals can display signs of self-mutilation, aggression to cagemates, hyperreactivity, and increased seizure susceptibility *(1,37,40,48,49)*. In mice, the behavioral manifestations of toxicity appear between 24 and 48 h in a dose-related manner. Although a dose of 2.0 mg/kg can produce hyperreactivity and tremor in a majority of the exposed animals, 1.5 mg/kg does not result in behavioral effects, and limited, if any, neurodegeneration. In rats, the dose range for such behaviors is observed at a high dose of 8 mg/kg, with less severe effects seen in fewer animals at 4–6 mg/kg. Although seizures can occur in the high-dosed animals, the lack of mitochondria changes such as swelling argues against the pathologic changes being related to TMT-induced seizures *(37)*. As would be expected with damage to the hippocampus, long-term deficits in learning and memory tasks are evident in both experimental animal models as well as accidental human exposure *(48,50–53)*.

2. RAT MODEL

2.1. Neuropathology

2.1.1. Neuronal Degeneration

In the rat, acute dosing results in a prominent neuronal necrosis in the pyramidal neurons of the hippocampal cornu ammonis. Neuronal loss occurs in the pyriform/entorhinal cortex, is the most severe in areas CA3 and CA4, is less severe in area CA1, and is minimal in area CA2 *(54)*. Neuronal necrosis is also observed, with time, in the basal ganglia, brainstem, spinal cord, dorsal root ganglia, and various other sites *(55,56)*. It has been previously reported that the neuronal necrosis is associated with an influx of microglial cells, neuronophagia, and astrocyte hypertrophy *(36,37,55,57,58,58a)*. Additionally, a highly predictable pattern of pathologic changes occurs over a relatively short period of time. Electron microscopic (EM) evaluation of TMT-induced pathology demonstrates no infiltration of circulating lymphocytes during these time periods.

In examining this model, we have focused on the hippocampus because of both its predilection for TMT-induced neurodegeneration and its well-characterized structural organization. In our laboratory, an acute intraperitoneal injection of TMT (6 mg/kg) produces a mild individual neuronal necrosis that is evident within the hippocampal pyramidal cell and granule cell region (Fig. 2). The necrosis observed in these regions is characterized by loss of Nissl substance, intense cytoplasmic eosinophilia, pyknosis and karyorrhexis of nuclei, and cell body shrinkage. A distinct loss of neuronal cell bodies is also evident in the CA3–4 pyramidal cell region of the hippocampus and occurs over a 30-d time frame. Unlike other models of hippocampal damage (e.g., excitotoxicity), there is little evidence to support compensatory reinnervation or reactive sprouting in the CA3 region in response to TMT-induced damage. Additional experiments performed in this laboratory also failed to demonstrate induction of neurotrophin mRNA or

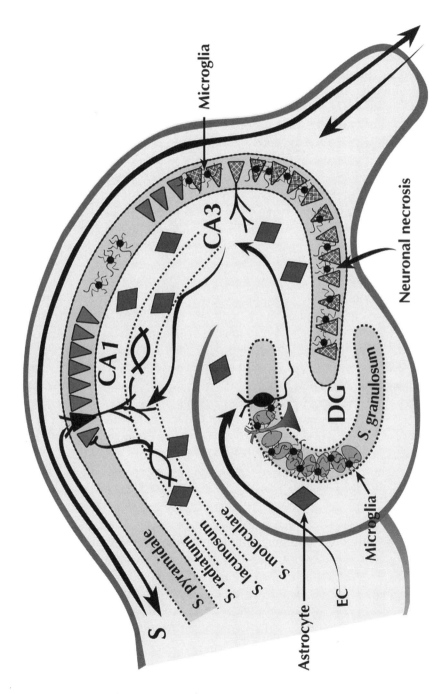

Fig. 2. Schematic of neuronal and glia responses in the rat hippocampus 1 wk following injection of TMT. Neuronal necrosis is localized to the CA3–4 pyramidal cell layer, ramified microglia are localized around CA3–4 neurons; enlarged GFAP-positive astrocytes are seen throughout the hippocampus.

elevations in immunoreactive proteins usually associated with synaptic sprouting. Data in the literature, however, indicate that growth factor expression, in the form of astrocytic nerve growth factor and tyrosine kinase A, increases at later time-points, suggesting the possibility of delayed reactive response to TMT injury *(59)*.

2.1.2. Glia Response

The temporal response of glia cells in this pattern of TMT-induced neurodegeneration is of considerable interest to this laboratory. Within 4 d of TMT administration, astrocytes begin to display increased immunoreactivity to GFAP antibodies with an increase in the number of astrocytes showing dense processes. Pathological changes continue to occur over the next few weeks, and by d 30, the neuronal cell loss in the CA3–4 pyramidal region is accompanied by a significant increase in GFAP astrocyte reactivity. Although the neuronal degeneration is localized to the CA3–4 region, the astrocyte response is distributed throughout the hippocampus, with the exception of sparse staining in the region of the dentate neurons (Fig. 2). At 4 and 10 d post-TMT, astrocytes display increased immunoreactivity for GFAP; however, evidence of hypertrophy is limited. By 30 d, astrocytes show distinct signs of hypertrophy, as demonstrated by an enlargement of the cell bodies and a thickening of the glia processes *(60)*.

In comparison to the response of astrocytes throughout the hippocampus, microglia, as detected by lectin binding or OX-42 immunohistochemistry, were present primarily in the CA3–4 pyramidal cell region. A large majority of the cells are in close juxtaposition to neuronal cell bodies as early as 1 d following TMT injection and the number of cells present increases over the next 14 d. The cells display multiple spindly processes characteristic of reactive microglia; however, very few cells display ameboid morphology characteristic of activated microglia (Fig. 2). This pattern suggests that the early presence and biochemical response of microglia cells may be associated with signaling that leads to the subsequent neuronal degeneration. However, the role of phagocytosis as a contributory mechanism in the neuronal loss has yet to be established, given the distinct lack of an activated microglia phenotype.

2.2. Molecular Responses to TMT

The mRNA levels of IL-1α, IL-6, and TNF-α are elevated within 6 h of a systemic injection of TMT and continue to be elevated for approx 8 d (Fig. 3). Elevated cytokine mRNA levels typically result in enhanced protein production, although a lack of translation of mRNA has also been observed previously *(61,62)*. In such cases, cells that contain untranslated IL-1 mRNA can be maintained in a "primed" state such that minimal amounts of stimuli can produce a rapid secretion of active cytokines. Thus, the extended duration of cytokine elevation could be the result of either continued exposure to TMT or maintenance of the system in a "primed state."

Although the hippocampus is the primary target site for TMT-induced pathology, areas of the cortex associated with the hippocampal projection sites show signs of focal necrosis and associated reactive microglia. Associated with this cortical effect is a transient elevation of mRNA levels for TNF-α at 12 h and an increase in IL-6 and IL-1α at 24 h. This distinct regional patterns of cytokine elevation is consistent with differential morphological sensitivity and further suggests that the increase in cytokine mRNA levels and the temporal pattern of the response are both important in the final pathology. Inter-

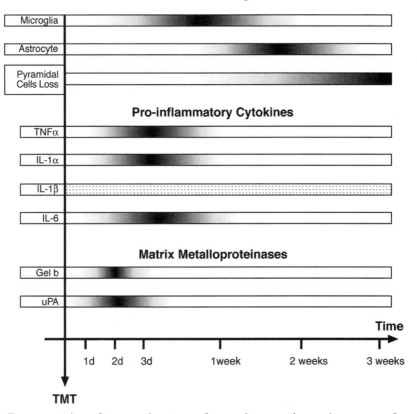

Fig. 3. Representation of temporal pattern of neurodegeneration and response of cytokines and matrix metalloproteinases in the hippocampus following TMT injection in the rat.

estingly, whereas IL-1α mRNA levels are elevated, IL-1β levels in the hippocampus remain within the range of control values. This may be the result of the nature of the insult because previous reports of elevated IL-1β mRNA in the brain have utilized more severe models of neural injury *(3,21,63)*.

In other models of hippocampal neurodegeneration (e.g., kainic acid neurotoxicity), neuronal degeneration occurs with seizure activity and is associated with a transient increase in pro-inflammatory cytokines *(21)*. In the case of TMT, one possibility for the difference in cytokine expression is the long half-life of the compound in the brain (10 d) *(45)* provides for a continued, low-grade stimulus. The continued presence of TMT may cause additional cellular injury, stimulate continued cytokine production, and recruit other cells into the inflammatory response. An opposing, or perhaps, additional factor in sustained cytokine production may involve the lack of induction in negative immunoregulators from resident cells of the brain. It is possible, for example, that TMT exposure yields increased protease activity that acts directly on the plasma membrane producing cytoskeletal changes, or propagates the cytokine response in part through proteolytic cleavage of cytokine proteins. Both gelatinase A and B, as well as, urokinase-type plasminogen activator (uPA), participate in the breakdown of the basal lamina around cerebral capillaries *(64)* and can be modulated by TNF-α in the brain *(65)*. Because

TMT causes a distinct elevation in TNF-α, it is logical to propose that such changes would occur following TMT exposure. In support of this, both gelatinase B and uPA are elevated in both the hippocampus and the cortex following a single injection of TMT. TMT exposure results in an increase in the 40-kDa isoform of uPA as early as 12 h following injection that was prolonged for 96 h, which is consistent with the pattern of cytokine elevation in the hippocampus *(66)* (Fig. 3). These data suggest that the early induction of cytokines by TMT is not an isolated event, but a step in a continuing cascade of neuro-inflammatory responses for which inhibitors of matrix metalloproteinases might offer a therapeutic intervention strategy.

3. MOUSE MODEL

3.1. Neuropathology

3.1.1. Neuronal Degeneration

As compared to the rat, acute exposure to TMT in the young adult mouse produces extensive damage to the granule cells of the fascia dentata *(38,39,43,44,66a)*. Damage to the hippocampal pyramidal neurons is minimal; however, a punctate pattern of isolated neurons in the CA1,2 and CA3 subfields of the Ammon's horn show vacuolar changes *(38)*. If mice are exposed to TMT prior to postnatal d 15, the dentate granule cells show a similar pattern of damage with the additional loss of pyramidal cells because of perturbation at the time of cell formation.

The mouse model that has been used in our laboratory relies on an acute (intraperitoneal injection; ip) exposure to TMT in postnatal d 19–24 CD-1 male mice. The steep dose response to TMT is displayed in the pathology at 24 h. At this time, minimal, if any, neuronal injury occurs at a dose of 1.5 mg/kg, a distinct necrosis of the dentate granule cells is seen with 2.0 mg/kg, and a severe necrosis manifests at 2.5 mg/kg. This pattern shifts somewhat with increasing animal age, resulting in the need to increase the dosing to 3–3.5 mg/kg to produce a similar level of damage in the adult. The acute ip injection of TMT produces apoptosis at 12 h, with a punctate pattern of terminal deoxynucleotidyl transferase dUTP Nick end labeling (TUNEL) positive cells evident in the dentate granule cell region; however, the number of TUNEL-positive cells is significantly less than the number of neurons undergoing necrosis at 24 h. mRNA levels for cyclooxygenase-2 and caspase 3 are elevated in dentate neurons *(66b)*. A small but distinct number of TUNEL-positive neurons is seen in the cerebellum at 6 h post-TMT administration, with no evidence of neuronal necrosis at later time-points. As the dentate granule cells degenerate over 72 h (Figs. 4 and 5), there is a loss of mossy fibers and a reduction in synaptic connections from the dentate to the pyramidal cell layer. This observation has also been reported in a rat model that produces damage to the dentate *(67)* and in the mouse model that demonstrated a decrease in synaptophysin immunoreactivity in the CA3–4 pyramidal cell layer *(68)*. Over a 1-wk period following exposure, immunohistochemical staining failed to demonstrate alterations in other neuronal-specific markers such as calbindin 28K and parvalbumin.

3.1.2. Glia Response

Within 24 h of a systemic injection of TMT, GFAP immunoreactivity in astrocytes is increased and the cells display dense immunoreactive cell bodies and processes char-

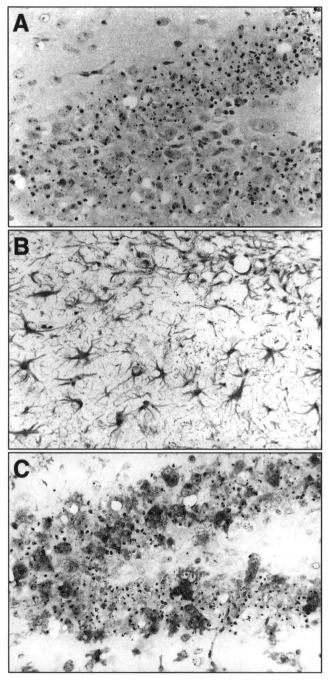

Fig. 4. Histopathology of dentate gyrus in the mouse 72 h following injection of TMT. (**A**) Neuronal necrosis (hematoxylin and eosin); (**B**) GFAP-positive astrocytes; (**C**) lectin-positive microglia.

acteristic of astrocyte reactivity. However, the swollen hypertrophy seen in the rat is not apparent in the mouse. In both models, the astrocyte response is distributed throughout the hippocampus and is not localized to the site of neuronal injury. The morphological

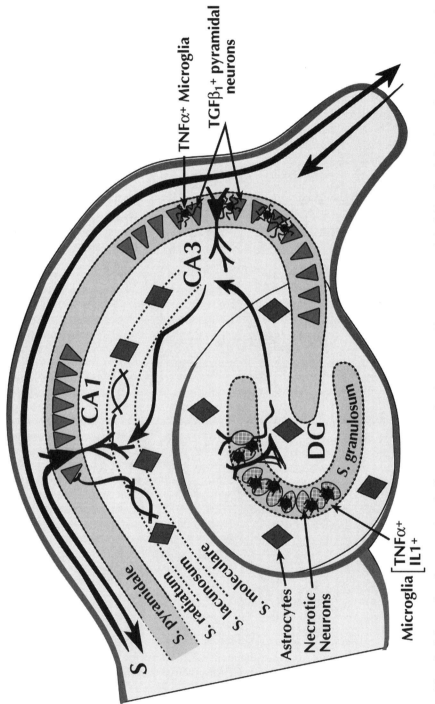

Fig. 5. Schematic of neuronal and glia responses in the mouse hippocampus 72 h following injection of TMT. Neuronal necrosis is localized to the dentate gyrus; ramified microglia are seen in the CA3–4 region, amoeboid microglia are seen in the dentate gyrus; GFAP-positive astrocytes are increased throughout the hippocampus. *In situ* hybridization demonstrates localization of IL-1α mRNA to microglia cells in the CA1 and dentate region; TNF-α mRNA is localized to microglia cells in the dentate and in the CA3–4 region. Transforming growth factor (TGF)-β1 mRNA is localized to CA3–4 neurons.

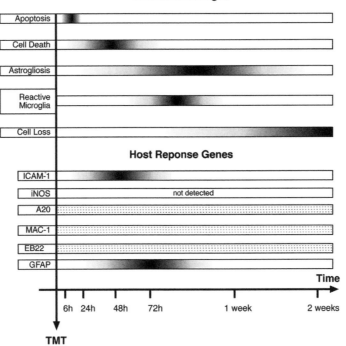

Fig. 6. Representation of temporal pattern of neurodegeneration and response of host response genes in the hippocampus following TMT injection in the mouse.

response of microglia in the mouse is much more pronounced than that seen in the rat, but it is temporally delayed and follows neuronal necrosis *(66a,68)*. While in the rat, the morphological response of microglia occurs early and appears to be associated with neurons in the injury area, in the mouse, the morphological response follows the onset of neuronal necrosis. Microglia first become evident at 3 d post-TMT exposure with reactive microglia present in both the pyramidal and dentate granule cell layers. Activated microglia with amoeboid phagocytic morphology are present only in the dentate as an area of neuronal cell death (Figs. 4 and 5). Similar to the distribution pattern seen in the rat, astrocyte and microglia responses are also seen, with time, in the various brain regions associated with the hippocampal formation such as septal and cortical regions.

3.1.3. Molecular Responses to TMT

In response to inflammatory injury, cells often respond with an elevation in various host response genes. The pattern of such genes in the hippocampus following TMT shows elevations in the mRNA levels for intercellular adhesion molecule-1 (ICAM-1) and GFAP in a dose-related manner with increases seen as early as 12 h, with the high dose of 2.5 and by 24 h at 2.0 mg/kg *(68)*. EB22/5 (a murine acute-phase response gene homologous to the α-1-antichymotrypsin gene) and Mac-1 (macrophage-1 antigen) are not elevated until 72 h and this occurs in the high-dose group only. At no time is iNOS (inducible nitric oxide synthase) or A20 (an antiapoptotic TNF-α-inducible early-response gene) elevated in either dose group (Fig. 6).

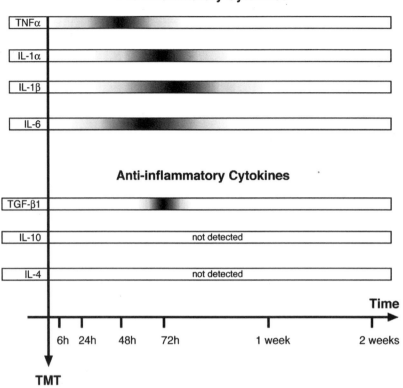

Fig. 7. Representation of temporal pattern of cytokine mRNA elevations in the hippocampus following TMT injection in the mouse.

In many cases, alterations in cytokine levels during the acute phase of central nervous system injury, disease processes, antibody-mediated responses, and chemical-induced injury have been attributed to microglia activation *(12,69–71)*. Within the hippocampus, increased lectin binding to microglia suggests that these cells may be responsible for the elevated cytokine mRNA levels seen with TMT exposure. The initial elevation in TNF-α mRNA occurs as early as 12 h following TMT administration and precedes any morphological evidence of glia reactivity. The elevated levels are maintained for a minimum of 72 h and return to within-control levels by 1 wk *(68)*. IL-1α, IL-1β, and IL-6 mRNA levels are elevated at 48 h (Fig. 7). Each transcript shows a dose-related pattern of elevation with either an increase in level or earlier onset and longer duration occurring in the higher dose of TMT (2.5 mg/kg). *In situ* hybridization for TNF-α and IL-1α indicates distinct cellular localization patterns (Fig. 8). mRNA for TNF-α is localized to both ramified microglia throughout the pyramidal cell layer and in amoeboid microglia localized in the dentate gyrus. Cellular localization of IL-1α mRNA is demonstrated by punctate staining of amoeboid microglia in the pyramidal cell layer and denser staining in the dentate region. This pattern of localization demonstrates the cojuxtaposition of TNF-α and IL-1α expressing microglia with both neurons that degenerate and those that survive the insult.

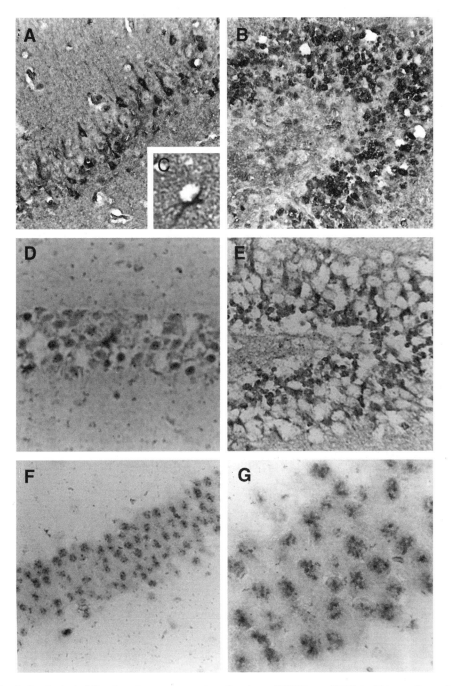

Fig. 8. *In situ* hybridization demonstrates localization of TNF-α mRNA to (**A**) reactive microglia in the CA-1, (**B**) activated microglia in the dentate region, and IL-1α mRNA localization to microglia cells in the CA3–4 region (**D**) and in the dentate (**E**). TGF-β1 mRNA localization to CA3–4 neurons (**F,G**). (From ref. *68.*)

One of the earliest changes following TMT is in the chemokine macrophage inflammatory protein-1α. Elevated mRNA levels are seen in both 2.0- and 2.5-mg/kg dose groups as early as 6 h with an associated dose response in the severity of the response

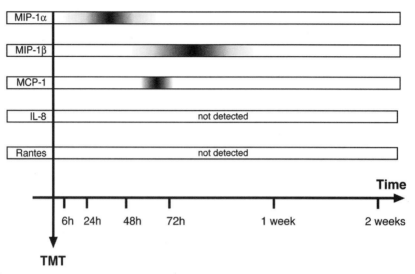

Fig. 9. Representation of temporal pattern of chemokine mRNA response in the hippocampus following TMT injection in the mouse.

(Fig. 9). Transforming growth factor-β1 (TGF-β1) is elevated by 72 h post-TMT injection. As determined by *in situ* hybridization, the cellular localization of TGF-β1 is primarily along the vascular lining at the early time of 24 h. This is accompanied by a punctate pattern of staining in cells representative of either a subpopulation of the microglia cells or necrotic neurons in the dentate region *(68)*. After 24 h, localization is no longer evident in the dentate; however, a prominent staining is seen in the pyramidal cells of the hippocampus (Fig. 8). The localization to the noninjured pyramidal cells suggests a response of the neurons to the deafferentation lesions in the hippocampus. This response occurs after the onset of necrosis of the dentate neurons and may be a specific response to the loss of synaptic contact.

Early-signaling components normally associated with inflammation and infiltrating cells are absent during TMT-induced neurodegeneration. The elevated cytokine levels do not appear to be caused by peripheral monocytes, as previous studies have demonstrated no such cellular infiltration *(37)*. The lack of infiltrating leukocytes is supported by the absence of elevations in the mRNA levels of IL-2, IL-4, and IL-10. The cytokine responses are not dependent on interferon (IFN)-γ, as levels are not elevated in any brain region. The elevation does not appear to be as a result of elevated TNF-α in the periphery, as no alteration is seen in plasma protein levels as a result of TMT administration. In fact, TMT is thought to be nonimmunotoxic, as compared to the effects of both triphenyltin and dibutyltin, TMT has no effect on F-actin depolymerization in thymocytes *(73)*.

4. RESPONSE OF CULTURED GLIAL CELLS

The glia response can be elicited by a number of factors that are not easily experimentally controlled in the intact animal, therefore, in vitro studies using mixed glial cell cultures have been conducted to understand glia responses to various injuries. For example, scratch wounds to cultured astrocytes share many of the major characteristics of astrogliosis observed in vivo (e.g., astrocyte hyperplasia, hypertrophy, and elevated GFAP content) *(74)*. Additionally, scratched glial cells in culture display similar morphology and patterns of pro-inflammatory cytokine elevations, as seen with reactive astrocytes in vivo following a penetrating stab wound *(23,75)*. In the intact hippocampus, the pattern of neuronal response to TMT differs between mice and rats, and in the rat, it can vary depending on the dosing regimen. Although the glia response is similar in both location and temporal manifestation, there are some differences. In the rat, the astrocytes progress to show a distinct hypertrophy and the microglia show thickened processes but do not progress to a phagocytic phenotype. In the mouse, the astrocyte response is characterized by an increased density of processes and no distinct sign of hypertrophy. In addition, the microglia demonstrated both ramified reactive and amoeboid activated morphological phenotypes. However, in both species, the morphological response of glia was evident within the first 24 h. In rats, it precedes neuronal necrosis, whereas in mice, it is a subsequent event. In culture however, glial cells from both species act relatively similar in response to TMT exposure.

4.1. Morphological Response of Glia

Previous studies have shown direct effects of TMT on astrocytes in culture *(76–79)*. Maier and colleagues *(60)* continued this work and examined the morphological and cytokine response of glia cultures following TMT exposure. Dose levels used in vitro were based on those predicted to occur in brain tissue following in vivo exposure. Exposure of cultured primary mixed glial cells with TMT (10 μM) produces a morphological response within 6 h, with astrocytes displaying an enlarged and rounded cell body with distinct cell processes becoming evident. At this time, the microglia become enlarged and display phagocytic activity *(80)*. Upon further examination, such morphological changes in glia display both a dose and temporal response following TMT. By 24 h, the severity of the astrocyte response is characterized by enlarged and rounded cell bodies and a retraction of the thick monolayer processes into long, thin, glial fibrillary acidic protein-dense processes *(60,80,80a)* (Fig. 10). These changes progress over the next 24 h, with suggestions of cell loss. The microglia in these cultures show an early increase in phagocytic activity, as evidenced by fluorescent bead uptake. Within 24 h, the structural changes of microglia of the ramified phenotype show increased fragmentation of processes *(80)*.

4.2. Biochemical Changes

A number of biochemical indicators of cell function have been examined in cultured glial cells following TMT exposure. At an early time-point of 6 h, no changes are seen in general astrocyte biochemical function or cell viability with doses of TMT less than 10 μM. At later time-points, a dose-response effect can be detected in a number of biochemical end points. Within 24 h of exposure to 10 μM TMT, cell viability remains within the normal range. Various biochemical features of astrocyte functioning also remain within normal levels such as ^{3}[H]-glutamate uptake and ^{3}[H]-thymidine and ^{3}[H]-leucine incorporation, whereas glutamine synthetase activity is decreased by approx 40%. No changes were seen at lower doses. When other measures of cell function are examined, distinct changes occur in the mRNA levels of structural proteins in the cultures

Control

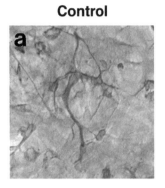

TMT

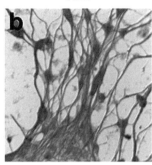

Anti-TNFα + TMT

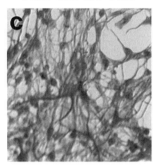

Anti-IL-1α + TMT

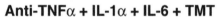

Anti-IL-6 + TMT

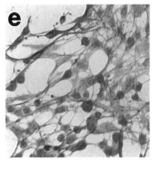

Anti-TNFα + IL-1α + IL-6 + TMT

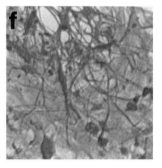

Fig. 10. Representative morphological response of cultured mixed glia cells to 24-h exposure to TMT (10 μ*M*) in the absence and presence of pro-inflammatory cytokine antibodies.

following TMT exposure. For example, both mRNA levels for vimentin and GFAP are elevated as early as 3 h of exposure. When the cultures are examined at 24 h, β-actin, vimentin, and GFAP mRNA levels are decreased. Proper expression and state of assembly of GFAP is critical in modulating astrocyte motility and shape, especially extension of astrocytic processes *(14,81)*. The alterations in mRNA levels for these structurally related proteins may be related to an effect on the cell membrane *(76,77,82)*, a direct interference with the expression of the proteins, or a disruption in precise packaging of structural proteins as a result of inflammatory cytokine signaling *(83)*. A microglia cytokine response has often been linked to a release of nitric oxide *(84–87)*. Although the nitric oxide synthase inhibitor L-NAME is ineffective in preventing hippocampal neurodegeneration induced by TMT in rats, an in vitro nitric oxide response can be evident. When mixed glia cultures are exposed to TMT, a nitric oxide response can be seen to increase within 24 h until 48 h following exposure. However, when cultures were purified to either astrocytes or microglia, TMT is unable to induce a nitric oxide response. This suggests an interaction between the cell types in the initiation and maintenance of such a response to TMT.

4.3. Pro-Inflammatory Cytokines

Because pro-inflammatory cytokine mRNA levels are elevated in the hippocampus following a systemic injection of TMT, the question remained as to whether this is a direct effect on the glia cells or in response to a neuronal signal. When primary glial cell cultures are exposed to TMT, mRNA levels for pro-inflammatory cytokines are elevated in a similar manner as that seen in the hippocampus. In the rat hippocampus, TNF-α mRNA levels remained elevated for up to 8 d following TMT injection, possibly the result of the prolonged recruitment of cells into the injury response. In culture, the elevation is short-lived and returns to within-control levels by 24 h possibly because of the lack of a bystander effect. mRNA levels for IL-1α, IL-1β, and IL-6 are increased for up to 48 h *(87a,87b)*. Mixed glial cells exposed to TMT secrete biologically active TNF protein in a dose-related manner *(60)*. Elevations are seen as early as 6 h and protein production is sustained for up to 48 h. Protein levels of TNF and IL-1 were significantly elevated at 24 and 48 h, respectively. In contrast to the in vivo response of a delayed elevation in TGF-β1 mRNA, no elevation is seen in culture. This is consistent with the *in situ* hybridization localization of TGF-β1 in pyramidal neurons of the hippocampus following TMT exposure. Similar to the pattern seen in the hippocampus, levels of mRNA for IL-2, IL-3, IL-4, IL-5, IL-6, IL-10, and IFN-γ were not detected in the glia cell cultures.

5. MODULATION OF GLIA RESPONSE

5.1. Cytokine Manipulations In Vitro

Although the morphological response of both astrocytes and microglia to TMT exposure appears to coincide with cytokine elevation, a direct causal relationship remains to be determined. When cells are exposed to a dose of lipopolysaccharide (LPS), one that would produce equivalent or greater amounts of TNF-α than that produced by 10 μ*M* TMT, only a slight change in cell morphology is evident. A 10-fold increase in concentration of LPS is required to produce as severe a morphological change as seen with TMT. Based on this comparison, it would appear that morphological changes produced by TMT are not entirely due to cytokine production and may involve additional biochemical processes. The influence of cytokine proteins on the morphological response to TMT has been examined by the coadministration of recombinant proteins with various doses of TMT. When mixed glia cultures are incubated with recombinant proteins for TNF-α, IL-1α, IL-1β, or IL-6 for a total of 25 h, no distinct morphological responses are evident. When cells are exposed to TMT in the presence of either IL-1α, IL-1β, or IL-6, an exacerbation of the response is seen with lower doses (1 and 5 μ*M* TMT) *(80)*. It is possible that this is the result of a direct effect of IL-1; however, it may also be in response to the induction of IL-6 by IL-1, as has been demonstrated by Aloisi and colleagues *(88)*. At higher concentrations of TMT, a distinct morphological response is seen. Coexposure to either IL-1α or IL-1β does not exacerbate this TMT response; however, coexposure with IL-6 produces a pronounced exacerbation.

In the in vitro model of TMT exposure, the addition of neutralizing antibodies to pro-inflammatory cytokines offers a level of protection in the morphological responses. However, the level and extent of protection is restricted. In cells pre-exposed to neutralizing antibodies for TNF-α, the glia monolayer remains intact for 24 h of TMT exposure; however, the antibody is unable to block the response when exposure continues for 48 h. This suggested that the neutralizing antibody for TNF-α is able to delay the onset but not prevent the morphological response. Preincubation with neutralizing antibodies to IL-1 or IL-6 are unable to prevent the normal morphological response to TMT. It is only when cells are preincubated with a combination of neutralizing antibodies to TNF-α, IL-1, and IL-6 that a moderate level of protection is seen in the glia monolayer morphology (Fig. 10). Full protection however, is not evident.

Although neutralizing antibodies to cytokines offer some ability to modulate the morphological response, effects on the induction of cytokine mRNA levels show a somewhat different pattern *(80)*. When mixed glia cultures are incubated with antibodies to TNF-α or IL-1α, there is no alteration in the relatively low basal levels of mRNA for any of the pro-inflammatory cytokines. However, exposure of control cells to anti-IL-6 results in a significant elevation in mRNA levels for IL-1α, IL-1β, and IL-6. A slight elevation can be seen in TNF-α mRNA levels. The combination of neutralizing antibodies to TNF-α, IL-1, and IL-6 show a pattern of elevation in mRNA levels for IL-1α, IL-1β, IL-6, and TNF-α similar to that seen with anti-IL-6 alone. At 6 h post-TMT, elevations are seen in TNF-α, IL-1α, and IL-1β. The coexposure to TMT for 6 h, in the presence of neutralizing antibodies, produces elevations in mRNA levels for IL-1α, IL-1β, IL-6, and TNF-α similar to those seen with TMT alone. Thus, whereas morphological changes induced by TMT can be altered by the combination of neutralizing antibodies, the initiation of the cytokine cascade is not modified.

The response of glia to increase the synthesis of certain cytokines has been previously reported to vary depending on the presence or absence of neurons in the culture system *(89)*. Although it has been clearly demonstrated that TMT exposure can elicit a cytokine response in cultured glial cells, whether this response would occur in the presence of neurons remains a question. Following a 24-h exposure to TMT, cultured hippocampal neurons display shrunken cell bodies and retracted neuronal processes. When cocultured with glial cells however, neurons retain their normal appearance during exposure to TMT. However, the glia cells display their standard morphological response to TMT characterized by process retraction *(80)*. These data are consistent with previous observations that neurons plated alone are more sensitive to TMT than cocultures of neurons and glia or glia alone *(90)*. In cocultures of hippocampal neurons and glia, a 6-h exposure to TMT produces a slight elevation in mRNA levels for TNFα, IL-1α, and IL-6 *(80)*. The magnitude of the elevation is not as great as that seen in the isolated mixed glia cultures. This differential level of response suggests an interdependency of the cells in the cytokine response. In addition to the general trophic support provided to neurons by the astrocytes, stimulation with IL-1, TNF-α, and IL-6 can elevate the production of nerve growth factor *(91–93)* and may contribute to the observed neuroprotection to cultured neurons. The altered expression of cytokines in mixed neuronal/glia cultures demonstrates a complex interaction between neurons and glia. It suggests that neurons may significantly influence the activation threshold of astrocytes to a cytokine-mediated response. Whether this modulation is the result of a soluble factor or a factor associated with membrane contact remains to be determined. It emphasizes the need for caution in attempts to extrapolate results from in vitro culture systems to in vivo situations where extensive cell–cell communication may alter the tissue response to injury.

5.2. In Vivo Responses

5.2.1. Pharmacological Manipulations

Glucocorticoids (GCS) are potent anti-inflammatory steroids that can inhibit both central and peripheral cytokine synthesis and action *(94–98)*. Anti-inflammatory actions of glucocorticoid administration are usually evident in acute injury and short-term pretreatment with drugs such as dexamethasone (DEX). Whereas glucocorticoids have limited if any demonstrated effects in the uninjured brain, they have the capability to suppress many functions of activated monocyte/macrophages in the injured brain, including production and release of cytokines. DEX has been reported to inhibit the release of TNF-α, IL-1, and IL-6 protein in the periphery *(99–101)*. Such acute effects of glucocorticoids may be beneficial in the downregulation of cytokine secretion and the inflammatory response.

Pretreatment with DEX does not alter the morphological responses seen with TMT exposure, in that neuronal necrosis and astrocyte reactivity are both present. Consistent with the morphological changes in astrocytes, mRNA levels for the astrocyte-specific protein GFAP are elevated by TMT, and pretreatment with DEX shows no modulation of this effect *(102)*. This finding is consistent with a previous study by O'Callaghan et al. *(103)* that demonstrated the inability of glucocorticoids to regulate the expression of GFAP in rat brain following either a stab wound or systemic injection of TMT. Pretreatment with DEX (10 mg/kg) significantly increases mRNA levels in the hippocampus for ICAM, EB22, TNF-α, TNF-β, and IL-1α over increases seen with TMT alone. mRNA levels of iNOS and A-20 are not changed under these conditions *(102)*. Consistent with previously reported findings *(68)*, no elevations are seen in mRNA levels for IL-4, IL-2, IL-6, or IFN-γ following TMT administration, and pretreatment with dexamethasone did not alter this profile.

In an injury response of the brain, an elevation in GFAP mRNA is often accompanied by an elevation in the mRNA for the astrocyte-associated serine proteinase inhibitor EB22/5.3. Such proteinase inhibitors can be induced by oxidative damage, cytokines, and inflammatory reactions and serve to neutralize proteolytic enzymes released from inflammatory cells *(104,105)*. EB22/5.3 mRNA levels are only slightly elevated by TMT, and a low dose of DEX does not change this level of elevation; however, with 10 mg/kg DEX, a significant elevation can be seen in the TMT-exposed hippocampus. A similar pattern is seen for TNF-β, whereas TNF-α and IL-1α are elevated in all TMT-dosed conditions, suggesting a specific association of EB22/5.3 with the cytokine TNF-β. mRNA levels for ICAM-1, a receptor for the integrins LFA-1 and Mac-1 required for leukocyte migration and extravasation *(106–108)* are altered by TMT exposure. ICAM-1 is elevated and Mac-1 is decreased. A low dose of DEX does not alter this response; however, the higher dose of DEX elevated this response substantially *(68)*. Within the central nervous system, a number of cell types can express ICAM-1 (neurons, astrocytes, microglia, and brain endothelial cells) and expression can be induced by TNF-α, IL-1α, and IFN-γ *(109–111)*. One interpretation of these data, from TMT-induced hippocampal injury, is that DEX can interfere with the processing and production of the pro-inflammatory cytokine proteins. Available data suggests that many of the effects on cytokines act through both transcriptional and posttranscriptional mechanisms *(112)*. It has been reported that inhibition of cytokine expression by glucocorticoids occurs at the transcriptional level, resulting in an inhibition of cytokine mRNA expression and a parallel decrease in cytokine secretion *(113,114)*. Whereas these reports strongly suggest that GCS act proximally by reducing cytokine availability via transcriptional repression and/or destabilization of cytokine mRNA, other reports suggest that GCS do not alter cytokine mRNA stability or affect cytokine-mediated cellular activation *(115)*.

In addition to the general cytokine-mediated effects on tissue injury, glucocorticoid receptor activation is attenuated by cytokines in immune and nonimmune tissue; DEX is often unable to suppress microglial functions under pathological conditions. The response seen with DEX could be related to the cytokine-induced downregulation of microglia glucocorticoid receptor expression and mineralocorticoid receptor activation. The activity in the hippocampus of DEX and cytokines is possibly mediated via the glucocorticoid response element (GRE) that could account for the altered AP-1 binding evident in both the DEX- and the TMT-treated hippocampus *(116)*.

5.2.2. Genetic Mutant Models

5.2.2.1. OP/OP MOUSE

The osteopetrotic (op/op) mutant mouse is deficient in biologically active colony-stimulating factor 1 (CSF-1) and has often been used in an attempt to understand the role of microglia proliferation and phagocytosis in brain injury. CSF-1 stimulates the proliferation and differentiation of monocytes and macrophages that have important trophic and/or scavenger roles in the function of the tissues in which they reside. We used this mutant mouse to examine the role of microglia in chemically induced trauma. The hypothesis was that if the neuronal degeneration following TMT is dependent on a microglia inflammatory response, a depletion of microglia would result in an attenuation of the response. Following injection of TMT, normal mice display neuronal necrosis, neuronal loss, astrocyte hypertrophy, and microglia activation. In op/op homozygous mice, there is an increased vulnerability for neuronal necrosis, as demonstrated by a shift in the dose response with a lack of microglia activation and astrocyte hypertrophy. TMT injury induces elevations in mRNA levels for the host response genes; ICAM-1, EB22, and GFAP. TNF-α mRNA levels are significantly elevated in microglia in both the nonhomozygous and op/op homozygous mice. In the op/op homozygous mouse, a dramatic increase is seen in both TNF-α and TNF-β that is significantly elevated over both saline-dosed and nonhomozygous mice receiving TMT *(72)*.

Chemokines are chemoattractant cytokines that play an important role in early events of inflammation and brain injury. As early as 24 h post-TMT, mRNA levels of macrophage inflammatory protein (MIP)-1α and MIP-1β are significantly elevated in the hippocampus of nonhomozygous mice. The op/op homozygous mice show a slight elevation in MIP-1α, as compared to the basal level in the homozygous mice. However, the level of induction is significantly less than that seen with a TMT injection in the nonhomozygous mice. Interferon-inducible protein (IP-10) and monocyte chemoattractant protein (MCP-1) are elevated in the TMT-dosed op/op homozygous mice. These data suggest a differential regulation of the chemokine pathway in the CSF-1-deficient mice and may represent alterations in the astrocyte population and contribute to the altered functional response of microglia.

5.2.2.2. CYTOKINE RECEPTOR KNOCKOUT MICE

There are two distinct TNF receptors of 55 and 75 kDa that play roles in the biological actions of TNF; however, the 55-kDa receptor appears to be responsible for many biological actions associated with cytotoxicity. Expression levels of TNF receptors are under regulatory control by a wide array of endogenous factors, including cytokines (e.g., IL-1, IL-2, IL-6, IL-8, granulocyte macrophage colony stimulating factor [GM-CSF]), protein kinases C and A, steroids, and calcium ionophores. These factors influence the number of surface receptors for TNF and, consequently, play a role in the biological response. For astrocytes, this can include β-amyloid, iNOS, and various cytokines *(84)*. Because TMT administration produces a dramatic elevation in TNF-α, it was hypothesized that the activation of TNF receptors would be an initiating event in the neurodegenerative process. The elevation of IL-1 mRNA in both the in vivo and the in vitro systems following TMT suggest a role for this pro-inflammatory cytokine in the signaling process associated with neurodegeneration. Because it has been reported that IL-1 receptors are present on dentate granule cells, it could be hypothesized that the pattern of neuronal death in the mouse following TMT is related to the elevation of IL-1 and possible activation of the IL-1 receptor. In our attempts to address these various hypotheses, we have examined both the neuropathology and the cytokine response in genetically altered mice. The outcome of these preliminary studies raise more questions than they answer. In each of the knockout animal models TNF-α receptor I, TNF-α receptor II, TNF-α receptor I/II, and IL-1 receptor, no significant modulation is evident in the TMT-induced neuropathology. At the current time, there is little evidence of an alteration in the prominent cytokine response. Although earlier studies suggested a critical role for these cytokines in the neurodegenerative process, these preliminary data suggest a more limited role for these cytokines in the actual manifestation of the damage.

5.2.2.3. *APOE* KNOCKOUT AND TRANSGENIC MICE

In the pathogenesis of Alzheimer's disease (AD), there is evidence for both a neuronal component and an "inflammatory" response *(69)*. These inflammatory responses are marked by astrocyte reactivity and microglia activation throughout the cortex *(117)*. In particular, activated microglia are found to be clustered near amyloid plaques *(118–121)* that contain other components such as β-amyloid protein, thrombin, and apolipoprotein E (ApoE) *(122)*. These data have led to the supposition that microglia and their immune-like secreted factors play a prominent role in the onset and progression of disease-related neurodegeneration. A variety of factors and processes have been implicated in the initiation and progression of AD pathology (for review, *see* refs. *103, 123,* and *124*). Such factors include amyloid fragment deposition, reactive gliosis, α-1-antichymotrypsin, and apolipoprotein E (ApoE).

To examine possible differential susceptibility associated with the apolipoprotein genotype we have used both *APOE* knockout mice, which have been reported to have an increased sensitivity to brain injury *(125–128)*, and knockouts containing the human *APOE* allele 4 transgene. Animals received TMT (2 mg/kg body wt.) at either 21 d or 8 mo of age. Similar to what is seen in the CD-1 mice, within 24 h, TMT produces cell death in hippocampal dentate granule neurons and mild astrogliosis in wild-type (C57BL6J), *APOE* knockout, and *APOE*4 mice. At both ages, mRNA levels of GFAP are elevated by TMT in all genotypes. In 21-d-old mice, TNF-α, IL-1α, MIP-1α, and TGF-β1 mRNA levels are elevated in all genotypes *(129)*. Although the patterns are similar in young animals, when the animals reach 8 mo of age, genotype-specific differences can be observed. In both wild-type and *APOE*4 mice, chemical injury produces an elevation in mRNA levels for TNF-α, IL-1α, and MIP-1α. In the *APOE* knockout mice, the normal chemical-induced elevation in TNF-α and IL-1α is not present and the induction of MIP-1α is decreased. TMT produces a slight increase of TGF-β1 mRNA levels only in wild-type mice, whereas no elevation is seen in *APOE* knockout or *APOE4* mice *(129)*. These data suggest that there is an age- and *APOE*-genotype-dependent effect in both microglia and neuronal response. They also suggest a regulatory role for ApoE in maintaining a critical balance of various pro-inflammatory cytokines, as is demonstrated by the differences in basal levels of TNF-β, TNF-α, MIP-1α, and TGF-β1 mRNA in the older *APOE* knockout and the *APOE4* mice. This shift seen with both *APOE* genotypes may represent altered systems that are unable to properly respond to injury. This is supported by the ApoE-genotype-dependent influence of EB-22, TNF-α, and MIP-1α mRNA levels in response to chemically induced injury.

6. SUMMARY

Acute exposure to the hippocampal neurotoxicant TMT produces localized neuronal necrosis that is accompanied by both a microglia and an astrocyte response. Elevations in specific chemokines, pro-inflammatory cytokines, and host response genes are temporally related to this process and show a dose dependency. Whereas glia are evident throughout the hippocampus, a distinct distribution exists for amoeboid microglia in the region of neuronal necrosis. The lack of significant infiltrating cells into the brain upon injury allows one to examine the reaction of the resident microglia in an intact system as well as efficacy of therapeutic intervention. The tight temporal response to TMT exposure will allow for the further study of a number of factors in brain injury and to demonstrate impact within the framework of dose-response relationships. The specific pattern of neuronal damage and microglia reaction in the hippocampus makes this an excellent model for examining the primary and interactive roles of cytokines and associated signaling proteins as they relate to the process of neuronal cell death, cell loss, and survival.

REFERENCES

1. Chao, C. C., Hu, S., and Peterson, P. K. (1995) Glia, cytokines, and neurotoxicity. *Crit. Rev. Neurobiol.* **9,** 189–205.
2. Eng, L. F. (1988) Regulation of glial intermediate filaments in astrogliosis, in *Biochemical Pathology of Astrocytes* (Norenberg, M. D., Hertz, L., and Schousboe, A., eds.), Liss, New York, pp. 70–90.

3. Yabuuchi, K., Minami, M., Katsumata, S., and Satoh, M. (1993) In situ hybridization study of interleukin-1 beta mRNA induced by kainic acid in the rat brain. *Brain Res. Mol. Brain Res.* **20,** 153–161.

4. Eng, L. F. (1985) Glial fibrillary acidic protein: the major protein of glial intermediate filaments in differentiated astrocytes. *J. Neuroimmunol.* **8,** 203–214.

5. Banati, R. B., Gehrmann, J., Schubert, P., and Kreutzberg, G. W. (1993) Cytotoxicity of microglia. *Glia* **7,** 111–118.

6. Giulian, D., Vaca, K., and Corpuz, M. (1993) Brain glia release factors with opposing actions on neuronal survival. *J. Neurosci.* **13,** 29–37.

7. Panek, R. B. and Benveniste, E. N. (1995) Class II MHC gene expression in microglia. *J. Immunol.* **154,** 2846–2854.

8. Sei, Y., Vitakovic, L., and Yokoyama, M. M. (1995) Cytokines in the central nervous system: regulatory roles in neuronal function, cell death, and repair. *Neuroimmunomodulation* **2,** 121–133.

9. Gadient, R. A. and Otten, U. H. (1997) Interleukin-6 (IL-6)—a molecule with both beneficial and destructive potentials. *Prog. Neurobiol.* **52,** 379–390.

10. Gelbard, H. A., James, H. J., Sharer, L. R., Perry, S. W., Saito, Y., Kazee, A. M., et al. (1995) Apoptotic neurons in brains from pediatric patients with HIV-1 encephalitis and progressive encephalopathy. *Neuropathol. Appl. Neurobiol.* **21,** 208–217.

11. Giulian, D. and Corpuz, M. (1993) Microglial secretion products and their impact on the nervous system. *Adv. Neurol.* **59,** 315–320.

12. Giulian, D., Li, J., Li, X., George, J., and Rutecki, P. A. (1994) The impact of microglia-derived cytokines upon gliosis in the CNS. *Dev. Neurosci.* **16,** 128–136.

13. Laping, N. J., Teter, B., Nichols, N. R., Rozovsky, I., and Finch, C. E. (1994) Glial fibrillary acidic protein: regulation by hormones, cytokines, and growth factors. *Brain Pathol.* **1,** 259–275.

14. Lees, G. (1993) The possible contribution of microglia and macrophages to delayed neuronal death after ischemia. *J. Neurol. Sci.* **114,** 119–122.

15. Hauser, S. L., Doolittle, T. H., Lincoln, R., Brown, R. H., and Dinarello, C. A. (1990) Cytokine accumulations in CSF of multiple sclerosis patients: frequent detection of interleukin-1 and tumor necrosis factor but not interleukin-6. *Neurology* **40,** 1735–1739.

16. Hofman, F. M., Hinton, D. R., Johnson, K., and Merrill, J. E. (1989) Tumor necrosis factor identified in multiple sclerosis brain. *J. Exp. Med.* **170,** 607–612.

17. Selmaj, K., Raine, C. S., Cannella, B., and Brosnan, C. F. (1991) Identification of lymphotoxin and tumor necrosis factor in multiple sclerosis lesions. *J. Clin. Invest.* **87,** 949–954.

18. Mrak, R. E., Sheng, J. G., and Griffin, W. S. (1995) Glial cytokines in Alzheimer's disease: review and pathogenic implications. *Hum. Pathol.* **26,** 816–823.

19. Giulian, D. and Lachman, L. B. (1985) Interleukin-1 stimulation of astroglial proliferation after brain injury. *Science* **228,** 497–499.

20. Knoblach, S. M. and Faden, A. I. (1998) Interleukin-10 improved outcome and alters pro-inflammatory cytokine expression after experimental traumatic brain injury. *Exp. Neurol.* **153,** 143–151.

21. Minami, M., Kuraishi, Y., and Satoh, M. (1991) Effects of kainic acid on messenger RNA levels of IL-1beta, IL-6, TNF alpha and LIF in the rat brain. *Biochem. Biophys. Res. Commun.* **176,** 593–598.

22. Shohami, E., Novikov, M., Bass, R., Yamin, A., and Gallily, R. (1994) Closed head injury triggers early production of TNF alpha and IL-6 by brain tissue. *J. Cereb. Blood Flow Metab.* **14,** 615–619.

23. Taupin, V., Toulmond, S., Serrano, A., Benavides, J., and Zavala, F. (1993) Increase in IL-6, IL-1, and TNF levels in rat brain following traumatic lesion. Influence of pre-and post-traumatic treatment with Ro5 4864, a peripheral-type (p site) benzodiazepine ligand. *J. Neuroimmunol.* **42,** 177–185.

24. Woodroofe, M. N., Sarna, G. S., Wadhwa, M., Hayes, G. M., Loughlin, A. J., Tinker, A., et al. (1991) Detection of interleukin-1 and interleukin-6 in adult rat brain, following mechanical injury, by in vivo microdialysis: evidence of a role for microglia in cytokine production. *J. Neuroimmunol.* **33,** 227–236.

25. Benveniste, E. N. (1998) Cytokine actions in the central nervous system. *Cytokine Growth Factor Rev.* **9,** 259–275.

26. Feuerstein, G. Z., Liu, T., and Barone, F. C. (1994) Cytokines, inflammation, and brain injury: Role of tumor necrosis factor-α. *Cerebrovasc. Brain Metab. Rev.* **6,** 341–360.

27. Arvin, B., Neville, L. F., Barone, F. D., and Feuerstein, G. Z. (1996) The role of inflammation and cytokines in brain injury. *Neurosci. Biobehav. Rev.* **20,** 445–452.

28. Benveniste, E. N. (1993) Astrocyte-microglia interactions. in *Astrocytes Pharmacology and Function* (Murphy, S., ed.), Academic, San Diego, CA, pp. 355–377.

29. Lemke, R., Hartiage-Rubsamen, M., and Schliebs, R. (1999) Differential injury-dependent glial expression of interleukins-1 alpha, beta, and interleukin-6 in rat brain. *Glia* **27,** 75–87.

30. Raivich, G., Jones, L. L., Werner, A., Bluthmann, H., Doetschmann, T., and Kreutzberg, G. W. (1999) Molecular signals for glial activation: pro-and anti-inflammatory cytokines in the injured brain. *Acta Neurochir. (Wien)* **73(Suppl.),** 21–30.

31. Fathallah-Shaykh, H. M., Gao, W., Cho, M., and Herrera, M. A. (1998) Priming in the brain, an immunologically privileged organ, elicits anti-tumor immunity. *Int. J. Cancer* **75,** 266–276.

32. Hallett, M. B. and Lloyds, D. (1995) Neutrophil priming: the cellular signals that say 'amber' but not green. *Immunol. Today* **16,** 264–268.

33. Renz, H., Hanke, A., Hofmann, P., Wolff, L. J., Schmidt, A., Ruschoff, J., et al. (1992) Sensitization of rat alveolar macrophages to enhanced TNF alpha release by in vivo treatment with dexamethasone. *Cell Immunol.* **144,** 249–257.

34. Arai, K., Lee, F., Miyajima, A., Miyatake, S., Arai, N., and Yokota, T. (1990) Cytokines: coordinators of immune and inflammatory responses. *Annu. Rev. Biochem.* **59,** 783–836.

35. Aschner, N., Allen, J. W., Kimelberg, H. K., LoPachin, R. M., and Streit, W. J. (1999) Glial cells in neurotoxicity development. *Annu. Rev. Pharmacol. Toxicol.* **39,** 151–173.

36. Balaban, C. D., O'Callaghan, J. P., and Billingsley, M. L. (1988) Trimethyltin-induced neuronal damage in the rat brain: comparative studies using silver degeneration stains, immunocytochemistry and immunoassay for neuronotypic and gliotypic proteins. *Neuroscience* **26,** 337–361.

37. Bouldin, T. W., Goines, N. D., Bagnell, C. R., and Krigman, M. R. (1981) Pathogenesis of trimethyltin neuronal toxicity: ultrastructural and cytochemical observations. *Am. J. Pathol.* **104,** 237–249.

38. Chang, L. W. (1986) Neuropathology of trimethyltin: a proposed pathogenetic mechanism. *Fundam. Appl. Toxicol.* **6,** 217–232.

39. Chang, L. W., Tiemeyer, T. M., Wenger, G. R., and McMillan, D. E. (1982) Neuropathology of mouse hippocampus in acute trimethyltin intoxication. *Neurobehav. Toxicol. Teratol.* **4,** 149–156.

40. Dyer, R., Deshields, T., and Wonderlin, W. (1982) Trimethyltin-induced changes in gross morphology of the hippocampus. *Neurobehav. Toxicol. Teratol.* **4,** 141–147.

41. Kutscher, C. L. (1992) A morphometric analysis of trimethyltin-induced changes in rat brain using the Timm technique. *Brain Res. Bull.* **28,** 519–527.

42. Reuhl, K. R. (1987) Neuropathology of organometallic compounds: review of selected literature. in *Neurotoxicants and Neurobiological Function* (Tilson, H. A. and Sparber, S. B., eds.), Wiley, New York, pp. 118–136.

43. Reuhl, K. R. and Cramner, J. M. (1984) Developmental neuropathology of organotin compounds. *Neurotoxicology* **5,** 187–204.

44. Reuhl, K. R., Smallridge, E. A., Chang, L. W., and Mackenzie, B. A. (1983) Developmental effects of trimethyltin intoxication in the neonatal mouse. I. Light microscopic studies. *Neurotoxicology* **4,** 19–28.
45. Cook, L. L., Stine, K. E., and Reiter, L. W. (1986) Tin distribution in adult rat tissues after exposure to trimethyltin and triethyltin. *Toxicol. Appl. Pharmacol.* **76,** 344–348.
46. Mushak, P., Krigman, M. R., and Mailman, R. B. (1982) Comparative organotin toxicity in the developing rat: somatic and morphological changes and relationship to accumulation of total tin. *Neurobehav. Toxicol. Teratol.* **4,** 209–215.
47. Doctor, S. V., Sultatos, L. G., and Murphy, S. D. (1983) Distribution of trimethyltin in various tissues of the male mouse. *Toxicol. Lett.* **17,** 43–48.
48. Ishida, N., Akaike, M., Tsutsumi, S., Kanai, H., Masui, A., Sadamatsu, M., et al. (1997) Trimethyltin syndrome as a hippocampal degeneration model: temporal changes and neurochemical features of seizure susceptibility and learning impairment. *Neuroscience* **81,** 1183–1191.
49. Sloviter, R. S., von-Knebel-Doeberitz, C., Walsh, T. J., and Dempster, D. W. (1986) On the role of seizure activity in the hippocampal damage produced by trimethyltin. *Brain Res.* **367,** 169–182.
50. Alessandri, B., FitzGerald, R. E., Schcaeppi, U., Krinke, G. J., and Classen, W. (1994) The use of an unbaited tunnnel maze in neurotoxicology: I. Trimethyltin-induced brain lesions. *Neurotoxicology* **15,** 349–357.
51. Feldman, R. G., White, R. F., and Eriator, I. I. (1993) Trimethyltin encephalopathy. *Arch. Neurol.* **50,** 1320–1324.
52. Fortemps, E., Amand, G., Bomboir, A., Lauwerys, R., and Laterre, E. C. (1978) Trimethyltin poisoning: report of two cases. *Int. Arch. Occup. Environ. Health* **41,** 1–6.
53. Wenger, G. R., McMillan, D. E., and Chang, L. W. (1982) Behavioral toxicology of acute trimethyltin exposure in the mouse. *Neurobehav. Toxicol. Teratol.* **4,** 157–161.
54. Robertson, D. G., Gray, R. H., and de-la-Iglesia, F. A. (1987) Quantitative assessment of trimethyltin induced pathology of the hippocampus. *Toxicol. Pathol.* **15,** 7–17.
55. McCann, M. J., O'Callaghan, J. P., Martin, P. M., Bertram, T., and Streit, W. J. (1996) Differential activation of microglia and astrocytes following trimethyl tin-induced neurodegeneration. *Neuroscience* **72,** 273–281.
56. O'Shaughnessy, D. J. and Losos, G. J. (1986) Peripheral and central nervous system lesions caused by triethyl- and trimethyltin salts in rats. *Toxicol. Pathol.* **14,** 141–148.
57. Maier, W. E., Brown, H. W., Tilson, H. A., Luster, M. L., and Harry, G. J. (1995) Trimethyltin increased interleukin (IL)-1α, IL-6 and tumor necrosis factor-α mRNA levels in rat hippocampus. *J.Neuroimmunol.* **59,** 65–75.
58. Monnet-Tschudi, R., Zurich, M. G., Pithon, E., van Melle, G., and Honegger, P. (1995) Microglial responsiveness as a sensitive marker for trimethyltin (TMT) neurotoxicity. *Brain Res.* **690,** 8–14.
58a. Haga, S., Haga, C., Aizawa, T., and Ikeda, K. (2002) Neuronal degeneration and glia cell-responses following trimethyltin intoxication in the rat. *Acta Neuropathol. (Berl)* **103,** 575–582.
59. Koczyk, D. and Oderfeld-Nowak, B. (2000) Long-term microglial and astroglial activation in the hippocampus of trimethyltin-intoxicated rat: stimulation of NGF and TrkA immunoreactivities in astroglia but not in microglia. *Int. J. Dev. Neurosci.* **18,** 591–606.
60. Maier, W. E., Bartenbach, M. J., Brown, H. W., Tilson, H. A., and Harry, G. J. (1997) Induction of tumor necrosis factor alpha in cultured glial cells by trimethyltin. *Neurochem. Int.* **30,** 385–392.
61. Schindler, R., Clark, B. D., and Dinarello, C. A. (1990) Dissociation between interleukin-1 beta mRNA and protein synthesis in human peripheral blood mononuclear cells. *J. Biol. Chem.* **265,** 10,232–10,237.

62. Schindler, R., Lonnemann, G., Shaldon, S., Koch, K. M., and Dinarello, C. A. (1990) Transcription, not synthesis, of interleukin-1 and tumor necrosis factor by complement. *Kidney Int.* **37,** 85–93.

63. Minami, M., Kuraishi, Y., Yabuuchi, K., Yamazaki, A., and Satoh, M. (1992) Induction of interleukin-1 beta mRNA in rat brain after transient forebrain ischemia. *J. Neurochem.* **58,** 390–392.

64. Reich, R., Thompson, E. W., Iwamoto, Y., Martin, G. R., Denson, J. R., Fuller, G. C., et al. (1988) Effects of inhibitors of plasminogen activator, serine proteinases, and collagenase IV on the invasion of basement membranes by metastatic cells. *Cancer Res.* **48,** 3307–3312.

65. Rosenberg, G. A, Estrada, E. Y, Dencoff, J. E., and Stetler-Stevenson, W. G. (1995) Tumor necorsis factor-alpha-induced gelatinase B caused delayed opening of the blood-brain barrier: an expanded therapeutic window. *Brain Res.* **703,** 151–155.

66. Dencoff, J. E. Rosenberg, G. A., and Harry, G. J. (1997) Trimethyltin induces gelatinase B and urokinase in rat brain. *Neurosci. Lett.* **228,** 147–150.

66a. Fiedorowicz, A., Figiel, I., Kaminska, B., Zaremba, M., Wilk, S., and Oderfeld-Nowak, B. (2001) Dentate granule neuron apoptosis and glia activation in murine hippocampus induced by trimethyltin exposure. *Brain Res.* **912,** 116–127.

66b. Geloso, M. C., Vercelli, A., Corvine, V., Repici, M., Boca, M., Haglid, K., et al. (2002) Cyclooxygenase-2 and caspase 3 expression in trimethylin-induced apoptosis in the mouse hippocampus. *Exp. Neurol.* **175,** 152–160.

67. Harry, G. J., Goodrum, J. F., Krigman, M. R., and Morell, P. (1985) The use of Synapsin I as a biochemical marker for neuronal damage. *Brain Res.* **326,** 9–18.

68. Bruccoleri, A., Brown, H., and Harry, G. J. (1998) Cellular localization and temporal elevation of tumor necrosis factor-α interleukin-1α, and transforming growth factor-β1 mRNA in hippocampal injury response induced by trimethyltin. *J. Neurochem.* **71,** 1577–1578.

69. Giulian D. (1999) Microglia and the immune pathology of Alzheimer's disease. *Am. J. Hum. Genet.* **65,** 13–18.

70. Kreutzberg, G. W. (1996) Microglia: a sensor for pathological events in the CNS. *Trends Neurosci.* **19,** 312–318.

71. Merrill J. E. and Benveniste E. N. (1996) Cytokines in inflammatory brain lesions: helpful and harmful. *Trends Neurosci.* **19,** 331–338.

72. Bruccoleri, A. and Harry, G. J. (2000) Chemical-induced hippocampal neurodegeneration and elevations in TNFα, TNFβ, IP-10, and MCP-1 mRNA in osteopetrotic (op/op) mice. *J. Neurosci. Res.* **62,** 146–155.

73. Chow, S. C. and Orrenius, S. (1994) Rapid cytoskeleton modification in thymocytes induced by the immunotoxicant tributyltin. *Toxicol. Appl. Pharmacol.* **127,** 19–26.

74. Yu, A. C. H., Lee, Y. L., and Eng, L. F. (1993) Astrogliosis in culture: I. The model and the effect of antisense oligonucleotides on glial fibrillary acidic protein synthesis. *J. Neurosci. Res.* **34,** 295–303.

75. Eng, L. F., Lee, Y. L., Murphy, G. M., and Yu, A. C. (1995) A RT-PCR study of gene expression in a mechanical injury model. *Prog. Brain Res.* **105,** 219–229.

76. Aschner, M. and Aschner, J. L. (1992) Cellular and molecular effects of trimethyltin and triethyltin: relevance to organotin neurotoxicity. *Neurosci. Biobehav. Rev.* **16,** 427–435.

77. Aschner, M., Gannon, M., and Kimelberg, H. K. (1992) Interactions of trimethyl tin (TMT) with rat primary astrocyte cultures: altered uptake and efflux of rubidium, L-glutamate, and D-aspartate. *Brain Res.* **582,** 181–185.

78. Dawson, R. Jr., Patterson, T. A., and Epler, B. (1995) Endogenous excitatory amino acid release from brain slices and astrocyte cultures evoked by trimethyltin and other neurotoxic agents. *Neurochem. Res.* **20,** 847–858.

79. Richter-Landsberg, C. and Besser, A. (1994) Effects of organotins on rat brain astrocytes in culture. *J. Neurochem.* **63**, 2202–2209.

80. Harry, G. J., Tyler, K., Tilson, H. A., and Maier, W. E. (2002) Morphological alterations and elevations in TNFα, IL-1α, and IL-6 in mixed glia cultures following exposure to trimethyltin (TMT): modulation of toxicity by pro-inflammatory cytokine recombinant proteins and neutralizing antibodies. *Toxicol. Appl. Pharmacol.* **180**, 205–218.

80a. Figiel, I. and Fiedorowicz, A. (2002) Trimethyltin-evoked neuronal apoptosis and glia response in mixed cultures of rat hippocampal dentate gyrus: a new model for the study of the cell type-specific influence of neurotoxins. *Neurotoxicology* **23**, 77–86.

81. Laping, N. J., Morgan, T. E., Nichols, N. R., Rozovsky, I., Young-Chan, C. S., Zarow, C., et al. (1994) Transforming growth factor-beta 1 induces neuronal and astrocyte genes: tubulin alpha 1, glial fibrillary acidic protein and clusterin. *Neuroscience* **58**, 563–572.

82. Harkins, A. B. and Armstrong, D. L. (1992) Trimethyltin alters membrane properties of CA1 hippocampal neurons. *Neurotoxicology* **13**, 569–582.

83. Liu, W., Shafit-Zagardo, B., Aquino, D. A., Zhao, M. L., Dickson, D. W., Brosnan, C. F., et al. (1994) Cytoskeletal alterations in human fetal astrocytes induced by interleukin-1 beta. *J. Neurochem.* **63**, 1625–1634.

84. Akama, K. T. and Van Eldik, L. J. (2000) Beta-amyloid stimulation of inducible nitric-oxide synthase in astrocytes is interleukin-1 beta and tumor necrosis factor-alpha (TNF alpha)-dependent, and involves a TNF alpha receptor-associated factor and NFKappa B-inducing kinase-dependent signaling mechanism. *J. Biol. Chem.* **275**, 7918–7924.

85. Hu, S., Ali, H., Sheng, W. S., Ehrlich, L. C., Peterson, P. K., and Chao, C. C. (1999) Gp-41-mediated astrocyte inducible nitric oxide synthase mRNA expression: involvement of inter-leukin-1beta production by microglia. *J. Neurosci.* **19**, 6468–6474.

86. Nomura, Y. and Kitamura, Y. (1993) Inducible nitric oxide synthase in glial cells. *Neurosci. Res.* **18**, 103–107.

87. Romero, L. I., Tatro, J. B., Field, J. A., and Reichlin, S. (1996) Roles of IL-1 and TNF-alpha in endotoxin-induced activation of nitric oxide synthase in cultured rat brain cells. *Am. J. Physiol.* **270**, R326–R332.

87a. Viviani, B., Corsini, E., Pesenti, M., Galli, C. L., and Marinovich, M. (2001) Trimethyl-tin-activated cyclooxygenase stimulates tumor necrosis factor-alpha release from glial cells through reactive oxygen species. *Toxicol. Appl. Pharmacol.* **172**, 93–97.

87b. Jahnke, G. D., Brunssen, S., Maier, W. E., and Harry, G. J. (2001) Neurotoxicant-induced elevation of adrenomedullin expression in hippocampus and glia cultures. *J. Neurosci. Res.* **66**, 464–474.

88. Aloisi, R., Care, A., Borsellino, G., Gallo, P., Rosa, S., Bassani, A., et al. (1992) Production of hemolymphopoietic cytokines (IL-6, IL-8, colony-stimulating factors) by normal human astrocytes in response to IL-1β and TNF-α. *J. Immunol.* **149**, 2358–2366.

89. Aloisi, F., Borsellino, A., Care, A., Testa, U., Gallo, P., Russo, G., et al. (1995) Cytokine regulation of astrocyte function: *in vitro* studies using cells from the human brain. *Int. J. Dev. Neurosci.* **13**, 265–274.

90. Thompson, T. A., Lewis, J. M., Dejneka, N. S., Severs, W. B., Polavarapu, R., and Billingsley, M. L. (1996) Induction of apoptosis by organotin compounds in vitro: neuronal protection with antisense oligonucleotides directed against stannin. *J. Pharmacol. Exp. Ther.* **276**, 1201–1216.

91. Gadient, R. A., Cron, K. C., and Otten, U. (1990) Interleukin-1 beta and tumor necrosis factor-alpha synergistically stimulate nerve growth factor (NGF) release from cultured rat astrocytes. *Neurosci. Lett.* **117**, 335–340.

92. Kossmann, T., Volkmar, H., Imhof, H.-G., Trentz, O., and Morganti-Kossman, M. C. (1996) Interleukin-6 released in human cerebrospinal fluid following traumatic brain injury may trigger nerve growth factor production in astrocytes. *Brain Res.* **713**, 143-152.

93. Spranger, M., Lindholm, D., Bandtlow, C., Heumann, R., Gnahn, H., Naher-Noe, M., et al. (1990) Regulation of nerve growth factor (NGF) synthesis in the rat central nervous system. Comparison between the effects of interleukin-1 and various growth factors in astrocyte cultures and in-vivo. *Eur. J. Neurosci.* **2,** 69–76.

94. Almawi, W. Y., Bayhum, H. N., Rahme, A. A., and Reider, M. J. (1996) Regulation of cytokine and cytokine receptor expression by glucocorticoids. *J. Leukocyte Biol.* **60,** 563–572.

95. Bateman, A., Singh, A., Kral, T., and Solomon, S. (1989) The immune-hypothalamic-pituitary-adrenal axis. *Endocr. Rev.* **10,** 92–112.

96. Cato, A. C. B. and Wade, E. (1996) Molecular mechanisms of anti-inflammatory action of glucocorticoids. *BioEssays* **18,** 371–378.

97. Goujon, E., Laye, S., Parnet, P., and Dantzer, R. (1997) Regulation of cytokine gene expression in the central nervous system by glucocorticoids: mechanisms and functional consequences. *Psychoneuroendrocrinology* **22,** S75–S80.

98. Guyre, P. M., Girard, M. T., Morganelli, P. M., and Manganiello, P. D. (1988) Glucocorticoid effects on the production and actions of immune cytokines. *J. Steroid Biochem.* **30,** 89–93.

99. Chao, C. C., Hu, S., Close, K., Choi, C. S., Molitor, T. W., Novick, W. J., et al. (1992) Cytokine release from microglia: differential inhibition by pentoxifylline and dexamethasone. *J. Infect. Dis.* **166,** 847–853.

100. Leu, W., Oppenheim, J. J., and Matsushima, K. (1988) Analysis of the suppression of IL-1α and IL-1β production in human peripheral blood mononuclear cells by a glucocorticoid hormone. *J. Immunol.* **140,** 1895–1902.

101. Zuckerman, S. H., Shellhaas, J., and Butler, L. D. (1989) Differential regulation of lipopolysacharide-induced interleukin-1 and tumor necrosis factor synthesis: effects of endogenous glucocorticoids and the role of pituitary-adrenal axis. *Eur. J. Immunol.* **19,** 301–305.

102. Bruccoleri, A., Pennypacker, K. R., and Harry, G. J. (1999) Effect of dexamethasone on elevated cytokine mRNA levels in chemical-induced hippocampal injury. *J. Neurosci. Res.* **57,** 916–926.

103. O'Callaghan, J. P., Brinton, R. E., and McEwen, B. S. (1991) Glucocorticoids regulate the synthesis of glial fibrillary acidic protein in intact and adrenalectomized rats but do not affect its expression following brain injury. *J. Neurochem.* **57,** 860–869.

104. Abraham, C. R. (1992) The role of the acute-phase protein α1-antichymotrypsin in brain dysfunction and injury. *Res. Immunol.* **143,** 631–636.

105. Inglis, J. D., Lee, M., Davidson, D. R., and Hill, R. E. (1991) Isolation of two cDNAs encoding novel alpha 1-antichymotrypsin-like proteins in a murine chondrocytic cell line. *Gene* **106,** 213–220.

106. Marlin, S. D. and Springer, T. A. (1987) Purified intercellular adhesion molecule-1 (ICAM-1) is a ligand for lymphocyte function-associated antigen 1 (LFA-1). *Cell* **51,** 813–819.

107. Rothlein, R., Dustin, M. L., Marlin, S. D., and Springer, T. A. (1986) A human intercellular adhesion molecule (ICAM-1) distinct from LFA-1. *J. Immunol.* **137,** 1270–1274.

108. Smith, C. W., Marlin, S. D., Rothlein, R., Toman, C., and Anderson, D. C. (1989) Cooperative interactions of LFA-1 and Mac-1 with intercellular adhesion molecule-1 infacilitating adherence and transendothelial migration of human neutrophils in vitro. *J. Clin. Invest.* **83,** 2008–2017.

109. Frohman, E. M., Frohman, T. C., Dustin, M. L., Vayavegula, B., Choi, A., Gupta, S., et al. (1989) The induction of intercellular adhesion molecule-1 (ICAM-1) expression on human fetal astrocytes by interferon-γ, tumor necrosis factor-α lymphotoxin, and interleukin-1: relevance to intercerebral antigen presentation. *J. Neuroimmunol.* **23,** 117–124.

110. Shrikant, P., Chung, I. Y., Ballestas, M., and Benveniste, E. N. (1994) Regulation of intercellular adhesion molecule-1 gene expression by tumor necrosis factor-α, interleukin-1β, and interferon-γ in astrocytes. *J. Neuroimmunol.* **51,** 209–222.

111. Shrikant, P., Weber, E., Jilling, T., and Benveniste, E. N. (1995) Intercellular adhesion molecule-1 gene expression by glial cells. Differential mechanisms of inhibition by IL-10 and IL-6. *J. Immunol.* **155,** 1489–1501.

112. Han, J., Thompson, P., and Beutler, B. (1990) Dexamethasone and pentoxifylline inhibit endotoxin-induced cachectin/tumor necrosis factor synthesis at separate points in the signaling pathway. *J. Exp. Med.* **172,** 391–394.

113. Amano, Y., Lee, S. W., and Allison, A. C. (1993) Inhibition by glucocorticoids of the formation of interleukin-1α, interleukin-1β, and interleukin-6: mediation by decreased mRNA stability. *Mol. Pharmacol.* **43,** 176–182.

114. Boumpas, D. T., Anastassiou, E. D., Older, S. A., Tsokos, G. C., Nelson, D. L., and Balow, J. E. (1991) Dexamethasone inhibits human interleukin 2 but not interleukin 2 receptor gene expression in vitro at the level of nuclear transcription. *J. Clin. Invest.* **87,** 1739–1747.

115. Villafuerte, B. C., Koop, B. L., Pao, C. I., and Phillips, L. S. (1995) Glucocorticoid regulation of insulin-like growth factor-binding protein-3. *Endocrinology* **136,** 1928–1933.

116. Xiao, Y., Harry, G. J., and Pennypacker, K. R. (1999) Expression of AP-1 transcription factors in rat hippocampus and cerebellum after trimethyltin neurotoxicity. *Neurotoxicology* **20,** 761–766.

117. Egensperger, R., Kosel, S., von Eitzen, U., and Graeber, M. B. (1998) Microglial activation in Alzheimer disease: association with APOE genotype. *Brain Pathol.* **8,** 439–447.

118. Griffin, W. S., Sheng, J. G., Roberts, G. W., and Mrak, R. E. (1995) Interleukin-1 expression in different plaque type in Alzheimer's disease: significance in plaque evolution. *J. Neuropathol. Exp. Neurol.* **54,** 276–281.

119. Licastro, R., Mallory, M., Hansen, L. A., and Masliah, E. (1998) Increased levels of alpha-1-antichymotrypsin in brains of patients with Alzheimer's disease correlate with activated astrocytes and are affected by APOE genotype. *J. Neuroimmunol.* **88,** 105–110.

120. Saitoh, T., Kang, D., Mallory, M., DeTeresa, R., and Masliah, E. (1997) Glial cells in Alzheimer's disease: preferential effect of APOE risk on scattered microglia. *Gerontology* **43,** 109–118.

121. Uchihara, T., Duyckaerts, C., He, Y., Kobayashi, K., Seilhean, D., Amouyel, P., et al. (1995) ApoE immunoreactivity and microglial cells in Alzheimer's disease brain. *Neurosci. Lett.* **195,** 5–8.

122. McGeer, P. L., Klegeris, A., Walker, D. G., Yasuhara, O., and McGeer, E. G. (1994) Pathological proteins in senile plaques. *Tohoku J. Exp. Med.* **174,** 269–277.

123. Cruts, M. and Van Broeckhoven, C. (1998) Molecular genetics of Alzheimer's disease. *Ann. Med.* **30,** 560–565.

124. Higgins, G. A., Large, C. H., Rupniak, H. T., and Barnes, J. C. (1997) Apolipoprotein E and Alzheimer's disease: a review of recent studies. *Pharmacol. Biochem. Behav.* **56,** 675–685.

125. Chen, Y., Lomnitski, L., Michaelson, D. M., and Shohami, E. (1997) Motor and cognitive deficits in apolipoprotein E-deficient mice after closed head injury. *Neuroscience* **80,** 1255–1262.

126. Horsburgh, K., Kelly, S., McCulloch, J., Higgins, G. A., and Nicoll, J. A. R. (1999) Increased neuronal damage in apolipoprotein E-deficient mice following global ischaemia. *NeuroReport* **10,** 837–841.

127. Laskowitz, D. T., Sheng, H., Bart, R. D., Joyner, K. A., Roses, A. D., and Warner, D. S. (1997) Apolipoprotein E-deficient mice have increased susceptibility to focal cerebral ischemia. *J. Cereb. Blood Flow Metab.* **17,** 753–758.

128. Lomnitski, L., Kohen, R., Chen, Y., Shohami, E., Trembovler, V., Vogel, T., et al. (1997) Reduced levels of antioxidants in brains of apolipoprotein E-deficient mice following closed head injury. *Pharmacol. Biochem. Behav.* **56,** 669–673.

129. Harry, G. J., Lefebvre d'Hellencourt, C., Bruccoleri, A., and Schmechel, D. (2000) Age-dependent cytokine responses: trimethyltin hippocampal injury in wild-type, APOE knockout, and APOE4 mice. *Brain Behav. Immunol.* **14,** 288–304.

17

Inflammation and Cyclo-Oxygenase in Alzheimer's Disease

Experimental Approaches and Therapeutic Implications

Patrick Pompl, Tara Brennan, Lap Ho, and Giulio Maria Pasinetti

1. ALZHEIMER'S DISEASE: AN INFLAMMATORY DISEASE?

There is clear evidence of inflammation in the degenerating Alzheimer's disease (AD) brain, however, it is not presently known whether inflammatory mechanisms are a cause or consequence of this disease (Fig. 1) *(1,2)*. The uncertainty stems from the fact that inflammatory events and pathological hallmarks of AD (such as senile plaques [SPs] and neurofibrillary tangles [NFTs]) occur in close temporal and physical proximity *(3)*. For example, it has been shown that the majority of mature senile plaques are surrounded by reactive microglia that have the characteristics of antigen-presenting tissue macrophages, including HLA-DR surface markers, suggestive of an autoimmune response. It has been proposed that this is a remedial circumstance in which microglia are attempting to clear the amyloid debris from the neuropil. However, experimental studies show that suppression of microglia by treatment with ibuprofen (a nonsteroidal anti-inflammatory drug [NSAID]) can actually diminish the presence of amyloid deposits in the brain *(4)* and suggests a contradictive hypothesis. This latter hypothesis is further supported by retrospective epidemiological studies that have found decreased incidence of AD in populations who regularly take NSAIDs.

Although a lucid cause–effect relationship for inflammation and AD pathology is not easily ascertained, a thorough cataloging of inflammatory indices continues so that meaningful associations can eventually be derived. Thus far, the inflammatory molecules found to be altered in AD include (1) upregulation of cytokines, such as interleukin (IL)-1, IL-6, and tumor necrosis factor (TNF)-α, and -acute phase proteins, such as α1-antichymotrypsin (ACT) and α2 macroglobulin (α2 MAC) *(5)*; (2) the presence of an active complement system in the AD brain *(6)* with generation of the lytic membrane attack complex *(7)* and, presumably, a release of anaphylatoxins [which under certain conditions, may be neuroprotectant *(8)*]; (3) an upregulation of cyclo-oxygenase-2 (COX)2, an enzyme involved in inflammation and neuronal functions in the brain of AD *(9)*.

From: *Neuroinflammation, 2nd Edition: Mechanisms and Management*
Edited by: P. L. Wood © Humana Press Inc., Totowa, NJ

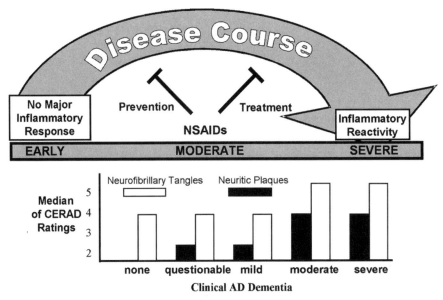

Fig. 1. Potential role of inflammation as a function of the progression of Alzheimer's disease dementia. Relationship with the Consortium to Establish a Registry for Alzheimer's Disease (CERAD) rating of AD neuropathology.

Several of these findings have been identified as having causative potential in AD neuropathology. For example, in mice, it has been shown that overexpression of the inflammatory cytokine IL-6 in the brain leads to neurodegeneration *(10)*. As well, complement components *(11)* can potentiate amyloid neurotoxicity in vitro, and neuronal COX-2 overexpression can potentiate amyloid deposition *(12)*. However, despite the growing list of potential degenerative effects of inflammatory mediators whose expression is altered in the AD brain, the AD-specific role of these pro-inflammatory agents remains elusive. Recent evidence suggests that polymorphisms in inflammatory genes influence the risk of AD *(13)* and may potentially explain some of the existing complications in reconciling pathologic findings among different cases. It is for this reason that it is essential not only to identify (early) biomarkers of brain inflammation but also to derive profiles of the heterogeneity of these features in the populations under study. Elucidation of the functional relationship of early molecular biomarkers to more advanced degenerative changes will yield significant insight for future treatments of AD and provide potential selection criterion and covariant factors for clinical trials of anti-inflammatory drugs.

2. INFLAMMATION AND ALZHEIMER'S DISEASE CLINICAL PROGRESSION

The inflammatory events present in AD are dynamic and evolve with the stimuli that elicited their activation (Fig. 1). However, the above-mentioned inflammatory events have only recently been observed to change with the clinical progression of AD. This discovery was facilitated through the development of an accurate longitudinal assessment of the progression of AD across populations. These assessments, the Clinical Dementia

Rating (CDR) scale and the consortium to establish a registry for AD (CERAD) *(3,14)*, employ systematic analyses of psychometric tests and postmortem impressions, respectively, to assign a numeric value to individual cases that covers the spectrum of AD clinical progression. The CDR scale encompasses cognitively normal (CDR 0.0), to mild cognitive impairment (MCI)/questionable AD (CDR 0.5), mild to moderate dementia (CDR 1–2), and severe dementia (CDR 5.0) cases, the CERAD evaluates patients on a basis of SPs and NFTs on a scale of 1–5 (postmortem). Interestingly, pro-inflammatory activities have been found to be evident in AD as early as the CDR 0.5 stage and become increasingly evident as the disease progresses to later stages *(14–16)*.

The CDR scale has provided researchers with the opportunity to correlate findings from different patient populations and rapidly validate clinical impressions. For illustration, it has been reported that there is a significant elevation in prostaglandin (PG) E_2 content in the cerebrospinal fluid of probable AD cases *(15)*. This evidence was substantiated by reports showing that COX-2, which functions to convert arachidonic acid to PG, is clearly upregulated in the hippocampus of AD cases. A more recent study utilizing CDR-rated cases found that the expression of COX-2 is elevated as early as CDR 0.5, and significantly increases between CDR 0.5 and 1.0 cases *(16)*. In light of previous data, this study strongly suggested that COX-2-related activity may play a significant role in the conversion from MCI to frank AD dementia. This is particularly interesting considering the lack of biological markers that represent this most fundamental phase of AD progression. Further, these data are supportive of a large body of epidemiological data showing a protective effect for NSAIDs, a large number of which function by blocking COX-2.

Other classical markers of inflammation such as cytokine expression and HLA-DR immunoreactive microglia are also apparent in AD, but only in later, more severe phases of the disease (CDR 5). These stages are also characterized by extensive oxidative damage and widespread elevation of upstream and downstream caspases *(49)*. Thus, based on a collective evaluation of the progression of inflammation in AD as a function of CDR, it may be the case that COX inhibitors or other NSAIDs will prove successful in early stages of AD; however, conjunctive treatment with antioxidants and caspase inhibitors may be more useful in later stages of the disease. The design of drugs that target inflammatory cascades specific to each stage of AD will allow us to more precisely address the progression of this disease and optimize therapeutic efficacy.

3. THE NEED FOR CHARACTERIZING EARLY INFLAMMATORY CASCADES IN ALZHEIMER'S DISEASE

Although a great deal of clinical and experimental data have been gathered on the pathologic and inflammatory changes in AD, current clinical investigations of therapeutics aimed at mitigating these changes are having difficulty. For example, a meta-analysis review of 17 epidemiological studies *(17)* suggests that anti-inflammatory treatments, such as NSAIDs and steroids, may decrease the relative risk (RR) for AD by as much as 50%, and a recent prospective study found that the RR for AD fell with increasing duration of NSAID use *(18)*—indicating that 2 or more years of NSAID treatment resulted in a 50% decrease in RR. However, in a 6-mo study of the NSAID indomethacin (100–150 mg/d) patients showed only minor improvement on a battery of cognitive tests *(19)*. Further, this finding could not be reproduced in a study using NSAID compounds such as diclofenac–misoprostol (a high dropout rate prevented statistically significant results) *(20)* or prednisone *(21)*.

A potential confound responsible for these discrepancies may be that systematic characterization of disease stage by methods such as CDR has only recently begun to receive needed attention and that the mechanisms of action of the large and diverse class of NSAID compounds currently available are not fully understood. However, of greater consideration is that the preponderance of epidemiological studies are performed on patients receiving NSAID treatment well before clinical manifestations of AD are observed. In contrast, therapeutic studies are conducted on patients with illness that is severe enough to exceed the clinical detection threshold (CDR 1–2). Thus, this latter case presents a significantly different pathophysiologic environment than earlier presymptomatic stages (CDR 0.5) *(3)* and, as mentioned previously, may require additional/different therapeutic treatment.

Although the underlying profile of neuropathology and inflammation in AD has not yet been fully elucidated, it is likely that late-stage AD will yield a far more complex challenge for drug intervention, possibly with less potential for amelioration. Thus, it is imperative that the influence of inflammatory activity in early AD or cases at high risk to develop AD is understood, so that effective anti-inflammatory treatment strategies can be developed for the prevention and/or attenuation neurodegeneration. As with any disease, therapeutic–preventive intervention is most effective when administered early in the disease's progression (Fig. 1).

4. CYCLO-OXYGENASE: A TARGET FOR NSAIDs IN ALZHEIMER'S DISEASE

Based on the early upregulation of COX-2 in neurons, prior to an elevation of the cytokines interleukin (IL)-6 transforming growth factor (TGF)- β1 expression *(14)* (discussed in Section 1), it has been proposed that the increase in neuronal COX-2 seen during the earliest detectable phase of AD dementia (mild dementia; CDR 1) lays a precedence for later inflammatory neurodegeneration. This is further supported by a great deal of existing evidence that NSAIDs (most of which block COX) provide a protective effect for AD and suggests that COX inhibition may be involved in the attenuation of neurodegeneration. Thus, there is currently enormous interest in clinical trials of NSAIDs in AD.

The mechanism of action of NSAIDs is not entirely clear, however, it is generally believed that their effects are attributable to the competitive inhibition of COX catalytic activity and the subsequent reduction of inflammatory PG production. COX exists in two isoforms that are coded by distinct genes on different chromosomes *(22–24)*. There is an approx 50% homology between these enzymes, as well as similar catalytic activities. However, the COX-1 and -2 isoforms are physiologically distinct. For example, COX-2 is inducible in response to inflammatory signals such as cytokines and lipopolysaccharide and is downregulated by glucocorticoids. In contrast, COX-1 expression is generally constitutive. It appears that the COX-2 isoform is responsible for the mediation of inflammatory activity, whereas COX-1 maintains housekeeping functions (including gastric cytoprotection and platelet aggregation). Further, traditional NSAIDs are nonselective COX inhibitors, but the effects of these compounds on inflammation are believed to derive from the inhibition of COX-2 activity, whereas selective COX-1 inhibition is implicated in gastrointestinal, renal, and platelet toxicity *(25)*. Thus, the use of newly developed, selective COX-2 inhibitors in AD treatment may hold promise for anti-inflammatory action and significantly reduce toxicity *(25,26)*.

Major efforts are currently under way to determine whether COX inhibitors can indeed help control the destructive progression of AD. However, there is a growing number of candidate NSAIDs with widely divergent activities. In order to optimize drug selection and clinical trial design, it is vital that NSAIDs are further characterized for their actions and administered in a manner specific to each clinical phase of AD.

5. NOVEL ROLE(S) FOR NSAIDs IN AD

Several National Institute of Health (NIH)-supported trials are currently investigating whether selective or nonselective NSAID COX inhibitors delay the onset of AD and/or provide supplemental treatment for patients with confirmed AD diagnosis. Many industry-sponsored trials of selective COX-2 inhibitors are also targeting individuals with mild cognitive impairment.

Much of the research on NSAIDs in early AD supports the theory that the COX-inhibiting properties of these drugs contribute to their apparent efficacy in protecting against AD *(1)*. However, a recent study of transgenic mice and cell cultures *(27)* offers an alternative explanation. A study by Weggen et al. found that nonselective, rather than selective, COX-2 inhibitors slow the production of the fibril-forming β-amyloid $(A\beta)_{1-42}$, even in the absence of COX-2 and COX-1 gene expression. Aβ peptides $(A\beta_{1-42}, A\beta_{1-40},$ and $A\beta_{1-38}$, among others) are composite elements of amyloid plaques that are produced in excess during AD and are believed to be central to AD pathogenesis. It is supposed that the higher production of $A\beta_{1-42}$, compared to $A\beta_{1-40}$, and $A\beta_{1-38}$, throughout the progression of AD could be seminal to amyloid (Aβ) plaque formation. Thus, the data presented by Weggen et al. suggest that NSAIDs, in addition to their anti-inflammatory activities, may also function by targeting Aβ peptide production. However, it is of note that these findings were demonstrated to be COX-independent with only one highly distinct NSAID (sulindac) at very high, and perhaps toxic, concentrations *(48)*. If a mechanism for this novel explanation can be identified, a new class of NSAID compounds for the treatment of AD may emerge.

6. COX-2: A NOVEL TARGET FOR ANTIAMYLOIDOSIS STRATEGIES

In a recent study from our laboratories, we found evidence for a direct link between COX-2 expression in the brain and AD-type amyloidosis in vivo *(12)*. In this study, mice expressing neuronal human (h)COX-2 were backcrossed with the $APP_{swe}/PS1$-A246E mouse model of AD-type neuropathology. Consistent with previous studies showing a relationship between COX-2 expression and amyloidosis in AD *(10,16,28–30)*, we found that neuronal hCOX-2 expression induced a significant potentiation of amyloid burden in the brain of 24-mo-old mice *(12)*. Using mass spectrometry proteomic technology, we found that coincidental with increased amyloidosis, a significant elevation of $A\beta_{1-38}, A\beta_{1-40},$ and $A\beta_{1-42}$ in the brains of COX-2/APP_{swe}/PS1-A246E animals occurred. These data suggest a novel and direct role for COX-2 in amyloid deposition (Fig. 2) and indicate that under disease conditions, COX-2 activities may be an elemental therapeutic target. Moreover, it argues for further study of the involvement of COX in AD pathogenesis. Thus, as discussed, future studies investigating the pharmacological relevance of NSAIDs to amyloidosis and the role of COX are necessary in

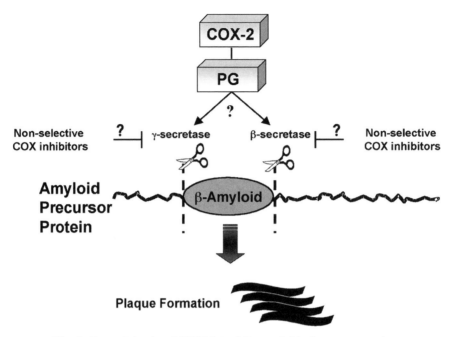

Fig. 2. Potential role of COX-2 in AD amyloid plaque generation.

order to further clarify the potential efficacy of COX-inhibiting drugs as therapeutic treatments for AD.

7. COX-2 AND CELL CYCLE ACTIVITIES: IMPLICATIONS FOR AD INFLAMMATORY NEURODEGENERATION

Recently, components of the cell division cycle have been examined for potential involvement in AD neuropathology. In support of these investigations, a study from our laboratory found that expression of the cell division cycle component CDK4 and COX-2 in hippocampal neurons were coregulated during the clinical progression of AD dementia.

The four phases that comprise the cell division cycle include G1 (growth), S (DNA synthesis and replication), G2 (growth for cell division), and M (mitosis). The main regulators of the progression of the cell cycle are the cyclin/cyclin-dependent kinase (CDK) complexes. The sequential expression/activation of these proteins not only orchestrates the transition from one phase to the next, but can also serve as a marker of different stages of the cell cycle *(31)*. The transition of resting cells from the G0 to G1 phase is controlled by the cyclin D1/CDK4,6 complexes. Activation of these complexes is triggered by mitogenic growth factors and inhibited by the CDK4,6 inhibitor protein (p)18.

An important target substrate for the cyclinD1/CDK4,6 complex is the tumor suppressor retinoblastoma protein (pRb), which is phosphorylated by activated cyclinD1/Cdk4,6. Recent evidence shows that pRB can also be phosphorylated in response to amyloid (Aβ)-mediated neurotoxicity *(32)*. Hyperphosphorylated pRb is then released from the transcription factor complex E2F and can, subsequently, activate genes required for S-phase transition. We hypothesize that COX-2, which can be induced by aggregated Aβ peptides *(9)*, may facilitate transition of resting neurons from the G0 phase into G1 phase by decreasing the expression of the CDK4,6 inhibitor p18 *(33)*. This activity may then possibly lead to an unsuccessful attempt to re-enter the cell cycle and, subsequently, influence caspase activity and neuronal death.

To explore this hypothesis, we used DNA microarray technology to screen select brain regions susceptible to AD-type neuropathology in transgenic hCOX-2/APP$_{swe}$/PS1-A246E (discussed in Section 6). We found that COX-2 overexpression in these mice resulted in decreased mRNA expression of the CDK4,6 inhibitor p18 when compared to APP$_{swe}$/PS1-A246E transgenic mice. Interestingly, this hCOX-2-mediated decrease of p18 expression was found to be reversible by treatment with the NSAID nimesulide (a preferential COX-2 inhibitor) in the feed of the mice *(33)*. This further substantiated our hypothesis that COX-2 may facilitate the re-entry of neurons into the cell cycle and proposed yet another potential therapeutic activity for NSAIDs in AD.

Currently, little is known about the role of p18 and other CDK4,6 inhibitors in the brain. However, given the obvious involvement of COX-2 in the expression of the cell division cycle components in neurons and the evidence demonstrating modulation of these responses by NSAID treatment, these studies clearly warrant further examination of cell division cycle activities in AD.

8. THE ROLE OF COX-2 IN ALZHEIMER'S DISEASE NEUROPATHOLOGY: A CELL CYCLE CONNECTION?

The search for factors responsible for the formation of NFTs and amyloid plaque pathology has yielded support for the hypothesis that the cell cycle may play an important role in AD neurodegeneration. For example, there is significant accumulation of mitogenic growth factors (e.g., epidermal growth factor) in amyloid plaques, suggesting that a trigger for re-entry of neurons into the cell cycle may be upregulated in AD. Interestingly, COX-2, whose expression is also regulated by growth factors and tumor promoters *(34)*, may represent an important link between amyloid pathology and the cell cycle.

With regard to neurofibrillary pathology, it has been reported that COX-2 expression is preferentially elevated in NFT-positive neurons in AD *(29)* and other neurodegenerative disorders *(35)*. There is also in vitro evidence that the cell division cycle molecule CDK 5 actively phosphorylates the microtuble binding protein tau and induces AD-type NFT formation *(36)*. Although destabilization of microtubules and coincidental activation of kinases that are able to phosphorylate tau are features of cell division, we hypothesize that COX-2 by promoting cell cycle activities (described in Section 7) might also participate in abnormal tau phosphorylation and further influence the clinical progression of AD. This hypothesis is consistent with evidence showing that COX inhibitors may also arrest the progression of cell division *(37)*.

9. Aβ VACCINATION IMMUNOTHERAPY IN ALZHEIMER'S DISEASE

As there is currently no cure or effective treatment for AD, the search for novel therapeutics to combat the disorder is receiving massive attention from both industry and academia. The most recent therapeutic intervention to show great promise was the Elan Corporation and American Home Products (Elan) Aβ vaccination immunotherapy. Investigators at Elan studying neuropathologic changes in a murine model of AD found that immunization with β-amyloid (Aβ) peptides could evoke an antibody response that prevented amyloidosis and ameliorated the behavioral impairment characteristic of this model *(38–41)*. Unfortunately, it was reported in January 2002 that the inflammation associated with this treatment, although tolerable in animals, exceeded expected potential and was detrimental to AD-affected humans. In March 2002, Elan indicated that a number of subjects had developed cerebral inflammation. Upon this observation, the trial was suspended and patients were immediately treated with anti-inflammatory drugs that succeeded in taming the inflammatory contraindication. However, these events have subsequently resulted in the termination of Aβ vaccination clinical trials.

The mechanism of action proposed for this therapy was determined from several studies of murine models of AD-type neuropathology. Investigators found that both passive (intraperitoneal injection of prederived Aβ antibodies) and active (ip injection of Aβ peptides) immunization with Aβ could evoke similar responses. Treatment was determined to function by activating pro-inflammatory microglial cells through immunoglobulin Fc receptor signaling, which facilitated clearance and degradation of Aβ plaques *(38,41)*—presenting for the first time a potentially beneficial role for features of inflammation in AD. These studies are undoubtedly the most significant leap in understanding the role of Aβ in AD neuropathology over the past decade. However, the induction of pro-inflammatory cascades [e.g., microglial cytokine synthesis and generation of free radicals *(42)*], is little understood, and has been implicated in neuronal injury *(43)*. Additionally, as discussed throughout this chapter, inflammatory phenomena are characteristic of the pathology of AD. Thus, there are significant concerns about the pro-inflammatory consequences of Aβ immunization (for discussion, *see* ref. *2*).

Although the current protocol for Aβ immunization will no longer be used in humans, the data generated have established the dangers of this therapy, and also suggested that precautionary procedures, such as coadministration of anti-inflammatory drugs, may not be a straightforward alternative. For example, if the induction of specific cytokines or other unknown microglial proinflammatory factors are integral to microglial scavenging of Aβ plaques, the use of anti-inflammatory drugs that interfere with microglial cytokine activity may aberrantly impact immunization-based Aβ plaque clearance. Alternatively, more selective drugs that attenuate inflammatory responses not integral to microglial function may prove to be effective.

Unfortunately, given the abrupt and unexpected failure of this therapy, we can neither verify positive (cognitive) nor deny additional negative consequences of this treatment in humans to justify further study. Further, given the irreversible nature of vaccination of the immune system for a self-protein, it remains difficult to predict whether this treatment will have additional problems in the long term. However, we believe that further and more extensive laboratory investigation of the actions of this treatment and, potentially, alternative modes for eliciting similar modulation of amyloid pathology will undoubtedly bring us closer to successfully treating AD dementia.

10. NSAID STUDIES IN ALZHEIMER'S DISEASE

In light of the multifarious nature of the data supporting the use of NSAID compounds for the treatment of AD, it is without question that these compounds will continue to be evaluated exhaustively. However, because a subset of these compounds is implicated in iatrogenic illness, toxicity remains a major roadblock for the study of anti-inflammatory dosages of traditional NSAIDs in AD.

The Alzheimer's Disease Cooperative Study (ADCS) has opted to study two NSAID-type regimens that are expected to have substantially less toxic side effects than the indomethacin and diclofenac regimens previously reported. The first is a new selective COX-2 inhibitor called rofecoxib. As discussed earlier, COX-2 may be the target of action of NSAIDs in the AD brain, and because COX-2 inhibitors appear to carry a reduced risk of serious gastrointestinal toxicity, rofecoxib is a favorable candidate for therapeutic trials. The second active drug regimen is low-dose naproxen (200 mg twice daily). Naproxen is a nonselective COX inhibitor and, therefore, carries the risk of toxicity at full dose, similar to indomethacin and diclofenac. However, if, in fact, COX-1 is an important target for NSAID action in AD, naproxen may be prove to be more effective than a selective COX-2 inhibitor because of its actions on both isoforms. It is important to note that 200 mg/d is the over-the-counter analgesic dose for this drug, which is substantially less than the typically prescribed anti-inflammatory dose and may prove insufficient for altering AD progression. Nevertheless, short-term studies suggest that this regimen is reasonably well tolerated in the elderly *(44)* and several epidemiologic studies suggest that casual use of NSAIDs obtained over-the-counter is sufficiently neuroprotective *(45)*. Further, in rodent studies *(46)*, it has been found that systemic administration of naproxen suppresses COX activity in the brain *(47)* (indicating effective brain penetration), further supporting the study of a low-dose regimen *(48)*.

In conclusion, elucidating the role of inflammatory processes has undoubtedly provided an impetus for further development of anti-inflammatory therapies for AD. However, there is a critical need to improve the diagnosis of patients in the earliest stages of the disease so that drug therapies can be administered as early as possible and optimize their efficacy.

REFERENCES

1. Aisen, P. S. (1997) Inflammation and Alzheimer's disease: mechanisms and therapeutic strategies. *Gerontology* **43,** 143–149.
2. Pasinetti, G. M., Ho, L., Pompl, P. (2002) Amyloid immunization in Alzheimer's disease: do we promote amyloid scavenging at the cost of inflammatory degeneration? *Neurobiol. Aging* **23,** 667–668.
3. Haroutunian, V., Perl, D. P., Purohit, D. P., Marin, D., Khan, K., Lantz, M., et al. (1998) Regional distribution of neuritic plaques in the nondemented elderly and subjects with very mild Alzheimer disease. *Arch. Neurol.* **55,** 1185–1191.
4. Lim, G. P., Yang, F., Chu, T., Chen, P., Beech, W., Teter, B., et al. (2000) Ibuprofen suppresses plaque pathology and inflammation in a mouse model for Alzheimer's disease. *J. Neurosci.* **20,** 5709–5714.
5. Vandenabeele, P. and Fiers, W. (1991) Is amyloidogenesis during Alzheimer's disease due to an IL-1/IL-6-mediated "acute phase response" in the brain? *Immunol. Today* **12,** 217–219.
6. Pasinetti, G. M. (1996) Inflammatory mechanisms in neurodegeneration and Alzheimer's disease: the role of the complement system. *Neurobiol. Aging* **17,** 707–716.
7. Webster, S., Bonnell, B., and Rogers, J. (1997) Charge-based binding of complement component C1q to the Alzheimer amyloid beta-peptide. *Am. J. Pathol.* **150,** 1531–1536.
8. Mukherjee, P. and Pasinetti, G. M. (2000) The role of complement anaphylatoxin C5a in neurodegeneration: implications in Alzheimer's disease. *J. Neuroimmunol.* **105,** 124–130.

9. Pasinetti, G. M. and Aisen, P. S. (1998) Cyclooxygenase-2 expression is increased in frontal cortex of Alzheimer's disease brain. *Neuroscience* **87,** 319–324.

10. Campbell, I. L., Abraham, C. R., Masliah, E., Kemper, P., Inglis, J. D., Oldstone, M. B., et al. (1993) Neurologic disease induced in transgenic mice by cerebral overexpression of interleukin 6. *Proc. Natl. Acad. Sci. USA* **90,** 10,061–10,065.

11. Oda, T., Lehrer-Graiwer, J., Finch, C. E., and Pasinetti, G. M. (1995) Complement and beta-amyloid neurotoxicity in vitro: a model for Alzheimer's disease. *Alzheimer's Res.* **1,** 29–34.

12. Xiang, Z., Ho, L., Yemul, S., Zhao, Z., Pompl, P., et al. (2002) Cyclooxygenase-2 promotes amyloid plaque deposition in a mouse model of Alzheimer's disease neuropathology. *Gene Expression* **10,** 271–278.

13. McGeer, P. L. and McGeer, E. G. (2001) Polymorphisms in inflammatory genes and the risk of Alzheimer disease. *Arch. Neurol.* **58,** 1790–1792.

14. Luterman, J. D., Haroutunian, V., Yemul, S., Ho, L., Purohit, D., Aisen, P. S., et al. (2000) Cytokine gene expression as a function of the clinical progression of Alzheimer disease dementia. *Arch. Neurol.* **57,** 1153–1160.

15. Ho, L., Luterman, J. D., Aisen, P. S., Pasinetti, G. M., Montine, T. J., and Morrow, J. D. (2000) Elevated CSF prostaglandin E2 levels in patients with probable AD. *Neurology* **55,** 323.

16. Ho, L., Purohit, D., Haroutunian, V., Luterman, J. D., Willis, F., Naslund, J., et al. (2001) Neuronal cyclooxygenase 2 expression in the hippocampal formation as a function of the clinical progression of Alzheimer disease. *Arch. Neurol.* **58,** 487–492.

17. McGeer, P. L., Schulzer, M., and McGeer, E. G. (1996) Arthritis and anti-inflammatory agents as possible protective factors for Alzheimer's disease: a review of 17 epidemiologic studies. *Neurology* **47,** 425–432.

18. In't Veld, B. A., Ruitenberg, A., Hofman, A., Launer, L. J., van Duijn, C. M., Stijnen, T., et al. (2001) Nonsteroidal antiinflammatory drugs and the risk of Alzheimer's disease. *N. Engl. J. Med.* **345,** 1515–1521.

19. Rogers, J., Kirby, L. C., Hempelman, S. R., Berry, D. L., McGeer, P. L., Kaszniak, A. W., et al. (1993) Clinical trial of indomethacin in Alzheimer's disease. *Neurology* **43,** 1609–1611.

20. Scharf, S., Mander, A., Ugoni, A., Vajda, F., and Christophidis, N. (1999) A double-blind, placebo-controlled trial of diclofenac/misoprostol in Alzheimer's disease. *Neurology* **53,** 197–201.

21. Aisen, P. S., Davis, K. L., Berg, J. D., Schafer, K., Campbell, K., Thomas, R. G., et al. (2000) A randomized controlled trial of prednisone in Alzheimer's disease. Alzheimer's Disease Cooperative Study. *Neurology* **54,** 588–593.

22. Kujubu, D. A., Fletcher, B. S., Varnum, B. C., Lim, R. W., and Herschman, H. R. (1991) TIS10, a phorbol ester tumor promoter-inducible mRNA from Swiss 3T3 cells, encodes a novel prostaglandin synthase/cyclooxygenase homologue. *J. Biol. Chem.* **266,** 12,866–12,872.

23. Cao, C., Matsumura, K., Yamagata, K., and Watanabe, Y. (1995) Induction by lipopolysaccharide of cyclooxygenase-2 mRNA in rat brain; its possible role in the febrile response. *Brain Res.* **697,** 187–196.

24. O'Banion, M. K., Winn, V. D., and Young, D. A. (1992) cDNA cloning and functional activity of a glucocorticoid-regulated inflammatory cyclooxygenase. *Proc. Natl. Acad. Sci. USA* **89,** 4888–4892.

25. Warner, T. D., Giuliano, F., Vojnovic, I., Bukasa, A., Mitchell, J. A., and Vane, J. R. (1999) Nonsteroid drug selectivities for cyclo-oxygenase-1 rather than cyclo-oxygenase-2 are associated with human gastrointestinal toxicity: a full in vitro analysis. *Proc. Natl. Acad. Sci. USA* **96,** 7563–7568.

26. Vane, J. R. and Botting, R. M. (1995) New insights into the mode of action of anti-inflammatory drugs. *Inflam. Res.* **44,** 1–10.

27. Weggen, S., Eriksen, J. L., Das, P., Sagi, S. A., Wang, R., Pietrzik, C. U., et al. (2001) A subset of NSAIDs lower amyloidogenic Abeta42 independently of cyclooxygenase activity. *Nature* **414,** 212–216.
28. Ho, L., Pieroni, C., Winger, D., Purohit, D. P., Aisen, P. S., and Pasinetti, G. M. (1999) Regional distribution of cyclooxygenase-2 in the hippocampal formation in Alzheimer's disease. *J. Neurosci. Res.* **57,** 295–303.
29. Oka, A. and Takashima, S. (1997) Induction of cyclo-oxygenase 2 in brains of patients with Down's syndrome and dementia of Alzheimer type: specific localization in affected neurones and axons. *Neuroreport* **8,** 1161–1164.
30. Yasojima, K., Schwab, C., McGeer, E. G., and McGeer, P. L. (1999) Distribution of cyclo-oxygenase-1 and cyclooxygenase-2 mRNAs and proteins in human brain and peripheral organs. *Brain Res.* **830,** 226–236.
31. Grana, X. and Reddy, E. P. (1995) Cell cycle control in mammalian cells: role of cyclins, cyclin dependent kinases (CDKs), growth suppressor genes and cyclin-dependent kinase inhibitors (CKIs). *Oncogene* **11,** 211–219.
32. Giovanni, A., Keramaris, E., Morris, E. J., Hou, S. T., O'Hare, M., Dyson, N., et al. (2000) E2F1 mediates death of B-amyloid-treated cortical neurons in a manner independent of p53 and dependent on Bax and caspase 3. *J. Biol. Chem.* **275,** 11,553–11,560.
33. Mirjany, M., Ho, L., and Pasinetti, G. M. (2002) Role of cyclooxygenase-2 in neuronal cell cycle activity and glutamate-mediated excitotoxicity. *J. Pharmacol. Exp. Ther.* **301,** 494–500.
34. Fletcher, B. S., Lim, R. W., Varnum, B. C., Kujubu, D. A., Koski, R. A., and Herschman, H. R. (1991) Structure and expression of TIS21, a primary response gene induced by growth factors and tumor promoters. *J. Biol. Chem.* **266,** 14,511–14,518.
35. Oka, A., Takashima, S. (1997) Induction of cyclo-oxygenase 2 in brains of patients with Down's syndrome and dementia of Alzheimer's type: specific localization in affected neurons and axons. *Neuroreport* **8,** 1161–1164.
36. Baumann, K., Mandelkow, E. M., Biernat, J., Piwnica-Worms, H., and Mandelkow, E. (1993) Abnormal Alzheimer-like phosphorylation of tau-protein by cyclin-dependent kinases cdk2 and cdk5. *FEBS Lett.* **336,** 417–424.
37. Shiff, S. J., Koutsos, M. I., Qiao, L., and Rigas, B. (1996) Nonsteroidal antiinflammatory drugs inhibit the proliferation of colon adenocarcinoma cells: effects on cell cycle and apoptosis. *Exp. Cell Res.* **222,** 179–188.
38. Bard, F., Cannon, C., Barbour, R., Burke, R. L., Games, D., Grajeda, H., et al. (2000) Peripherally administered antibodies against amyloid beta-peptide enter the central nervous system and reduce pathology in a mouse model of Alzheimer disease. *Nat. Med.* **6,** 916–919.
39. Janus, C., Pearson, J., McLaurin, J., Mathews, P. M., Jiang, Y., Schmidt, S. D., et al. (2000) A beta peptide immunization reduces behavioural impairment and plaques in a model of Alzheimer's disease. *Nature* **408,** 979–982.
40. Morgan, D., Diamond, D. M., Gottschall, P. E., Ugen, K. E., Dickey, C., Hardy, J., et al. (2000) A beta peptide vaccination prevents memory loss in an animal model of Alzheimer's disease. *Nature* **408,** 982–985.
41. Schenk, D., Barbour, R., Dunn, W., Gordon, G., Grajeda, H., Guido, T., et al. (1999) Immunization with amyloid-beta attenuates Alzheimer-disease-like pathology in the PDAPP mouse. *Nature* **400,** 173–177.
42. Levi, G., Minghetti, L., and Aloisi, F. (1998) Regulation of prostanoid synthesis in microglial cells and effects of prostaglandin E2 on microglial functions. *Biochimie* **80,** 899–904.
43. Ulvestad, E., Williams, K., Matre, R., Nyland, H., Olivier, A., and Antel, J. (1994) Fc receptors for IgG on cultured human microglia mediate cytotoxicity and phagocytosis of antibody-coated targets. *J. Neuropathol. Exp. Neurol.* **53,** 27–36.

44. DeArmond, B., Francisco, C. A., Lin, J. S., Huang, F. Y., Halladay, S., Bartziek, R. D., et al. (1995) Safety profile of over-the-counter naproxen sodium. *Clin. Ther.* **17,** 587–601.
45. Geczy, M., Peltier, L., and Wolbach, R. (1987) Naproxen tolerability in the elderly: a summary report. *J. Rheumatol.* **14,** 348–354.
46. Abdel-Halim, M. S., Sjoquist, B., and Anggard, E. (1978) Inhibition of prostaglandin synthesis in rat brain. *Acta Pharmacol. Toxicol. (Copenh.)* **43,** 266–272.
47. Ferrari, R. A., Ward, S. J., Zobre, C. M., Van Liew, D. K., Perrone, M. H,, Connell, M. J., et al. (1990) Estimation of the in vivo effect of cyclooxygenase inhibitors on prostaglandin E2 levels in mouse brain. *Eur. J. Pharmacol.* **179,** 25–34.
48. Pasinetti, G. M. and Pompl, P. (2002) Inflammation and Alzheimer's disease: are we well-ADAPTed? *Lancet Neurology* **1,** 403–404.
49. Pasinetti, G. M., Ho, L., and Pompl, P. (2002) AN1792 vaccination immunotherapy in Alzheimer's disease: the case of a therapy before its time. *Neurobiol. Aging* **23,** 685–686.

IV
Multiple Sclerosis

18

Experimental Autoimmune Encephalomyelitis

Hans-Peter Hartung and Bernd C. Kieseier

1. INTRODUCTION

Much of our present understanding of the pathogenesis in multiple sclerosis (MS) and other disorders of the central nervous system (CNS) is based on information gained through studying experimental animal models. Whereas the investigation of human tissue samples—obtained by biopsies or postmortem—can provide only a focused view at one given time-point in the disease process of which the precise pathobiological history cannot be ascertained, such animal models offer an excellent tool to study the pathogenesis and evolution of the entire course of the disease.

Experimental autoimmune (allergic) encephalomyelitis (EAE) is one of the most intensively studied models in immunology and has contributed profoundly to our understanding of autoimmunity in general and helped to clarify basic aspects of neuroimmunology, such as immune surveillance in the nervous system, physiological regulation of inflammation, and effector mechanisms of neural damage.

Histomorphologically, EAE is characterized by inflammation, demyelination, and axonal damage, and therefore is used widely as an animal model for MS *(1)*.

2. HISTORICAL ASPECTS OF EXPERIMENTAL AUTOIMMUNE ENCEPHALOMYELITIS

Experimental autoimmune encephalomyelitis was first described in 1933 by Rivers and co-workers, who induced inflammatory demyelinating lesions after active immunization with brain tissue in monkeys and rabbits *(2)*. In the following 50 yr basic biological concepts and detailed pathological analysis of this model disease were established and this disorder was assigned to the class of cell-mediated immune reactions, primarily driven by CD4+ Th1 lymphocytes *(3)*. During this time, it also became obvious that macrophages found within the lesion do not just represent intrinsic CNS phagocytes or microglia cells responding to tissue destruction by cleaning up debris; in fact, macrophages were identified as primary cells of tissue destruction, an observation constituting a fundamental shift in the understanding of cellular immunobiology of the central nervous system at that time.

In the mid- and late 1950s, the discovery of single homogenous proteins of myelin as effective encephalitogens was initiated *(4,5)*. In the following 40 yr, various encephalitogenic proteins have been identified. Moreover, the in vitro generation of T-cell lines

From: *Neuroinflammation, 2nd Edition: Mechanisms and Management*
Edited by: P. L. Wood © Humana Press Inc., Totowa, NJ

against individual proteins and their peptides as specific epitopes in the induction of the disease paved the way to the identification of putative autoantigens. It also became obvious that the distribution and character of lesions vary with different encephalitogens (6).

Genetic studies in EAE started in the early 1970s and identified major histocompatibility complex (MHC) class II as a major susceptibility gene locus. However, the genetic determinants of immune regulation still remain elusive at present.

The 1980s and 1990s were dominated by detailed analysis of cell subtypes and immune mediators, such as cytokines, complement components, or antibodies, in the pathogenesis of the disorder (7). On the one hand, these studies revealed the complexity of immune mechanisms relevant to the course of the disease in greater detail; on the other hand, the increased understanding of cell-specific responses during the disease course and the discovery of regulatory capacities of Th2 type cytokines opened, at least theoretically, new avenues for specific therapeutic intervention (8,9).

During the last 15 yr the availability of transgenic and gene knockout animals facilitated the possibility to obtain deeper insights into the specific immunopathogenic functions of distinct immune mediators. However, the initial euphoria that deletion or overexpression of a specific gene of interest could answer many remaining questions on functional aspects in the pathogenesis of inflammatory demyelination has already given way to a more critical view of such animal models.

3. ACUTE EXPERIMENTAL AUTOIMMUNE ENCEPHALOMYELITIS

Experimental autoimmune encephalomyelitis can be reliably elicited in a number of different species by immunization with either CNS tissue or CNS myelin in complete Freund's adjuvant (10). This type of model is termed actively induced or active EAE because an active immune response in the recipient animal is required to cause the disease.

Alternatively, EAE can be induced by the transfer of large numbers of primed T-lymphocytes from actively immunized animals to the untreated recipient (11). This transfer of lymphoid cells from an immune donor to a normal syngeneic recipient is termed "adoptive transfer." Such cell transfers must take place between genetically identical donors and recipients, such as members of the same inbred strain of rodents, so that the donor lymphocytes are not rejected by the recipient. In the context of inflammatory demyelination of the CNS, this model disease is termed "adoptive transfer EAE (AT-EAE)" and it provides conclusive proof of the pivotal role of T-lymphocytes in this group of disorders (see Fig. 1).

The striking advantage in studying animal models is the synchronous course of a disease that can be examined at various time-points. This advantage becomes even more valuable in the evaluation of new therapeutic strategies (12). Whereas in clinical trials hundreds of patients need to be included in order to provide statistically significant effects, experimental therapies can be evaluated in the inbread rodent models with only some 6–12 animals per treatment group, depending on the antigen and strain of rodent used, the mode of immunization, and the respective treatment (13).

The full neurological picture of EAE requires formation of round-cell infiltrates that are composed of autoreactive activated T-cells, nonspecifically recruited CD4+ T-lymphocytes, and activated macrophages (14). Depletion of macrophages in susceptible rodents interferes with the induction of EAE by active immunization of T-cell transfer

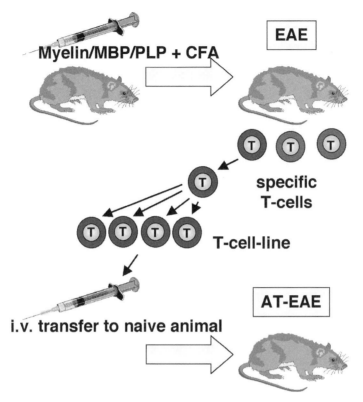

Fig. 1. Experimental autoimmune encephalomyelitis can actively induced by immunization with CNS myelin or myelin components, such as myelin basic protein (MBP), myelin oligodendrocyte glycoprotein (MOG), proteolipid protein (PLP), and complete Freund's adjuvant (CFA). At the peak of clinical disease lymphoid cells can be obtained and, after generation of encephalitogenic T-cell lines and restimulation, induce the disease in a normal syngeneic recipient, a model termed AT-EAE.

(15). Furthermore, EAE variants with a fulminant course are characterized by a high content of activated macrophages in the CNS infiltrates, whereas those with mild disease have infiltrates with remarkably few activated macrophages *(6)*.

3.1. EAE in Lewis Rats

Lewis rat EAE is a monophasic disease. After immunization, weight loss represents one of the earliest markers of the disease, followed by neurological deficits, which can be quantitated on predefined scales. In AT-EAE, maximum disease severity is normally found 3–4 d after the intravenous injection of encephalitogenic T-lymphocytes. In actively induced EAE disease onset usually is detectable around d 10, with maximum clinical severity around d 12–13 after myelin injection. The clinical course of the disease runs more synchronously in the adoptive-transfer model compared to the actively induced disease, thus even lower numbers of animals are required in AT-EAE.

As indicated earlier, clinical disease severity correlates with the number of infiltrating mononuclear cells in the spinal cord and brainstem. During the recovery phase of the disease increased numbers of apoptotic T-lymphocytes are detectable, a mechanism

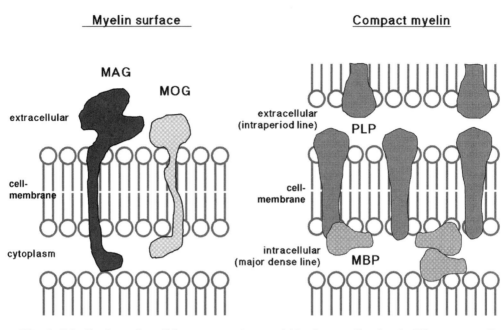

Fig. 2. Distribution of candidate auto-antigens within the myelin-sheath. Whereas myelin associated glycoprotein (MAG) and myelin-oligodendrocyte glycoprotein (MOG) are expressed on the surface of the myelin sheaths, other myelin proteins, such as proteolipid protein (PLP) and myelin basic protein (MBP) are found within compact myelin.

apparently important in containing the immune reaction within the inflamed CNS (reviewed in ref. *16*). The limitation of Lewis rat EAE as a model for the pathogenesis of MS is (1) the lack of spontaneous relapses and (2) the absence of primarily demyelinating lesions.

The monophasic course of this model disease can be overcome by immunopharmacological modulation. For example, by application of low-dose cyclosporine A, relapses of active EAE in Lewis rats can be achieved. However, such additional immunomodulation makes it difficult to evaluate specific effects of therapeutic strategies tested in this model. Animal models with chronic relapsing EAE have been established in mice and represent an alternative to relapsing EAE in Lewis rats at present.

In contrast to MS, the histopathology of myelin basic protein (MBP)-induced EAE in Lewis rats, either by the protein itself or by specific T-cell lines, is characterized by inflammation and axonal degeneration in the spinal cord. Demyelination, however, a characteristic finding of MS pathology, is only observable when cotransfer of antibodies specific for myelin-oligodendrocyte glycoprotein (MOG), together with encephalitogenic T-cells, is performed *(17)*. This model underlines the important role of antibodies in the pathogenesis of inflammatory demyelination (*see* Fig. 2).

3.2. Encephalitogenic T-Lymphocytes

Encephalitogenic T-cell lines can be obtained from immune organs or CNS infiltrates, as described in previously established protocols *(18)*. Encephalitogenic T-cell lines were commonly assumed to be CD4[+] Th cells and, as such, would recognize target autoantigens in the context of MHC class II proteins. However, in recent reports, the encephalitogenic potential of myelin-specific CD8[+] T-lymphocytes was demonstrated. With these cells, severe CNS autoimmunity was inducible in mice, indicating that CD8[+] T-cells might also function as effector cells in the pathogenesis of inflammatory demyelination of the CNS *(19,20)*. Moreover, neurons are susceptible to perforin-mediated T-cell cytotoxicity, and CD8[+] T-lymphocytes can cause local transsection of axons, a major disabling factor in MS *(21,22)*. On the other hand, in previous studies, immunization against MBP in the absence of CD8[+] T-cells resulted in an enhanced encephalitogenic response in mouse EAE *(23)*. Also, transgenic mice lacking CD8[+] T-lymphocytes suffer from increased severity of relapse, although the disease course appeared to be milder *(24)*. Thus, the precise role of CD8[+] T-lymphocytes in the pathogenesis of inflammatory CNS demyelination is not elaborated; however, it becomes obvious that the immunologically relevant T-cell component in EAE and MS is not only CD4[+].

The range of disease-associated T-cell receptor (TCR) variable regions used by encephalitogenic T-lymphocytes has been found to be limited in different rodent strains. This finding raised hopes of developing TCR-specific immunotherapies. However, during recent years, it became obvious that TCR usage is much more heterogeneous than originally suspected *(25)*; thus, the concept of targeting specific TCRs has been abandoned by most research groups.

Encephalitogenic T-cells are able to transfer EAE to naive syngeneic recipient animals *(11)*. However, these myelin-specific T-lymphocytes need to be activated before transfer, because the same cells are harmless in the resting state. On the other hand, encephalitogenic T-cells can also be isolated from completely naive, nonimmunized animals (and humans). It has been demonstrated that MBP-specific T-lymphocytes from naive Lewis rats exhibit the same functional and structural properties as T-cell lines from preimmunized animals *(26)*. These observations strongly support the concept that potentially autoaggressive T-cell clones are part of the physiological immune repertoire.

Taken together, these observations underline the crucial role of T-lymphocytes in the pathogenesis of inflammatory demyelinating disorders of the CNS *(27,28)*.

3.3. The Relevance of Different Antigens

The present literature suggests that most brain-derived proteins are capable to elicit a T-cell-mediated immune response and, as such, could function as targets for encephalitogenic T-lymphocytes. It has been demonstrated that T-cells respond against a large variety of myelin antigens, such as MBP *(4)*, proteolipid protein (PLP) *(5)*, MOG *(29)*, and myelin-associated glycoprotein (MAG) *(6)*. However, also T-cell responses against non-CNS-specific antigens, such as S100 *(30)* or the astrocytic glial fibrillary acidic protein (GFAP) *(6)*, have been shown to be effective in inducing an inflammatory reaction within the CNS (*see* Table 1 and Fig. 2). The histopathology of these various diseases induced by T-lymphocytes targeting different antigens is surprisingly similar. They are all characterized by perivascular infiltrates, consisting of T-cells and activated macrophages. Differences can be seen in the distribution pattern within the central and peripheral nervous system (*see* Table 2). Even within the CNS, the topography apparently depends on the availability and distribution of the antigen *(31)*.

Table 1
Candidate Autoantigens

CNS myelin protein	Percent of total myelin protein
Proteolipid protein (PLP)	50
Myelin basic protein (MBP)	30
Myelin-associated glycoprotein (MAG)	1
Myelin-oligodendrocyte glycoprotein (MOG)	1
Nonmyelin-specific antigens	
S100	
Astrocytic glial fibrillary acidic protein (GFAP)	

Table 2
Characteristics of Antigen-Induced EAE

Animal model	Similarities to MS	Differences to MS
Lewis rat		
Active EAE (CNS Myelin, MBP, MOG, PLP)	T-Cell inflammation, minor Ab reponse	Monophasic, axonal damage, secondary demyelination
AT-EAE (MBP, S100)	T-Cell inflammation	Axonal damage
Active EAE + Cotransfer Anti-MOG-AbS	T-Cell inflammation, demyelination	Only transient demyelination
Congenic Lewis (DA, BN)		
Active EAE (MOG)	Relapsing–remitting course, similar histopathology	Relapses increase in severity
Murine EAE (SJL, PL/J, C57/BI6)		
Active EAE (MBP, MOG, PLP)	Relapsing–remitting and chronic course, demyelination, axonal damage	Animals severly affected from non-specific systemic disease

Note: Ab = antibody; DA = dark agouti; BN = Brown Norway.
Source: Adapted from ref. *13.*

Interestingly, T-cell responses mounted against compact myelin proteins, such as MBP or PLP, precipitate inflammatory lesions primarily in areas of thick myelin (e.g., the spinal cord or medulla oblongata). In contrast, antigens localized to the oligodendrocyte membrane, such as MOG or MAG, elicit predominantly T-cell responses in areas of thin myelin sheaths (e.g., the forebrain or the cerebellum). T-cells directed against S100 precipitate lesions in the cerebral cortex, a localization only rarely seen after transfer of myelin-specific T-cells (*see* Table 3).

Table 3
Distribution of Inflammatory Reaction Within the Nervous System in AT-EAE

Topography	Antigen				
	MBP	MOG	MAG	S100	GFAP
CO	+/–	+/–	+/–	++	–
CSO	+/–	+	+	++	–
ON	+/–	++		+/–	–
MES	+/–	+/–	+/–	++	+
CWM	+	++	+	+/–	+/–
MO	++	+	+/–	++	+
SC	+++	+	+/–	++	+
PNS	+	–	+/–	++	+/–
Retina	–	–	–	+	–
Uvea	–	–	–	+	–

Note: CO = cerebral cortex; CSO = centrum semiovale; ON = optic nerve; MES = mesencephalon; CWM = cerebellar white matter; MO = medulla oblongata; SC = spinal cord; PNS = peripheral nervous system.
Source: Adapted from ref. *23*.

4. CHRONIC EXPERIMENTAL AUTOIMMUNE ENCEPHALOMYELITIS

Whereas the immunopathogenesis of acute monophasic EAE is well characterized, many more questions remain regarding factors that operate in the induction of relapses or chronic disease. Most animal species and strains are resistant to the reinduction of EAE. However, certain strains, such as SJL-J or PJ mice, react with new inflammatory episodes after each antigenic challenge *(32)*. Moreover, after a single transfer of MBP-reactive T-lymphocytes, these animals develop a spontaneous chronic relapsing disease. In this transfer model, the encephalitogenic reaction shifts from the originally transferred MBP-reactive T-cells to an immune response against other cryptic MBP determinants or even PLP *(33,34)*. The precise mechanisms underlying this shift in antigenicity still remain elusive.

In active EAE, chronic disease can be induced in a large variety of different animal species. The persistence of the antigen within the CNS is important in maintaining chronicity. This can be achieved by initial sensitization only or repeated antigen challenge during the course of the disease. The precise immunomechanism resulting in relapse and chronicity remain speculative at present. Whether changes in the local cytokine pattern might result in active disease or remission in chronic EAE in analogy to acute EAE *(35)* still needs to be assessed.

5. OTHER EAE MODELS

5.1. Virus-Induced Inflammatory Demyelination

A large number of different viruses have been used in animals to induce inflammatory demyelination of the CNS. However, their pathogenetic complexity renders them problematic as models for MS. In order to produce an inflammatory and demyelinating lesion, the virus has to be neurotropic, persist in the CNS, and elicit a pathogenic antiviral immune response. In susceptible animals, such as SJL or DBA/1 mice, persistent infection with Theiler's murine encephalomyelitis viruses (TMEV) leads to chronic progressive, immune-mediated demyelination of the CNS. In this model, disease induction of autoimmunity through epitope spreading has recently been demonstrated *(36)*. The pathomorphology in TMEV infection and MS is similar, although chronic demyelination in TMEV is induced in the absence of antimyelin responses by B-cells. This model has been employed to study therapeutic strategies in remyelination *(37)*.

5.2. Experimental Autoimmune Encephalomyelitis in Primates

To mimic more closely human MS and to overcome the species barrier between rodents and men, nonhuman primate models were developed. The fundamental principles obtained in rodent EAE could be confirmed in the primate model and histopathological analysis of lesions revealed strong similarities between primate EAE and human MS *(38)*. As such, this model represents an excellent tool to explore novel therapeutic strategies for inflammatory demyelination of the CNS.

5.3. Gene Control in EAE

Experimental autoimmune encephalomyelitis does not only offer insights into the pathomechanisms of inflammatory demyelination, it also might help to define specific genes regulating pathogenetic pathways of autoimmune neuroinflammation. In the animal model, genetic heterogeneity can be minimized using inbred strains. To study MHC and non-MHC regulatory mechanisms congenic inbred strains in which either the non-MHC background genome is kept constant and the MHC haplotypes are varied or vice versa, are used *(39)*. Studies in these models revealed that various aspects of inflammatory responses are regulated independently of non-MHC genes and that the proneness to mount inflammatory responses in the CNS only partially overlaps with the susceptibility for EAE *(39)*.

Another approach to identify alterations in gene expression profiles in EAE is the application of the microarray technique. Screening of inflamed CNS tissue with oligonucleotide microarrays revealed 213 out of around 11,000 different genes to be regulated differentially and defined at least 51 genes with chromosomal location as putative candidate genes for susceptibility to EAE *(40)*. Such studies can be used to identify potential new candidates in the pathogenesis of MS that clearly need to be validated.

6. EAE IN TRANSGENIC ANIMALS

Progress in molecular biology generated a large number of transgenic mouse models in which specific genes are overexpressed or deleted (knockout). These models made it possible to study the physiological and pathogenetic role of distinct immune molecules in vivo. Especially proinflammatory cytokines, such as tumor necrosis factor (TNF)-α, lymphotoxin, and interferon (IFN)-γ have recently been the focus of several extensive studies *(41)*.

However, these studies generated controversial results. For example, it could be demonstrated that selective overexpression of TNF-α in the CNS can lead to a spontaneous inflammatory demyelinating disease *(42)* or, as shown in a different model, can augment, in the absence of spontaneous CNS pathology, the process of inflammatory demyelination after induction of EAE *(43)*. In contrast, deletion of the TNF gene did not alter the development of EAE *(44,45)*. Paradoxically, other groups could demonstrate that upon immunization with MOG, TNF-deficient mice develop severe inflammatory demyelination of the CNS. Interestingly, treatment with TNF in both TNF-deficient and TNF-overexpressing mice reduced disease severity, pointing to potential anti-inflammatory properties of this cytokine *(46)*. Although TNF-α, as well as other molecules, has been incriminated as an important effector molecule in the pathogenesis of inflammatory demyelination, studies in transgenic animals clearly demonstrate that such molecules are not exclusively essential for initiating the lesions of EAE. Several possible explanations can be offered for these apparent discrepancies. First, the effect of the genetic manipulation on the development and cellular function of the animal is not always completely understood. Thus, a close evaluation of the animals over their entire life-span is necessary to exclude developmental changes that might alter cellular responsiveness in these organisms. Second, the immune system is highly redundant, thus gene-targeted deletion of functional molecules consequently induces compensatory upregulation of molecules with similar functions and properties. The role of such compensatory mechanisms has to be assessed to analyze which effector mechanisms act in parallel during the evolution of inflammatory demyelination. This is of critical importance because any therapeutic intervention targeting single-effector pathways might have only limited efficacy if such subsidiary mechanisms are at work. Conditional knockout models might be helpful to overcome this problem and to increase our current understanding *(47)*.

7. EAE AS A MODEL FOR DESIGNING NOVEL THERAPIES

Experimental models not only provide deeper insights into the relevant pathomechanisms involved in the disease (*see* Fig. 3), but they also represent tools to evaluate the safety and efficacy of novel therapeutic strategies *(12,48)*.

Acute EAE is an excellent model to study the effect of anti-inflammatory strategies, because the clinical course of the disease as well as type and severity of the inflammatory reaction can be clearly documented.

However, such approaches do not address all pathogenetically relevant aspects of MS. Progression of the disease is not only the result of an acute inflammatory reaction. As such, acute EAE provides only limited information if a long-term treatment for MS needs to be evaluated.

Chronic EAE, in contrast, follows a clinical course more similar to human MS. Course and outcome are, however, rather unpredictable, and hence large numbers of animals are required to reliably assess clinical efficacy of a novel therapeutic approach.

Finally, based on experience during recent years, caution is warranted when translating results obtained from experimental studies in EAE to MS. EAE-based therapeutic strategies have been found ineffective or detrimental in clinical trials in MS. Examples are therapeutic approaches with soluble TNF-receptor, IFN-γ in MS, or oral tolerization with myelin.

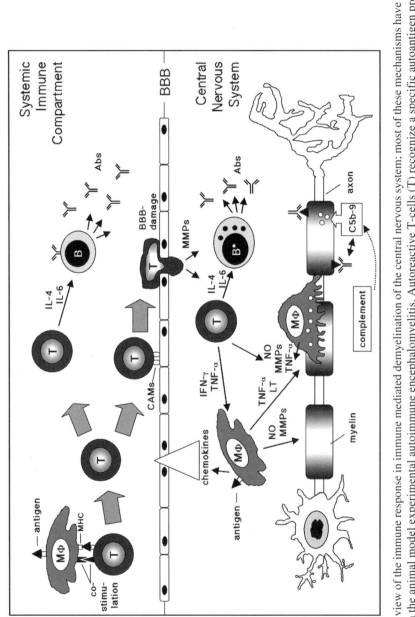

Fig. 3. Synoptic view of the immune response in immune mediated demyelination of the central nervous system; most of these mechanisms have been identified or clarified in the animal model experimental autoimmune encephalomyelitis. Autoreactive T-cells (T) recognize a specific autoantigen presented by MHC class II molecules and the simultaneous delivery of costimulatory signals (CD28, B7-1) on the cell surface of antigen-presenting cells (APC), such as macrophages (Mφ), in the systemic immune compartment. Activated T-lymphocytes can cross the blood brain barrier (BBB) in order to enter the CNS. The mechanisms of transendothelial migration is mediated by the complex interplay of cellular adhesion molecules (CAMs), chemokines, and matrix metalloproteinases (MMPs). Within the CNS T-cells activate microglia cells/Mφ to enhanced phagocytic activity, production of cytokines, such as IFN-γ, TNF-α, various interleukins (IL), and the release of toxic mediators, such as nitric oxide (NO), propagating demyelination, and axonal loss. Autoantibodies (Abs) crossing the BBB or locally produced by B-cells or mast cells (B*) contribute to this process. Autoantigens activate the complement cascade resulting in the formation of the membrane-attack complex (C5b-9) and subsequent lysis of target structures.

REFERENCES

1. Lassmann, H., Zimprich, F., Rössler, K., and Vass, K. (1991) Inflammation in the nervous system. Basic mechanisms and immunological concepts. *Rev. Neurol. (Paris)* **147**, 763–781.
2. Rivers, T. M., Sprunt, D. H., and Berry, G. P. (1933) Observations on attempts to produce acute disseminated encephalomyelitis in monkeys. *J. Exp. Med.* **58**, 39–53.
3. Raine, C. S. (1994) The Dale E. McFarlin Memorial lecture: the immunology of the multiple sclerosis lesion. *Ann. Neurol.* **36**, S61–S72.
4. Kies, M. W., Murphy, J. B., and Alvord, E. C. (1960) Fractionation of guinea pig brain proteins with encephalitogenic activity. *Fed. Proc.* **19**, 207.
5. Waksman, B. H., Porter, H., Lees, M. B., and Adams, R. D. (1954) A study of the chemical nature and components of bovine white matter effective in producing allergic encephalomyelitis in the rabbit. *J. Exp. Med.* **100**, 451–471.
6. Berger, T., Weerth, S., Kojima, K., Linington, C., Wekerle, H., and Lasssmann, H. (1997) Experimental autoimmune encephalomyelitis: the antigen specificity of T-lymphocytes determines the topography of lesions in the central and peripheral nervous system. *Lab. Invest.* **76**, 355–364.
7. Wekerle, H. (1993) Experimental autoimmune encephalomyelitis as a model of immune-mediated CNS disease. *Curr. Opin. Neurobiol.* **3**, 779–784.
8. Hartung, H.-P., Jung, S., Stoll, G., Zielaseck, J., Schmidt, B., Archelos, J. J., et al. (1992) Inflammatory mediators in demyelinating disorders of the CNS and PNS. *J. Neuroimmunol.* **40**, 197–210.
9. Hohlfeld, R., Meinl, E., Weber, F., Zipp, F., Schmidt, S., Sotgiu, S., et al. (1995) The role of autoimmune T lymphocytes in the pathogenesis of multiple sclerosis. *Neurology* **45**, S33–S38.
10. Wekerle, H., Kojima, K., Lannes-Vieira, J., Lassmann, H., and Linington, C. (1994) Animal models. *Ann. Neurol.* **36**, S47–S53.
11. Paterson, P. Y. (1960) Transfer of allergic encephalomyelitis in rats by means of lymph node cells. *J. Exp. Med.* **111**, 205–208.
12. Hohlfeld, R. (1997) Biotechnological agents for the immunotherapy of multiple sclerosis. Principles, problems and perspectives. *Brain* **120**, 865–916.
13. Gold, R., Hartung, H.-P., and Toyka, K. V. (2000) Animal models for autoimmune demyelinating disorders of the nervous system. *Mol. Med. Today* **6**, 88–91.
14. Hartung, H.-P. and Rieckmann, P. (1997) Pathogenesis of immune-mediated demyelination in the CNS. *J. Neural Transm.* **50(Suppl.)**, 173–181.
15. Huitinga, I., Van Rooijen, N., de Groot, C. J. A., Uitdehaag, B. M. J., and Dijkstra, C. D. (1990) Suppression of experimental allergic encephalomyelitis in Lewis rats after elimination of macrophages. *J. Exp. Med.* **172**, 1025–1033.
16. Gold, R., Hartung, H. P., and Lassmann, H. (1997) T-cell apoptosis in autoimmune diseases: termination of inflammation in the nervous system and other sites with specialized immune-defense mechanisms. *Trends Neurosci.* **20**, 399–404.
17. Storch, M. K., Stefferl, A., Brehm, U., Weissert, R., Wallström, E., Kerschensteiner, M., et al. (1998) Autoimmunity to myelin oligodendrocyte glycoprotein (MOG) in rats mimics the spectrum of multiple sclerosis pathology. *Brain Pathol.* **8**, 681–694.
18. Ben-Nun, A., Wekerle, H., and Cohen, I. R. (1981) The rapid isolation of clonable antigen-specific T lymphocyte lines capable of mediating autoimmune encephalomyelitis. *Eur. J. Immunol.* **11**, 195–199.
19. Sun, D., Whitaker, J. N., Huang, Z., Liu, D., Coleclough, C., Wekerle, H., et al. (2001) Myelin antigen-specific CD8+ T cells are encephalitogenic and produce severe disease in C57BL/6 mice. *J. Immunol.* **166**, 7579–7587.
20. Huseby, E. S., Liggitt, D., Brabb, T., Schnabel, B., Ohlen, C., and Goverman, J. (2001) A pathogenic role for myelin-specific CD8(+) T cells in a model for multiple sclerosis. *J. Exp. Med.* **194**, 669–676.

21. Medana, I., Li, Z., Flugel, A., Tschopp, J., Wekerle, H., and Neumann, H. (2001) Fas ligand (CD95L) protects neurons against perforin-mediated T lymphocyte cytotoxicity. *J. Immunol.* **167,** 674–681.

22. Medana, I., Martinic, M. A., Wekerle, H., and Neumann, H. (2001) Transection of major histocompatibility complex class I-induced neurites by cytotoxic T lymphocytes. *Am. J. Pathol.* **159,** 809–815.

23. Jiang, H., Zhang, S.-L., and Pernis, B. (1992) Role of CD8+ T cells in murine experimental allergic encephalomyelitis. *Science* **256,** 1213–1215.

24. Koh, D.-R., Fung-Leung, W.-P., Ho, A., Gray, D., Acha-Orbea, H., and Mak, T. W. (1992) Less mortality but more relapses in experimental allergic encephalomyelitis in CD8–/– mice. *Science* **256,** 1210–1213.

25. Hafler, D., Saadeh, M. G., Kuchroo, V. K., Milford, E., and Steinman, L. (1996) TCR usage in human and experimental demyelinating disease. *Immunol. Today* **17,** 152–159.

26. Schlüsener, H. and Wekerle, H. (1985) Autoaggressive T lymphocyte lines recognize the encephalitogenic region of myelin basic protein; in vitro selection from unprimed T cell populations. *J. Immunol.* **157,** 5249–5253.

27. Martino, G. and Hartung, H.-P. (1999) Immunopathogenesis of multiple sclerosis: the role of T cells. *Curr. Opin. Neurol.* **12,** 309–321.

28. Steinman, L. (2001) Myelin-specific CD8 T cells in the pathogenesis of experimental allergic encephalitis and multiple sclerosis. *J. Exp. Med.* **194,** 669–676.

29. Linington, C., Berger, T., Perry, L., Weerth, S., Hinze-Selch, D., Zhang, Y., et al. (1993) T Cells specific for the myelin oligodendrocyte glycoprotein (MOG) mediate an unusual autoimmune inflammatory response in the central nervous system. *Europ. J. Immunol.* **23,** 1364–1372.

30. Kojima, K., Berger, T., Lassmann, H., Hinze-Selch, D., Zhang, Y., Gehrmann, J., et al. (1994) Experimental autoimmune panencephalitis and uveoretinitis transferred to Lewis rat by T-lymphocytes specific for the S100b molecule, a calcium binding protein of astroglia. *J. Exp. Med.* **180,** 817–829.

31. Lassmann, H. and Wekerle, H. (1998) Experimental models of multiple sclerosis, in *McAlpine's Multiple Sclerosis* (Compston, A., Ebers, G., Lassmann, H., McDonald, I., Matthews, B., and Wekerle, H., eds.), London, Churchill Livingston.

32. Kozlowski, P. B., Schuller-Levis, G. B., and Wisniewski, H. M. (1987) Induction of synchronized relapses in SJL/J mice with chronic relapsing experimental allergic encephalomyelitis. *Acta Neuropathol.* **74,** 163–168.

33. Lehmann, P. V., Forsthuber, T., Miller, A., and Sercarz, E. E. (1992) Spreading of T-cell autoimmunity to cryptic determinants of an antigen. *Nature* **358,** 155–157.

34. Cross, A. H., Tuohy, V. K., and Raine, C. S. (1993) Development of reactivity to new myelin antigens during chronic relapsing autoimmune demyelination. *Cell Immunol.* **146,** 261–269.

35. Kennedy, M. K., Torrance, D. S., Picha, K. S., and Mohler, K. M. (1992) Analysis of cytokine mRNA expression in the central nervous system of mice with experimental autoimmune encephalomyelitis reveals that IL-10 mRNA expression correlates with recovery. *J. Immunol.* **149,** 2946–2505.

36. Miller, S. D., Vanderlught, C. L., Smith Begolka, W., Pao, W., Yauch, R. L., Neville, K. L., et al. (1997) Persistent infection with Theiler's virus leads to CNS autoimmunity via epitope spreading. *Nat. Med.* **10,** 1133–1136.

37. Miller, D. J., Asakura, K., and Rodriguez, M. (1996) Central nervous system remyelination. Clinical application of basic neuroscience principles. *Brain. Pathol.* **6,** 331–344.

38. 't Hart, B. A., van Meurs, M., Brok, H. P., Massacesi, L., Bauer, J., Boon, L., et al. (2000) A new primate model for multiple sclerosis in the common marmoset. *Immunol. Today* **21,** 290–297.

39. Olsson, T., Dahlman, I., Wallstrom, E., Weissert, R., and Piehl, F. (2000) Genetics of rat neuroinflammation. *J. Neuroimmunol.* **107,** 191–200.

40. Ibrahim, S. M., Mix, E., Böttcher, T., Koczan, D., Gold, R., Rolfs, A., et al. (2001) Gene expression profiling of the nervous system in murine experimental autoimmune encephalomyelitis. *Brain* **124,** 1927–1938.
41. Kieseier, B. C., Storch, M. K., Archelos, J. J., Martino, G., and Hartung, H.-P. (1999) Effector pathways in immune mediated central nervous system demyelination. *Curr. Opin. Neurol.* **12,** 323–336.
42. Probert, L., Akassoglu, K., Pasparakis, M., Kontogeorgos, G., and Kollias, G. (1995) Spontaneous inflammatory demyelintaing disease in transgenic mice showing central nervous system-specific expression of tumor necrosis factor α. *Proc. Natl. Acad. Sci. USA* **29,** 11,294–11,298.
43. Taupin, V., Renno, T., Bourbonnierre, L., Peterson, A. C., Rodriguez, M., and Owens, T. (1997) Increased severity of experimental autoimmune encephalomyelitis, chronic macrophage/microglia reactivity and demyelination in transgenic mice producing tumor necrosis factor-alpha in the central nervous system. *Eur. J. Immunol.* **27,** 905–913.
44. Frei, K., Eugster, H.-P., Bopst, M., Constantinescu, C. S., Lavi, E., and Fontana, A. (1997) Tumor necrosis factor a and lymphotoxin a are not required for induction of acute experimental autoimmune encephalomyelitis. *J. Exp. Med.* **185,** 2177–2182.
45. Körner, H., Riminton, D. S., Strickland, D. H., Lemckert, F. A., Pollard, J. D., and Sedgwick, J. D. (1997) Critical points in tumor necrosis factor action in central nervous system autoimmune inflammation defined by gene targeting. *J. Exp. Med.* **186,** 1585–1590.
46. Liu, J., Marino, M. W., Wong, G., Grail, D., Dunn, A., Bettadapura, J., et al. (1998) TNF is a potent anti-inflammatory cytokine in autoimmune-mediated demyelination. *Nat. Med.* **4,** 78–83.
47. Steinman, L. (1997) Some misconceptions about understanding autoimmunity through experiments with knockouts. *J. Exp. Med.* **185,** 2039–2041.
48. Steinman, L. (1999) Assessment of animal models for MS and demyelinating disease in the design of rational therapy. *Neuron* **24,** 511–514.

19

Neuroimmunologic Mechanisms
in the Etiology of Multiple Sclerosis

Claudia F. Lucchinetti, W. Brück, and Hans Lassmann

1. INTRODUCTION

Multiple sclerosis (MS) is a common chronic inflammatory demyelinating disorder of the central nervous system (CNS) *(1,2)* that is heterogeneous with respect to its clinical course, response to therapy, radiology, structural pathology, and immunopathogenesis. The factors that contribute to this heterogeneity are largely unknown, although it is likely a complex trait with genetic and environmental components. MS is a disease of young people with a median age of onset of 28 yr, and affects approx 0.1% of the population in temperate climates. The disorder is the most common cause of nontraumatic disability in young adults, with 50% of patients requiring a cane to walk 15 yr after disease onset *(3)*.

Approximately 85% of patients present with relapsing-remitting disease (RRMS), in which neurological symptoms and signs develop over several days, plateau, and then usually improve over days to weeks *(4)*. Inflammatory infiltrates and demyelination in the brain and spinal cord white matter usually accompany these clinical exacerbations. Periods of clinical quiescence (remissions) typically occur between these exacerbations; however, remissions vary in length and are rarely permanent. The remaining 15% of patients continuously progress from disease onset without associated relapses. This disease course is defined as "primary progressive" (PPMS). In PPMS patients, the disease usually begins after the age of 40, males and females are equally affected, and the prognosis is often worse. Approximately two-thirds of patients with RRMS eventually undergo a similar fate, as relapse frequency decreases over time and progressive neurological dysfunction ensues, signaling the development of secondary progressive disease (SPMS) *(3)*. Some patients who convert to a secondary progressive course continue to experience superimposed relapses.

The cause of MS and its genetic basis are not yet defined. The pathological hallmark of MS is the demyelinated plaque in the CNS white matter with relative axonal sparing and glial scar formation *(5,6)*. These lesions are scattered throughout the CNS with a predilection for the optic nerves, brainstem, spinal cord, and periventricular white matter. Traditionally, MS has been considered an autoimmune disorder consisting of myelin autoreactive T-cells that drive an inflammatory process, leading to secondary macrophage

From: *Neuroinflammation, 2nd Edition: Mechanisms and Management*
Edited by: P. L. Wood © Humana Press Inc., Totowa, NJ

recruitment and subsequent myelin and oligodendrocyte destruction. However, recent detailed studies on a large collections of MS lesions have indicated that structural features of the plaques are extremely variable and the events involved in the immunopathogenesis of MS may be more complicated *(7–9)*. This is not surprising given the inherent heterogeneity observed with respect to the clinical, radiographic, genetic, and morphological features of this disease.

2. THE SPECTRUM OF HUMAN INFLAMMATORY DEMYELINATING DISEASES

The classical clinico-pathological pattern of chronic MS represents only one member of a family of closely related inflammatory demyelinating leukoencephalitides that include acute MS (Marburg variant) Balo's concentric sclerosis, acute disseminated encephalomyelitis (ADEM), and neuromyelitis optica (Devic's disease). The literature on the classification of these syndromes is often confusing. Some studies emphasize specific clinical or pathologic features to distinguish between these syndromes. However, there are examples of transitional cases that defy a specific terminology. For example, the typical concentric lesions of Balo's concentric sclerosis can be found adjacent to more typical MS plaques. In addition, some patients have lesions with histologic features of both ADEM and MS. Although the clinical and pathological characteristics of these diseases are diverse, the presence of transitional forms suggests a spectrum of inflammatory diseases that may share a pathogenic relationship.

2.1. Marburg Variant of MS

Acute MS was recognized as a subtype of the disease by Otto Marburg in 1906 *(10)*. Clinically, this entity is characterized by rapid progression and an exceptionally severe course, which typically ends in death within a year from presentation. The course is generally monophasic and relentlessly progressive, with death usually secondary to brainstem involvement. Although for most patients this is the presenting episode of demyelination, there are a several recorded cases of well-documented MS that subsequently progressed to a fulminant terminal stage.

Pathologically, the lesions are more destructive than typical MS or ADEM lesions and are characterized by massive macrophage infiltration, acute axonal injury, and necrosis. Multiple small lesions may be disseminated throughout the brain and spinal cord and may coalesce to form large confluent white matter plaques. In some cases, there is widespread diffuse demyelination throughout the white matter. Despite the destructive nature of these lesions, small areas of remyelination are often observed.

One study suggests that this acute form of MS may be associated with immature myelin basic protein *(11)*. An autopsy study on a single case of Marburg's disease documented pronounced posttranslational changes that converted mature myelin basic protein to an extensively citrullinated and poorly phosphorylated immature form. These changes were thought to render myelin more susceptible to breakdown.

More recent neuropathological studies suggest that these fulminant destructive lesions are associated with deposition of immunoglobulins (mainly IgG) and pronounced complement activation at sites of active myelin destruction. These observations suggest an important role for demyelinating antibodies *(8,12,13)*.

2.2. Balo's Concentric Sclerosis

Balo's concentric sclerosis is another acute variant of MS that is distinguished by its unique pattern of pathology. Similar to the Marburg variant, the clinical course is typically fulminant with an acute monophasic progressive course over weeks to months that often ends in death within 1 yr. Death is usually the result of cerebral herniation or pneumonia. Occasionally, the course may be subacute and there are rare reports of long-term survival. Age of onset is between 20 and 50 yr, and patients often present with predominantly cerebral symptoms, including headache, disturbances in consciousness, aphasia, cognitive decline, psychiatric symptoms, seizures, and signs of raised intracranial pressure.

The pathology of Balo's concentric sclerosis reveals areas of focal necrosis of white matter with mass effect and often spares the cerebellum, brainstem, optic chiasm, and spinal cord. The essential features are large demyelinated plaques that show peculiar alternating rims of myelin preservation and loss, giving the lesions the macroscopic and microscopic structure of onion bulbs. The cause for the peculiar alternating pattern is still a matter of debate. Several investigators argue that Balo's concentric sclerosis is a rare variant of MS, because transitional forms have been observed where the presence of both concentric lesions and MS plaques are found within the same patient. The distinction between Balo's sclerosis and the Marburg variant of MS is vague because Marburg's original case of fulminant MS from 1906 had pathological features resembling Balo's sclerosis. These observations underscore the difficulty in classifying the acute leukoencephalitides.

2.3. Perivenous Encephalomyelitis

Several syndromes can be grouped under the term "perivenous encephalomyelitis." These include acute disseminated encephalomyelitis (ADEM), postinfectious encephalomyelitis, postvaccinial encephalomyelitis, and the more severe hyperacute syndrome acute hemorrhagic leukoencephalomyelitis (AHLE) *(14,15)*.

Acute disseminated encephalomyelitis is generally a monophasic disorder that typically begins within 6 d to 6 wk following an antigenic challenge. At least 70% of patients report a precipitating event (infection or vaccine) during the prior few weeks. However, the only epidemiologically and pathologically proven association is with the rabies vaccination. ADEM has also been described following the administration of antisera and some drugs.

Acute disseminated encephalomyelitis can occur at any age, but is more common in childhood. The clinical course is rapidly progressive, leading to focal or multifocal neurologic deficits. A prodrome of headache, low-grade fever, myalgias, and malaise often precedes the onset of ADEM by a few days. Headache, fever, seizures, myelopathy, optic neuritis, and brainstem or cerebellar disturbances are common. Drowsiness and lethargy are frequent and may progress to coma. The overall mortality in ADEM is 10–25%. Although ADEM is considered a monophasic disorder, there are rare cases of this disorder relapsing up to 18 mo following an infection. Occasionally ADEM evolves over a few months and there may be a second clinical deterioration or subacute progression over a period of time. In these cases, distinction from MS is difficult *(15)*.

The pathology of ADEM reveals inflammation, predominantly in the Virchow Robin spaces and diffuse, often symmetric perivenular demyelination *(14)*. In the more severe cases of AHLE, the inflammatory reaction is associated with perivascular hemorrhages, fibrin deposits, and severe brain edema. The limited extent and pattern of demyelination help distinguish ADEM from MS lesions. However, as in the other acute leukoencephalitides, there are also transitional cases with histologic features of both ADEM and MS, suggesting that these conditions may represent a continuous spectrum of inflammatory demyelinating disease.

Initial theories speculated that ADEM represented a delayed but direct invasion of the nervous system by virus or by reactivation of latent virus in the central nervous system (CNS). However, the pathology is unlike typical viral encephalitides, and viral antigens or particles have not been identified in the lesions. It is generally believed that ADEM represents a transient autoimmune response toward myelin or other self-antigens, possibly via the mechanism of molecular mimicry or the activation of autoreactive T-cell clones in a nonspecific manner. This is based on observations that the neurologic syndrome was obtained in patients vaccinated with Semple rabies vaccine that had been grown in rabbit spinal cord. In addition, ADEM closely resembles experimental allergic encephalomyelitis (EAE), which is produced by injection of rodents with brain homogenate and myelin basic protein (MBP). Lymphocytic reactivity toward MBP has been identified in blood and cerebrospinal fluid (CSF) from patients with ADEM, but its absence in others indicates a role for other antigens. Because ADEM shows extensive inflammation in the absence of significant demyelination, it suggests that there may be a need for additional demyelinating amplification factors in order to develop the demyelinated MS plaque.

2.4. Neuromyelitis Optica (Devic's Disease)

Neuromyelitis optica (NMO) was originally considered a monophasic syndrome consisting of acute, severe transverse myelitis and bilateral simultaneous optic neuritis in close temporal association, leading to paraplegia and blindness *(16)*. However, both monophasic and relapsing variants have been described. The most recent and comprehensive retrospective review defined NMO as a monophasic or relapsing disorder characterized by severe transverse myelitis associated with either unilateral or bilateral optic neuritis *(17)*. Typically, symptoms of optic neuritis and myelitis develop over hours to days and are often preceded by headache, nausea, somnolence, fever, or malaise. Most patients develop bilateral optic neuritis (>80%). Patients with NMO can follow one of several possible courses. Approximately 35% have a monophasic illness, 55% develop relapses restricted to the optic nerves and spinal cord, and, rarely, patients may have a relentlessly progressive course to death. Occasional rare individuals who satisfy clinical criteria for NMO pursue a clinical course consistent with MS. NMO occurs in patients ranging from 1 to 73 yr of age, with the monophasic form occurring more commonly in younger patients (mean age = 27 yr). There is no sex predilection in the monophasic syndrome; however, relapsing NMO is more frequent in women (F:M, 3.8:1). Most reports of NMO involving predominantly adult patients emphasize the poor prognosis. Patients with relapsing NMO typically have very aggressive disease with frequent and severe exacerbations and a poor prognosis. However, many patients with a monophasic course may recover remarkably, with little to no residual neurologic deficit.

The pathogenic mechanisms that result in selective localization of inflammatory demyelinating lesions to the optic nerves and spinal cord are unknown. Serologic and clinical evidence of B-cell autoimmunity has been observed in a high proportion of patients with NMO. Pathologically, active NMO lesions are characterized by extensive demyelination across multiple spinal cord levels, associated with cavitation, necrosis, and acute axonal pathology (spheroids), in both gray and white matter. There is a pronounced loss of oligodendrocytes within the lesions. The inflammatory infiltrates in active lesions are characterized by extensive macrophage infiltration associated with large numbers of perivascular granulocytes and eosinophils and rare CD3+ and CD8+ T-cells. There is a pronounced perivascular deposition of immunoglobulins (mainly IgM) and complement C9neo antigen in active lesions associated with prominent vascular fibrosis and hyalinization in both active and inactive lesions. The extent of complement activation, eosinophilic infiltration, and vascular fibrosis supports a role for humoral immunity in the pathogenesis of NMO *(18)*. The relationship between NMO and chronic MS is controversial.

3. THE MS PLAQUE

The basic pathologic feature unique to MS is the presence of multifocal demyelinated plaques. These focal areas of myelin destruction occur on a background of an inflammatory reaction consisting predominantly of macrophages and T-lymphocytes. MS lesions can be characterized as active or inactive. Macrophage activation and phagocytosis of myelin proteins in the lesions are reliable indicators of ongoing demyelinating activity *(19–21)*. Active lesions are heavily infiltrated by macrophages containing myelin debris, which are often closely associated with the disintegrating myelin sheath.

The chronic inactive MS plaque is a sharply circumscribed hypocellular plaque with no evidence of active myelin breakdown. Fibrillary gliosis is prominent, axonal density is often markedly reduced, variable inflammation may be present, and remyelination is typically sparse. Mature oligodendrocytes are markedly diminished or absent from chronic inactive lesions.

4. INFLAMMATION IN MULTIPLE SCLEROSIS

Although the pathological hallmark of the MS lesion is focal demyelination, these focal areas of myelin destruction occur on a background of an inflammatory reaction consisting of T-lymphocytes, a few B-lymphocytes and plasma cells, and extensive macrophage/microglial activation *(22)*. This pathology is similar to that found in experimental autoimmune encephalomyelitis (EAE), an animal model of MS, which can be induced in susceptible animals by active sensitization with CNS tissue, myelin, myelin proteins, or by the transfer of auto-reactive T-cells *(23)*. The inflammatory reaction in MS is associated with the upregulation of a variety of Th1 cytokines, including interleukin (IL)-2, interferon (IFN)-γ, and tumor necrosis factor (TNF)-α. In addition, the pattern of chemokine expression is also compatible with a Th1-mediated process *(24–26)*. However, the pathogenesis of MS lesions is more complex, compared to that of a pure Th1-mediated CNS autoimmune disease. Evidence is accumulating that cells other than the classical Th1 T-cells may contribute to the inflammatory process in autoimmune diseases *(27)*. CD8+ class I restricted T-cells outnumber CD4+ T-cells in MS lesions and they are clonally expanded in MS lesions *(28)*. Axonal destruction in MS lesions correlates better with CD8+ T-cells and macrophages rather than CD4+ T-cells *(29)*. There is also evidence that Th2 cells can participate in pathologic autoimmune processes. Antibodies to both myelin-oligodendrocyte glycoprotein (MOG) and MBP have been demonstrated in MS lesions and in the serum of MS patients *(13,30)*, and circulating Th2 cells could drive this antibody formation *(31)*. The exact pathogenic role of the inflammatory response in MS is not clear. Neuropathologic studies reveal that inflammatory cells are not always present in areas of active demyelination and persistent inflammation is a frequent and typical feature of chronic inactive MS lesions *(31a)*. Finally, recent observations on the local production of neurotrophic factors by leukocytes may indicate an important role for inflammation in the repair of MS lesions *(32)*. Nonetheless, the emphasis on the inflammatory aspects of the MS lesion continues to be the major impetus for therapeutic strategies to date, despite yielding disappointing results.

Although inflammatory mechanisms seem to be an important aspect contributing to tissue injury in MS, whether it is a primary or secondary event in lesion formation is still not clear. Furthermore, the degree to which demyelination and axonal injury are a direct consequence of inflammation remains unclear.

4.1. Mechanisms of T-Cell Entry into the CNS; Pathogeneic Relevance to MS

A limited trafficking of antigen nonspecific T-cells occurs across an intact blood-brain barrier (BBB) via the interaction of adhesion molecules expressed on the surface of lymphocytes and integrins present on the endothelial surface of blood vessels. Rolling, adherence, and diapedesis of T-lymphocytes are mediated by VCAM-1/VLA-4 and ICAM-1/LFA-1 interactions. Genetic and environmental factors (i.e., viral infection, bacterial lipopolysaccharides, superantigens, reactive metabolites, or metabolic stress) may facilitate the entry of potentially pathogenetic autoreactive T-cells and antibodies into the CNS via BBB disruption *(2)*. In the CNS, local factors may also upregulate the expression of endothelial adhesion molecules (ICAM-1, VCAM-1, E-selectin), which further enhance the movement of pathogenetic cells into the CNS *(2)*. Circulating levels of ICAM-1 and VCAM-1 are elevated in RRMS *(33)*.

Once autoreactive T-cells have entered the CNS, matrix metalloproteinases (MMPs), especially MMP9, are thought to contribute to degrading extracellular matrix macromolecules *(34)*. In addition, MMPs are involved in other functions including proteolysis of myelin components, regulating cytokine production (i.e., TNF-α), and may play a role in regulating apoptotic cell death by disrupting cell–matrix contacts with the subsequent loss of integrin signaling *(35)*. MMPs and tissue inhibitors of MMPs (TIMPs) are present in the serum and CSF of MS patients and expressed in plaques *(36,37)*. Serum MMP9 levels may be higher in RRMS patients and may correlate with brain MRI markers of inflammation *(38)*. Beta-interferons, which reduce relapse rate and severity, are potent MMP9 inhibitors *(39)* and may limit T-cell infiltration and cytokine production *(40)*.

Within the CNS, pro-inflammatory cytokines activate resident and hematogenous macrophages. Recruitment and attraction of these cells occurs via integrins and chemokines and is believed to contribute to tissue injury and demyelination. Selective expression of individual chemokines may influence the cellular composition of inflammatory lesions because chemokine receptors are associated with either Th1 or Th2 responses. Th1 pro-inflammatory cells may be associated with CCR5 (receptors for chemokines RANTES, microphage-inflammatory protein [MIP]-1α and MIP-1B) and CXCR3 (receptors for

IP-10 and MIG), whereas Th2 inflammatory cells may shift toward the display of CCR 3 (receptors for MCP-3, MCP-4, and RANTES) and CCR8 *(41)*. In MS, CCR5 and CXCR3 are overexpressed in peripheral and lesional T-lymphocytes *(42)*, and the CSF may contain elevated levels of the chemokines IP-10, RANTES, and MIG *(24)*.

4.2. The Role of Demyelinating Amplification Factors in MS Pathogenesis

In most experimental animal models (rats, pigs, primates), T-cell-mediated immune responses against brain antigens result in brain inflammation, but only limited demyelination. This resembles the pathology of acute disseminated encephalomyelitis in which perivascular inflammation dominates, with minimal if any perivenular demyelination. These observations suggest that additional pathogenetic factors are necessary in order to produce the widespread demyelination classically seen in MS. These factors may include demyelinating antibodies, cytokines and other soluble mediators, cytotoxic T-cells, reactive oxygen and nitrogen species, excitotoxic mechanisms, or primary oligodendrocyte injury *(43)*.

4.2.1. B-Cells and Antibodies in MS

Despite the emphasis on T-cells in MS, over the last several years there has been growing interest in the potential importance of B-cells and antibodies in MS pathogenesis *(44)*. Although B-cells are often found in chronic MS plaques, some acute lesions have demonstrated a predominance of B-cells *(6)*. The potential demyelinating ability of antibodies in culture has also been demonstrated, as has the identification of autoantibodies against myelin antigens in MS lesions, CSF, and blood. Receptor-mediated phagocytosis of myelin by macrophages has been demonstrated both in culture and within MS lesions *(45,46)*. In addition, pure inflammatory T-cell-mediated EAE can be transformed into a massively demyelinating disease when specific antibodies directed against MOG antigen expressed on the surface of the myelin sheath are administered simultaneously *(47)*. In this model, T-cell-mediated brain inflammation results in the local activation of macrophages and microglia, with disruption of the blood-brain barrier (BBB). This facilitates the entry of circulating demyelinating antibodies and complement components to enter the CNS and to destroy myelin either via complement activation or via an antibody-dependent cellular cytotoxicity reaction (ADCC). These experimental lesions are characterized by the local precipitation of the lytic terminal complement complex on the surface of myelin sheaths and oligodendrocytes. The deposition of complement components, including C1q, C3d, C5-9, as well as the terminal lytic component C9neo, has also been demonstrated within active MS lesions *(8,12)*. Oligodendrocytes are also known to be particularly susceptible to attack by complement components *(48)*. Recent studies by Raine et al. on MS tissue and the marmoset model of EAE have also suggested the possible involvement of anti-MOG antibodies in contributing to myelin destruction *(13,49)*. Finally, the beneficial clinical response of plasmapheresis in some MS patients following a fulminant MS exacerbation further supports a potential role for humoral immunity in MS pathogenesis *(50)*.

4.2.2. Cytokines

Cytokines are soluble molecules whose functions include mediating "pro-inflammatory" and "anti-inflammatory" effects upon the immune system. Studies on changes in serum and CSF cytokines and on cytokine expression in MS lesions suggest they may play an essential role in the pathophysiology of this disease. An extensive profile of cytokine protein and mRNA expression has been described in MS lesions *(51,52)*, including IL-1, -2, -4, -6, and –10, INF-γ, TNF-α, transforming growth factor (TGF), perforin, and lymphotoxin. A positive correlation between some of these pro-inflammatory cytokines and MS disease activity has also been reported. Lymphotoxin, TNF-α, and perforin *(53)* are known to be oligodendrogliotoxic, as is the Fas pathway *(54)* and IFN-γ. Several cytokine transgenic experimental models have been generated which produce spontaneous demyelination (TNF-α, IFN-γ, IL-3) *(55,56)*. Finally, cytokines may directly act on CNS tissue to induce functional deficit. This has especially has shown for the effect of nitric oxide on axons.

4.2.3. Cytotoxic T-Cells

Several lines of evidence suggest that MHC class I-restricted CD8+ T-cell responses may play a role in MS pathogenesis. CD8+ lymphocytes are present in MS lesions and, in fact, numerically predominate in many lesions. Clonal expansion of CD8+ T-cells has been shown in multiple sclerosis *(28)*. Oligodendrocytes express MHC class I antigens when stimulated by IFN-γ and, therefore, could be potential targets of a CD8+ class I MHC-mediated cytotoxic response *(57)*. Recent studies have demonstrated that self-peptides derived from human myelin proteins can induce autoreactive CD8+ cytotoxic T-lymphocyte responses which produce TNF-α and IFN-γ *(58)*. Little is known so far on the functional role of cytotoxic, MHC class I-restricted T-lymphocytes in inflammatory brain lesions. Previous studies suggested a major regulatory role, because blockade by specific antibodies *(59)* or genetic deletion of CD8+ T-cells *(60)* was associated with poor recovery from EAE and an increased incidence of relapses of the disease. Furthermore, TCR peptide-specific CD8+ T-cells were able to selectively destroy MBP-reactive encephalitogenic CD4+ T-cells in vitro and suppress EAE *(61)*. Two recent reports directly addressed the potential role of self-reactive class I-restricted T-cells as effector cells in brain lesions. Passive transfer of MBP-specific CD8+ T-cell clones into irradiated recipients, simultaneously treated with IL-2, induced brain inflammation with extensive vascular damage and unselective perivascular ischemic tissue damage *(62)*. In another study *(63)*, recipient animals developed severe inflammatory and destructive lesions in the CNS following passive transfer of an enriched population of MOG-reactive CD8+ T-cells. These studies suggest a potential important role for CD8+ T-cells in mediating tissue injury in MS and may become an attractive target for new therapeutic strategies.

Oligodendrocytes can also be destroyed in vitro directly by activated CD4+ lymphocytes independent of TNF-α, possibly via the interaction of Fas antigen with Fas ligand *(54,64)*. Soluble products from T-cells such as perforin could also mediate oligodendrocyte cytotoxicity, via the formation of pores in the target cell resulting in cell death via necrosis *(65)*. The target antigen for cytotoxic T-cell reactions in MS lesions is unknown; however, some studies suggest that stress proteins expressed in MS lesions may be one potential target.

4.2.4. Direct Oligodendrocyte Injury

Certain neurotropic viruses are known to infect oligodendrocytes *(66)*. Numerous studies demonstrating the presence of viral antigens or nucleotide sequences have been reported in MS lesions. Although no specific virus has been implicated in the cause of MS, several case reports document brain virus infections leading to a pathology mimicking MS *(67,68)*. Furthermore, human herpesvirus 6 (HHV-6) antigen expression was reported to localize to oligodendrocytes in MS tissue, suggesting a potential pathogenic role for this virus *(69)*. However, a direct causal link remains to be confirmed.

4.2.5. Other Potential Mediators of Tissue Destruction in MS Lesions

Reactive oxygen and nitrogen species are known potent oligodendrogliotoxic molecules produced by activated macrophages and thought to cause cell injury via necrosis and lysis rather than apoptosis. Increased levels of NO metabolites have been reported in the CSF and serum of MS patients *(70)*, and reactive oxygen and nitrogen intermediates (ROI/RNI) have been demonstrated within active MS lesions *(71)*. Excitotoxic mechanisms have also been postulated to contribute to tissue injury in MS secondary to abnormal glutamate mechanisms *(72)*.

5. OLIGODENDROCYTE PATHOLOGY IN MS

The interaction between oligodendroglial cells and the immune system triggers multiple distinct molecular and cellular processes that may be relevant to MS pathogenesis. It is therefore unlikely that the oligodendrocyte is strictly destroyed as a secondary consequence to primary myelin injury. The mechanisms of oligodendrocyte death in MS have not been fully elucidated. Three primary potential mechanisms have been considered and they include necrosis, apoptosis, and dying-back oligodendrogliopathy. Necrosis is the result of acute cellular injury and is characterized by disruption of the plasma membrane, organelle and cytoplasmic swelling, and random DNA cleavage. Apoptosis, on the other hand, is the result of controlled autodigestion of the cell leading to cytoskeletal disruption, cell shrinkage, and membrane blebbing, with nuclear condensation and endonuclease fragmentation leading to the formation of apoptotic bodies and the loss of mitochondrial function. Several MS pathological studies have demonstrated the presence of apoptotic cells bearing myelin markers at nearly all stages of the disease *(7,8,73)*; however, other studies have failed to confirm these observations. Furthermore, other potential candidates in MS pathogenesis are thought to result in necrosis of oligodendrocytes *(74–76)*. Although the same etiological factor may lead to either apoptosis or necrosis, it is important to try to distinguish between the two processes, because the strategies for treating them may be entirely different. The destruction of oligodendrocytes may be prevented by the administration of growth factors such as insulin-like growth factor (IGF) and ciliary neurotrophic factor (CNTF), which has been shown to protect oligodendrocytes from natural or TNF-α-induced apoptosis, but not from complement or nitric oxide induced necrosis *(77)*.

Another potential mechanism of oligodendrocyte damage in MS is the concept of a dying-back process. This was first introduced as a mechanism of cell damage in experimental models of axonal degeneration induced by a variety of toxic and metabolic insults. Dying-back neuropathies are a group of clinical and experimental conditions characterized by the inability of the cell body to support the metabolic processes necessary to maintain the distal axon. This results in degeneration of the distal processes before degeneration of the cell body. In 1981, Ludwin suggested that a similar dying-back process occurred in oligodendrocytes in the experimental model of cuprizone toxicity *(78)*. In this model, degeneration of the distal, periaxonal oligodendrocyte processes was observed initially, followed by demyelination and oligodendrocyte loss. More recently, similar ultrastructural changes were described in brain biopsies from MS lesions *(79)*. Similar morphologic alterations in the distal, periaxonal oligodendrocyte processes were noted at the edge of demyelinating lesions of Theiler's murine encephalomyelitis (TMEV)-infected mice, a viral animal model of multiple sclerosis *(80)*. This dying-back oligodendrogliopathy was associated with widespread demyelination and clinical deficit.

Pathological studies of MS lesions have reported a variable degree of oligodendrocyte preservation in actively demyelinating lesions *(80a)*. A detailed analysis of oligodendrocyte density relative to demyelinating activity in a series of 113 MS cases demonstrated 2 principal patterns of oligodendrocyte pathology in MS lesions *(7)*. In the first pattern, oligodendrocytes were variably reduced during active stages of myelin destruction, but reappeared within inactive or remyelinating areas. The other pattern was characterized by extensive destruction of myelinating cells at active sites of demyelination in the absence of progenitor cell recruitment in inactive plaque areas. In these cases, remyelination was sparse or absent. The profound heterogeneity in extent and topography of oligodendrocyte destruction in active demyelinating lesions suggested that in subsets of MS patients, myelin, mature oligodendrocytes, and, possibly, oligodendrocyte progenitors were differentially affected. These observations suggested that different mechanisms of myelin and/or oligodendrocyte injury may be operating in individual MS patients and may influence the likelihood of effective remyelination with the MS lesion.

6. REMYELINATION IN MULTIPLE SCLEROSIS

Neuropathologic studies of MS lesions have clearly demonstrated the presence of remyelination. Although lesions remyelination in chronic MS is incomplete and generally restricted to the edge of the demyelinated plaques, examination of plaques from acute and early MS lesions may show extensive remyelination and are referred to as "shadow plaques." These shadow plaques are sharply demarcated areas of myelin pallor and gliosis. Although, initially, these plaques were considered incompletely demyelinated areas, subsequent studies confirmed that these represented complete remyelination of a previously demyelinated plaque. The ultrastructural characteristic of these remyelinated areas are uniformly thin myelin sheaths in relation to their axon diameter.

Remyelination during the early stages of some MS lesions can be extensive *(79,81, 82)*, and may occur simultaneously with ongoing demyelination. During these early stages of myelin sheath formation, there may be a pronounced inflammatory infiltrate present within the lesions. Remyelinated lesions may also become targets of new demyelinating attacks *(83)*. These studies suggest there is an ongoing dynamic interaction between pathogenic and reparative factors within the evolving MS lesion.

The extent of remyelination appears to depend on the availability of oligodendrocytes or their progenitor cells in the lesions *(82,84,85)*. However, the presence of cells in very early stages of oligodendrocyte development have been identified in completely demyelinated plaques devoid of mature oligodendrocytes *(80a)*. Furthermore, a recent study suggests that premyelinating oligodendrocytes may be present in chronic MS lesions, suggesting that remyelination at these late states may not be limited by an absence of oligodendrocyte progenitors, but rather that damaged axons may not be receptive to remyelination signals *(86)*. To what extent these cells can be stimulated to divide, repopulate the lesions, and initiate remyelination must still be demonstrated.

7. AXON PATHOLOGY IN MS

Axonal damage in MS lesions has already been recognized more than 100 yr ago *(87)*. Axonal damage can be determined histopathologically by the extent of axonal loss relative to the normal white matter. In addition, acute axonal damage can be detected by immunohistochemistry for the amyloid precursor protein (APP) *(88–91)*. APP is found in neurons and undergoes anterograde axonal transport. In case of an axonal transection, the transport is interrupted and APP accumulates in the proximal axonal ends. As a consequence, so-called APP-positive spheroids are formed that persist for less than 30 d. A variety of different histopathological studies exist that investigated acute axonal damage in multiple sclerosis *(29,92–94)*, and suggest that axonal injury may underlie permanent neurological deficits in MS patients *(94)*. Magnetic resonance spectroscopy (MRS) of MS plaques revealed a correlation of disability and reduced levels of *N*-acetylaspartate (NAA) *(95)*, which is a biochemical marker found exclusively in neurons and their processes *(96)*. Also, the volume of hypointense lesions ("black holes") in T1-weighted magnetic resonance imaging (MRI) and the extent of brain atrophy, both putative markers of axonal damage, correlated with the degree of disability *(97–100)*. In histopathological studies, a significant reduction of axon density was found in MS lesions *(101)*. An average axonal loss of 59–82% was observed in demyelinated and remyelinated MS plaques compared to the periplaque white matter *(102)*. The acute axonal damage determined by APP staining or axonal spheroids was investigated in acute *(92,93)*, early *(29)*, and late chronic *(93)* MS cases. From these studies, it is well known that axonal damage is an early event during development of MS plaques and that the highest degree of acute axonal damage is associated with active demyelination *(29,92,93)*. In chronic disease stages, acute axonal pathology is much less prominent *(92)*. Damage to axons may occur partly independent of active demyelination and there is a slow burning axonal destruction ongoing, especially in lesions that are completely demyelinated. Reports of as much as 50% axonal loss in the corpus callosum outside of macroscopic MS lesions suggests that extensive loss of axons in the normal-appearing white matter (NAWM) may be the result of the effects of axonal transsection of lesions the actions of diffusable neurotoxic factors from lesions *(103)*. A recent detailed analysis of the NAWM of a patient with acute MS also described axonal changes in detail *(104)*.

The pathogenesis of axon destruction is unknown. An association was found between the numbers of CD8 positive cells and the extent of axon damage *(29)*. A CD8–MHC class I-mediated pathway of axon destruction has also been suggested from experimental studies *(105)*. Nitric oxide is another candidate molecule mediating functional and structural axon damage *(106)*. Axons can also be directly damaged by a cellular *(107)* or antibody-mediated *(108,109)* inflammatory reaction. The genetic background of axonal susceptibility to damage may be affected by a gene that prevents axon degeneration in the mouse mutant WLD^s *(110–112)*. This gene encodes for a chimeric protein that is not yet characterized in detail *(113,114)*.

8. IMMUNOPATHOLOGIC HETEROGENEITY IN MS

In addition to the heterogeneity observed in the structural aspects of MS lesions, there is an important degree of interindividual variability in the immunopathological features of MS lesions. In a large study of actively demyelinating MS lesions (based on 51 biopsies and 32 autopsies), quite diverse patterns of myelin destruction were observed *(8)*. The majority of active MS plaques were characterized by the precipitation of immunoglobulins and complement components at sites of active myelin breakdown. These lesions resembled the model of MOG-induced autoimmune encephalomyelitis. However not all cases of MS followed this pathway. The other cases demonstrated signs suggestive of a primary oligodendrocyte dystrophy. This was reflected either by a disproportionate loss of myelin-associated glycoprotein (MAG) and oligodendrocyte apoptosis or degeneration of oligodendrocytes in a small rim of periplaque white matter adjacent to active sites of demyelination. The patterns of demyelination were heterogeneous between patients, but homogenous within multiple active plaques from the same patients. Therefore, it is possible that different pathogenic mechanisms of demyelination may operate in different subgroups of MS patients.

9. MRI AND MS

Magnetic resonance imaging is not only used for the diagnosis of MS, but has become increasingly important in order to monitor disease activity, establish prognostic parameters, and find the radiological correlate of key morphological features of the MS lesion such as inflammation, demyelination, remyelination, or axonal loss *(115–117)*. There is a poor relationship between MR measures and clinical findings, and T2-weighted images, in particular, lack pathologic specificity. General radiological markers include total lesion load (T1 or T2), T1 hypointense lesion load and gadolinium enhancement. Although it has generally been accepted that breakdown of the blood-brain barrier is the initial event in lesion formation, a detailed and careful diffusion MRI study suggested the opposite, namely that subtle changes in the NAWM may precede and perhaps trigger the formation of demyelinated plaques *(119)*. These different pathways of lesion formation may have their substrate in the heterogenous pathology of MS plaques observed in recent pathological studies *(8,9)*.

Axon pathology in MS is considered an important feature of MS plaques and may represent the substrate of disability *(94)*. T1 hypointense lesions are thought to represent axonal damage in MS lesions and to correlate with clinical disability *(119)*. A comparative pathologic–radiologic study in postmortem autopsy cases showed that MTR and T1 contrast ratio significantly correlated with axonal density in the lesions *(120)*. These hypointense T1 lesions revealed significantly lower concentrations of NAA in proton magnetic resonance spectroscopy (MRS) *(121)*. NAA is regarded as a biochemical marker of axon integrity and, therefore, considered the most reliable current marker for monitoring axonal pathology in vivo *(122)*. Finally, axonal damage may eventually lead to loss of brain tissue and brain atrophy, a potential new surrogate marker of MS *(123)*. Whether there is a specific radiologic correlate for remyelination is questionable. A recent experimental study observed an increase in MTR during demyelination and remyelination, suggesting that a candidate marker for this process may be available *(124)*.

10. CONCLUSIONS

Since Charcot's original description in 1868, detailed analysis of the MS lesion has led to increased understanding of the complex and potentially diverse immunopathological processes that may be involved in lesion pathogenesis. The MS lesion, the hallmark of which is demyelination, remains the target for therapy. Although clinical deficits are dependent in part on demyelination, axonal injury is likely a major cause of irreversible neurologic damage. The degree to which demyelination and axonal injury are a direct consequence of inflammation is still uncertain. Although current therapies have been directed toward modulation of inflammatory components, future therapies will require novel approaches that inhibit demyelination, prevent neuronal death, protect axons, and promote repair. Variability in the pathological features of the MS lesion has been previously recognized, but was largely thought to reflect the varying intensity of the inflammatory process. Recent studies, however, suggest an interindividual pathologic heterogeneity in the patterns of inflammation, demyelination, oligodendrocyte damage, axonal injury, and remyelination in different subgroups of MS patients. Four distinct immunopathological patterns have been identified, with all active lesions from a single MS patient demonstrating the same pattern of damage. These observations suggest that alternate pathogenic mechanisms of tissue injury may underlie what previously was considered a single disease, and raise the possibility that different therapeutic strategies may be required for different MS patients. A major challenge now is to confirm these observations and to identify paraclinical markers that reliably predict specific pathological features of the MS lesion. Determining to what extent heterogenous therapeutic responses in MS patients may be the result of distinct immunopathological subtypes of the disease may ultimately lead to individualized and more effective therapies for MS.

REFERENCES

1. Noseworthy, J. H. (1999) Progress in determining the causes and treatment of multiple sclerosis. *Nature* **399(Suppl.),** A40–A47.
2. Noseworthy, J. H., Lucchinetti, C. F., Rodriguez, M., and Weinshekner, B. G. (2000) Medical progress: multiple sclerosis. *N. Engl. J .Med.* **343,** 938–952.
3. Weinshenker, B. G., Bass, B., Rice, G. P., Noseworthy, J., Carriere, W., Baskerville, J., et al. (1989) The natural history of multiple sclerosis: a geographically based study. I. Clinical course and disability. *Brain* **112,** 133–146.
4. Schumacher, G. A., Beebe, G., Kilber, R. F., et al. (1965) Problems of experimental trials of therapy in MS: report of the panel on evaluation of experimental trials in MS. *Ann. NY Acad. Sci.* **122,** 552–568.
5. Prineas, J. W. (1985) The neuropathology of multiple sclerosis, in *Handbook of Clinical Neurology Volume 47: Demyelinating Diseases* (Koetsier, J. C., ed.), Elsevier Science, Amsterdam, pp. 337–395.
6. Lassmann, H. (1998) Pathology of multiple sclerosis, in *McAlpine's Multiple Sclerosis* (Compston, A., Ebers, G., Lassmann, H., McDonald, I., Matthews, B., and Wekerle, H., eds.), Churchill Livingstone, London, pp. 323–358.
7. Lucchinetti, C., Bruck, W., Parisi, J., Scheithauer, B., Rodriguez, M., and Lassmann, H. (1999) A quantitative analysis of oligodendrocytes in multiple sclerosis lesions: a study of 113 cases. *Brain* **122,** 2279–2295.
8. Lucchinetti, C., Bruck, W., Parisi, J., Scheithauer, B., Rodriguez, M., and Lassmann, H. (2000) Heterogeneity of multiple sclerosis lesions: implications for the pathogenesis of demyelination. *Ann. Neurol.* **47,** 707–717.
9. Lassmann, H., Brück, W., and Lucchinetti, C. (2001) Heterogeneity of multiple sclerosis pathogenesis: implications for diagnosis and therapy. *Trends Mol. Med.* **7,** 115–121.
10. Marburg, O. (1906) Die sogenannte "akute multiple sklerose." *Jahrb. Psychiatrie* **27,** 211–312.

11. Wood, D. D., Bilbao, J. M., O'Connors, P., and Moscarello, M. A. (1996) Acute multiple sclerosis (Marburg type) is associated with developmentally immature myelin basic protein. *Ann. Neurol.* **40,** 18–24.

12. Storch, M. K., Piddlesden, S., Haltia, M., Iivanainen, M., Morgan, P., and Lassmann. H. (1998) Multiple sclerosis: in situ evidence for antibody and complement mediated demyelination. *Ann. Neurol.* **43,** 465–471.

13. Genain, C. P., Cannella, B., Hauser, S. L., and Genain, C. P. (1999) Identification of autoantibodies associated with myelin damage in multiple sclerosis. *Nat. Med.* **5,** 170–175.

14. Hart, M. and Earle, K. (1975) Haemorrhagic and perivenous encephalitis: a clinical-pathological review of 38 cases. *J. Neurol. Neurosurg. Psych.* **38,** 585–591.

15. Kesselring, J., Miller, D. H., Robb, S. A., Kendall, B. E., Moseley, I. F., Kingsley, D., et al. (1990) Acute disseminated encephalomyelitis. MRI findings and the distinction from multiple sclerosis. *Brain* **113,** 291–302.

16. Devic. C. (1894) Myelite subaigue complique de nevrite optique. *Bull Med.* **5,** 18–30.

17. Wingerchuk, D. M., Hogancamp, W. E., O'Brien, P. C., and Weinshenker, B. G. (1999) The clinical course of neuromyelitis optica (Devic's syndrome). *Neurology* **53,** 1107–1114.

18. Lucchinetti, C. F., Mandler, R., McGavern, D., Bruck, W., Gleich, G., Ransohoff, R., et al. (2002) A role for humoral mechanisms in the pathogenesis of Devic's neuromyelitis optica. *Brain* **125,** 1450–1461.

19. Brück, W., Porada, P., Poser, S., et al. (1995) Monocyte/macrophage differentiation in early multiple sclerosis lesions. *Ann. Neurol.* **38,** 788–796.

20. Brück, W., Sommermeier, N., Bergmann, M., Zettl, U., Goebel, H. H., Kretzschmar, H. A., et al. (1996) Macrophages in multiple sclerosis. *Immunobiology* **195,** 588–600.

21. Lassmann, H., Raine, C. S., Antel, J., and Prineas, J. W. (1998) Immunopathology of multiple sclerosis: report on an international meeting held at the Institute of Neurology of the University of Vienna. *J. Neuroimmunol.* **86,** 213–217.

22. Prineas, J. W. and Wright, R. G. (1978) Macrophages, lymphocytes, and plasma cells in the perivascular compartment in chronic multiple sclerosis. *Lab. Invest.* **38,** 409–421.

23. Lassmann, H. (1983) *Comparative Neuropathology of Chronic Experimental Allergic Encephalomyelitis and Multiple Sclerosis.* Springer-Verlag, Heidelberg.

24. Sorensen, T. L., Tani, M., Jensen, J., Pierce, V., Lucchinetti, C., Folcik, V. A., et al. (1999) Expression of specific chemokines and chemokine receptors in the central nervous system of multiple sclerosis patients. *J. Clin. Invest.* **103,** 807–815.

25. Simpson, J. E., Newcombe, J., Cuzner, M. L., and Woodroofe, M. N. (2000) Expression of the interferon-γ-inducible chemokines IP-10 and Mig and their receptor, CXCR3, in multiple sclerosis lesions. *Neuropathol. Appl. Neurobiol.* **26,** 133–142.

26. Simpson, J., Rezaie, P., Newcombe, J., Cuzner, M. L., Male, D., and Woodroofe, M. N. (2000) Expression of β-chemokine receptors CCR2, CCR3 and CCR5 in multiple sclerosis central nervous systen tissue. *J. Neuroimmunol.* **108,** 192–200.

27. Scotet, E., Peyrat, M.-A., Saulquin, X., Retiere, C., Couedel, C., Davodeau, F., et al. (1999) Frequent enrichment for CD8 T cells reactive against common herpes viruses in chronic inflammatory lesions: towards a reassessment of the physiopathological significance of T cell clonal expansions found in autoimmune inflammatory processes. *Eur. J. Immunol.* **29,** 973–985.

28. Babbe, H., Roers, A., Waisman, A., Lassmann, H., Goebels, N., Hohlfeld, R., et al. (2000) Clonal expansions of CD8(+) T cells dominate the T cell infiltrate in active multiple sclerosis lesions as shown by micromanipulation and single cell polymerase chain reaction. *J. Exp. Med.* **192,** 393–404.

29. Bitsch, A., Schuchardt, J., Bunkowski, S., Kuhlmann, T., and Bruck, W. (2000) Acute axonal injury in multiple sclerosis: correlation with demyelination and inflammation. *Brain* **123,** 1174–1183.

30. Reindl, M., Linington, C., Brehm, U., Egg, R., Dilitz, E., Deisenhammer, F., et al. (1999) Antibodies against the myelin oligodendrocyte glycoprotein and the myelin basic protein in multiple sclerosis and other neurological diseases: a comparative study. *Brain* **122,** 2047–2056.

31. Lindert, R.-B., Haase, C. G., Brehm, U., Linington, C., Wekerle, H., Hohlfeld, R. (1999) Multiple sclerosis: B- and T-cell responses to the extracellular domain of the myelin oligodendrocyte glycoprotein. *Brain* **122,** 2089–2099.

31a. Guseo, A. and Jellinger, K. (1975) The significance of perivascular infiltrates in multiple sclerosis. *J. Neurol.* **211,** 51–60.

32. Kerschensteiner, M., Gallmeier, E., Behrens, L., et al. (1999) Activated human T cells, B cells and monocytes produce brain-derived neurotrophic factor (BDNF) in vitro and in brain lesions: a neuroprotective role for inflammation? *J. Exp. Med.* **189,** 865–870.

33. Giovannoni, G., Lai, M., Thorpe, J., Kidd, D., Chamoun, V., Thompson, A. J., et al. (1997) Longitudinal study of soluble adhesion molecules in multiple sclerosis: correlation with gadolinium-enhanced magnetic resonance imaging. *Neurology* **48,** 1557–1565.

34. Yong, V. W., Krekoski, C. A., Forsyth, P. A., Bell, R., and Edwards, D. R. (1998) Matrix metalloproteinases and diseases of the CNS. *Trends Neurosci.* **21,** 75–80.

35. Chandler, S., Miller, K. M., Clements, J. M., Lury, J., Corkill, D., Anthony, D. C., et al. (1997) Matrix metalloproteinases, tumour necrosis factor and multiple sclerosis: an overview. *J. Neuroimmunol.* **72,** 155–161.

36. Cuzner, M. L., Gveric, D., Strand, C., Loughlin, A. J., Paeman, L., Opdenakker, G., et al. (1996) The expression of tissue-type plasminogen activator, matrix metalloproteinases and endogenous inhibitors in the central nervous system in multiple sclerosis: comparison of stages in lesion evolution. *J. Neuropathol. Exp. Neurol.* **55,** 1194–1204.

37. Maeda, A. and Sobel, R. A. (1996) Matrix metalloproteinases in the normal human central nervous system, microglial nodules, and multiple sclerosis lesions. *J. Neuropathol. Exp. Neurol.* **55,** 300–309.

38. Lee, M. A., Palace, J., Stabler, G., Ford, J., Gearing, A., and Miller, K. (1999) Serum gelatinase B, TIMP-1 and TIMP-2 levels in multiple sclerosis. A longitudinal clinical and MRI study. *Brain* **122,** 191–197.

39. Uhm, J. H., Dooley, N. P., Stuve, O., Francis, G. S., Duquette, P., Antel, J. P., et al. (1999) Migratory behavior of lymphocytes isolated from multiple sclerosis patients: effects of interferon beta-1b therapy. *Ann. Neurol.* **46,** 319–324.

40. Yong, V. W., Chabot, S., Stuve, O., and Williams, G. (1998) Interferon beta in the treatment of multiple sclerosis: mechanisms of action. *Neurology* **51,** 682–689.

41. Bonecchi, R., Bianchi, G., Bordignon, P. P., D'Ambrosio, D., Lang, R., Borsatti, A., et al. (1998) Differential expression of chemokine receptors and chemotactic responsiveness of type 1 T helper cells (Th1s) and Th2s. *J. Exp. Med.* **187,** 129–134.

42. Zhang, G. X., Baker, C. M., Kolson, D. L., and Rostomi, A. M. (2000) Chemokines and chemokine receptors in the pathogenesis of multiple sclerosis. *Multiple Sclerosis* **6,** 3–13.

43. Lucchinetti, C. F., Bruck, W., Rodriguez, M., and Lassmann, H. (1998) Multiple sclerosis: lessons learned from neuropathology. *Semin. Neurol.* **18,** 337–349.

44. Lovas, G., Szilágyi, N., Majtényi, K., Palkovits, M., and Kolomy, S. (2000) Axonal changes in chronic demyelinated cervical spinal cord plaques. *Brain* **123,** 308–317.

45. Prineas, J. W. and Graham, J. S. (1981) Multiple sclerosis: capping of surface immunoglobulin G on macrophages engaged in myelin breakdown. *Ann. Neurol.* **10,** 149–158.

46. Prineas, J. W., Kwon, E. E., Cho, E. S., and Sharer, L. R. (1984) Continual breakdown and regeneration of myelin in progressive multiple sclerosis. *Ann. NY Acad. Sci.* **436,** 11–32.

47. Linington, C., Bradl, M., Lassmann, H., Brunner, C., and Vass, K. (1988) Augmentation of demyelination in rat acute allergic encephalomyelitis by circulating mouse monoclonal

antibodies directed against a myelin/oligodendrocyte glycoprotein. *Am. J. Pathol.* **130,** 443–454.

48. Zajicek, J. P., Wing, M., Scolding, N. J., and Compston, D. A. (1992). Interactions between oligodendrocytes and microglia. A major role for complement and tumour necrosis factor in oligodendrocyte adherence and killing. *Brain* **115,** 1611–1631.

49. Raine, C. S., Cannella, B., Hauser, S. L., and Genain, C. P. (1999) Demyelination in primate autoimmune encephalomyelitis and acute multiple sclerosis lesions: a case for antigen-specific antibody mediation. *Ann. Neurol.* **46,** 144–160.

50. Weinshenker, B. G., O'Brien, P. C., Petterson, T. M., et al. (1999) A randomized trial of plasma exchange in acute central nervous system inflammatory demyelinating disease. *Ann. Neurol.* **46,** 878–886.

51. Canella, B. and Raine, C. S. (1995) The adhesion molecule and cytokine profile of multiple sclerosis lesions. *Ann. Neurol.* **37,** 424–435.

52. Link, H. (1998) The cytokine storm in multiple sclerosis. *Multiple Sclerosis* **4,** 12–15.

53. Murray, P. D., Pavelko, K. D., Leibowitz, J., Lin, X., and Rodriguez, M. (1998) CD4+ and CD8+ T-cells make discrete contributions to demyelination and neurologic disease in a viral model of multiple sclerosis. *J. Virol.* **72,** 7320–7329.

54. D'Souza, S. D., Bonetti, B., Balasingam, V., Cashman, N. R., Barker, P. A., Troutt, A. B., et al. (1996) Multiple sclerosis: Fas signaling in oligodendrocyte cell death. *J. Exp. Med.* **184,** 2361–2370.

55. Probert, L., Akassoglou, K., Pasparakis, M., Kontogeorgos, G., and Kollias, G. (1995) Spontaneous inflammatory demyelinating disease in transgenic mice showing central nervous system-specific expression of tumor necrosis factor α. *Proc. Natl. Acad. Sci. USA* **92,** 11,294–11,298.

56. Chiang, C.-S., Powell, H. C., Gold, L. H., Samimi, A., and Campbell, I. L. (1996) Macrophage/microglial-mediated primary demyelination and motor disease induced by the central nervous system production of interleukin-3 in transgenic mice. *J. Clin. Invest.* **97,** 1512–1524.

57. Grenier, Y., Ruijs, T. C., Robitaille, Y., Olivier, A., and Antel, J. P. (1989) Immunohistochemical studies of adult human glial cells. *J. Neuroimmunol.* **21,** 103–115.

58. Tsuchida, T., Parker, K. C., Turner, R. V., McFarland, H. F., Coligan, J. E., and Biddison, W. E. (1994) Autoreactive CD8+ T cell responses to human myelin protein-derived peptides. *Proc. Natl. Acad. Sci. USA* **91,** 10,859–10,863.

59. Jiang, H., Zhang, S. L., and Pernis, B. (1992) Role of CD8+ T cells in murine experimental allergic encephalomyelitis. *Science* **256,** 1213–1215.

60. Koh, D. R., Fung-Leung, W. P., Ho, A., et al. (1992) Less mortality but more relapses in experimental allergic encephalomyelitis in CD8$^{-/-}$ mice. *Science* **256,** 1210–1213.

61. Sun, D., Qin, Y., Chluba, J., Epplen, J. T., and Wekerle, H. (1988) Suppression of experimentally induced autoimmune encephalomyelitis by cytolytic T–T-cell interactions. *Nature* **332,** 843–845.

62. Huseby, E. S., Liggitt, D., Brabb, T., et al. (2001) A pathogenic role for myelin-specific CD8 (+) T-cells in a model for multiple sclerosis. *J. Exp. Med.* **194,** 669–676.

63. Sun, D., Whitaker, J. N., Huang, Z., et al. (2001) Myelin antigen specific CD8+ T cells are encephalitogenic and produce severe disease in C57BL/6 mice. *J. Immunol.* **166,** 7579–7587.

64. D'Souza, S., Alinauskas, K., McCrea, E., Goodyer, C., and Antel, J. P. (1995) Differential susceptibility of human CNS-derived cell populations to TNF-dependent and independent immune-mediated injury. *J. Neurosci.* **15,** 7293–7300.

65. Scolding, N., Jones, J., Compston, D. A., and Morgan, B. P. (1990) Oligodendrocyte susceptibility to injury by T-cell perforin. *Immunology* **70,** 6–10.

66. Fazakerley, J. K. and Buchmeier, M. J. (1993) Pathogenesis of virus-induced demyelination. *Adv. Virus Res.* **42,** 249–324.

67. Carrigan, D. R., Harrington, D., and Knox, K. K. (1996) Subacute leukoencephalitis caused by CNS infection with human herpesvirus-6 manifesting as acute multiple sclerosis. *Neurology* **47,** 145–148.
68. Sanders, V. J., Waddell, A. E., Felisan, S. L., Li, X., and Conrad, A. J. (1996) Herpes simplex virus in postmortem multiple sclerosis brain tissue. *Arch. Neurol.* **53,** 123–124.
69. Challoner, P. B., Smith, K. T., Parker, J. D., MacLeod, D. L., Coulter, S. N., Rose, T. M., et al. (1995) Plaque-associated expression of human herpesvirus 6 in multiple sclerosis. *Proc. Natl. Acad. Sci. USA* **92,** 7440–7444.
70. Johnson, A. W., Land, J. M., Thompson, E. J., et al. (1995) Evidence for increased nitric oxide production in multiple sclerosis. *J. Neurol. Neurosurg. Psychiatry* **58,** 107.
71. Bo, L., Dawson, T. M., Wesselingh, S., Mork, S., Choi, K., Kong, P. A., et al. (1994) Induction of nitric oxide synthase in demyelinating regions of multiple sclerosis brains. *Ann. Neurol.* **35,** 778–786.
72. Pitt, D., Werner, P., and Raine, C. (2000) Glutamate excitotoxicity in a model of multiple sclerosis. *Nat. Med.* **6,** 67–69.
73. Ozawa, K., Suchanek, G., Breitschopf, H., Brück, W., Budka, H., Jellinger, K., et al. (1994) Patterns of oligodendroglia pathology in multiple sclerosis. *Brain* **117,** 1311–1322.
74. Freedman, M. S., Ruijs, T. C., Selin, L. K., and Antel, J. P. (1991) Peripheral blood gamma/delta T cells lyse fresh human brain derived oligodendrocytes. *Ann. Neurol.* **30,** 794–780.
75. Griot, C., Vandevelde, M., Richard, A., Peterhans, E., and Stocker, R. (1993) Selective degeneration of oligodendrocytes mediated by reactive oxygen species. *Free Radical Res. Commun.* **11,** 181–193.
76. Mitrovic, B., Ignarro, L. J., Montestruque, S., et al. (1994) Nitric oxide as a potential pathological mechanism in demyelination: its differential effects on primary glial cells in vitro. *Neuroscience* **61,** 575–585.
77. Louis, J. C., Magal, E., Takayama, S., and Varon, S. (1993) CNTF protection of oligodendrocytes against natural and tumor necrosis factor-induced death. *Science* **259,** 689–692.
78. Ludwin, S. K. and Johnson, E. S. (1981) Evidence of a "dying-back" gliopathy in demyelinating disease. *Ann. Neurol.* **9,** 301–305.
79. Rodriguez, M. and Scheithauer, B. W. (1994) Ultrastructure of multiple sclerosis. *Ultrastruct. Pathol.* **18,** 3–13.
80. Rodriguez, M. (1985) Virus-induced demyelination in mice: "dying-back" of oligodendrocytes. *Mayo Clin. Proc.* **3,** 433–438.
80a. Wolswijk, G. (1998) Chronic stage multiple sclerosis lesions contain a relatively quiescent population of oligodendrocyte precursor cells. *J. Neurosci.* **18,** 601–609.
81. Lassmann, H., Brück, W., Lucchinetti, C., and Rodriguez, M. (1997) Remyelination in multiple sclerosis. *Multiple Sclerosis* **3,** 133–136.
82. Prineas, J. W., Barnard, R. O., Kwon, E. E., Sharer, L. R., and Cho, E. S. (1993) Multiple sclerosis: remyelination of nascent lesions. *Ann. Neurol.* **33,** 137–151.
83. Prineas, J. W., Barnard, R. O., Revesz, T., Kwon, E. E., Sharer, L., and Cho, E. (1993) Multiple sclerosis: pathology of recurrent lesions. *Brain* **116,** 681–693.
84. Raine, C. S., Scheinberg, L., and Waltz, J. M. (1981) Multiple sclerosis: oligodendrocyte survival and proliferation in an active established lesion. *Lab. Invest.* **45,** 534–546.
85. Brück, W., Schmied, M., Suchanek, G., Brück, Y., Breitschopf, H., Poser, S., et al. (1994) Oligodendrocytes in the early course of multiple sclerosis. *Ann. Neurol.* **35,** 65–73.
86. Chang, A., Tourtellotte, W. W., Rudick, R., and Trapp, B. D. (2002) Premyelinating oligodendrocytes in chronic lesions of multiple sclerosis. *N. Engl. J. Med.* **346,** 165–173.
87. Kornek, B. and Lassmann, H. (1999) Axonal pathology in multiple sclerosis: a historical note. *Brain Pathol.* **9,** 651–656.

88. Li, G. L., Farooque, M., Holtz, A., and Olsson, Y. (1995) Changes of β-amyloid precursor protein after compression trauma to the spinal cord: an experimental study in the rat using immunohistochemistry. *J. Neurotrauma* **12,** 269–277.

89. Pelletier, J., Suchet, L., Witjas, T., Habib, M., Guttmann, C. R. G., Salamon, G., et al. (2001) A longitudinal study of callosal atrophy and interhemispheric dysfunction in relapsing-remitting multiple sclerosis. *Arch. Neurol.* **58,** 105–111.

90. Bramlett, H. M., Kraydieh, S., Green, E. J., and Dietrich, W. D. (1997) Temporal and regional patterns of axonal damage following traumatic brain injury: a beta-amyloid precursor protein immunocytochemical study in rats. *J. Neuropathol. Exp. Neurol.* **56,** 1132–1141.

91. Yam, P. S., Takasago, T., Dewar, D., Graham, D. I., and McCulloch, J. (1997) Amyloid precursor protein accumulates in white matter at the matgin of a focal ischaemic lesion. *Brain Res.* **760,** 150–157.

92. Ferguson, B., Matyszak, M. K., Esiri, M., and Perry, V. H. (1997) Axonal damage in acute multiple sclerosis lesions. *Brain* **120,** 393–399.

93. Kornek, B., Storch, M., Weissert, R., Wallstroem, E., Stefferl, A., Olsson, T., et al. (2000) Multiple sclerosis and chronic autoimmune encephalomyelitis: a comparative quantitative study of axonal injury in active, inactive and remyelinated lesions. *Am. J. Pathol.* **157,** 267–276.

94. Trapp, B. D., Peterson, J., Ransahoff, R. M., Rudick, R., Sverre, M., and Bo, L. (1998) Axonal transection in the lesions of multiple sclerosis. *N. Engl. J. Med.* **338,** 278–285.

95. Matthews, P. M., De Stefano, N., Narayanan, S., Francis, G. S., Wolinsky, J. S., Antel, J. P., et al. (1998) Putting magnetic resonance spectroscopy studies in context: axonal damage and disability in multiple sclerosis. *Semin. Neurol.* **18,** 327–336.

96. Birken, D. L. and Oldendorf, W. H. (1989) *N*-Acetyl-L-aspartic acid: a literature review of a compound prominent in ^{1}H-NMR spectroscopic studies of brain. *Neurosci. Behav. Rev.* **13,** 23–31.

97. Fisher, E., Rudick, R. A., Cutter, G., Baier, M., Miller, D., Weinstock-Guttman, B., et al. (2000) Relationship between brain atrophy and disability: an 8-year follow-up study of multiple sclerosis patients. *Multiple Sclerosis* **6,** 373–377.

98. Pesini, P., Kopp, J., Wong, H., Walsh, J. H., Grant, G., and Hökfelt, T. (1999) An immunohistochemical marker for Wallerian degeneration of fibers in the central and peripheral nervous system. *Brain Res.* **828,** 41–59.

99. Grimaud, J., Barker, G. J., Wang, L., Lai, M., MacManus, D. G., Webb, S. L., et al. (1999) Correlation of magnetic resonance imaging parameters with clinical disability in multiple sclerosis: a preliminary study. *J. Neurol.* **246,** 961–967.

100. Paolillo, A., Pozzilli, C., Gasperini, C., Giugni, E., Mainero, C., Giuliani, S., et al. (2000) Brain atrophy in relapsing-remitting multiple sclerosis: relationship with "black holes," disease duration and clinical disability. *J. Neurol. Sci.* **174,** 85–91.

101. Lovas, G., Szilagyi, N., Majtenyi, K., et al. (2000) Axonal changes in chronic demyelinated cervical spinal cord plaques. *Brain* **123,** 308–317.

102. Mews, I., Bergmann, M., Bunkowski, S., Gullotta, F., and Bruck, W. (1998) Oligodendrocyte and axon pathology in clinically silent multiple sclerosis lesions. *Multiple Sclerosis* **4,** 55–62.

103. Evangelou, N., Esiri, M. M., Smith, S., Palace, J., and Matthews, P. M. (2000) Quantitative pathological evidence for axonal loss in normal appearing white matter in multiple sclerosis. *Ann. Neurol.* **47,** 391–395.

104. Bjartmar, C., Kinkel, P. R., Kidd, G., Rudick, R. A., and Trapp, B. D. (2001) Axonal loss in normal-appearing white matter in a patient with acute MS. *Neurology* **57,** 1248–1252.

105. Rivera-Quinones, C., McGavern, D., Schmelzer, J. D., Hunter, S. F., Low, P. A., and Rodriguez, M. (1998) Absence of neurological deficits following extensive demyelination in a class I-deficient murine model of multiple sclerosis. *Nat. Med.* **4,** 187–193.

106. Smith, K. J., Kapoor, R., and Felts, P. A. (1999) Demyelination: the role of reactive oxygen and nitrogen species. *Brain Pathol.* **9,** 69–92.

107. Gimsa, U., Peter, S. V. A., Lehmann, K., Bechmann, I., and Nitsch, R. (2000) Axonal damage induced by invading T cells in organotypic central nervous system tissue in vitro: involvement of microglial cells. *Brain Pathol.* **10,** 365–377.

108. Rawes, J. A., Calabrese, V. P., Khan, O. A., and DeVries, G. H. (1998) Antibodies to the axolemma-enriched fraction in the cerebrospinal fluid and serum of patients with multiple sclerosis and other neurological diseases. *Multiple Sclerosis* **3,** 363–369.

109. Sadatipour, B. T., Greer, J. M., and Pender, M. P. (1998) Increased circulating antiganglioside antibodies in primary and secondary progressive multiple sclerosis. *Ann. Neurol.* **44,** 980–983.

110. Perry, V. H., Brown, M. C., and Lunn, E. R. (1991) Very slow retrograde and Wallerian degeneration in the CNS of C57BL/Ola mice. *Eur. J. Neurosci.* **3,** 102–105.

111. Lyon, M. F., Ogunkolade, B. W., Brown, M. C., Atherton, D. J., and Perry, V. H. (1993) A gene affecting Wallerian nerve degeneration maps distally on chromosome 4. *Proc. Natl. Acad. Sci. USA* **90,** 9717–9720.

112. Brück, W., Brück, Y., Maruschak, B., and Friede, R. L. (1995) Mechanisms of macrophage recruitment in Wallerian degeneration. *Acta Neuropathol.* **89,** 363–367.

113. Coleman, M. P., Confort, L., Buckmaster, E. A., Tarlton, A., Ewing, R. E., Browns, M. C., et al. (1998) An 85-kb tandem triplication in the slow Wallerian degeneration (*Wld^s*) mouse. *Proc. Natl. Acad. Sci. USA* **95,** 9985–9990.

114. Conforti, L., Tarlton, A., Mack, T. G. A., Mi, W., Buckmaster, E. A., Wagner, D., et al. (2000) A Ufd2/D4Cole1e chimeric protein and overexpression of Rbp7 in the slow Wallerian degeneration (Wld^s) mouse. *Proc. Natl. Acad. Sci. USA* **97,** 11,377–11,382.

115. Castelijns, J. A. and Barkhof, F. (1999) Magnetic resonance (MR) imaging as a marker for multiple sclerosis. *Biomed. Pharmacother.* **53,** 351–357.

116. McFarland, H. F. (1999) Correlation between MR and clinical findings of disease activity in multiple sclerosis. *AJNR* **20,** 1777–1778.

117. Rovaris, M. and Filippi, M. (1999) Magnetic resonance techniques to monitor disease evolution and treatment trial outcomes in multiple sclerosis. *Curr. Opin. Neurol.* **12,** 337–344.

118. Werring, D. J., Brassat, D., Droogan, A. G., Clark, C. A., Symms, M. R., Barker, G. J., et al. (2000) The pathogenesis of lesions and normal appearing white matter changes in multiple sclerosis. A serial diffusion MRI study. *Brain* **123,** 1667–1676.

119. Truyen, L., van Waesberghe, J. H. T. M., van Walderveen, M. A. A., van Oosten, B. W., Polman, C. H., Hommes, O. R., et al. (1996) Accumulation of hypointense lesions ("black holes") on T_1 spin-echo MRI correlates with disease progression in multiple sclerosis. *Neurology* **47,** 1469–1476.

120. van Waesberghe, J. H. T. M., Kamphorst, W., De Groot, C. J. A., van Walderveen, M. A. A., Castelijns, J. A., Ravid, R., et al. (1999) Axonal loss in multiple sclerosis lesions: magnetic resonance imaging insights into substrates of disability. *Ann. Neurol.* **46,** 747–754.

121. Brex, P. A., Parker, G. J. M., Leary, S. M., Molyneux, P. D., Barker, G. J., Davie, C. A., et al. (2000) Lesion heterogeneity in multiple sclerosis: a study of the relations between appearances on T1 weighted images, T1 relaxation times, and metabolite concentrations. *J. Neurol. Neurosurg. Psychiatry* **68,** 627–632.

122. De Stefano, N., Narayanan, S., Matthews, P. M., Mortilla, M., Dotti, M. T., Federico, A., et al. (2000) Proton MR spectroscopy to assess axonal damage in multiple sclerosis and other white matter diseases. *J. Neurovirol.* **6(Suppl. 2),** S121–S129.

123. Jagust, W. J. and Noseworthy, J. H. (2000) Brain atrophy as a surrogate marker in MS. Faster, simpler, better? *Neurology* **54,** 782–783.

124. Deloire-Grassin, M. S. A., Brochet, B., Quesson, B., Delalande, C., Dousset, V., Canioni, P., et al. (2000) In vivo evaluation of remyelination in rat brain by magnetization transfer imaging. *J. Neurol. Sci.* **178,** 10–16.

20

In Vivo Imaging of Neuroinflammation in Neurodegenerative Diseases

Annachiara Cagnin, Alexander Gerhard, and Richard B. Banati

1. THE CONCEPT OF NEUROINFLAMMATION

Microglia are normally quiescent, mesoderm-derived brain macrophages and are the resident immunocompetent cells of the central nervous system (CNS). In conditions of intact blood-brain barrier when blood-borne cells are largely absent, microglia, together with perivascular cells, are the first line of the brain's immune defense system. Any even subtle or subacute neuronal insult/damage induces an activation of resting microglial cells. Activated microglia produce a widespread variety of pro-inflammatory molecules, change their morphology, and, if cell death occurs, finally mature into full-blown macrophages (1).

Rapid local activation of microglia can occur without the lymphocytic infiltrations characteristic of classical inflammatory brain disease, such as in multiple sclerosis. This observation has lead to the concept of "neuroinflammation" in a variety of primarily non-inflammatory neurological condition, including neurodegenerative diseases, such as Alzheimer's disease and Parkinson's disease (2), metabolic encephalopathies, such as hepatic encephalopathy, and processes involving neuronal remodeling and plasticity after brain or peripheral nerve injury (3,4).

Here, we briefly outline the rational for employing positron emission tomography (PET) using the ligand [^{11}C](R)-PK11195, which binds with relative cellular selectivity to activated microglia in order (1) to detect in vivo neuroinflammatory changes occurring in a variety of brain diseases and at different disease stages and (2) to monitor the progression of neuroinflammation as generic in vivo marker of "disease activity." The use of [^{11}C](R)-PK11195 PET is described as a systematic attempt at measuring the emerging phenomenology tissue pathology itself—as opposed to measuring, for example, the loss of neuronal function or structure. Although methodological issues, such sensitivity or absolute quantification of the specific ligand, may remain, [^{11}C](R)-PK11195 PET provides a first proof of principle for the clinical utility of imaging glial cells in vivo and its potential toward establishing a cell-biology-based in vivo neuropathology.

2. THE PK11195 BINDING SITE AND ITS CELLULAR SOURCE IN THE BRAIN

PK11195 [1-(2-chlorophenyl)-N-methyl-N-(1-methylpropyl)-3-isoquinoline carboxamide] is a specific ligand for the "peripheral benzodiazepine binding site" (PBBS),

From: *Neuroinflammation, 2nd Edition: Mechanisms and Management*
Edited by: P. L. Wood © Humana Press Inc., Totowa, NJ

which is particularly abundant on cells of mononuclear phagocyte lineage. Labeled with carbon-11, PK11195 can be used as a ligand for PET studies *(5–7)*. The name of PBBS reflects the fact that certain benzodiazepines, such as diazepam, bind this receptor, which is distinct from the central benzodiazepine receptor associated with γ-aminobutyric acid (GABA)-regulated channels. The PBBS is a heteromeric complex that appears to be largely, although not exclusively, localized in the outer membrane of mitochondria *(8–10)*. The gene for the PBBS is remarkably well preserved across species, implying an important housekeeping function. A role in the steroidogenesis and regulation of immunological response has been observed *(11)*.

In normal brain parenchyma, PBBS expression is low or absent, except for some constitutive expression in certain areas, such as those without blood-brain barrier (BBB) or high synaptic turnover. In brain diseases with disrupted blood-brain barrier, expression of PBBS colocalizes with infiltrating hematogenous cells *(5,12)*. In conditions of intact BBB, the PK11195 binding is the result of the de novo expression of PBBS in glial cells. Although astrocytes clearly express the PBBS at high levels when grown in cell culture, their contribution to the overall in vivo binding in tissue pathology is unclear *(13)*. Numerous experimental models have shown that autoradiographically detected PK11195-binding colocalizes better with the temporo-spatial profile of microglial activation than that of reactive astrocytes *(12,14–17)*. Histological data obtained from multiple sclerosis and experimental autoimmune encephalomyelitis tissue using high-resolution microautoradiography {with the *R*-enantiomer [^{3}H](R)-PK11195 PET *(7)*}, combined with immunohistochemical cell identification *(18)* has shown that increased binding of [^{3}H](R)-PK11195 is found on infiltrating blood-borne cells and activated microglia but not on reactive astrocytes. Remaining discrepancies to localization data using a polyclonal antibody may *(19)* may hint to differences between the sites or PBBS subunits recognized by immunostaining and those to which (R)-PK11195 binds on tissue sections and in vivo. For the purpose of interpreting [^{11}C](R)-PK11195 PET, however, there is no indication for any significant contribution to the signal by astrocytes. In vivo findings from [^{11}C](R)-PK11195 PET study performed in stable (i.e., few seizures) epilepsy patients with hippocampal sclerosis, a condition characterized by marked astrogliosis, showed no increases of [^{11}C](R)-PK11195 binding, supporting the view that the in vivo [^{11}C](R)-PK11195 signal is preferentially, if not exclusively, the result of the presence of activated microglia *(20)*.

3. IMAGING NEUROINFLAMMATION IN BRAIN DISEASES INVOLVING A DISTRIBUTED NEURAL NETWORK

In vitro experiments of microautoradiography using [^{3}H]PK11195 after peripheral axotomy has demonstrated that activated [^{3}H]PK11195-binding microglial cells are also found remote from the primary lesion in regions that are connected via retrograde and anterograde neuronal projections *(21)*. Axotomy of the facial nerve results in an activation of microglial cells in the facial nucleus as a response to the retrograde neuronal reaction. Likewise, axotomy of the sciatic nerve induces an anterograde microglial activation in the ipsilateral nucleus gracilis, the nucleus receiving the ipsilaterally projecting fibers ascending from the dorsal root ganglion. Similarly, an experimental focal ischemic lesion in the rat motor cortex leads to an increased [^{3}H]PK11195, binding not only at the site of the primary lesion but also secondarily in the ipsilateral thalamus reflecting damage to the cortico-thalamic projections *(22)*. In vivo, significantly increased [^{11}C]*(R)*-PK11195 signals in the thalamus secondary to pathology elsewhere have been observed in patients with cerebrovascular cortical stroke *(3)*.

Optic neuritis in patients with multiple sclerosis is a white matter tract pathology associated with anterograde glial reactions in the projection areas of the optic nerves (i.e., the lateral geniculate bodies). In this condition of optic tract pathology, increased [^{11}C]*(R)*-PK11195 binding was found in the lateral geniculate bodies, an observation that can clinically be related to the loss of vision *(18)*. Because volumetric T1-weighted (with and without gadolinium enhancement) and T2-weighted magnetic resonance images (MRIs) did not reveal disruption of the blood-brain barrier or obvious structural abnormalities in the lateral geniculate bodies, the increased [^{11}C]*(R)*-PK11195 binding is, indeed, indicative of a subtle microglial activation secondary to the degeneration of the optic tract.

Secondary neuroinflammatory reactions along fiber tracts may also occur throughout an entire neuronal circuitry. Such distributed and projected neuroinflammation along neuronal networks can be seen subacutely in patients with herpes simplex encephalitis. We studied clinically stable patients with unilateral or asymmetric herpes simplex encephalitis between *x* and *y* months after disease onset and conducted serial follow-up scans ([^{11}C]*(R)*-PK11195 PET and volumetric MRIs over a period of 2 yr *(23)*. Increased [^{11}C]*(R)*-PK11195 signals in the cerebral regions most damaged overlapped with those areas that subsequently developed the most marked atrophic changes. Additionally, microglial activation was found widely distributed beyond the recognizable structural damage. In fact, [^{11}C]*(R)*-PK11195 binding was increased also in regions contralateral to the affected site and in areas spared from structural damage but connected to the lesioned areas through white matter tracts. For example, in one patient, increased [^{11}C]*(R)*-PK11195 binding was present in structurally normal-appearing regions all belonging to the right limbic circuitry and along the superior longitudinal fasciculus and right angular gyrus. That latter observation appeared particularly relevant, because this patient presented with a peculiar cognitive deficit characterized by a decreased ability to recognize facial emotions, a task in which right angular gyrus seems critically involved *(24)*. These findings illustrate that focal damage can lead to a widespread microglial activation along an entire neural network (limbic structures and associated areas) and that the pattern of glial activation thus closely reflects neuronal function or deficit.

Activated microglia are also seen in the basal ganglia and brainstem nuclei of patients with idiopathic Parkinson's disease (PD) and multiple system atrophy (MSA) *(25,26)*. Increased [^{11}C]*(R)*-PK11195 binding in the substantia nigra and the globus pallidus *(27)* has been measured in PD patients (Fig. 1) whereas MSA patients show more widespread increases in [^{11}C]*(R)*-PK11195 binding, including nigra, pons, pallidum, caudate, putamen, and dorsolateral prefrontal cortex *(28)* (Fig. 2). These studies also reveal marked ingroup variations with a certain degree of overlap between the cases of typical and atypical parkinsonism. Although imaging of activated microglia with [^{11}C]*(R)*-PK11195 PET does not serve to override an a priori clinical classification, it provides a generic measurement of disease location and progression that may help to understand the heterogeneity in the clinical presentation of patients with parkinsonian or other neurodegenerative disorders.

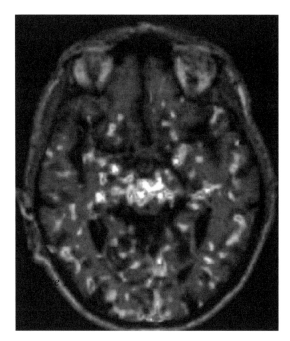

Fig. 1. Binding potential maps of PD patients coregistered to the individual MRI. Increased binding can be seen especially in the nigral and brainstem area. The binding potential maps has been masked for extracerebral binding in the scull.

4. IMAGING NEUROINFLAMMATION IN BRAIN DISEASES INVOLVING A LARGE-SCALE NEURAL NETWORK

Normal brain aging is clinically associated with a subtle cognitive decline characterized by a nonspecific pattern of impairment of higher cognitive functions involving mainly the attention. It has been hypothesized that the neuronal substrate of such cognitive impairment is a diffuse derangement of networks leading to a reduce ability on focus the neural activity on the task (disconnectivity theory) *(29)*. The neuropathological correlates are a widespread mild loss of cortical neurons and dendritic and synaptic compensatory changes. We have studied microglial activation in "normal" brain aging in 14 healthy cognitively intact subjects between the ages of 32 and 80 yr by means of magnetic resonance imaging (MRI) and [^{11}C](R)-PK11195 PET scans *(30)*. An age-related increase of [^{11}C](R)-PK11195 binding was found only in the thalamus but not in cortical areas (Fig. 3A). The absence of increased [^{11}C](R)-PK11195 binding in the aging cerebral cortex might be the result of a lack of sensitivity of [^{11}C](R)-PK11195 PET for the detection of the very low-grade and, consequently, subthreshold cortical pathology in normal aging brains. However, in the thalamus, the high density of direct corticothalamic projections converging into the thalamus appears to lead to a cumulative signal that essentially represents a regional amplification of the widespread but subthreshold cortical pathology.

The process of neurodegeneration in Alzheimer's disease (AD) is associated with local glial responses. Activated microglial cells have been found in the core and around the amyloid plaques in brains of AD patients. Some studies have implied an active role

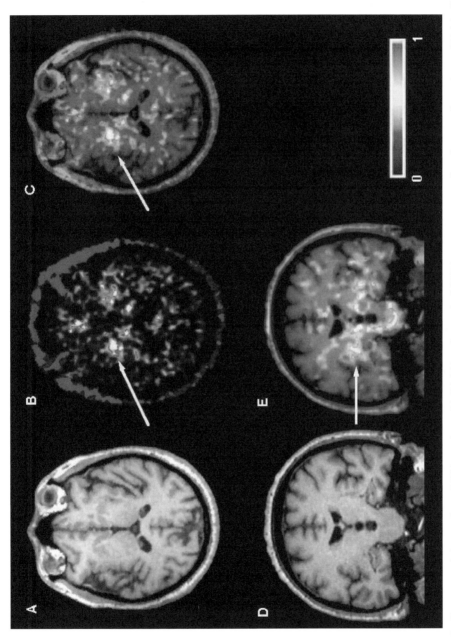

Fig. 2. Binding potential map for a MSA patient: (**A,D**) T1 volumetric MRI in transverse and and coronal sections; (**B**) binding potential map in transverse section; (**C,E**) coregistered binding potential maps to MR sections. The intensity scale is callibrated for binding potential values from 0 to 1. The binding potential map has been masked for extracerebral binding. Increased [^{11}C](R)-PK11195-binding can be seen in the nigra, putamen, pallidum, and pons.

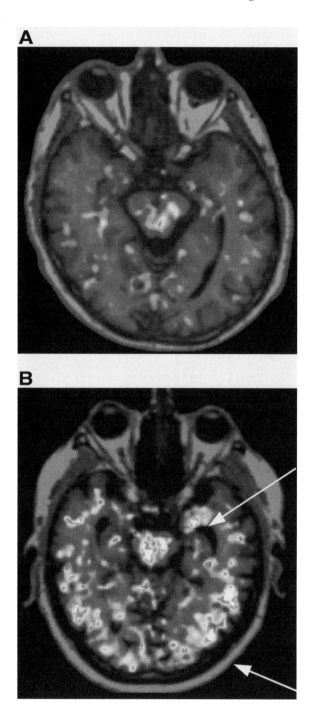

Fig. 3. $[^{11}C](R)$-PK11195-binding-potential maps in transverse sections of the brain of a normal elderly subject (**A**) and an Alzheimer's disease (AD) patient (**B**) with mild dementia. $[^{11}C]$ (R)-PK11195 PET images were coregistered to the individual MRI images. Increased $[^{11}C]$ (R)-PK11195 binding was detected in the amygdala (top arrow) and in the inferior lateral (bottom arrow) and mesial temporal lobe in the AD patient but not in the brain of the elderly control.

of microglia in the mediation of amyloid toxicity and subsequent secondary tissue damage via the release of cytokines and cytotoxic molecules *(31)*. This observation had led to the hypothesis that a chronic, self-perpetuating neuroinflammatory stimulation of microglia in AD pathology is an important pathogenetic mechanism. Consequently, the suppression of the brain's inflammatory immune reaction to neuronal degeneration (e.g., by the use of nonsteroidal anti-inflammatory therapy) may reduce the risk of developing AD and slow the progression of disease *(32)*.

In our $[^{11}C](R)$-PK11195 PET study performed in a group of AD patients with mild to moderate disease, increased signals were measured in the temporal cortex (particularly the fusiform, the prahippocampal, and the inferior temporal gyri), the inferior parietal cortex, the posterior cingulated, and the amygdala *(33)*. (Fig. 3B) The findings agree well with the known spatial and histopathological distribution of AD pathology *(34)*. In addition, we were able do demonstrate that areas with high $[^{11}C](R)$-PK11195 signals subsequently underwent the most marked atrophic changes within the following year, as shown by longitudinal serial volumetric MRI scans. This suggests that an in vivo measure of activated microglia provides an indirect index of disease activity. There is also first indications that in patients with minimal cognitive impairment (defined as isolated objective memory impairment and not yet demented) $[^{11}C](R)$-PK11195 PET may help to detect active neuroinflammatory pathology in specific areas, such as the inferior temporal cortex before conversion into clinically manifest dementia *(33)*. The patient with minimal cognitive impairment included in our study showed an increased $[^{11}C](R)$-PK11195 binding in subregions of the temporal lobe similar to those found in AD patients but not in the inferior parietal lobe. Interestingly, the areas with high $[^{11}C](R)$-PK11195 binding underwent marked atrophic changes 23 mo after the PET scanning although the cognitive performance remained stable over the same period of time. This finding supports the view that pathological process can remain unrecognized for several years before reaching the threshold of clinical manifestation.

ACKNOWLEDGMENTS

R.B.B. is supported by the Medical Research Council, the 5th European Framework Programme (MANAD), the Multiple Sclerosis Society of Great Britain and Northern Ireland, the Max-Planck-Institute of Neurobiology (Martinsried, Germany), and through the Deutsche Forschungsgemeinschaft grant "The mitochondrial benzodiazepine receptor as indicator of early CNS pathology, clinical application in PET." A.C. was supported by a Fellowship from the European Community (BMH4/CT98/5100) in the Training and Mobility of Researchers Programme in Biomedicine. A.G. is currently supported by an Action Research (UK) project grant.

REFERENCES

1. Kreutzberg, G. W. (1996) Microglia: a sensor for pathological events in the CNS. *TINS* **19,** 312–318.
2. Graeber, B. M. (2001) Glial inflammation in neurodegenerative diseases. *Immunology* **101 (Suppl. 1),** 52.
3. Pappata, S., Levasseur, M., Gunn, R. N., Myers, R., Crouzel, C., Syrota, A. et al. (1999) Thalamic microglial activation in ischemic stroke detected in vivo by PET and $[^{11}C](R)$-PK11195. *Neurology* **55,** 1052–1054.

4. Banati, A. R., Cagnin, A., Brooks, D. J., Gunn, R. N., Myers, R., et al. (2001) Long-term trans-synaptic glial responses in the human thalamus after peripheral nerve injury. *Neuroreport* **12,** 3439–3442.

5. Benavides, J., Cornu, P., Dennis, T., et al. (1988) Imaging of human brain lesions with an omega-3 site radioligand. *Ann. Neurol.* **24,** 708–712.

6. Cremer, J. E., Hume, S. P., Cullen, B. M., Myers, R., Manjil, L. G., Turton, D. R., et al. (1992) The distribution of radioactivity in brains of rats given [N-methyl-^{11}C]PK 11195 in vivo after induction of a cortical ischaemic lesion. *Int. J. Radiat. Appl. Instrum. B* **19,** 159–166.

7. Shah, F., Pike, V. W., Ashworth, S., and McDermott, J. (1994) Synthesis of the enantiomer of [N-methyl-^{11}C]PK11195 and comparison of their behaviours as PK (peripheral benzodiazepine) binding site radioligands in rats. *Nuclear Med. Biol.* **21,** 573–581.

8. Hertz, L. (1993) Binding characteristics of the receptor and coupling to transport proteins, in *Peripheral Benzodiazepine Receptors* (Giessen-Crouse, E., ed.), Academic, London, pp. 27–51.

9. Anholt, R. R., Pedersen, P. L., DeSouza, E. B., and Snyder, S. H. (1986) The peripheral-type benzodiazepine receptor. Localisation to the mitochondrial outer membrane. *J. Biol. Chem.* **261,** 776–783.

10. Hardwick, M., Fertikh, D., Culty, M., Li, H., Vidic, B., and Papadopoulos, V. (1999) Peripheral-type benzodiazepine receptor (PBR) in human breast tissue: correlation of breast cancer cell aggressive phenotype with PBR expression, nuclear localization and PBR-mediated cell proliferation and nuclear transport of cholesterol. *Cancer Res.* **59,** 831–842.

11. Gavish, M., Bachman, I., Shoukrun, R., Katz, Y., Veenman, L., Weisinger, G., et al. (1999) Enigma of the peripheral benzodiazepine receptor. *Pharmacol. Rev.* **51,** 629–650.

12. Dubois, A., Benavides, J., Peny, B., et al. (1988) Imaging primary and secondary ischaemic and excitotoxic brain lesions. An autoradiographic study of peripheral type benzodiazepine binding sites in the rat and cat. *Brain Res.* **445,** 77–90.

13. Itzhak, Y., Baker, L., and Norenberg, M. D. (1993) Characterization of the peripheral-type benzodiazepine receptors in cultured astrocytes: evidence for multiplicity. *Glia* **9,** 211–218.

14. Myers, R., Manjil, L. G., Cullen, B. M., Price, G. W., Frackowiak, R. S. J., and Cremer, J. E. (1991) Macrophage and astrocyte populations in relation to [3H]PK 11195 binding in rat brain cortex following a local ischaemic lesion. *J. Cereb. Blood Flow Metab.* **11,** 314–332.

15. Stephenson, D. T., Schober, D. A., Smalstig, E. B., Mincy, R. C., Gehlert, D. R., and Clemens, J. A. (1995) Peripheral benzodiazepine receptors are colocalized with activated microglia following transient global forebrain ischemia in the rat. *J. Neurosci.* **15,** 5263–5274.

16. Conway, E. L., Gundlach, A. L., and Craven, J. A. (1998) Temporal changes in glial fibrillary acidic protein messenger RNA and [3H] PK11195 binding in relation to imidazoline-I2-receptor and alpha 2-adrenoceptor binding in the hippocampus following transient global forebrain ischaemia in the rat. *Neuroscience* **82,** 805–817.

17. Rao, V. L., Dogan, A., Bowen, K. K., and Dempsey, R. J. (2000) Traumatic brain injury leads to increased expression of peripheral-type benzodiazepine receptors, neuronal death, and activation of astrocytes and microglia in rat thalamus. *Exp. Neurol.* **16,** 102–114.

18. Banati, R. B., Newcombe, J., Gunn, R. N., Cagnin, A., Turkhemer, F., Heppner, F., et al. (2000) The peripheral benzodiazepine binding site in the brain in multiple sclerosis: quantitative in vivo-imaging of microglia as a measure of disease activity. *Brain* **123,** 2321–2337.

19. Kuhlmann, A. C. and Guilarte, T. R. (2000) Cellular and subcellular localization of peripheral benzodiazepine receptors after trimethyltin neurotoxicity. *J. Neurochem.* **4,** 1694–1704.

20. Banati, R. B., Goerres, G. W., Myers, R., et al. (1999) [^{11}C](R)-PK11195 PET-imaging of activated microglia in vivo in Rasmussen's encephalitis. *Neurology* **53,** 2199–2203.

21. Banati, R. B., Myers, R., and Kreutzberg, G. W. (1997) PK ("peripheral benzodiazepine")-binding sites in the CNS indicate early and discrete brain lesions: microautoradiographic detection of [3H]PK11195 binding to activated microglia. *J. Neurocytol.* **26,** 77–82.

22. Myers, R., Manjil, L. G., Frackowiak, R. S., Cremer, J. E., et al. (1991) (3H) PK 11195 and the localisation of secondary thalamic lesions following focal ischemia in rat motor cortex. *Neurosci. Lett.* **133,** 20–24.

23. Cagnin, A., Myers, R., Gunn, R. N., Lawrence, A. D., Stevens, T., Kreutzberg, G. W., et al. (2001) In vivo-visualisation of activated glia by [11C](R)-PK11195 PET following herpes encephalitis reveals projected neuronal damage beyond the primary focal lesion. *Brain* **124,** 2014–2027.

24. Hamann, S. B., Stefanacci, L., Squire, L. R., Adolphs, R., Tranel, D., Damasio, H., et al. (1996) Recognizing facial emotion. *Nature* **379,** 497.

25. McGeer, P. L., Itagaki, S., Boyes, B. E., and McGeer, E. G. (1988) Reactive microglia are positive for HLA-DR in the substantia nigra of Parkinson's and Alzheimer's disease brains. *Neurology* **38,** 1285–1291.

26. Probst Cousin, S., Rickert, C. H., Schmid, K. W., and Gullotta, F. (1998) Cell death mechanisms in multiple system atrophy. *J. Neuropathol. Exp. Neurol.* **57,** 814–821.

27. Banati, R. B., Cagnin, A., Myers, R., Gunn, R., Piccini, P., Olanow, C. W., et al. (1999) In vivo detection of activated microglia by [11C] PK11195-PET indicates involvement of the globus pallidum in idiopathic Parkinson's disease. *Parkinsonism Related Disord.* **5,** 56–57.

28. Gerhard, A., Banati, R. B., Cagnin, A., Meyers, R., Gunn, R., Kreutzberg, G. W., et al. (2000) In vivo imaging of activated microglia with [11C]PK11195 positron emission tomography in patients with multiple system atrophy. *Mov. Disord.* **15,** 215.

29. Esposito, G., Kirkby, B. S., Van Horn, J. D., et al. (1999) Context dependent, neural system specific neurophysiological concomitants of ageing: mapping PET correlates during cognitive activation. *Brain* **122,** 77–90.

30. Cagnin, A., Myers, R., Gunn, R. N., Turkheimer, F. E., Cunningham, V. J., Brooks, D. J., et al. (2001) Imaging activated microglia in the ageing human brain, in *Physiological Imaging of the Brain with PET* (Gjedde, A., Hansen, S. B., Knudsen, G.M., and Paulson, O. B., eds.), Academic, San Diego, CA, pp. 361–367.

31. Kalaria, R. N., Harshbarger-Kelly, M., Cohen, D. L., and Premkumar, D. R. (1996) Molecular aspects of inflammatory and immune responses in Alzheimer's disease. *Neurobiol. Aging* **17,** 687–693.

32. McGeer, P. L. and McGeer, E. G. (1995) The inflammatory response system of brain: implications for therapy of Alzheimer and other neurodegenerative diseases. *Brain Res. Rev.* **21,** 195–218.

33. Cagnin, A., Brooks, D. J., Kennedy, A. M., Gunn, R. N., Myers, R., Turkeimer, F. E., Jones, T., et al. (2001) In-vivo measurement of activated microglia in dementia. *Lancet* **358,** 461–467.

34. Esiri, M. M., Hyman, B. T., Beyreuther, K., and Asters, C. L. (1997) Ageing and dementia, in *Greenfield's Neuropathology* (Graham, D. I. and Lantos, P. L., eds.), Oxford University Press, New York, pp. 153–231.

V

Parkinson's and Huntington's Diseases

.

21

Inflammatory Mechanisms in Parkinson's Disease

Joseph Rogers and Carl J. Kovelowski

1. INTRODUCTION

Although other neurotransmitter systems may also be involved, the primary pathology underlying Parkinson's disease (PD) is damage to and loss of pigmented dopamine (DA) neurons in the lateral and ventral substantia nigra pars compacta (SN). Whether inflammation causes, contributes to, or is a minor consequence of DA degeneration in PD remains undetermined, but there is clear evidence that it occurs, including numerous reports of focal gliosis, which forms extensive SN scars in severe cases of neuron damage *(1–4)*, as well as increases in classic markers of inflammatory attack.

2. MOLECULAR MEDIATORS OF PARKINSON'S DISEASE INFLAMMATION

At the molecular level, a wide range of inflammatory mediators has been implicated in PD pathophysiology. For example, complement proteins C1–C9 and their respective mRNAs are upregulated in PD compared to normal elderly SN (reviewed in ref. *5*). The proteins are not merely present, but they are fully activated for inflammatory actions, as indicated by the presence of anaphylatoxin and opsonizing split products *(5,6)*. Our preliminary data also show that microglia in the PD SN express much more CD11B, the receptor for the pivotal complement opsonins C3b/iC3b, than do human elderly control microglia (Fig. 1). The source for complement activation in PD is uncertain. PD serum antibodies, when injected into rodent SN, cause a significantly greater loss of DA neurons than control serum antibodies *(7)*, a result that is presumably mediated by antibody-dependent complement activation *(8)*. Alternatively, multiple antibody-independent activators of complement have been demonstrated *(9–12)*, and some of these (e.g., naked DNA, neurofilaments) *(11,12)* would be expected to be present as the detritus from prior DA losses in the PD SN. Whatever its source, complement activation can be one of the most toxic mechanisms in the inflammatory arsenal and is fully capable of damaging not only targeted cells but also "innocent bystander" cells *(13)*. Indeed, it has been estimated that if regulation by complement inhibitors such as CD59 were rendered inoperative, spontaneous activation of complement present in the circulation would be sufficient to lyse virtually every cell in the body in a matter of minutes (S. Meri and U. Helsinki, personal communication).

From: *Neuroinflammation, 2nd Edition: Mechanisms and Management*
Edited by: P. L. Wood © Humana Press Inc., Totowa, NJ

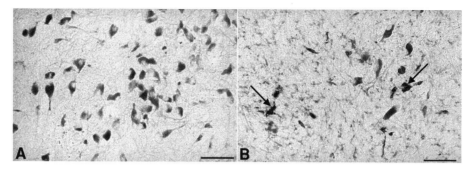

Fig. 1. Relationships of activated microglia (CD11b antibody with black reaction product) and dopamine neurons (tyrosine hydroxylase antibody with brown reaction product) in normal elderly control (**A**) and PD (**B**) substantia nigra. Control nigra shows only weak immunoreactivity for activated microglia, whereas immunoreactivity is profuse in the PD nigra. Moreover, in the PD nigra activated microglia cluster around and on (arrows) deteriorating dopamine neurons.

The triad of classic pro-inflammatory cytokines, interleukin-1 (IL-1) *(14–16)*, IL-6 *(14)*, and tumor necrosis factor-α (TNF-α) *(14,16–18)*, is reportedly increased in PD SN, as is the transcription factor NF-κB *(19)*, which helps regulate IL-1, IL-6, and TNF-α expression. IL-6 is also increased in the striatum of mice treated with MPTP (1-methyl-4-phenyl-1,2,3,6-tetrahydropyridine), an animal model of PD *(20)*. Human elderly microglia express these cytokines in vitro *(21)* and in PD SN *(15)*. TNF-α-immunoreactive microglia are particularly prominent around intact and deteriorating SN DA neurons in PD, but are not detected in normal elderly SN *(15)*. Type 1 TNF-α receptors are found on human SN DA neurons *(18)*. Activation of the TNF-α signal transduction pathway in cultured mesencephalic neurons results in NF-κB translocation, production of oxygen free radicals, and apoptosis of the cells *(19)*. In addition to the SN, IL-1β, IL-6, and TNF-α concentrations are elevated in PD cerebrospinal fluid (CSF) *(14,22,23)*, caudate, and putamen *(14)*. These molecules amplify and sustain inflammatory responses, and are generally considered to be pathogenic elements in numerous inflammatory disorders such as rheumatoid arthritis and inflammatory bowel syndrome. Interferon (INF)-γ, which can modulate glial major histocompatibility complex (MHC) expression *(24)*, midkine, a chemokine mediating chemotaxis and activation of inflammatory cells, and chromogranin A, another activator *(25)*, are elevated in PD SN.

Increased lipid peroxidation and oxidative stress are widely reported in the PD SN *(26–34)*. Although there are many possible sources for these pathophysiologic processes, including mitochondrial dysfunction, it is worth remembering that reactive oxygen and nitrogen species constitute a primary means of attack for inflammatory cells (i.e., the respiratory burst). Nitric oxide (NO) reacts with superoxide to form peroxynitrite, a potent oxidant *(35)*, and glial increases in NO expression are believed to have detrimental consequences for compromised neurons *(36–39)*. NO-dependent neuronal *(40)* and oligodendrocyte *(37)* injury has been demonstrated in culture, and inhibition of inducible nitric oxide synthase (iNOS) improves neuronal survival in mixed primary rodent *(41)* and human fetal *(40)* cortical cultures. Elevated iNOS expression has also been reported in brain microvessels *(42)* and neurons bearing neurofibrillary tangles *(49)* in Alzheimer's disease (AD), and in amoeboid microglia in demyelinating multiple sclerosis lesions *(37,43,44)*. Increased iNOS immunoreactivity is observed in ameboid microgia in PD but not control SN *(4)*. Because melanogenesis in DA SN neurons may lead to production of reactive oxygen intermediates, it has been suggested that these cells may be especially vulnerable to oxidative stress (reviewed in ref. *33*).

Prostaglandins are another set of inflammatory mediators that have been extensively investigated for their potential role in neurodegenerative disorders such as PD. These molecules are synthesized from arachidonic acid by cyclo-oxygenase (COX), of which there are two types. COX-I is a predominantly constitutive isoform that is involved both in inflammation and cellular homeostasis *(45)*, whereas COX-2 is an inducible isoform that is upregulated by inflammatory cytokines, mitogens, and reactive oxygen intermediates *(46–48)*. Inhibition of COX in rodent brain protects against ischemic damage *(49,50)*, kianic acid-induced seizures, and neurotoxicity *(51)*. Cytotoxic effects of COX are thought to be mediated through reactive oxygen species generated by peroxidase conversion of prostaglandin (PG)-G, to PGH *(50)*. Reactive astrocytes express COX-2 following damage *(52,53)*, and astrocytes cultured from rodent cortex, cerebellum, and spinal cord synthesize and release PGD_2, PGE_2, prostacyclin, thromboxane A_2, and PGF_2 *(54)*. Neuronal COX-1 and COX-2 immunoreactivity is observed in human elderly SN *(4,55)*, with no apparent difference in staining when control and PD samples are compared. Likewise, neuronal COX-2 staining remains unchanged after cerebellar kianic acid *(52)* or hippocampal bacillus Calmette–Guerin *(56)* lesions. By contrast, COX-1 and COX-2 immunoreactivity of reactive astrocytes and activated microglia increases in affected areas under neurodegenerative conditions *(52,55)*, including PD SN *(55)*.

3. CELLULAR MEDIATORS OF PARKINSON'S DISEASE INFLAMMATION

At the cellular level, activated glia, particularly activated microglia, have been widely implicated as cellular effectors of brain inflammation in a number of neurodegenerative disorders, including PD (reviewed in ref. *5*). Degenerating PD DA neurons are initially covered by astrocyte processes and surrounded by ramified microglia *(4)*. Subsequently, astrocyte processes withdraw from the degrading neuronal somata and amoeboid microglia accumulate within the astrocyte envelope *(4)*. The amoeboid morphology is one that is taken on by microglia when they become highly activated under inflammatory conditions. Notably, the SN is reported to have one of the highest concentrations of microglia in the brain *(57)*.

The change of glia from resting to activated states could be doubly pathologic, because, under normal conditions, resting astrocytes and microglia subserve important support functions for neurons. In culture, for example, medium conditioned either by rodent astrocytes *(58,59)* or microglia *(60)* supports DA neuron survival. These beneficial functions can even extend, in part, to brain injury. For example, neurotrophic factors are synthesized by astrocytes cultured from rodent hemiparkinsonian-lesioned brain *(61)*, and activated microglia express neurotrophic factors and induce DA sprouting in the injured rodent striatum *(62)*. Hirsch et al. *(15)* report that DA vulnerability in SN pars compacta, ventral tegmental area, and the catecholaminergic A8 cell group varies inversely with the number of astrocytes normally present in each, implying that the more astrocytes, the more protection of DA neurons *(15)*. Glial-derived neurotrophic factor (GDNF) administration has protective effects against 6-hydroxydopamine (6-OHDA) SN lesions at pharmacologic and behavioral levels (reviewed in ref. *63*). Transforming growth factor-β1 (TGF-β1) and TGF-β2 levels are increased in the PD striatum and CSF

(64–66), consistent with attempts to quench inflammation and to induce new neuritic growth.

Many of the normal trophic functions of glia may be lost or overwhelmed when the cells become chronically activated in progressive neurodegenerative disorders, for there is abundant evidence that in such disorders, activated glia play destructive roles by direct and indirect inflammatory attack *(67–73)*. For example, virtually all of the inflammatory mediators that have been shown to be upregulated in PD SN can be produced by glia, especially activated microglia *(21; reviewed in refs. 5 and 73)*. Activated microglia also express Fc receptors, CD4 antigen, and MHC classes I and II antigens *(3,74–77)*. Both astrocytes and microglia in culture can generate the calcium-independent, readily inducible form of nitrogen oxide synthase (iNOS) *(40,78–80)*. In turn, iNOS is upregulated following exposure to various cytokines *(81)*, many of which are elevated in PD SN. Knott and colleagues *(55)* have characterized the cellular sources of iNOS in human elderly control and PD SN, first demonstrating that process-bearing iNOS immunoreactive cells in both conditions colocalized either with a glial fibrillary acidic protein (GFAP) marker for astrocytes or an EBM II marker for microglia. In control SN, the percentage of iNOS-immunopositive astrocytes and microglia was less than 10–20% of the glial population. In PD SN gliotic foci, however, amoeboid microglia surrounding fragmented neurons were found to be the major iNOS-immunoreactive cell type. Virtually all these cells were positive for iNOS, as previously shown by Hunot et al. *(82)*. In contrast, the astrocyte contribution to total glial iNOS was small, consistent with findings in culture that microglia produce more NO per cell than astrocytes *(36)*.

Smeyne et al. *(83)* used novel chimeric murine SN cultures to show that C57Bl/6J and SWR/J neurons were significantly more sensitive to the effects of MPTP when cocultured with C57Bl/6J glia than SWR/J glia, suggesting to the authors that glia are a critical cell type in the genesis of experimental parkinsonism. Lipopolysaccharide (LPS) activation of astrocytes reportedly causes extracellular accumulation of hydrogen peroxide, glutamate, and NO *(84)*, and enhances neurotoxicity to MPTP *(85)*.

4. INFLAMMATORY MECHANISMS IN ANIMAL MODELS OF PARKINSON'S DISEASE

At the whole animal level, MPTP treatment has been shown to result in a wide range of inflammatory responses *(86)*, including enhanced production of IL-6 *(20)*, lipid peroxidation *(31,32)*, generation of reactive oxygen intermediates *(31,87)*, activation of microglia *(88,89)* and astrocytes *(90–94)*, upregulation of endothelial ICAM *(90)*, activation of the c-Jun N-terminal kinase (JNK) pathway *(95–97)*, and infiltration by CD4+ and CD8+ lymphocytes *(90)*. Conversely, specific COX-1 and COX-2 inhibitors *(98)*, aspirin *(99)*, and sodium salicylate *(99,100)* have been found to block MPTP-mediated DA depletion and motor deficits. 6-OHDA lesions of the SN also result in cytokine upregulation *(101)*, lipid peroxidation *(34,102)*, generation of reactive oxygen species *(103)*, and microglial activation/proliferation *(104)*.

Perhaps one of the most compelling studies of PD SN inflammation is the demonstration that a single injection of LPS into rodent SN activates microglia/macrophages and induces selective death of SN DA neurons and depletion of DA and its metabolites without significant damage to serotonergic, cholinergic, or GABAergic neurons or their neurotransmitter products. The loss of DA function is still detectable 1 yr postinjection *(105)*. Lu et al. *(106)* injected LPS just above the rodent SN and observed widespread SN glial activation followed by death of SN neurons. The importance of these findings and those on amelioration of PD symptoms by anti-inflammatory drugs cannot be overemphasized, as they suggest that PD inflammation is pathogenic rather than a mere detritus-clearing process. Moreover, these studies demonstrate that inflammation selectively targets the DA system, a specificity that is not encountered in any other neurodegenerative disorder.

5. MECHANISMS FOR INFLAMMATORY ATTACK ON DA NEURONS

From this brief review, it is clear at multiple levels of analysis that inflammation targets DA neurons in the PD SN. Four relatively global possibilities might be considered in attempts to account for this phenomenon. The first and most obvious is simple detritus removal; that is, when cells of the body sustain damage and die, innate inflammatory responses arise to remove the detritus. The brain is no exception, as has been amply documented in a number of neurodegenerative disorders. Because DA neurons are specifically vulnerable in PD, inflammatory mechanisms may naturally be directed toward DA neurons or their remnants. Although this view relegates PD SN inflammation to an effect of the specific vulnerability of PD DA neurons, rather than a cause of such vulnerability, it may still have substantial pathogenic relevance for two reasons. First, the inflammatory mechanisms that are invoked to remove detritus may sometimes cause secondary bystander damage. This is typically not a problem in organ systems where new cells can rapidly replace those that have been inadvertently damaged by the inflammatory response, but it is a difficult problem for the nervous system. Second, novel interactions of inflammatory mediators with pathologic elements of various neurodegenerative disorders have been observed. Nearly a dozen pro-inflammatory molecules in AD, for example, have been found to bind amyloid β peptide (Aβ) and to alter its tertiary structure, its metabolism, or both (reviewed in ref. *73*). In multiple sclerosis, certain components of myelin have been reported to activate complement in an antibody-independent fashion *(107)*. Thus, the initiation of innate inflammatory mechanisms at sites of neuron damage in the context of detritus removal can be seen to carry with it the inherent risk of causing further damage through "innocent bystander" mechanisms and through unexpected interractions of inflammatory molecules with pathologic elements characteristic of the neurodegenerative disorder. These problems may be further compounded by the fact that the activation of glia from their resting states deprives neurons of nurturing or protective factors generated by the resting glia.

A second possible explanation for the selective inflammatory attack on SN DA neurons in PD is that PD SN DA neurons may possess some novel attribute that invites the attack. Such a mechanism is strongly suggested by the finding that a single injection of the inflammatory stimulant LPS into rodent SN results in loss of DA markers but not markers for other SN neurotransmitters *(105)*. Moreover, LPS stimulation in rodent SN appears to cause pronounced astrocytic reactivity and microglial activation that colocalizes with DA neurons *(105,106)*, consistent with the glial responses observed in the PD SN *(4,5)*. Although it is possible that LPS directly damages DA neurons and therefore stimulates inflammatory attack in the context of detritus removal, we have not seen any direct toxicity of LPS on neurons in our cell culture models, nor do we know of any reports of direct LPS neurotoxicity in the literature. A more parsimonious explanation for the collective findings from animal models is that DA neurons possess or secrete some novel factor or factors that invites inflammatory attack by activated microglia and/or reactive astrocytes.

A novel attribute mechanism may also extend to PD glia; that is, there may be some novel, inherent difference in PD SN astrocytes and/or microglia that leads them to attack DA neurons under conditions where normal elderly glia would not do so. Superficially, one might suspect that this attribute(s) would arise only after the astrocytes have become reactive and/or the microglia have become activated; otherwise, damage to DA neurons would begin at birth (when glia are presumably unactivated) and continue throughout life. However, PD can strike as early as the twenties and the activation state of glia in PD patients prior to the onset of clinical symptoms is not known. Moreover, because resting microglia and astrocytes are well established to play important supportive roles for neurons in the normal brain, it could be that some constitutive defect in normal PD SN glia makes them less able to nurture DA neurons. To our knowledge, none of these possibilities has been previously explored.

A fourth possible mechanism for selective inflammatory damage to PD SN DA neurons is that various inflammatory toxins are generated in the PD SN without specific targets, but DA neurons are more vulnerable to these toxins than other cells *(108)*. In this view, the inflammatory attack itself is not selective; rather, selectivity is the byproduct of a heightened sensitivity to inflammatory attack on the part of DA neurons. For example, melanogenesis in SN DA neurons may increase their susceptibility to oxidative stress *(33)*. Alternatively, rodent SN DA neurons are not pigmented, yet they are still selectively lost after stimulation of inflammation by LPS *(105,106)*. Special vulnerability of DA neurons to inflammation may also not adequately explain glial distributions in the PD SN. Inflammatory attack in the PD SN must be mounted by glia, either through secretion of toxic intermediates or by direct cell–cell interactions, both of which require proximity for selective operation. PD SN glia do indeed specifically cluster around DA neurons, but not other SN neurotransmitter phenotype neurons, a selective chemotaxis that is unexplained by the hypothesis of heightened DA vulnerability to inflammatory attack. Even if DA neurons are ultimately more susceptible, some novel attribute of DA neurons must draw activated glia specifically to them before their hypothetical heightened vulnerability to attack becomes a factor. These and other scenarios to explain the relationship between PD inflammation and PD DA loss warrant substantial further research.

6. COMPARISON OF PARKINSON'S DISEASE AND ALZHEIMER'S DISEASE INFLAMMATION

The molecules, cell types, and mechanisms that mediate inflammation in PD and AD appear to be highly similar, with the main difference being that AD inflammation research has been ongoing for many more years, permitting an even more expansive list of players. Complement, cytokines, chemokines, inflammation-related growth factors, inflammation-related transcription factors, microglia, and astrocytes have all been implicated in both disorders (for a complete review in AD, *see* ref. *73*). The underlying cause of the symptoms of PD and AD is neuron and neurite loss, both of which occur over a disease course that can last a decade or more and which have pathologic antecedents that arise many years prior to the onset of clinical signs. Just as the basic principles that characterize peripheral inflammatory attack can typically be applied to inflammation in AD, so, also, many of the lessons learned in AD may help inform research into PD inflammation, and vice versa.

What perhaps distinguishes PD and AD inflammation most are their initiating stimuli and eventual targets. In AD, Aβ deposits and neurofibrillary tangles represent chronic, highly insoluble, β-pleated, abnormal protein aggregates—classic qualities for stimulating inflammation. Consistent with this, the inflammatory molecules and cells in AD brain virtually always find their highest concentrations or reactivity associated with Aβ and tangles. Clusters of MHCII immunoreactive microglia within Aβ deposits, for example, were the first indication that inflammation might be occurring in AD *(109, 110)*, and the full range of classical pathway complement molecules can be visualized in adjacent sections of a single Aβ plaque *(112)*. Neighboring neurites are most likely attacked as innocent bystanders in AD, with little known specificity for the neuron types that are damaged. By contrast, PD inflammation appears to specifically target DA neurons. Although Lewy bodies do constitute abnormal, relatively inert aggregates in PD, a glial and inflammatory focus clearly extends to SN DA neurons that do not exhibit overt Lewy body inclusions.

7. SUMMARY AND CONCLUSIONS

There is now compelling evidence at molecular, cellular, and whole animal levels that inflammation plays at least some pathophysiologic role in PD. This would not be inconsistent with many of the putative etiologies of PD such as repeated head trauma, central nervous system viral infections, and even environmental toxins such as MPTP, all of which have been shown to have inflammation as one of their final common denominators. Given recent results from animal models of PD, it may soon be time to entertain clinical studies with anti-inflammatory drugs to delay the onset or, as an adjunct to existing symptomatic treatments, slow the progression of PD.

ACKNOWLEDGMENTS

This work was supported by NIA grant no. AGO7367, the Alzheimer's Association, and the Arizona Alzheimer's Research Center.

REFERENCES

1. Forno, L. S. (1996) Neuropathology of Parkinson's disease. *J. Neuropathol. Exp. Neurol.* **55,** 259–272.
2. Forno, L. S., DeLanne, L. E., Irwin, I., Di Monte, D., and Langston, J. W. (1992) Astrocytes and Parkinson's disease. *Prog. Brain Res.* **94,** 429–436.
3. McGeer, P. L., Itagaki, S., Boyes, B. E., and McGeer, E. G. (1988) Reactive microglia are positive for HLA-DR in the substantia nigra of Parkinson's and Alzheimer's disease brains. *Neurology* **38,** 1285–1291.
4. Knott, C., Wilkin, G. P., and Stern, G. (1999) Astrocytes and microglia in the substantia nigra and caudete-putamen in Parkinson's disease. *Parkinson's Rel. Dis.* **5,** 115–122.
5. McGeer, P. L., Yasojima, K., and McGeer, E. G. (2001) Inflammation in Parkinson's disease. *Parkinson's Dis. Adv. Neurol.* **86,** 83–89.
6. Togo, T., Iseki, E., Marui, W., Akiyama, H., Ueda, K., and Kosaka, K. (2001) Glial involvement in the degeneration process of Lewy body-bearing neurons and the degradation process of Lewy bodies in brains of dementia with Lewy bodies. *J. Neurol. Sci.* **184,** 71–75.

7. Chen, S., Le, W. D., Xie, W. J., Alexianu, M. E., Engelhardt, J. I., Siklos, L., et al. (1998) Experimental destruction of substantia nigra Initiated by Parkinson disease immunoglobulins. *Arch. Neurol.* **55,** 1075–1080.

8. Defazio, G., Daltoso, R., Benvegnu, D., Minozzi, M. C., Cananzi, A. R., and Leon, A. (1994) Parkinsonian serum carries complement-dependent toxicity for rat mesencephalic dopaminergic neurons in culture. *Brain Res.* **633,** 206–212.

9. Rogers, J., Cooper, N. R., Schultz, J., McGeer, P. L., Webster, S., Styren, S. D., et al. (1992) Complement activation by β-amyloid in Alzheimer's disease. *Proc. Natl. Acad. Sci. USA* **89,** 10,016–10,020.

10. Johns, T. G. and Bernard, C. C. (1997) Binding of complement component C1q to myelin oligodendrocyte glycoprotein: a novel mechanism for regulating CNS inflammation. *Mol. Immunol.* **34(1),** 33–38.

11. Messmer, B. T. and Thaler, D. S. (2000) C1q-Binding peptides share sequence similarity with C4 and in complement activation. *Mol. Immunol.* **37(7),** 343–350.

12. Linder, E., Lehto, V. P., and Stenman, S. (1979) Activation of complement by cytoskeletal intermediate filaments. *Nature* **278(5700),** 176–178.

13. Cooper, N. R. (1999) The complement system, in *Inflammation: Basic Principles and Clinical Correlates* (Gatlin, J. and Snyderman, R., eds.), Lippincott–Raven, Philadelphia, pp. 281–315.

14. Mogi, M., Harada, M., Kondo, T., Riederer, P., Inagaki, H., Minami, M., et al. (1994) Interleukin-1β, Interleukin-6, epidermal growth factor and transforming growth factor-β are elevated in the brain from Parkinsonian patients. *Neurosci. Lett.* **180,** 147–150.

15. Hirsch, E. D., Hunot, S., Damier, P., and Faucheux, B. (1998) Glial cells and inflammation in Parkinson's disease: a role in neurodegeneration? *Ann. Neurol.* **44(3 Suppl. 1),** 115–120.

16. Hunot, S., Dugas, N., Faucheux, B., Hartmann, A., Tardieu, M., Debre, P., et al. (1999) FcγRII/CD23 is expressed in Parkinson's disease and induces, *in vitro*, production of nitric oxide and tumor necrosis factor-α in glial cells. *J. Neurosci.* **19,** 3440–3447.

17. Boka, G., Anglade, P., Wallach, D., Javoy-Agid, F., Agid, Y., and Hirsch, E. C. (1994) Immunocytochemical analysis of tumor necrosis factor and its receptor in Parkinson's disease. *Neurosci. Lett.* **172,** 151–154.

18. Boka, G., Anglade, P., Wallach, D., et al. (1994) Immunocytochemical analysis of tumor necrosis factor and its receptors in Parkinson's disease. *Neurosci. Lett.* **172,** 151–154.

19. Hunot, S., Brugg, B., Ricard, D., Michel, P. P., Muriel, M.-P., Ruberg, M., et al. (1997) Nuclear translocation of NF-kB is increased in dopaminergic neurons of patients with Parkinson's disease. *Proc. Natl. Acad. Sci, USA* **94,** 7531–7536.

20. Kaku, K., Shikimi, T., Kamisaki, Y., Shinozuka, K., Ishino, H., Okunishi, H., et al. (1999) Elevation of striatal interleukin-6 and serum corticosterone contents in MPTP-treated mice. *Clin. Exp. Pharmacol. Physiol.* **26,** 680–683.

21. Lue, L. F., Rydel, R., Brigham, E. F., Yang, L. B., Hampel, H., Murphy, G. M., et al. (2001) Inflammatory repertoire of Alzheimer's disease and nondemented elderly microglia in vitro. *Glia* **35,** 72–79.

22. Blum-Degen, D., Muller, T., Kuhn, W., Gerlach, M., Przuntek, H., and Riederer, P. (1995) Interleukin-1β and interleukin-6 are elevated in the cerebrospinal fluid of Alzheimer's and *de novo* Parkinson's disease patients. *Neurosci. Lett.* **202,** 17–20.

23. Mogi, M., Harada, M., Riederer, P., et al. (1994) Tumor necrosis factor-α (TNF-α) increases both in the brain and in the cerebrospinal fluid from Parkinsonian patients. *Neurosci. Lett.* **180,** 147–150.

24. Hunot, S., Becard, C., Faucheux, B., Agid, Y., and Hirsch, E. C. (1997) Immuno-histochemical analysis of interferon-γ and interleukin-1β in the substantia nigra of Parkinsonian patients. *Mov. Disord.* **12(Suppl. 1),** 20 (abstract).

25. Yasuhara, O., Kawamata, T., Aimi, Y., et al. (1994) Expression of chromogranin A in lesions of the central nervous system from patients with neurological diseases. *Neurosci. Lett.* **170,** 13–16.

26. Torreilles, F., Salman-Tabcheh, S., Guerin, M.-C., and Torreilles, J. (1999) Neuro-degenerative disorders: the role of peroxynitrite. *Brain Res. Rev.* **30,** 153–163.

27. Dexter, D. T., Carter, C. J., Wells, F. R., et al. (1989) Basal lipid peroxidation in substantia nigra is increased in Parkinson's disease. *J. Neurochem.* **52,** 381–389.

28. Dexter, D. T., Holley, A. E., Flitter, W. D., et al. (1994) Increased levels of lipid hydroperoxides in the Parkinsonian substantia nigra: an HPLC and ESR study. *Mov. Disord.* **9,** 92–97.

29. Stoessl, A. J. (1999) Etiology of Parkinson's disease. *Can. J. Neurol. Sci.* **26(Suppl. 2),** S5–S12.

30. Rajput, A. H. (1992) Frequency and cause of Parkinson's disease. *Can. J. Neurol. Sci.* **19(Suppl. 1),** 103–107.

31. Jenner, P. (1998) Oxidative mechanisms in nigral cell death in Parkinson's disease. *Mov. Disord.* **13,** 24–34.

32. Youdim, M. B. H., Ben-Shachar, D., and Riederer, P. (1993) The possible role of iron in the etiopathology of Parkinson's disease. *Mov. Disord.* **8,** 1–12.

33. Hegedus, Z. L. (2000) The probable involvement of soluble and deposited melanins, their intermediates and the reactive oxygen side-products in human diseases and aging. *Toxicology* **145,** 85–101.

34. Kumar, R., Agarwal, A. K., and Seth, P. K. (1995) Free radical-generated neurotoxicity of 6-hydroxydopamine. *J. Neurochem.* **64,** 1703–1707.

35. Dawson, T. M., Dawson, V. L., and Synder, S. H. (1992) A novel neuronal messenger molecule in brain: the free radical, nitric oxide. *Ann. Neurol.* **32,** 297–311.

36. Boje, K. M. and Arora, P. K. (1992) Microglial-produced nitric oxide and reactive nitrogen oxides mediate neuronal cell death. *Brain Res.* **587,** 250–256.

37. Merrill, J. D., Ignarro, L. J., Sherman, M. P., Melinek, J., and Lane, T. E. (1993) Microglial cell cytotoxicity of oligodendrocytes is mediated via nitric oxide. *J. Immunol.* **151,** 2132–2141.

38. Jenner, P. (1994) Oxidative damage in neurodegenerative disease. *Lancet* **344,** 796–798.

39. Wilkin, G. P. and Knott, C. (1999) Glia: a curtain raiser. *Parkinson's Dis. Adv. Neurol.* **80,** 3–7.

40. Chao, C. C., Hu, S., Sheng, W. S., Bu, D., Bukrinsky, M. I., and Peterson, P. K. (1996) Cytokine-stimulated astrocytes damage human neurons via a nitric oxide mechanism. *Glia* **16,** 276–284.

41. Dawson, V. L., Bahmblatt, H. P., Mong, J. A., and Dawson, T. M. (1994) Expression of inducible nitric oxide synthase causes delayed neurotoxicity in primary mixed neuronal-glial cortical cultures. *Neuropharmacology* **33,** 1425–1430.

42. Dorheim, M. A., Tracey, W. R., Pollock, J. S., and Grammas, P. (1994) Nitric oxide synthase activity is elevated in brain microvessels in Alzheimer's disease. *Biochem. Biophys. Res. Commun.* **205,** 659–665.

43. Bagasra, O., Michaelis, F. H., Mu Zheng, Y., Bopbroski, L. E., Spitsin, S. V., Fang Fu, Z., et al. (1995) Activation of inducible form of nitric oxide synthase in the brains of patients with multiple sclerosis. *PNAS* **92,** 12,041–12,045.

44. DeGroot, C. J. A., Ruuls, S. R., Theeuwes, J. W. M., Dijkstra, C. D., and Van der Valk, P. (1997) Immunocytochemical characterization of the expression of inducible and constitutive isoforms of nitric oxide synthase in demyelinating multiple sclerosis lesions. *J. Neuropathol. Exp. Neurol.* **56,** 10–20.

45. O'Neill, G. and Hutchinson, A. F. (1993) Expression of mRNA for cyclooxygenase-1 and cyclooxygenase-2 in human tissue. *FEBS Lett.* **330,** 156–160.

46. Williams, C. W. and DuBois, R. N. (1996) Prostoglandin endoperoxide synthase: why two isoforms? *Am. J. Physiol.* **270,** G393–G400.

47. Herschman, H. R. (1996) Prostaglandin synthase 2. *Biochim. Biophys. Acta* **1299,** 125–140.

48. Smith, W. L. and DeWitt, D. L. (1996) Prostaglandin endoperoxide H synthases-1 and -2. *Adv. Immunol.* **62,** 167–215.

49. Yamamoto, N., Uokotoa, K., Yoshidomi, M., Yamashita, A., and Oda, M. (1995) Protective effect of KBT-3022, a new cyclo-oxygenase inhibitor, in cerebral hypoxia and ischemia. *Jpn. J. Pharmacol.* **69,** 421–428.

50. Nogawa, S., Zhang, F., Ross, M. E., and Iadecola, C. (1997) Cyclooxygenase-2 gene expression in neurons contributes to ischemic brain damage. *J. Neurosci.* **17,** 2746–2755.

51. Baran, H., Vass, K., Lassmann, H., and Hornykiewicz, O. (1994) The cyclooxygenase and lipoxygenase inhibitor BW755C protects rats against kainic acid-induced seizures and neurotoxicity. *Brain Res.* **646,** 201–206.

52. Hirst, W. D., Young, K. A., Newton, R., Allport, V. C., Marriott, D. R., and Wilkin, G. P. (1999) Expression of Cox-2 by normal and reactive astrocytes in the adult rat brain central nervous system. *Mol. Cell. Neurosci.* **13,** 57–68.

53. Sandhya, T. L., Ong, W. Y., Horrocks, L. A., and Farooqui, A. A. (1998) An electron microscopic study of cytoplasmic phospholipase A_2 and cyclooxygenase-2 in the hippocampus after kainate lesions. *Brain Res.* **788,** 223–231.

54. Wilkins, G. P. and Marriott, D. R. (1993) Biochemical responses of astrocytes to neuroactive peptides, in *Astrocytes: Pharmacology and Function* (Murphy, S., ed.), Academic, San Diego, CA, pp. 67–87.

55. Knott, C., Stern, G., and Wilkin, G. P. (2000) Inflammatory regulators in Parkinson's disease: iNOS, lipocortin-1, and cyclooxygenases-1 and -2. *Mol. Cell. Neurosci.* **16,** 724–739.

56. Minghetti, L., Hughes, P., and Perry, V. H. (1999) Restricted cyclooxygenase-2 expression in the central nervous system following acute and delayed-type hypersensitivity responses to bacillus calmette-guerin. *Neuroscience* **92,** 1405–1415.

57. Lawson, L. J., Perry, V. H., Dri, P., and Gordon, S. (1990) Heterogeneity in the distribution and morphology of microglia in the normal adult mouse brain. *Neuroscience* **39,** 151–170.

58. O'Malley, E. K., Black, I. B., and Dreyful, C. F. (1991) Local support cells promote survival of substantia nigra dopaminergic neurons in culture. *Exp. Neurol.* **112,** 40–48.

59. O'Malley, E. K., Sieber, B. A., Black, I. B., and Dreyful, C. F. (1992) The type I astrocyte subtype augments substantia nigra dopaminergic neuron survival. *Brain Res.* **582,** 65–70.

60. Nagata, A., Takei, N., Nakajima, K., Saito, H., and Kohsaka, S. (1993) Microglial conditioned medium promotes survival and development of cultured mesencephalic neurons from embryonic rat brain. *J. Neurosci. Res.* **34,** 357–363.

61. Langan, T. S., Plunkett, R. J., Asada, H., Kelly, K., and Kaseloo, P. (1995) Long-term production of neurotrophic factors by astrocyte cultures from hemiparkinsonian rat brain. *Glia* **14,** 174–184.

62. Batchelor, P. E., Liberatore, G. T., Wong, J. Y. F., Porritt, M. H., Frerichs, F., Donnan, G. A., et al. (1999) Activated macrophages and microglia induce dopaminergic sprouting in the injured striatum and express brain-derived neurotrophic factor and glial cell line-derived neurotrophic factor. *J. Neurosci.* **19(5),** 1708–1716.

63. Lapchak, P. A. (1998) A preclinical development strategy designed to optimize the use of glial cell line-derived neurotrophic factor in the treatment of Parkinson's disease. *Mov. Disord.* **13(Suppl. 1),** 49–54.

64. Lippa, C. F., Smith, T. W., and Flanders, K. C. (1995) Transforming growth factor-β: neuronal and glial expression in CNS degenerative diseases. *Neurodegeneration* **4,** 425–432.

65. Mogi, M., Harada, M., Kondo, T., Narabayashi, H., Reiderer, P., and Magatsu, T. (1995) Transforming growth factor-β1 levels are elevated in the striatum and in ventricular cerebrospinal fluid in Parkinson's disease. *Neurosci. Lett.* **193,** 129–132.

66. Vawter, M. P., Dillon-Carter, O., Tourtellotte, W. W., Carvey, P., and Freed, W. J. (1996) TGF-β2 and TGF-β1 concentrations are elevated in Parkinson's disease in ventricular cerebrospinal fluid. *Exp. Neurol.* **142,** 313–322.

67. Giulian, D. and Lachman, L. B. (1985) Interleukin-1 stimulation of astroglial proliferation after brain injury. *Science* **228,** 497–499.

68. Sawada, M., Kondo, N., Suzumara, A., and Marunouchi, T. (1989) Production of tumor necrosis factor-alpha by microglia and astrocytes in culture. *Brain Res.* **491,** 394–397.

69. Vaca, K. and Wendt, E. (1992) Divergent effects of astroglial and microglial secretions on neuron growth and survival. *Exp. Neurol.* **118,** 62–72.

70. Giulian, D., Corpuz, M., Chapman, S., Mansouri, M., and Robertson, C. (1993) Reactive mononuclear phagocytes release neurotoxins after ischaemic and traumatic injury to the central nervous system. *J. Neurosci. Res.* **36,** 381–393.

71. Banati, R. B., Gehrmann, J., Schubert, P., and Kreutzberg, G. W. (1993) Cytotoxicity of microglia. *Glia* **7(1),** 111–118.

72. Gehrmann, J., Banati, R. B., Wiessner, C., Hossmann, K. A., and Kreutzberg, G. W. (1995) Reactive microglia in cerebral ischaemia: an early mediator of tissue damage? *Neuropathol. Appl. Neurobiol.* **21,** 277–289.

73. Neuroinflammation Working Group. (2000) Inflammation and Alzheimer's disease. *Neurobiol. Aging* **21,** 383–421.

74. Tooyama, I., Kimura, H., Akiyama, H., and McGeer, P. L. (1990) Reactive microglia express class I and class II major histocompatibility complex antigens in Alzheimer's disease. *Brain Res.* **523,** 273–280.

75. Styren, S. D., Civin, W. H., and Rogers, J. (1990) Molecular, cellular, and pathologic characterization of HLA-DR immunoreactivity in normal elderly and Alzheimer's disease brain. *Exp. Neurol.* **110,** 93–104.

76. Luber-Narod, J. and Rogers, J. (1988) Immune system associated antigens expressed by cells of the human central nervous system. *Neurosci. Lett.* **94,** 17–22.

77. Rogers, J., Luber-Narod, J., Styren, S. D., and Civin, W. H. (1988) Expression of immune system associated antigens by cells of the human central nervous system: relationship to the pathology of Alzheimer's disease. *Neurobiol. Aging* **9,** 339–349.

78. Lee, S. C., Liu, W., Dickson, D. W., Brosnan, C. F., and Berman, J. W. (1993) Cytokine production by human fetal microglia and astrocytes. *J. Immunol.* **150,** 1517–1523.

79. Minc-Golomb, D. and Schwartz, J. P. (1994) Expression of both constitutive and inducible nitric oxide synthases in neuronal and astrocyte cultures. *Ann. NY Acad. Sci.* **738,** 462–467.

80. Murphy, S. (2000) Production of nitric oxide by glial cells: regulation and potential roles in the CNS. *Glia* **29,** 1–14.

81. Murphy, S. and Grzybicki, D. (1996) Glial NO: normal and pathological roles. *Neuroscientist* **2,** 91–100.

82. Hunot, S., Boissiere, F., Faucheux, B., et al. (1996) Nitric oxide synthase and neuronal vulnerability in Parkinson's disease. *Neuroscience* **72,** 366–363.

83. Smeyne, M., Goloubeva, O., and Smeyne, R. J. (2001) Strain-dependent susceptibility to MPTP and MPP(+)-induced toxicity is determined by glia. *Glia* **34(2),** 73–80.

84. Akaike, L. T., Yoshida, M., Miyamoto, Y., Sato, K., Kohno, M., Sasamoto, K., et al. (1993) Antagonistic action of imidazolineoxyl *N*-oxide against endothelium relaxing factor/NO through a radical reaction. *Biochemistry* **32,** 827–832.

85. McNaught, K. and Jenner, P. (1999) Altered glial function causes neuronal death and increases neuronal susceptibility to 1-methyl-4-phenylpyridinium- and 6-hydroxydopa-mine-induced toxicity in astrocytic/ventral mesencephalic co-cultures. *J. Neurochem.* **73,** 2469–2476.

86. Czlonkowska, A., Kurkowska-Jastrzebska, I., and Czlonkowski, A. (2000) Inflammatory changes in the substantia nigra and striatum following MPTP intoxication. *Ann. Neurol.* **48(1),** 127.

87. Riederer, P., Sofie, E., Rausch, W. D., Schmidt, B., Reynolds, G. P., Jellinger, K., et al. (1989) Transition metals, ferritin, glutathione, and ascorbic acid in Parkinsonian brain. *J. Neurochem.* **52,** 515–520.

88. Czlonkowska, A., Kohutnicka, M., Kurkowska-Jastrzebska, I., and Czlonkowski, A. (1996) Microglial reaction in MPTP (1-methyl-4-phenyl-1,2,3,6-tetrahydropyridine) induced Parkinson's disease mice model. *Neurodegeneration* **5,** 137–143.

89. McGeer, P. L. Itagaki, S., and McGeer, E. G. (1988) Expression of the histocompatibility glycoprotein HLA-DR in neurological disease. *Acta Neuropathol.* **76,** 550–557.

90. Kurkowska-Jastrzebska, I., Wronska, A., Kohutnicka, M., Czlonkowski, A., and Czlonkowska, A. (1999) The inflammatory reaction following 1-methyl-4-phenyl-1,2,3,6-tetrahydropyridine intoxication in mouse. *Exp. Neurol.* **156,** 50–61.

91. Czlonkowska, A., Kohutnicka, M., Kurkowska-Jastrzebska, I., and Czlonkowski, A. (1996) Microglial reaction in MPTP induced Parkinson's disease mice model. *Neurodegeneration* **5,** 137–143.

92. Francis, J. W., Von Visger, J., Markelonis, G. J., and Oh, T. H. (1995) Neurological responses to the dopaminergic neurotoxicant 1-methyl-4-phenyl-1,2,3,6-tetrahydropyridine in mouse striatum. *Neurotoxicol. Teratol.* **17,** 7–12.

93. Kohutnicka, M., Lewandowska, I., Kurkowska-Jastrzebska, I., Czlonkowski, A., and Czlonkowska, A. (1998) The microglial and astrocytic involvement in Parkinson's disease mice model induced by 1-methyl-4-phenyl-1,2,3,6-tetrahydropyridine (MPTP). *Immunopharmacology* **39,** 167–180.

94. O'Callaghan, J. P., Miller, D. R., and Reinhard, J. F. (1990) Characterization of the origins of astrocyte response to injury using the dopaminergic neurotoxicant, 1-methyl-4-phenyl-1,2,3,6-tetrahydropyridine. *Brain Res.* **521,** 73–80.

95. Luo, Y., Umegaki, H., Wang, X., Abe, R., and Roth, G. S. (1998) Dopamine induces apoptosis through an oxidation-involved SAPK/JNK activation pathway. *J. Biol. Chem.* **273,** 3756–3764.

96. Henchcliffe, T. F., James, D., and Burke, R. E. (1999) Expression of c-fos, c-jun, and c-jun N-terminal kinase (JNK) in a developmental model of induced apoptotic death in neurons of the substantia nigra. *J. Neurochem.* **72,** 557–564.

97. Perez-Otano, I., Mandelzys, A., and Morgan, J. I. (1998) MPTP-parkinsonism is accompanied by persistent expression of a delta-FosB-like protein in dopaminergic pathways. *Brain Res. Mol. Brain Res.* **53,** 41–52.

98. Teismann, P. and Ferger, B. (2001) Inhibition of the cyclooxygenase isoenzymes COX-1 and COX-2 provide neuroprotection in the MPTP-mouse model of Parkinson's disease. *Synapse* **39,** 167–174.

99. Aubin, N., Curet, O., Deffois, A., and Carter, C. (1998) Aspirin and salicylate protect against MPTP-induced dopamine depletion in mice. *J. Neurochem.* **71,** 1635–1642.

100. Mohanakumar, K. P., Muralikrishnan, D., and Thomas, B. (2000) Neuroprotection by sodium salicylate against 1-methyl-4-phenyl-1,2,3,6-tetrahydropyridine-induced neurotoxicity. *Brain Res.* **864,** 281–290.

101. Mogi, M., Togari, A., Tanaka, K., Ogawa, N., Ichinose, H., and Nagatsu, T. (2000) Increase in level of tumor necrosis factor-α in 6-hypdroxydopamine-lesioned striatum in rats is suppressed by immunosuppressant FK506. *Neurosci. Lett.* **289,** 165–168.

102. Ogawa, N., Asanuma, M., Kondo, Y., Hirata, H., Nishibayashi, S., and Mori, A. (1994) Changes in lipid peroxidation, Cu/Zn-superoxide dismutase and its mRNA following an intracerebroventricular injection of 6-hydroxydopamine in mice. *Brain Res.* **646,** 337–340.

103. Perumal, A. S., Tordzrio, W. K., Katz, M., Jackson-Lewis, V., Cooper, T. B., Fahn, S., et al. (1989) Regional effects of 6-hydroxydopamine on free radical scavengers in rat brain. *Brain Res.* **504,** 139–141.

104. Akiyama, H. and McGeer, P. L. (1989) Microglial response to 6-hydroxydopamine-induced substantia nigra lesions. *Brain Res.* **489,** 247–253.

105. Herrera, A. J., Castano, A., Venero, J. L., Cano, J., and Machado, A. (2000) The single intranigral injection of LPS as a new model for studying the selective effects of inflammatory reactions on dopaminergic system. *Neurobiol. Dis.* **7,** 429–447.

106. Lu, W., Bing, G., and Hagg, T. (2000) Naloxone prevents microglia-induced degeneration of dopaminergic substantia nigra neurons in adult rats. *Neurosci.* **97(2),** 285–291.

107. Vanguri, P. and Shin, M. L. (1986) Activation of complement by myelin: identification of C1-binding proteins of human myelin from central nervous tissue. *J. Neurochem.* **46(5),** 1535–1541.

108. Kim, W. G., Mohney, R. P., Wilson, B., Jeohn, G.-H., Liu, B., and Hong, J.-S. (2000) Regional difference in susceptibility to lipopolysaccharide-induced neurotoxicity in the rat brain: role of microglia. *J. Neurosci.* **20,** 6309–6316.

109. Luber-Narod, J. and Rogers, J. (1988) Immune system associated antigens expressed by cells of the human central nervous system. *Neurosci. Lett.* **94,** 17–22.

110. Rogers, J., Luber-Narod, J., Styren, S. D., and Civin, W. H. (1988) Expression of immune system associated antigens by cells of the human central nervous system: relationship to the pathology of Alzheimer's disease. *Neurobiol. Aging* **9,** 339–349.

112. Rogers, J., Cooper, N. R., Schultz, J., McGeer, P. L., Webster, S., Styren, S. D., et al. (1992) Complement activation by β-amyloid in Alzheimer's disease. *Proc. Natl. Acad. Sci. USA* **89,** 10,016–10,020.

22

Neuroinflammatory Components of the 3-Nitropropionic Acid Model of Striatal Neurodegeneration

Hideki Hida, Hiroko Baba, and Hitoo Nishino

1. INTRODUCTION

Disturbances in mitochondrial energy metabolism, oxidative stress, and microglial activation, among others, are common to neurodegenerative disorders *(1)*. Recently, accumulating data indicate the possible involvement of the immune system in the pathogenesis of neurodegenerative disorders, such as Alzheimer's disease, Parkinson's disease, multiple sclerosis, amylotrophic autoimmune sclerosis (ALS), and so on *(2–5)*. In some cases, neuronal cell death induces inflammatory responses, and in other cases, inflammation itself results in neuronal cell death. 3-Nitropropionic acid (3-NP), a mitochondria toxin that irreversibly inhibits succinate dehydrogenase (SDH), induces an acute encephalopathy targeting primarily the striatum whose symptom is similar to those of Huntington's disease. 3-NP encephalopathy is characterized by striatum-specific lesions, inflammatory responses, and gender-related difference in vulnerability *(6–11)*. Thus, the 3-NP model of Huntington's disease seems to be a good model for studying the interrelation between inflammation and neurodegeneration. In this chapter, neuroinflammatory components of the 3-NP model of striatal neurodegenaration will be discussed.

2. BEHAVIORAL SYMPTOM IN 3-NP INTOXICATION

An acute encephalopathy after ingestion of mildewed sugar cane was observed from 1972 to 1989 in China *(12–14)*. Mildewed sugar contains Arthrinium fungus that has aliphatic nitrotoxin (3-NP) (Fig. 1). People who accidentally ingested 3-NP manifested signs of an acute encephalopathy (headache, convulsions, opisthotonus, dystonia, somnolence, coma). Among 884 cases of the encephalopathy, 88 died and bilateral necrosis of the basal ganglia was detected in these victims *(14)*.

Animal models of 3-NP intoxication demonstrate a similar selective lesion in the striatum as well as motor symptoms resembling those of Huntington's disease. There is a species difference in vulnerability to 3-NP. Rats are susceptible (LD_{50}: 25–26 mg/kg), but mice are quite resistant to the toxin *(15)*. A single administration of 3-NP (20 mg/kg, sc) to adult rats induces no abnormal behavior except a reduction in motor activity,

From: *Neuroinflammation, 2nd Edition: Mechanisms and Management*
Edited by: P. L. Wood © Humana Press Inc., Totowa, NJ

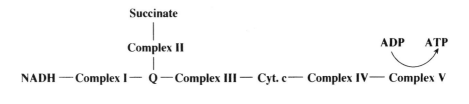

Fig. 1. Chemical structure of 3-NP and the action point of 3-NP as a mitochondria toxin that irreversibly inhibites succinate dehydrogenase.

but the second administration of same dose on the second day induces motor symptoms (catalepsy, lip smacking, abnormal gait, paddling, opisthotonus, rolling, somnolence, dyspnea, coma) in about one-third of animals *(8)*. The female siblings or young rats are rather resistant to the intoxicaton *(9)*. Chronic intoxication with 3-NP also induces striatal-specific lesions accompanying abnormal choreiform movement and asthenia in lower limbs but without acute motor symptoms. Inflammatory or immune responses are also evident in the chronic model *(6,16,17)*.

3. STRIATAL LESIONS

3.1. Magnetic Resonance Imaging

Following the second administration of 3-NP (20 mg/kg, sc), no marked changes were detected in the T1-weighted magnetic resonance image (MRI). However, during the 3–5 h after the second administration, the signal intensity of the T2-weighted image of both striata was higher in 3-NP-treated animals than that of controls. The diffusion-weighted image demonstrated more distinct hyperintensity in both striata for 3-NP-treated animals, suggesting striatal edema *(8)*.

3.2. Extravasation of IgG

When the dysfunction of the blood-brain barrier (BBB) was monitored by the extravasation of immunoglobulin G (Ig G), no IgG immunoreactivity was detected in sections made 24 h after the single administration of 3-NP. However, 2–3 h after the second administration of 3-NP, small immunoreactive deposits were detected in the lateral half of the striatum, particularly around vessels (Fig. 2A–C). During the 5–8 h after the second administration, the IgG immunoreactivity became stronger and extended to almost the entire striatum (except the medial end) (Fig. 3B) and the IgG immunoreaction was also detected in hippocampal CA1 to CA3 and the thalamus *(8)*.

3.3. Astrocytic Cell Death

When the histological damages were investigated using immunocytochemistry and TUNEL, during 2–3 h after the second administration of 3-NP, GFAP immunoreactive astrocytes were not visible in the center of the striatum (Figs. 2D and 3C), but they were present in normal fashion in other areas. MAP2 immunoreactive soma and dendrites

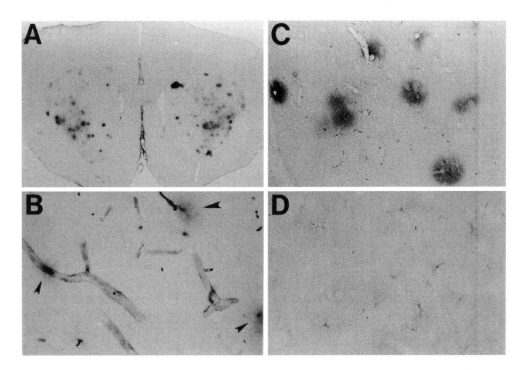

Fig. 2. Striatal extravasation of IgG at 3 h after the second administration of 3-NP (20 mg/ kg, sc). (**A**) IgG immunoreactive dots distributed in bilateral striata especially in centrolateral part; (**B**) IgG immunostaining figured out small branches of the lSTR artery, suggesting the invasion of IgG under the endothelial sheet. The extravasation of IgG has already started (arrowheads). (**C**) Higher magnification of (**A**). (**D**) At this moment, GFAP-positive astrocytes are already lost in the parenchyma of the striatum but remained among the fiber bundles. (**A–C**) IgG immunostaining; (**D**) GFAP immunostaining. (From ref. *11*.)

were present in apparently a normal amounts in all areas of the brain, including the striatum. TUNEL staining was negative in all areas of the brain at this early stage. However, 1 wk later, TUNEL-positive cells were visible around the lesions in the striatum and in the CA1 pyramidal cell layer but not in others *(8)*.

Ultrastructural analysis revealed a more precise mechanism of cell damages in the striatum (Fig. 4). One hour after the second administration of 3-NP, a marked edema in astrocytic soma and end-feet, endothelial cells, and pericytes was observed (Fig. 4C). Mitochondria in these structures became enlarged and spherical, lost their cristae, and, finally, ruptured. Myelin sheaths became rough and irregular; however, synaptic structures were less damaged. On the other hand, no detectable ultrastructural abnormalities were found in endothelial cells nor astrocytes in sections from the cortex (Fig. 4D,E). Similar but milder deficits in astrocytic end-feet but not in endothelial cells were detected in hippocampal CA1 after the third administration of 3-NP (Fig. 4F). In short, 3-NP intoxication results in severe damages in striatal artery (endothelial cells and astrocytic end-feet), milder damages in arteries of hippocampal CA1, and no damages in cortical arteries *(8)*.

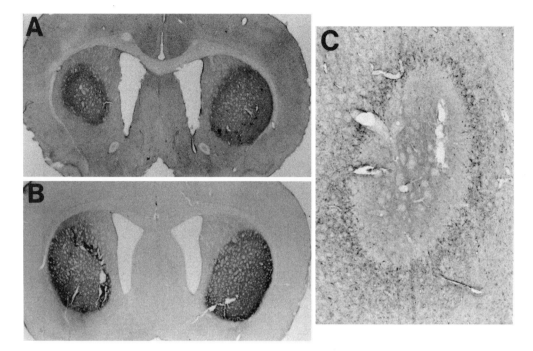

Fig. 3. Hematoxylin and eosin (H&E)-stained and immunostained sections from an animal for 1 wk without displaying any behavioral abnormalities. (**A**) H&E staining revealed striatum selective lesions with invasion of densely stained cells (neutrophils); (**B**) IgG extravasation; (**C**) loss of GFAP-positive astrocytes in the center and gliosis on the margin of the striatal lesion. (**A**) H&E staining; (**B**) IgG immunostaining; (**C**) GFAP immunostaining. (From ref. *11*.)

4. STRIATAL-SPECIFIC LESIONS

As discussed earlier, an acute strong intoxication with 3-NP induces BBB dysfunction in the striatum, in hippocampal CA1–CA3, and in the thalamus. However, an acute milder or chronic 3-NP intoxication results in striatum-specific lesions. Why does systemically administered 3-NP result in striatum-specific lesions? The specificity would be a composite of various vulnerabilities specific to the striatum.

4.1. Glutamate Excitotoxicity

The striatum is the only region where massive glutamatergic inputs and massive dopaminergic inputs converge. Glutamate and dopamine are important neurotransmitters in the brain, but once released concomitantly in excess, they act as potent neurotoxins *(18,19)*.

Striatum-specific lesions have been identified in chronic 3-NP intoxication models *(14,16,20,21)*. Medium-sized spiny neurons are more vulnerable to the toxin than aspiny NADPH-diaphorase-positive neurons or large cholinergic neurons. It is proposed that a subtle impairment of energy metabolism results in slow (secondary) excitotoxicity of neurons *(16,20,22,23)*. Prior ablation of the unilateral dorsal cerebral cortex had a little effect on the size of strital damage, but prior ablation of a wide area of the lateral cerebral cortex, resulting in the ablation of glutamatergic input to the lateral striatum, suppressed

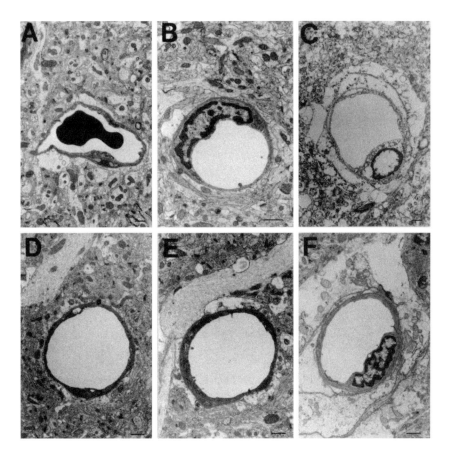

Fig. 4. Electron micrographs showing ultrastructural damage around capillaries in the striatum (**A,B,C**), cortex (**D,E**) and CA1 (**F**) after 3-NP administration. (**A**) control rat; (**B**) 24 h after the first injection of 3-NP; no damage of endothelial cells, astrocytic end-feet, or neighboring structures was observed; (**C**) 1 h after the second injection. Extensive damage of astrocytic end–feet, endothelial cells and pericytes was seen. Mitochondria became enlarged, lost cristae, and ruptured. Less damage in neurons, myelin and synapses was observed. (**D,E**) One hour after the second and third injections, respectively. Astrocytic end-feet were normal and no damage was detected. (**F**) One hour after the third injection. Astrocytic end-feet and pericytes were moderately swollen but mitochondria were still retained by the cells. Bars = 1 μm. (From ref. *8.*)

striatal damages induced by 3-NP *(11)*. It is reported that the expression of prostate apoptosis response-4 (Par-4), containing both a death domain and leucine zipper domain, that has been linked to apoptosis by various insults has been increased rapidly following administration of 3-NP. Par-4 acts upstream of caspase-3 activation in the cell death mechanism *(24)*. Recently, it has been reported that 3-NP intoxication produces a long-term potentiation (LTP) of the NMDA-mediated synaptic excitation in striatal spiny neurons, and the LTP involves increased intracellular calcium and activation of mitogen-activated protein (MAP) kinase and extracellular-regulated kinase (ERK) and is critically dependent on endogenous dopamine acting via D2 receptors *(25)*.

Thus, abundant data suggest that the glutamatergic excitotoxicity is the most plausible explanation underlying 3-NP-induced striatal neuronal death. However, other studies have yielded variable results on the protection of glutamate antagonists against the toxic effects of 3-NP *(26–28).*

4.2. Dopamine Toxicity

The striatum is most heavily innervated by dopaminergic inputs. Coactivation of D1 and D2 receptors results in activation of phosphlipase A2 *(29)* and leads to arachidonic acid formation and subsequent cascade, resulting in the production of prostaglandins and radicals *(30).* Oxygen radicals are also produced by oxidation of dopamine *(18).* Depletion of dopaminergic inputs to the striatum attenuates striatal cell death following global ischemia *(31).* Striatal lesions caused by 3-NP appear mainly in the centrolateral but not the medial part of the striatum, where D2 receptors are densely distributed *(32).* These observations prompted us to consider the involvement of dopaminergic toxicity in 3-NP-induced selective striatal damages.

Treatment of rats with a D2 agonist (quinpirole, 5 mg/kg) 15 min before each administration of 3-NP results in a decrease in incidence of motor symptoms, whereas pretreatment with a D2 antagonist (sulpiride, 80 mg/kg) results in an increase of motor symptoms. Damages in astrocytic end-feet, which cover cerebral vessels to establish the BBB, were attenuated and the extravasation of IgG was attenuated by the pretreatment with quinpirole.

The pattern of expression of D1 and D2 receptors varies among output neurons of the striatum *(33–35).* D2 receptors on dendrites and axon terminals of dopaminergic neurons play a key role in the autoregulation of dopamine activity and dopamine release *(35).* D2 receptors on corticostriatal terminals are involved in presynaptic inhibition of glutamate release *(36,37).* Thus, the decrease in incidence of motor symptoms and the attenuation of IgG extravasation following quinpirole treatment would be related to the suppression of dopamine overflow or overexcitation, whereas the potentiation by sulpiride could be explained by the enhancement of excitation. A moderate increase of dopamine and metabolites in striatal dialysates was detected for several hours following a single administration of 3-NP although no behavioral disturbances were detected. However, following the second administration, dopamine and metabolites increased within a few hours and accompanied motor symptoms. Thus, the increase in dopamine concentration and its turnover closely parallels behavioral disturbances *(8).*

4.3. Vulnerability of the Lateral Striatal Artery

Striatal lesions induced by 3-NP are located at the centrolateral part of the striatum and distal from the medial end of the structure. As mentioned earlier, the centrolateral part of the striatum is innervated by abundant glutamatergic as well as dopaminergic inputs. However, we must also consider that this area of the striatum is fed by the lateral striatal artery (lSTR-A). At the earliest stage of the striatal damage, when no significant damages are yet found in neurons and glias, small deposits of IgG appear just around capillaries and small branches of the lSTR-A. This means that dysfunction of the BBB of the lSTR-A marks a critical period for the striatal damage induced by 3-NP *(11,38,39).*

The lSTR-A arises from the middle cerebral artery and feeds the centrolateral part of the striatum. It has a high intraluminal pressure and a large diameter with a rather thin wall; thus, the resting tension on the wall of this artery is always high (Laplace law). Moreover, the lSTR-A often has acute angled or rectangled branches, which are hardly observed in other areas of the brain except the hippocampus and ventral pons *(9)*. In such branching areas, the blood flow would be slowed down, and a nonsteady turbulent flow would be expected, perhaps exposing the endothelial cells to stress. Thus, the lSTR-A may be exposed to a stressful condition even at the resting state. Accordingly, in a stressful state such as at 3-NP intoxication, ischemia, hypertension, and so forth, it is possible that the endothelial cells of the lSTR-A would be damaged most easily.

The lSTR-A, a perforate artery, is called an apoplexy artery that often causes functional disturbance leading to cerebral bleeding (putaminal bleeding) or lacunar infarction in human. The incidence of cerebral bleeding is highest with this artery and exhibits a significant gender difference *(11)*.

5. INFLAMMATORY COMPONENT

Amyloid angiopathy with vascular damage and inflammatory changes are hallmarks in the brain of Alzheimer's disease victims. In an experiment using the mesenteric artery, it has been found that the vascular actions of β-amyloid are distinct from the neurotoxic properties of the peptide and were prevented by the free-radical scavenging enzyme (SOD) *(40)*. In this model, circulating β-amyloid interacts with receptor sites on endothelial cells and leukocytes, resulting in the generation of reactive oxygen species (ROS). Then, the superoxide radicals would interact with NO to generate peroxynitrite, which leads to peroxidation and membrane degeneration, resulting in an increase in the BBB permeability *(40)*. In the cerebral vessel, induction of NO and cytokines alters the permeability of the BBB, increases the expression of intracellular adhesion molecule (ICAM)-1 and LFA-1, and induces another adhesion molecule vascular cell adhesion molecule [VCAM]-1) on endothelial cells and glial cells *(41)*.

Nitric oxide is normally metabolized to nitrate and nitrite; thus, the measurement of NO_x (nitrate and nitrite) level will provide clues on the approximate rate of NO metabolism. When the extracellular concentration of NO_x is measured using in vivo microdialysis in 3-NP-administered animals, it has been found that the NO_x level is higher in the striatum than in other parts of the brain (thalamus or cerebellum) at both basal (before 3-NP administration) and trailing (recovery phase after 3-NP administration) levels. Drug-responsive animals (rats that exhibited motor symptom following the second administration of 3-NP) had a higher striatal NO_x level than non-drug-responsive animals (rats that had no motor symptoms following the second administration of 3-NP) *(42,43)*. A parallel study from our laboratory revealed that an overexpression in eNOS message following the second administration of 3-NP was unique to the striatum and was not detected in other regions, suggesting selective derangement of the lSTR-A by 3-NP *(10)*. These data suggest that a high turnover in NO metabolism would underlie the striatal specific damages by 3-NP. NO is a biological messenger and neurotransmitter, but excessive NO is capable of disrupting the BBB permeability *(44)*.

A similar result has been reported in two other neuropathological animal models. Free-radical production in vascular endothelial cells and inflammatory responses in perivascular microglia accompany the selective neuronal death induced by thiamine deficiency *(44)*. Thiamine deficiency increases ferritin immunoreactivity in the wall of capillaries and larger vessels within the areas of neuronal damage and BBB breakdown but not in nonvulnerable areas. These data support the hypothesis that in thiamine deficiency, vascular factors constitute a critical part of a cascade of events leading to increases in BBB permeability to non-neuronal proteins and iron, leading to inflammation and oxidative stress. In the MPTP model of Parkinson's disease, microglial and astroglial reactions have been found around impaired neurons. Recently, an immune reaction including lymphocytic infiltration in the substantia nigra and striatum and an elevated MHC class I and II antigens expression on microglia were reported. In this model, ICAM-1 expression increased on the endothelial cells and appeared on microglia in the injured regions. Treatment with dexamethasone inhibited these immune responses and neuronal impairment, suggesting that an immune mechanism may contribute to the neuronal damage following MPTP administration *(45)*.

As was described earlier, the dysfunction of the BBB in the centrolateral part of the striatum is a unique feature of the 3-NP intoxication model. Complements and other biologically active substances enter the central nervous system (CNS) from the plasma when the BBB is damaged *(46)*, but, in many instances and particularly in neurodegeneration, complement proteins and activation products are found in the CNS despite the presence of an apparently intact BBB *(47,48)*. These findings lead to the proposal that local biosynthesis of complements may occur within the CNS and that local production of complements is a driving force in neurodegenerative diseases *(49–52)*. In the CNS, both microglias, which are macrophage derived and most abundantly located in the substantia nigra *(53)*, and astrocytes, the major glial cell in the CNS, have the capacity to produce some complements. In an inflamed CNS, astrocytes may produce all of the components necessary for the generation of a complete lytic complement system *(49)*.

The neuroprotective properties of cyclosporin A, an immunosuppressant, are mediated by its ability to prevent mitochondrial permeability transition during exposure to a high level of calcium or oxidative stress. Under conditions, in which cyclosporin A can gain access to striatal neurons through disruption of the BBB, significant neuroprotection from 3-NP toxicity was observed *(54)*.

In our previous studies, an expression of iNOS, which corresponded with IgG extravasation, was detected, and an expression of complements and complement receptors (C3bR and C4bR) was also detected in the centrolateral striatum of 3-NP-intoxicated animals *(6–8)*. Pretreatment with N-nitro-L-arginine methyl ester (L-NAME) along with each administration of 3-NP did not improve the behavioral disturbances but attenuated the extravasation of IgG as well as iNOS immunoreactivity. Pretreatment with aminoguanidine or FK506 attenuated the behavioral symptoms, extravasation of IgG and expression of iNOS immunoreactivity, and reduced the striatal damages *(7)*. Thus, in the pathophysiology of the 3-NP intoxication (striatum-specific lesions), NO production, eNOS/iNOS expression and involvement of immune/complement system are strongly suggested.

6. SUMMARY

Systemic administration of 3-NP induces striatum-selective lesions and motor symptoms similar to those of Huntington's disease. The striatal-selective lesions would be a composite of striatum specific vulnerabilities: glutamatergic excitotoxicity, dopaminergic toxicity, vulnerability of the lSTR-A, and inflammatory/immune response (Table 1).

Table 1
Various Risk Factors
Underlying Striatal Vulnerability

Glutamatergic toxicity
Dopaminergic toxicity
lSTR-A vulnerability
Anatomical weakness
NO metabolism (eNOS, iNOS)

Thus, the 3-NP-intoxication model might offer useful clues for understanding the inflammatory aspects and the role of the immune/complement system in the pathophysiology of brain insults.

REFERENCES

1. Park, L. C. H., Zhang, H., Sheu, K.-F. R., Calingasan, N. Y., Kristal, B. S., Lindsay, J. G., et al. (1999) Metabolic impairment induces oxidative stress, compromises inflammatory responses, and inactivates a key mitochondrial enzyme in microglia. *J. Neurochem.* **72,** 1948–1958.
2. Beal, M. F. (1996) Mitochondria, free radicals, and neurodegeneration. *Curr. Opin. Neurobiol.* **6,** 661–666.
3. Hirsch, E. C., Hunot, S., Damier, P., and Faucheux, B. (1998) Glial cells and inflammation in Parkinson's disease: a role in neurodegeneration? *Ann. Neurol.* **44(Suppl. 1),** S115–S120.
4. Siesjö B. K., Katsura K., Zhao, Q., Folbergrová, J., Pahlmark, K., Siesjö P., et al. (1995) Mechanisms of secondary damage in global and focal ischemia: a speculative synthesis. *J. Neurotrauma* **12,** 943–956.
5. Urabe, T., Hattori, N., Yoshikawa, M., Yoshino, H., Uchida, K., and Mizuno, Y. (1998) Colocalization of Bcl-2 and 4-hydroxynonenal modified proteins in microglial cells and neurons of rat brain following transient focal ischemia. *Neurosci. Lett.* **247,** 159–162.
6. Nishino, H., Shimano, Y., Kumazaki, M., and Sakurai, T. (1995) Chronically administered 3-nitropropionic acid induces striatal lesions attributed to dysfunction of the blood-brain barrier. *Neurosci. Lett.* **186,** 161–164.
7. Nishino, H., Fujimoto, I., Shimano, Y., Hida, H., Kumazaki, M., and Fukuda, A. (1996) 3-Nitropropionic acid produces striatum selective lesions accompanied by iNOS expression. *J. Chem. Neuroanat.* **10,** 209–212.
8. Nishino, H., Kumazaki, M., Fukuda, A., Fujimoto, I., Shimano, Y., Hida, H., et al. (1997) Acute 3-nitropropionic acid intoxication induces striatal astrocytic cell death and dysfunction of the blood-brain barrier: involvement of dopamine toxicity. *Neurosci. Res.* **27,** 343–355.
9. Nishino, H., Nakajima, K., Kumazaki, M., Fukuda, A., Muramatsu, K., Deshpande, S. B., et al. (1998) Estrogen protects against while testosterone exacerbates vulnerability of the lateral striatal artery to chemical hypoxia by 3-nitropropionic acid. *Neurosci. Res.* **30,** 303–312.
10. Nishino, H., Baba, H., Shimano, Y., Nakajima, K., Kumazaki, M., and Sakurai, T. (1999) Increase in eNOS expression well parallels with the dysfunction of the BBB in the striatum in 3-nitropropionate intoxication. *Soc. Neurosci. Abstr.* **25(1),** 1295.
11. Nishino, H., Hida, H., Kumazaki, M., Shimano, Y., Nakajima, K., Shimizu, H., et al. (2000) The striatum is the most vulnerable region in the brain to mitochondrial energy compromise: a hypothesis to explain its specific vulnerability. *J. Neurotrauma* **17,** 251–260.

12. He, F., Zhang, S., Zhang, C., Qian, F., Liu, X., and Lo, X. (1990) Mycotoxin induced encephalopathy and dystonia in children, in *Basic Science in Toxicology* (Volans, G. N., Sims, J., Sullivan, F. M., and Turner, P., eds.), Taylor & Francis, London, pp. 596–604.

13. Hu, W. (1986) The isolation and structure identification of a toxic substance, 3-nitropropionic acid, produced by *Arthrinium* from mildewed sugar cane. *Chin. J. Prev. Med.* **20,** 321–323.

14. Ludolph, A. C., He, F., Spencer, P. S., Hammerstad, J., and Sabri, M. (1991) 3-Nitropropionic acid—exogenous animal neurotoxin and possible human striatal toxin. *Can. J. Neurol. Sci.* **18,** 492–498.

15. Alexi, T., Hughes, P. E., Knusel, B., and Tobin, A. J. (1998) Metabolic compromise with systemic 3-nitropropionic acid produces striatal apoptosis in Sprague–Dawley rats but not in BALB/c ByJ mice. *Exp. Neurol.* **153,** 74–93.

16. Brouillet, M., Hantraye, P., Ferrante, R. J., Dolan, R., Leroy-Willig, A., Kowall, N. W., et al. (1995) Chronic mitochondrial energy impairment produces selective striatal degeneration and abnormal choreiform movements in primates. *Proc. Natl. Acad. Sci. USA* **92,** 7105–7109.

17. Shimano, Y., Kumazaki, M., Sakurai, T., Hida, H., Fujimoto, I., Fukuda, A., et al. (1995) Chronically administered 3-nitropropionic acid produces selective lesions in the striatum and reduces muscle tonus. *Obes. Res.* **3(Suppl. 5),** 779S–784S.

18. Ben-Shachar, D., Zuk, R., and Glinka, Y. (1995) Dopamine neurotoxicity: inhibition of mitochondrial respiration. *J. Neurochem.* **64,** 718–723.

19. Olney, J. W. (1978) Neurotoxicity of excitatory amino acids, in *Kainic Acid as a Tool in Neurobiology* (McGeer, E. G., Olney, J. W., and McGeer, P. L., eds.), Raven, New York, pp. 95–121.

20. Beal, M. F., Brouillet, E., Jenkins, B. G., Ferrante, R. J., Kowall, N. W., Miller, J. M., et al. (1993) Neurochemical and histologic characterization of striatal excitotoxic lesions produced by the mitochondrial toxin 3-nitropropionic acid. *J. Neurosci.* **13,** 4181–4192.

21. Borlongan, C. V., Nishino, H., and Sanberg, P. R. (1997) Systemic, but not intraparenchymal, administration of 3-nitropropionic acid mimics the neuropathology of Huntington's disease: a speculative explanation. *Neurosci. Res.* **28,** 185–189.

22. Beal, M. F. (1992) Does impairment of energy metabolism result in excitotoxic neuronal death in neurodegenerative illness? *Ann. Neurol.* **31,** 119–130.

23. Simpson, J. R. and Isacson, O. (1993) Mitochondrial impairment reduces the threshold for in vivo NMDA-mediated neuronal death in the striatum. *Exp. Neurol.* **121,** 57–64.

24. Duan, W., Guo, Z., and Mattson, M. P. (2000) Participation of par-4 in the degeneration of striatal neurons induced by metabolic compromise with 3-nitropropionic acid. *Exp. Neurol.* **165,** 1–11.

25. Calabresi, P., Gubellini, P., Picconi, B., Centonze, D., Pisani, A., Bonsi, P., et al. (2001) Inhibition of mitochondrial complex II induces a long-term potentiation of NMDA-mediated synaptic excitation in the striatum requiring endogenous dopamine. *J. Neurosci.* **21,** 5110–5120.

26. Behrens, M. I., Koh, J., Canzoniero, L. M., Sensi, S. L., Csernansky, C. A., and Choi, D. W. (1995) 3-Nitropropionic acid induces apoptosis in cultured striatal and cortical neurons. *Neuroreport* **6,** 545–548.

27. Fink, S. L., Ho, D. Y., and Sapolsky, R. M. (1996) Energy and glutamate dependency of 3-nitropropionic acid neurotoxicity in culture. *Exp. Neurol.* **138,** 298–304.

28. Weller, M. and Paul, S. M. (1993) 3-Nitropropionic acid is an indirect excitotoxin to cultured cerebellar granule neurons. *Eur. J. Pharmacol. Environ. Toxicol. Pharmacol.* **248,** 223–228.

29. Calabresi, P., Pisani, A., Mercuri, N. B., and Bernardi, G. (1996) The corticostriatal projection: from synaptic plasticity to dysfunctions of the basal ganglia. *Trends Neurosci.* **19,** 19–24.

30. Dumuis, A., Sebben, M., Haynes, L., Pin, J. P., and Bockaert, J. (1988) NMDA receptors activate the arachidonic acid cascade system in striatal neurons. *Nature* **336,** 68–70.

31. Globus, M. Y. T., Ginsbery, M. D., Dietrich, W. D., Busto, P., and Scheinberg, P. (1987) Substantia nigra lesion protects against ischemic damage in the striatum. *Neurosci. Lett.* **80,** 251–256.

32. Nishino, H., Hashitani, T., Mizukawa, K., Ogawa, N., and Shiosaka, S. (1990) Long-term survival of grafted cells, dopamine synthesis/release, receptor activity, and functional recovery after grafting of fetal nigral cells in model animals of hemi-Parkinson's disease, in *Basic, Clinical and Therapeutic Aspects of Alzheimer's and Parkinson's Disease* (Nagatsu, T., et al., eds.), Plenum, New York, Vol. 1, pp. 781–784.

33. Baik, J.-H., Picetti, R., Saiardi, A., Thiriet, G., Dierich, A., Depaulis, A., et al. (1995) Parkinsonian-like locomotor impairment in mice lacking dopamine D^2 receptors. *Nature* **377,** 424–428.

34. Gerfen, C. R. (1992) The neostriatal mosaic: multiple levels of compartmental organization. *Trends Neurosci.* **15,** 133–139.

35. Hersch, S. M., Ciliax, B. J., Gutekunst, C.-A., Rees, H. D., Heilman, C. J., Yung, K. K. L., et al. (1995) Electron microscopic analysis of D_1 and D_2 dopamine receptor proteins in dorsal striatum and their synaptic relationships with motor corticostriatal afferents. *J. Neurosci.* **15,** 5222–5237.

36. Garcia-Munoz, M., Young, S. J., and Groves, P. M. (1991) Terminal excitability of the corticostriatal pathway. I. Regulation by dopamine receptor stimulation. *Brain Res.* **551,** 195–206.

37. Maura, G., Giardi, A., and Raiteri, M. (1988) Release-regulating D2 dopamine receptors are located on striatal glutamatergic nerve terminals. *J. Pharmacol. Exp. Ther.* **247,** 680–684.

38. Hamilton, B. F. and Gould, D. H. (1987) Nature and distribution of brain lesions in rats intoxicated with 3-nitropropionic acid: a type of hypoxic (energy deficient) brain damage. *Acta Neurophathol.* **72,** 286–297.

39. Hamilton, B. F. and Gould, D. H. (1987) Correlation of morphologic brain lesions with physiologic alterations and blood-brain barrier impairment in 3-nitropropionic acid toxicity in rats. *Acta Neuropathol.* **74,** 67–74.

40. Thomas, T., Sutton, E. T., Bryant, M. W., and Rhodin, J. A. G. (1997) In vivo vascular damage, leukocyte activation and inflammatory response induced by β-amyloid. *J. Submicrosc. Cytol. Pathol.* **29,** 293–304.

41. Merrill, J. E. and Murphy, S. P. (1997) Inflammatory events at the blood brain barrier: regulation of adhesion molecules, cytokines, and chemokines by reactive nitrogen and oxygen species. *Brain Behav. Immun.* **11,** 245–263.

42. Hida, H., Masuda, T., Nakajima, K., and Nishino, H. (1999) Increase of nitric oxide metabolism in the striatum in 3-nitropropionic acid intoxication. *Jpn. J. Physiol.* **49(Suppl.),** S234.

43. Hida, H., Masuda, T., Nakajima, K., and Nishino, H. (1999) Effect of glial glutamate transporter on the regulation of NO metabolism in the striatum in 3-nitropropionic acid-treated rats. *Neurosci. Res.* **(Suppl. 23),** S64.

44. Calingasan N. Y. and Gibson, G. E. (2000) Vascular endothelium is a site of free radical production and inflammation in areas of neuronal loss in thiamine-deficient brain. *Ann. NY Acad. Sci.* **903,** 353–356.

45. Kurkowska-Jastrzebska, I., Wronska, A., Kohutnicka, M., Czlonkowski, A., and Czlonkowska, A. (1999) The inflammatory reaction following 1-methyl-4-phenyl-1,2,3,6-tetrahydropyridine intoxication in mouse. *Exp. Neurol.* **156,** 50–61.

46. Gay, D. and Esiri, M. (1991) Blood brain barrier damage in acute multiple sclerosis plaques. *Brain* **114,** 557–572.

47. Eikelenboom, P., Fraser, H., Rozemuller, J. M., and Stam, F. C. (1988) The blood-brain barrier in cerebral amyloidogenesis in Alzheimer's disease and scrapie, in *Immunology and*

Alzheimer's Disease (Poupland-Barthelaix, A., Emile, J., and Christen, Y., eds.), Springer-Verlag, Berlin, pp. 30–41.

48. Svensson, M. and Aldskogius, H. (1992) Evidence for activation of the complement cascade in the hypoglossal nucleus following peripheral nerve injury. *J. Neuroimmunol.* **40,** 99.

49. Gasque, P., Fontaine, M., and Morgan, B. P. (1995) Complement expression in human brain: biosynthesis of terminal pathway components and regulators in human glial cells and cell lines. *J. Immunol.* **154,** 4726–4733.

50. Johnson, S. A., Lampertetchells, M., Pasinetti, G. M., Rozovsky, I., and Finch, C. E. (1992) Complement messenger RNA in the mammalian brain: responses to Alzheimer's disease and experimental brain lesioning. *Neurobiol. Aging* **13,** 641–648.

51. Pasinetti, G. M., Johnson, S. A., Rozovsky, I., Lampert-Etchells, M., Morgan, D. G., Gordon, M. N., et al. (1992) Complement ClqB and C4 mRNAs responses to lesioning in rat brain. *Exp. Neurol.* **118,** 117–125.

52. Walker, D. G. and McGeer, P. L. (1992) Complement gene expression in human brain: comparison between normal and Alzheimer disease cases. *Mol. Brain Res.* **14,** 109–116.

53. Kim, W.-G., Mohney, R. P., Wilson, B., Jeohn, G.-H., Liu, B., and Hong, J.-S. (2000) Regional difference in susceptibility to lipopolysaccharide-induced neurotoxicity in the rat brain: role of microglia. *J. Neurosci.* **20,** 6309–6316.

54. Leventhal, L., Sortwell, C. E., Hanbury, R., Collier, T. J., Kordower, J. H., and Palfi, S. (2000) Cyclosporin a protects striatal neurons in vitro and in vivo from 3-nitropropionic acid toxicity. *J. Comp. Neurol.* **425,** 471–478.

Index